Materials Horizons: From Nature to Nanomaterials

Series Editor

Vijay Kumar Thakur, School of Aerospace, Transport and Manufacturing,
Cranfield University, Cranfield, UK

Materials are an indispensable part of human civilization since the inception of life on earth. With the passage of time, innumerable new materials have been explored as well as developed and the search for new innovative materials continues briskly. Keeping in mind the immense perspectives of various classes of materials, this series aims at providing a comprehensive collection of works across the breadth of materials research at cutting-edge interface of materials science with physics, chemistry, biology and engineering.

This series covers a galaxy of materials ranging from natural materials to nanomaterials. Some of the topics include but not limited to: biological materials, biomimetic materials, ceramics, composites, coatings, functional materials, glasses, inorganic materials, inorganic-organic hybrids, metals, membranes, magnetic materials, manufacturing of materials, nanomaterials, organic materials and pigments to name a few. The series provides most timely and comprehensive information on advanced synthesis, processing, characterization, manufacturing and applications in a broad range of interdisciplinary fields in science, engineering and technology.

This series accepts both authored and edited works, including textbooks, monographs, reference works, and professional books. The books in this series will provide a deep insight into the state-of-art of Materials Horizons and serve students, academic, government and industrial scientists involved in all aspects of materials research.

Review Process

The proposal for each volume is reviewed by the following:

1. Responsible (in-house) editor
2. One external subject expert
3. One of the editorial board members.

The chapters in each volume are individually reviewed single blind by expert reviewers and the volume editor.

Akarsh Verma · Hariome Sharan Gupta ·
Sushanta K. Sethi
Editors

Polymers and Composite Materials for Packaging: Smart Food Packaging and Solutions

Springer

Editors
Akarsh Verma
Department of Mechanical Engineering
UPES (University of Petroleum and Energy Studies)
Dehradun, Uttarakhand, India

Hariome Sharan Gupta
Department of Polymer and Process Engineering
Indian Institute of Technology Roorkee
Saharanpur, Uttar Pradesh, India

Sushanta K. Sethi
Department of Mechanical, Materials and Aerospace Engineering
Indian Institute of Technology (IIT) Dharwad
Dharwad, Karnataka, India

ISSN 2524-5384 ISSN 2524-5392 (electronic)
Materials Horizons: From Nature to Nanomaterials
ISBN 978-981-95-0676-7 ISBN 978-981-95-0677-4 (eBook)
https://doi.org/10.1007/978-981-95-0677-4

This Springer imprint is published by the registered company Springer Nature Singapore Pte Ltd.
The registered company address is: 152 Beach Road, #21-01/04 Gateway East, Singapore 189721, Singapore

Contents

Chapter 1
Edible Films as Sustainable Alternatives for Food Packaging

Mridul Pant and Swati Sharma

1 Introduction

Food packaging is vital in maintaining food safety, quality, and durability from its production to consumption stage. It serves as a physical and chemical barrier, and protects food from various environmental factors like oxygen exposure, moisture loss, microbial contamination, and physical damage, thus mitigating food spoilage and its wastage [1, 2]. Advanced packaging technologies now integrate active agents like oxygen scavengers, antimicrobial agents, and controlled-release systems to further enhance food shelf-life and preserve its quality [3–5]. In addition, intelligent packaging systems (time–temperature indicators, gas sensors, and RFID tracking) ensure food integrity all over the supply chain [4, 5]. Despite significant contributions to food safety and convenience, conventionally used packaging materials have limitations.

Generally, plastics rule the food packaging sector owing to their affordability and flexibility, still, their non-biodegradable nature leads to enormous waste accumulation in landfills and oceans, contributing to around 30% of municipal solid waste [6]. They face significant challenges during recycling, including polymer contamination and sorting inefficiencies, which restrict global recycling rates of plastic packaging to below 40% [7, 8]. Apart from plastics, metals, and glass are suitable choices in terms of recyclability but their production and transportation largely contribute to carbon emissions [9]. In the context of these environmental trade-offs, regulatory frameworks like the European Union's 65% recycling target for packaging by 2025 are driving innovations in sustainable packaging solutions [8].

Edible films have come up as a promising sustainable alternative to traditional packaging, addressing plastic waste limitations while maintaining food safety. These

M. Pant · S. Sharma (✉)
Department of Polymer and Process Engineering, Indian Institute of Technology, Roorkee, India
e-mail: Swati_s@pe.iitr.ac.in

A. Verma et al. (eds.), *Polymers and Composite Materials for Packaging: Smart Food Packaging and Solutions*, Materials Horizons: From Nature to Nanomaterials,
https://doi.org/10.1007/978-981-95-0677-4_1

films are made up of biodegradable biopolymers (polysaccharides, proteins, and lipids), and have good barrier properties [10, 11]. For instance, starch and chitosan-based films provide superior oxygen barrier properties, and enhanced moisture resistance is provided by lipid-based coatings [10, 11]. Figure 1 shows the detailed properties of edible films. Recent innovations in the field of edible film include the addition of various antimicrobial agents, and essential oil nanoemulsions, in biopolymer matrix to enhance food preservation while maintaining its quality. This chapter delves into the formulation, functional and physiochemical properties, and applications of edible films, mainly focusing on their potential in sustainable food systems. Moreover, it deliberates on challenges, future scope, and recommendations in commercialization, including processing limitations, economic feasibility, and regulatory concerns, while examining innovative trends that could enhance their commercial viability.

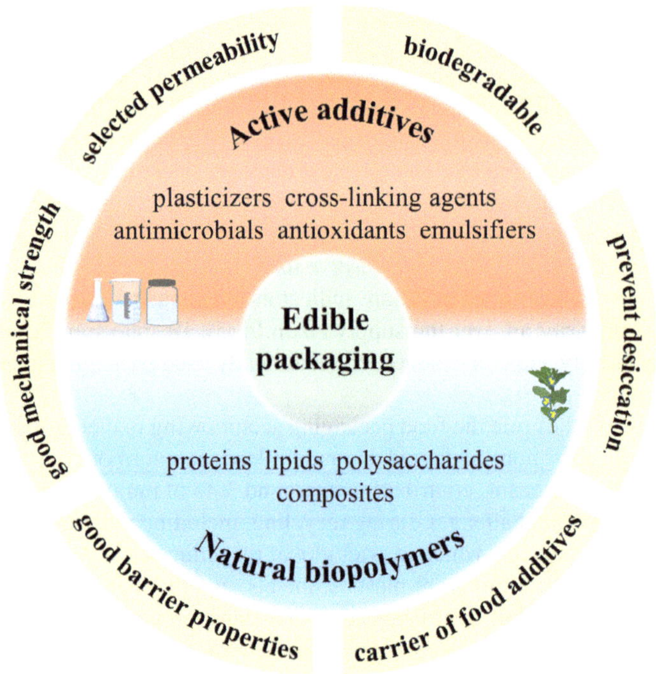

Fig. 1 Biopolymer matrix, active components, and properties of edible films (figure adapted from [12])

2 Composition and Formulation of Edible Films

2.1 *Biopolymer-Based Films*

Biopolymer-based films are sustainable and biodegradable options to traditionally used materials. Based on their sources these films can be classified as polysaccharide-based, protein-based, lipid-based, and composite as shown in Fig. 2.

2.1.1 Polysaccharide-Based Films (Starch, Cellulose, Chitosan)

Polysaccharides are non-toxic, edible, and abundantly present in nature, and owing to these properties they are extensively utilized in preparing edible films for food packaging to increase the shelf-life of food [13, 14]. Their glycosidic linkages allow monosaccharides and disaccharides to polymerize in particular ways to generate these natural polymers. The free OH groups can improve their functionality by easily forming hydrogen bonds with incorporated active ingredients [15]. The highly packed structure of polysaccharide-based film contributes to their good barrier properties (oxygen and aroma), and mechanical properties. However, in the case of highly humid environmental conditions, their hydrophilic nature could compromise the integrity of the packaging. This limitation can be overcome through the inclusion of lipophilic materials, such as oil and wax, to decrease hydrophilicity [14].

Polysaccharides used in edible films can be categorized based on their sources as plants (e.g., cellulose, starch, pectin, gum arabic), animals (e.g., chitosan), marine

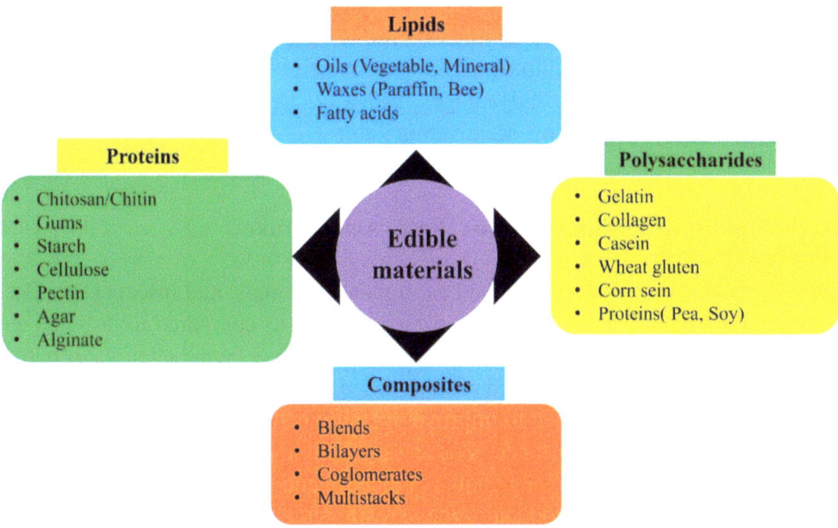

Fig. 2 Categorization of biopolymer-based films

life (e.g., alginate, agar, carrageenan), or microbial sources (e.g., pullulan, gellan, xanthan) [13, 15]. Each polysaccharide category possesses unique characteristics, allowing selection based on the specific requirements of the packaged food item. Alginate-based films can effectively regulate the respiration of fruits and vegetables owing to their excellent oxygen barrier properties. The adaptability and natural origin of polysaccharides make them a desirable option for the development of sustainable and efficient edible packaging solutions. Table 1 shows advantages, and disadvantages along with ways to improve the properties of polysaccharide-based films.

2.1.2 Protein-Based Films (Gelatin, Whey Protein, Soy Protein)

Protein is a heteropolymer made up of over 100 monomers, or amino acids, joined by peptide bonds. The core carbon of each of these amino acids is joined to (a) a carboxyl group, (b) a hydrogen atom, (c) an amino group, and (d) an R-group [13]. Natural polymers can exist in two different forms, i.e., fibrous and globular. Fibrous proteins like whey, soy, and zain maize, run parallel to the polypeptide chain, whereas globular proteins are spherically folded and resemble collagen proteins. Collagen, zein maize, soy, and whey protein have demonstrated good film-forming properties. Protein films have better mechanical properties than polysaccharides due to their distinct structure [13, 14]. In addition, antibacterial and antioxidant components can be incorporated into them and have good barriers to hydrophobic compounds (such as oil and aroma). However, in contrast to synthetic polymers, protein films have a higher water vapor permeability and a lower mechanical strength [15].

Soy protein and wheat gluten films offer good mechanical properties and resistance to oil and gas permeability, collagen and gelatin films are biocompatible and flexible, making them appropriate for food preservation [13, 14, 16]. Furthermore, casein-based edible films exhibit good oxygen barrier properties, enhancing food shelf-life. The advantages, disadvantages, and ways to improve the limitations of these protein-based edible films are given in Table 1.

2.1.3 Lipid-Based Films (Beeswax, Fatty Acids, Waxes)

Lipids are natural substances obtained from plants, animals, and insects [16]. They are not polymers and lack a self-supporting structure, as compared to proteins and polysaccharides. However, their hydrophobic nature offers them better water barrier properties over polysaccharides and protein-based films [14, 16]. Hence, lipid-based edible films could be employed to improve overall moisture barrier properties in the application where water resistance is necessary. Edible packaging uses a variety of lipids, such as edible waxes (long chains of alcohol and esters extracted from plants and animals), resins (like wood and terpene resin, which are known for their glossy texture), and fats and oils (mainly triglycerides extracted from plants and animals).

Table 1 Biopolymer-based films (biopolymer types, examples, advantages, limitations, and ways to improve the properties)

Biopolymer type	Example	Advantages	Disadvantages	Ways to improve properties	References
Polysaccharide-based films	Cellulose, starch, pectin, chitosan, alginate	Gas barrier property, biodegradable, good film-forming ability	Poor moisture resistance, brittle	Plasticizers addition, blending with proteins or lipids, cross-linking	[10, 13]
Protein-based films	Whey protein, casein, soy protein, zein	Good mechanical properties, moderate water resistance	Sensitive to humidity, potential allergenicity	Nanofillers incorporation (e.g., nanocellulose), cross-linking, blending with polysaccharides	[13, 17, 19]
Lipid-based films	Beeswax, waxes, fatty acids, shellac	Moisture barrier improves food appearance	Low mechanical strength, brittle, low gas barrier, greasy surface, impart a unique taste to packaged food	Emulsifiers and nanoparticles addition, polysaccharide blending	[13, 18]
Composite films	Starch-protein, chitosan-lipid, pectin-whey protein	Enhanced mechanical and barrier properties	Processing difficulty, and optimization required	Layer-by-layer deposition, nanocomposites formation, enzymatic cross-linking	[17–20]

Wax-based films in this category exhibit the highest moisture barrier properties. Additionally, the antimicrobial properties of biopolymer matrices can also be enhanced by the incorporation of essential oils, which are volatile and hydrophobic. Hence, the choice of lipid used for edible film preparation is dependent on the desired properties of packaging material and food type [14]. Some of the limitations of lipid-based films are shown in Table 1 along with methods to improve them.

2.1.4 Composite Films

Composite films also termed multi-component systems, typically have a protein, lipid layer, and hydrocolloid components supported by a polysaccharide or lipid material distributed within a polysaccharide or protein matrix [17, 18]. There are two types of composite edible films: binary and ternary. The former comprises a combination of two hydrocolloids, while the latter includes a combination of three hydrocolloids [19]. The combination of at least two ingredients is suggested in this edible packaging system, with the addition of the other component compensating for the weakness of the first component [18]. For example, the addition of lipids into the polysaccharides and proteins can improve their water vapor permeability while the mechanical properties of lipids are enhanced by proteins and polysaccharides addition. The advantages, limitations, and ways to overcome those limitations are provided in Table 1.

The molecular interactions and compatibility of the component ingredients and additives determine the characteristics of the composite edible films. The composite films enhance the required properties, based on the final application for various food products [20]. The addition of carbohydrates into the protein matrix for composite film preparation has shown great application in food packaging due to their water and gas barrier properties [17].

2.2 Functional Additives in Edible Films

2.2.1 Plasticizers, Emulsifiers, and Cross-linking Agents

Plasticizers are small molecular weight hydrophobic substances, generally added in film-forming solutions to increase their mechanical properties. Edible films typically contain plasticizers to increase their flexibility and durability. These consist of lipids, phospholipids, fatty acids, and surfactants; polyols such as glycerol, glyceryl derivatives, sorbitol, and polyethylene glycols; and mono-, di-, or oligosaccharides such as glucose, fructose-glucose syrups, and sucrose. Moreover, by choosing the right plasticizer for a particular biopolymer, it is possible to improve the flexibility and resilience of the film while optimizing its mechanical properties with a small degree of increase in permeability [15, 18].

Emulsifiers are chemicals with surface activity that can change the interfacial energy at the interface of immiscible systems (water–lipid) and have both polar and nonpolar properties. These substances are typically employed to create and stabilize evenly distributed lipid particles in composite emulsion films or to attain adequate surface wettability to guarantee appropriate surface coverage and adherence to the coated surface. Some of the commonly used emulsifiers are lecithin, glycerol monostearate, acetylated monoglyceride, polysorbates, etc. [21, 22].

Cross-linking agents are essential to improve the mechanical, thermal, and barrier properties of edible films used in food packaging [17]. Despite their primary function to enhance the mechanical properties of edible films, they can also impart active functionality to these films. Composite edible packaging systems with improved barrier properties and regulated release of bioactive components can be developed with the use of cross-linkers in conjunction with other active ingredients like plant extracts, essential oils, and nanomaterials [17, 21]. These substances improve the structural integrity and water resistance of biopolymer chains by forming covalent or hydrogen bonds between them. Various types of cross-linking agents are dialdehyde carboxymethyl cellulose (DCMC), alginate dialdehyde (ADA), tannic acid, and citric acid. Casein-based films can be effectively cross-linked by tannic acid, a phenolic substance obtained from plants, greatly improving their physicochemical characteristics for food packaging applications.

2.2.2 Antimicrobial and Antioxidant Agents

In edible food packaging, antioxidant compounds play a vital role in prolonging shelf-life, inhibiting oxidative deterioration, and improving food safety. These substances, when added to edible films, can minimize the development of off-flavors, color deterioration, and lipid oxidation in food items [23]. Some of the commonly used natural antioxidants are ascorbic acid, propolis, bee pollen, essential oils, plant extracts, α-tocopherol, and citric acid, they can be either used alone or in combination [18]. α-tocopherol (Vitamin E), catechin, quercetin, and caffeic acid derived from herbs and spices are effective in delaying lipid oxidation [23]. Plants-derived essential oils, such as citrus extracts, rosemary, oregano, and green tea, have been incorporated into edible films because of their strong antibacterial and antioxidant properties. These bioactive substances improve oxidative stability and inhibit microbial growth [24]. Moreover, the phenolic-rich plant extracts having antioxidant potential have been explored in edible films. For instance, Guava and Peepal leaf extract incorporated into carboxymethyl cellulose-based films exhibited substantial antioxidant activity, effectively reducing oxidative stress in stored food items [25].

Antibacterial compounds, both natural and synthetic, have been extensively studied for their potential use in edible packaging. The most common antimicrobial agents are polypeptides (lysozyme, peroxidase, lactoferrin, nisin), organic acids (acetic, benzoic, citric, fumaric, lactic, malic, propionic, sorbic, succinic, and tartaric acid), and essential oils (oregano, lemongrass, cinnamon, tea tree, clove, pimento, thyme, calendula, basil, rosemary, bergamot, sage, garlic, etc. [26]). Natural antibacterial agents like bacteriocins, essential oils, and plant-derived bioactive molecules have attracted a lot of interest because of their safety and biocompatibility. One of the studies by Moghimi et al. [27] on nanoemulsion *Thymus daneensis* essential oil-based edible films has good antibacterial activity against both gram-positive and gram-negative bacterial strains, hence can preserve perishable foods.

Apart from antibacterial agents, some natural polymers like chitosan also possess antibacterial activity and can disrupt bacterial cell membranes, hence it has been widely used in edible food packaging. Additionally, antibacterial agents (essential oils and metallic nanoparticles) can be also incorporated in chitosan-based edible coatings to enhance antimicrobial efficiency while sustaining eco-friendly properties. One another natural antimicrobial agent in this category is Lactoferrin, an antibacterial and immunomodulatory protein, Lactoferrin-infused bacterial cellulose films prevent foodborne pathogens including Escherichia coli and Staphylococcus aureus, making them ideal for antimicrobial active packaging [28].

2.2.3 Nanomaterials

Nanotechnology is an evolving technology in the food packaging industry. It is used to develop nano packaging that exhibits good functional and physiochemical properties, such as thermal, barrier, mechanical, antioxidant, and antimicrobial properties, which contribute to extending the shelf-life of food items. Different types of nanomaterials include nanoparticles, nanofibers, nanoemulsions, nanotubes, carbon dots, etc., [29, 30] are used in the preparation of edible films or used as coatings for the packaging of a variety of food items. Table 2 shows the properties of different nanomaterials and their application in food packaging. These nanomaterials are readily used in active packaging to improve confinement, convenience, protection, preservation, marketing, and environmental sustainability aspects of packaging material. These nanomaterials can be used in edible films and coating for the packaging of perishable food items such as meat, fruits, and cheese [14, 17].

3 Methods of Production of Edible Films

Figure 3 shows various methods used for edible film preparation along with the properties of films. This section briefly discusses all of the techniques that are used in the development of edible films.

Table 2 Different categories of nanomaterials, their examples, and applications in food packaging

Nanomaterial type	Examples	Applications in food packaging	References
Nanoparticles	Silver (Ag) nanoparticles, zinc oxide (ZnO) nanoparticles, titanium dioxide (TiO_2) nanoparticles, and silicon dioxide (SiO_2) nanoparticles	Antimicrobial properties, UV/oxygen barrier, photocatalytic activity, and improved mechanical properties	[31–34]
Nanoemulsions	Essential oil nanoemulsions (thyme, oregano, clove), curcumin nanoemulsions, and vitamin E nanoemulsions	Controlled release of antimicrobial and antioxidant compounds, extended shelf-life, improved stability, intelligent packaging, nutrient preservation	[35–37]
Nanofibers	Cellulose nanofibers, chitin/chitosan nanofibers, and poly (lactic acid) nanofibers	Mechanical reinforcement, gas barrier, antimicrobial properties, controlled release of active compounds	[38, 39]
Nanotubes	Carbon nanotubes, halloysite nanotubes, and boron nitride nanotubes	Electrical conductivity, gas barrier properties, controlled release of active agents, mechanical reinforcement, thermal stability	[40–42]
Carbon dots	Carbon quantum dots, graphene quantum dots	Food spoilage indicators, fluorescent detection of pathogens, oxygen scavenging, antioxidant and antimicrobial activity, biosensing, heavy metal detection, freshness indicators	[43, 44]

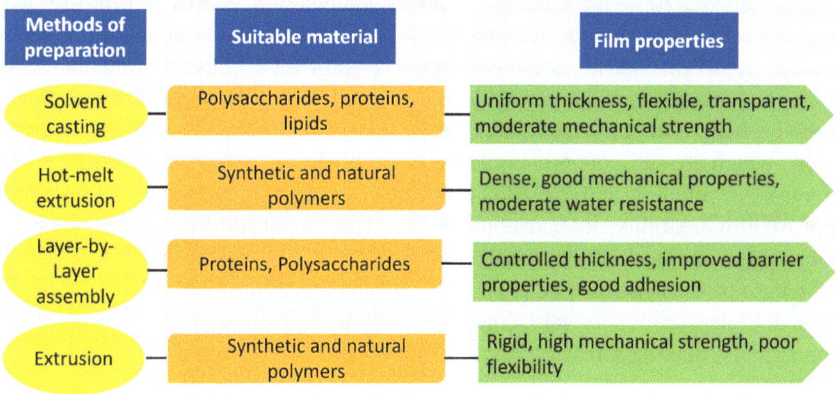

Fig. 3 Methods of preparation of edible films

3.1 Solvent Casting Method

Solvent casting is the most commonly employed method for the preparation of hydrocolloid-based edible films. It comprises dissolving polymers in suitable solvents (aqueous or organic solvents) to form a homogeneous solution, which is then casted into films and then dried [45, 46]. The process of solvent casting starts by dissolving film-forming polymers—like proteins, polysaccharides, or composite blends—in appropriate solvents (water, ethanol, or their mixtures). In addition, plasticizers (glycerol or sorbitol) are generally added to enhance flexibility, while bioactive components (e.g., antimicrobials, and antioxidants) can be incorporated to improve the functionality. Following the casting process, the film is dried at controlled temperatures (typically between 25 and 60 °C) and relative humidity levels (40–60%) to ensure uniform solvent evaporation [14, 45, 46]. This careful drying process allows for the proper realignment of polymer chains and the formation of a stable, well-structured film. The advantages of this method are its simplicity, low equipment costs, and compatibility with heat-sensitive additives. Nevertheless, some limitations like extended drying times, batch-to-batch variability, and challenges in scaling up limit their industrial utility [14].

3.2 Extrusion Technology

Edible films can also be prepared by a dry process (little to no solvent requirement) like extrusion, which may improve production on a commercial scale. Since there is little to no solvent, we don't have to wait for the solvent to evaporate and the drying time to be nullified [14]. This process involves combining raw materials (such as proteins and starch) with plasticizers and feeding them into an extruder, where shear pressures and high temperatures (90–150 °C) denature proteins or gelatinize starch to create a molten matrix. Some hydrophobic additives (lipids) can also be added to improve moisture resistance, while thermolabile substances may denature due to excessive heat. After that, the material is pushed through a die to create thin films or sheets, which are subsequently cooled and coiled. Twin-screw extruders are mainly effective for attaining homogeneous mixing and modifying polymer structures to improve film flexibility and barrier properties. The optical and mechanical properties of resulting films are dependent on the solvent used, extruder screw speed, heat output, and length of the heating zone [47]. This technique is better than solvent casting in terms of higher production rate, continuous output, and suitability for thermoplastic biopolymers, making it feasible for industrial-scale applications [46].

3.3 Layer-by-Layer Deposition

Layer-by-layer (LbL) deposition technique is based on electrostatic interactions. This method involves the sequential deposition of oppositely charged biopolymers, including chitosan (positively charged) and pectin (negatively charged), onto food surfaces or substrates through techniques such as dipping, spraying, or brushing [48, 49]. The thickness and composition of each layer can be precisely controlled to optimize barrier properties. LbL can improve the functionality of edible films via tailored compositional gradients. The issue of premature additive migration in films can also be prevented by the incorporation of active components (e.g., antimicrobials) within specific layers for controlled release. Moreover, the recent advancements in automated spraying systems and pH-adjusted formulations are broadening the applicability of LbL to diverse food matrices. However, some of its limitations are in terms of attaining uniform deposition on irregular food surfaces and time needs to be properly optimized for large-scale applications [48, 49].

3.4 Electrospinning and Other Emerging Techniques

Electrospinning is becoming an increasingly popular technique for developing edible nanofibrous films with high surface area-to-volume ratios, which allow for more effective encapsulation and regulated active ingredient release [50]. This method involves applying a high-voltage electric field to a polymer solution, which elongates the solution jet into 50–500 nm ultrafine fibers collected onto a grounded substrate [50]. Nonwoven mats often electrospun from materials like zein, chitosan, and pullulan have better mechanical and barrier properties than solvent-casted films. For example, thymol-containing electrospun zein fibers showed long-lasting antibacterial action against Listeria monocytogenes in cheese packaging [50]. Innovative methods, including three-dimensional (3D) printing and microfluidic spinning, are currently under investigation. 3D printing facilitates the accurate spatial arrangement of multilayered films containing bioactive compounds, whereas microfluidic systems generate core–shell fibers for the delivery of dual-active ingredients [48, 50]. Nonetheless, elevated equipment expenses and diminished production rates continue to impede their commercialization. Hybrid methods that integrate electrospinning with solvent casting or extrusion are being explored to achieve an optimal balance between performance and scalability [48, 50].

4 Physicochemical Properties of Edible Films

Edible films have recently emerged as a sustainable alternative to traditional plastic packaging, providing distinct physicochemical and functional properties geared toward food preservation. Films made from natural biopolymers (proteins, polysaccharides, and lipids), aim to extend shelf-life and reduce environmental impact. One of the factors affecting film properties can be mechanical factors like pressure (compression) or impact. Standard mechanical evaluations are employed to determine the strength of the edible film. Tensile strength, tearing strength, elongation at break, and abrasion resistance are essential tests for assessing film properties. Generally, edible films have lower tensile strength but good elongation at break as compared to traditional conventional plastic-based packaging films. The permeation of unwanted particles and aroma components can also degrade food quality and affect consumer acceptance of the item. In edible films, the coating serves as a moisture barrier and minimizes nutrient migration between food and its surroundings. These films establish a barrier that diminishes oxygen transmission between food and its external environment. This decreases the rate of food oxidation, hence extending the shelf-life of the food product. Edible films derived from polysaccharides and proteins are hydrophilic; when employed in high-humidity environments, their structural integrity diminishes, compromising the packaging's ability to protect the food product [14].

5 Active and Intelligent Edible Films

5.1 Antimicrobial Films for Food Safety

Antimicrobial-based packaging maintains food quality, ensures its safety, and enhances its shelf-life by decreasing and limiting the growth of food-spoiling microbes. In this packaging, antimicrobial components are fused in the biopolymer matrix of food packaging. The incorporation of antimicrobial agents into the biopolymer matrix substitutes the traditional method of directly applying antimicrobial compounds to food. Direct application on the food has a higher likelihood of dispersion. They inhibit microbial growth; however, these compounds may permeate the food product itself. Antimicrobial biopolymer edible films and coatings control the release of antimicrobial agents [51, 52].

5.2 Antioxidant-Enriched Films for Shelf-Life Extension

Edible films enriched with antioxidants represent a cutting-edge approach to extending shelf-life of food products, effectively merging the concepts of food preservation and sustainability. Biopolymer-based films are enhanced with natural antioxidants such as plant extracts, essential oils, or phenolic compounds. These materials function as dynamic packaging solutions by postponing lipid oxidation, capturing ROS, and suppressing pro-oxidative enzymes. These films help prevent spoilage and preserve sensory attributes like color, texture, and flavor [53, 54]. These films also offer gas and moisture barriers, as well as antimicrobial properties that prevent the growth of spoilage microorganisms [53, 55]. They are an environmentally friendly substitute for traditional packaging materials due to their biodegradable nature. Food waste can be decreased and food safety can be improved by using antioxidant-enriched edible films to preserve perishable goods like fruits, vegetables, meat, and aquatic products [55, 56].

5.3 Nutrient-Infused Edible Films

Nutrient-infused edible films offer a novel method for functional food packaging, by integrating preservation functions with improved nutritional delivery. These edible films are developed by incorporating bioactive compounds like vitamins, minerals, antioxidants, and probiotics into biopolymer base matrices at the time of production [55, 57]. In one of the studies, Oregon State University researchers prepared a chitosan-lysozyme composite film containing vitamin E and calcium to limit microbial growth and fortify coated foods with essential vitamins and minerals [57]. Similarly, protein extracts and bioactive agents (obtained from *Arthrospira sp.* (spirulina)) incorporated alginate-based films showcased high antioxidant activity (up to 180.8 mg TE/g) along with improved water vapor permeability, demonstrating dual functionality in food preservation as well as nutrient enrichment [58]. Further, nanoencapsulation techniques improve these films by incorporating lipid-soluble vitamins (A, D, E) or polyphenols into nanofibers or nanoparticles, thereby ensuring controlled release and stability during storage [55, 59]. Innovations in processed foods target micronutrient deficiencies along with dietary restrictions. For instance, gluten-free films made from cassava or potato starch, enriched with iron and vitamin B, provide alternatives for individuals with celiac disease. Regulatory compliance is achieved via the Generally Recognized as Safe (GRAS) designation for components such as chitosan and alginate [59, 60]. However, challenges persist in reconciling nutrient loading with mechanical integrity and sensory acceptability [57, 59].

5.4 Smart and Intelligent Packaging Technologies

Edible films play a fundamental role in smart and intelligent packaging technologies by incorporating functional properties that improve the shelf-life of the food items along with their quality, and safety. These films are generally incorporated with natural indicators, bio-sensors, and pH-sensitive components that enable real-time monitoring of food freshness. Recent innovation in this field includes anthocyanin-loaded polysaccharide-based films, which have the potential to show color changes in response to spoilage-related pH changes, hence providing a visual indication for food quality assessment [61, 62]. Another promising research in this field is the incorporation of betalain (red beetroot) into gelatin/hydroxypropyl methylcellulose (HPMC) films, which effectively monitor the meat and seafood item freshness by responding to total volatile basic nitrogen (TVB-N) changes [62]. These developments demonstrate the expanding possibilities of edible films in intelligent and active food packaging, providing a cost-effective and environmentally friendly substitute for traditional monitoring systems.

6 Applications in Food Packaging

The selection of edible packaging is primarily determined by the type of food that is packed as well as by storage time and conditions. Edible films made of proteins, incorporated with antioxidants and antimicrobial agents, like chitosan and essential oils, can successfully lower bacterial contamination and lipid oxidation in meat, poultry, and seafood products [63]. The shelf-life of both fresh and processed beef products is enhanced by these films, which improve food safety by preventing microbial growth. In fresh produce (fruits and vegetables), edible films delay ripening and microbiological spoiling while reducing oxidation and moisture loss in fresh produce. Fruits and vegetables can be kept fresh for long duration by using hydrocolloid-based coatings like pectin and alginate, which create a barrier that protects the food items from outside contaminants [64].

In dairy and bakery applications, edible films based on casein and whey protein act as oxygen barriers to keep cheese, butter, and baked goods from going rancid. These films also enhance moisture control and texture preservation, ensuring product stability over long storage time [65]. In addition, in the case of processed and ready-to-eat meals, edible films serve as carriers for bioactive ingredients like vitamins, probiotics, and antioxidants. New developments such as betalains incorporated gelatin or hydroxypropyl methylcellulose-based have been created as freshness markers for processed meals like meat and seafood [66]. Further innovations in edible film technology include improvements based on nanotechnology that increase the mechanical and barrier properties of films, making them more appropriate for use in the food industry on a commercial basis. With continuous development, edible films are transforming food packaging by providing useful and biodegradable options for a variety of food types.

7 Challenges and Future Perspectives

Despite several advantages offered by edible films, there are still a few challenges that need to be considered for their easy applicability. Firstly, edible film manufacturing from lab to industrial scale is tough. Solvent casting is a batch procedure mostly employed for edible films preparation, which is difficult to execute in continuous, high-speed operations. Large-scale development of films with uniform quality and drying or curing them online are challenges. Providing a stable supply of high-quality raw biopolymer ingredients at large scale is also difficult [67]. Food-contact safety guidelines require edible packaging to fulfill FDA "generally recognized as safe" criteria and maintain food-grade hygiene [68]. Technological innovation and investment are needed to overcome these obstacles [67]. Cost competitiveness is another substantial barrier for edible film preparation over conventional packaging (low material requirement and processing cost) [67]. Furthermore, edible films require special handling and protective wrappers to maintain cleanliness and integrity during transit, increasing the overall cost. Current research focuses on utilizing inexpensive feedstocks, such as agricultural by-products, and enhancing the manufacturing efficiency of the films. Attaining economies of scale and optimizing production processes will be essential for edible films to emerge as economically viable alternatives [69].

Recent advancements are enhancing the performance and functionality of edible packaging. Nanotechnology represents a significant domain; incorporating nanoscale fillers, including nano clays, nanocellulose, and metal nanoparticles, into edible biopolymer films can improve their mechanical and barrier properties. Another significant advancement is active packaging through the use of bioactive additives. Natural antimicrobials, including organic acids, enzymes, and essential oils, along with antioxidants, have been integrated into edible films to inhibit pathogens and oxidation, thereby prolonging shelf-life. Moreover, smart packaging features are emerging. Integrating pH-sensitive natural pigments, such as anthocyanins or betalains, develops color-changing films that signal food freshness and spoilage. These examples illustrate how nanocomposites, active compounds, and intelligent indicators enhance the functionality of edible films while mitigating certain limitations [3, 4, 10]. New composite formulations and multilayer edible films are expected to more effectively mimic the protective performance of plastics, particularly by enhancing water resistance and mechanical strength via innovative cross-linking or layering techniques. Efforts to streamline production, including the adaptation of extrusion processes for biopolymers, will improve scalability. Industry forecasts indicate significant growth for the edible packaging market, anticipated to surpass $6 billion in the next decade. With advancements in performance, edible films are likely to gain prevalence in specific applications, including coatings for fresh produce and single-serve soluble pouches. Ongoing innovation and collaboration among researchers, industry, and regulators will be essential for the integration of edible films into mainstream packaging practices.

8 Conclusion

Edible films are a sustainable substitute for conventional petroleum-based materials that address pertinent environmental concerns, representing a promising venture in food packaging. Various biopolymer matrices—polysaccharides, proteins, lipids, and their composites that form the foundation of these films have been discussed in this chapter, along with their potential benefits and constraints, which are influenced by different functional additives. Different production processes help in tailoring these edible films for various applications, whereas the addition of antibacterial and antioxidant chemicals has shown great potential for increasing shelf-life and maintaining food safety. Promising steps are being undertaken in developing these films, though major obstacles to general commercial adoption remain, including scalability, mechanical qualities, moisture sensitivity, and cost-effectiveness. Future research should focus on creating synergistic multi-component composite systems, improving manufacturing efficiency, and lowering manufacturing costs to close the difference between lab discovery and industrial applications.

Acknowledgements The author would like to thank MHRD for the financial support.

Conflicts of Interest There are no conflicts of interest to declare by the authors.

References

1. Summo C, De Angelis D (2022) The importance of edible films and coatings for sustainable food development". Foods 11(20):3221. https://doi.org/10.3390/FOODS11203221
2. Nehra A, Biswas D, Siracusa V, Roy S (2022) Natural gum-based functional bioactive films and coatings: a review. Int J Mol Sci 24:485. https://doi.org/10.3390/IJMS24010485
3. Yildirim S et al (2018) Active packaging applications for food. Compr Rev Food Sci Food Saf 17(1):165–199. https://doi.org/10.1111/1541-4337.12322
4. Mirza-Alizadeh A, Masoomian M, Shakooie M, Zabihzadeh-Khajavi M, Farhoodi M (2021) Trends and applications of intelligent packaging in dairy products: a review. Crit Rev Food Sci Nutr 62(2):383–397. https://doi.org/10.1080/10408398.2020.1817847
5. Westlake JR, Tran MW, Jiang Y, Zhang X, Burrows AD, Xie M (2022) Biodegradable active packaging with controlled release: principles, progress, and prospects. ACS Food Sci Technol 2(8):1166–1183. https://doi.org/10.1021/ACSFOODSCITECH.2C00070/ASSET/IMAGES/LARGE/FS2C00070_0005.JPEG
6. Marsh K, Bugusu B (2007) Food packaging—roles, materials, and environmental issues. J Food Sci 72(3):R39–R55. https://doi.org/10.1111/J.1750-3841.2007.00301.X

7. Blasenbauer D, Lipp AM, Fellner J, Tischberger-Aldrian A, Stipanović H, Lederer J (2024) Recovery of plastic packaging from mixed municipal solid waste: a case study from Austria. Waste Manag 180:9–22. https://doi.org/10.1016/J.WASMAN.2024.02.040
8. Roosen M et al (2020) Detailed analysis of the composition of selected plastic packaging waste products and its implications for mechanical and thermochemical recycling. Environ Sci Technol 54(20):13282–13293. https://doi.org/10.1021/ACS.EST.0C03371/ASSET/IMAGES/LARGE/ES0C03371_0005.JPEG
9. Kan M, Miller SA (2022) Environmental impacts of plastic packaging of food products. Resour Conserv Recycl 180:106156. https://doi.org/10.1016/J.RESCONREC.2022.106156
10. Kong I, Degraeve P, Pui LP (2022) Polysaccharide-based edible films incorporated with essential oil nanoemulsions: physico-chemical, mechanical properties and its application in food preservation—a review. Foods 11(4):555. https://doi.org/10.3390/FOODS11040555
11. Wei N et al (2024) Cassia seed gum films incorporated with partridge tea extract as an edible antioxidant food packaging film for preservation of chicken jerky. Polymers 16(8):1086. https://doi.org/10.3390/POLYM16081086
12. Wang H, Cao Z, Yao L, Feng T, Song S, Sun M (2023) Insights into the edible and biodegradable ulvan-based films and coatings for food packaging. Foods 12(8):1622. https://doi.org/10.3390/FOODS12081622
13. Mohamed SAA, El-Sakhawy M, El-Sakhawy MAM (2020) Polysaccharides, protein and lipid-based natural edible films in food packaging: a review. Carbohydr Polym 15:178
14. Kumar L, Ramakanth D, Akhila K, Gaikwad KK (2022) Edible films and coatings for food packaging applications: a review. Springer, New York
15. Pooja-Saklani PS, Nath S, Kishor-Das S, Singh SM (2019) A review of edible packaging for foods. Int J Curr Microbiol Appl Sci 8(07):2885–2895
16. Nurjanah H, Umam HI, Pambudi T (2023) Edible film for food packaging. AL ULUM JURNAL SAINS DAN TEKNOLOGI 9(3):127
17. Kumar N et al (2023) Recent trends in edible packaging for food applications: perspective for the future. Springer, New York
18. Trajkovska-Petkoska A, Daniloski D, D'Cunha NM, Naumovski N, Broach AT (2021) Edible packaging: sustainable solutions and novel trends in food packaging. Elsevier, Amsterdam
19. Dhumal CV, Sarkar P (2018) Composite edible films and coatings from food-grade biopolymers. Springer, New York
20. Basiak E, Galus S, Lenart A (2015) Characterisation of composite edible films based on wheat starch and whey-protein isolate. Int J Food Sci Technol 50(2):372–380. https://doi.org/10.1111/IJFS.12628
21. Janjarasskul T, Krochta JM (2010) Edible packaging materials. Annu Rev Food Sci Technol 1(1):415–448. https://doi.org/10.1146/ANNUREV.FOOD.080708.100836/CITE/REFWORKS
22. Taarji N et al (2020) Preparation of monodisperse O/W emulsions using a crude surface-active extract from argan by-products in microchannel emulsification. Colloids Surf A Physicochem Eng Asp 585:124050. https://doi.org/10.1016/J.COLSURFA.2019.124050
23. Sanches-Silva A et al (2014) Trends in the use of natural antioxidants in active food packaging: a review. https://doi.org/10.1080/19440049.2013.879215
24. Moghadas HC, Chauhan R, Smith JS (2024) Application of plant oils as functional additives in edible films and coatings for food packaging: a review. Multidisc Dig Publish Instit 90:997. https://doi.org/10.3390/foods13070997
25. Sharma S, Priyadarshi R, Mishra NC (2024) Carboxymethyl cellulose-based bioactive food packaging films incorporated with *Ficus religiosa*/*Psidium guajava* extracts: effect of additive ratio variation. J Vinyl Addit Technol 10:186. https://doi.org/10.1002/VNL.22186
26. Kamarudin SH et al (2022) A review on antimicrobial packaging from biodegradable polymer composites. Polymers 14(1):174. https://doi.org/10.3390/POLYM14010174
27. Moghimi R, Aliahmadi A, Rafati H (2017) Antibacterial hydroxypropyl methyl cellulose edible films containing nanoemulsions of *Thymus daenensis* essential oil for food packaging. Carbohydr Polym 175:241–248. https://doi.org/10.1016/J.CARBPOL.2017.07.086

28. Kraśniewska K, Gniewosz M (2012) Substances with antibacterial activity in edible films: a review. Pol J Food Nutr Sci 62(4):199–206. https://doi.org/10.2478/V10222-12-0059-3
29. Sharma C, Dhiman R, Rokana N, Panwar H (2017) Nanotechnology: an untapped resource for food packaging. Front Microbiol 8:243298. https://doi.org/10.3389/FMICB.2017.01735/PDF
30. Jeevahan J, Chandrasekaran M (2019) Nanoedible films for food packaging: a review. J Mater Sci 54(19):12290–12318. https://doi.org/10.1007/S10853-019-03742-Y/TABLES/3
31. Zhang W, Ahari H, Zhang Z, Jafari SM (2023) Role of silica (SiO_2) nano/micro-particles in the functionality of degradable packaging films/coatings and their application in food preservation. Trends Food Sci Technol 133:75–86. https://doi.org/10.1016/J.TIFS.2023.01.009
32. Zhang W, Rhim JW (2022) Titanium dioxide (TiO_2) for the manufacture of multifunctional active food packaging films. Food Packag Shelf Life 31:100806. https://doi.org/10.1016/J.FPSL.2021.100806
33. Carbone M, Donia DT, Sabbatella G, Antiochia R (2016) Silver nanoparticles in polymeric matrices for fresh food packaging. J King Saud Univ Sci 28(4):273–279. https://doi.org/10.1016/J.JKSUS.2016.05.004
34. Zhang W, Sani MA, Zhang Z, McClements DJ, Jafari SM (2023) High performance biopolymeric packaging films containing zinc oxide nanoparticles for fresh food preservation: a review. Int J Biol Macromol 230:123188. https://doi.org/10.1016/J.IJBIOMAC.2023.123188
35. Rehman A, Jafari SM, Aadil RM, Assadpour E, Randhawa MA, Mahmood S (2020) Development of active food packaging via incorporation of biopolymeric nanocarriers containing essential oils. Trends Food Sci Technol 101:106–121. https://doi.org/10.1016/J.TIFS.2020.05.001
36. Sanchez LT, Pinzon MI, Villa CC (2022) Development of active edible films made from banana starch and curcumin-loaded nanoemulsions. Food Chem 371:1121. https://doi.org/10.1016/J.FOODCHEM.2021.131121
37. Liu Q, Huang H, Chen H, Lin J, Wang Q (2019) Food-grade nanoemulsions: preparation, stability and application in encapsulation of bioactive compounds. Molecules 24:4242
38. Abdul-Khalil HPS et al (2016) A review on nanocellulosic fibres as new material for sustainable packaging: process and applications. Renew Sustain Energy Rev 64:823–836. https://doi.org/10.1016/J.RSER.2016.06.072
39. Tien ND, Lyngstadaas SP, Mano JF, Blaker JJ, Haugen HJ (2021) Recent developments in chitosan-based micro/nanofibers for sustainable food packaging, smart textiles, cosmeceuticals, and biomedical applications. Molecules 26:2683. https://doi.org/10.3390/MOLECULES26092683
40. Kumar L, Deshmukh RK, Hakim L, Gaikwad KK (2024) Halloysite nanotube as a functional material for active food packaging application: a review. Food Bioproc Tech 17(1):33–46. https://doi.org/10.1007/S11947-023-03092-3
41. Yadav M, Behera K, Chang YH, Chiu FC (2020) Cellulose nanocrystal reinforced chitosan based UV barrier composite films for sustainable packaging. Polymers 12(1):202. https://doi.org/10.3390/POLYM12010202
42. Bangar SP, Whiteside WS, Kajla P, Tavassoli M (2023) Value addition of rice straw cellulose fibers as a reinforcer in packaging applications. Int J Biol Macromol 243:5320. https://doi.org/10.1016/J.IJBIOMAC.2023.125320
43. Ezati P, Priyadarshi R, Rhim JW (2022) Prospects of sustainable and renewable source-based carbon quantum dots for food packaging applications. Sustain Mater Technol 33:e00494. https://doi.org/10.1016/J.SUSMAT.2022.E00494
44. Sharma AS, Ali S, Sabarinathan D, Murugavelu M, Li H, Chen Q (2021) Recent progress on graphene quantum dots-based fluorescence sensors for food safety and quality assessment applications. Compr Rev Food Sci Food Saf 20(6):5765–5801. https://doi.org/10.1111/1541-4337.12834
45. Chandrababu A, Ep B (2023) A comprehensive review on classification and application of edible films. Pharma Innov J 11:1240–1244
46. Prakoso FAH, Indiarto R, Utama GL (2023) Edible film casting techniques and materials and their utilization for meat-based product packaging. Polymers 15:2800. https://doi.org/10.3390/POLYM15132800

47. Suhag R, Kumar N, Petkoska AT, Upadhyay A (2020) Film formation and deposition methods of edible coating on food products: a review. Food Res Int 136:109582. https://doi.org/10.1016/J.FOODRES.2020.109582
48. Wang Q, Chen W, Zhu W, McClements DJ, Liu X, Liu F (2022) A review of multilayer and composite films and coatings for active biodegradable packaging. NPJ Sci Food 6(1):1–16. https://doi.org/10.1038/s41538-022-00132-8
49. Adhikari M, Koirala S, Anal AK (2023) Edible multilayer coating using electrostatic layer-by-layer deposition of chitosan and pectin enhances shelf life of fresh strawberries. Int J Food Sci Technol 58(2):871–879. https://doi.org/10.1111/IJFS.15704
50. Zhao L, Duan G, Zhang G, Yang H, Jiang S, He S (2020) Electrospun functional materials toward food packaging applications: a review. Nanomaterials 10(1):150. https://doi.org/10.3390/NANO10010150
51. Chawla R, Sivakumar S, Kaur H (2021) Antimicrobial edible films in food packaging: current scenario and recent nanotechnological advancements—a review. Carbohydr Polym Technol Appl 2:100024. https://doi.org/10.1016/J.CARPTA.2020.100024
52. Singh S, Gaikwad KK, Lee YS (2018) Antimicrobial and antioxidant properties of polyvinyl alcohol bio composite films containing seaweed extracted cellulose nano-crystal and basil leaves extract. Int J Biol Macromol 107:1879–1887. https://doi.org/10.1016/J.IJBIOMAC.2017.10.057
53. Moura-Alves M, Esteves A, Ciríaco M, Silva JA, Saraiva C (2023) Antimicrobial and antioxidant edible films and coatings in the shelf-life improvement of chicken meat. Foods 12(12):2308. https://doi.org/10.3390/FOODS12122308
54. Shahidi F, Hossain A (2022) Preservation of aquatic food using edible films and coatings containing essential oils: a review. Crit Rev Food Sci Nutr 62(1):66–105. https://doi.org/10.1080/10408398.2020.1812048
55. Gürdal AA, Çetinkaya T (2024) Advancements in edible films for aquatic product preservation and packaging. Rev Aquac 16(3):997–1020. https://doi.org/10.1111/RAQ.12880
56. Chettri S, Sharma N, Mohite AM (2023) Edible coatings and films for shelf-life extension of fruit and vegetables. Biomater Adv 154:213632. https://doi.org/10.1016/J.BIOADV.2023.213632
57. Edible coatings for added vitamins (n.d.) https://www.nutraingredients-usa.com/Article/2004/08/09/Edible-coatings-for-added-vitamins/. Accessed on 05 March 2025
58. Martins VFR, Poças F, Pintado M, Morais RMSC, Morais AMMB (2025) Edible films with protein and bioactive compounds from *Arthrospira* sp. Biol Life Sci Forum 40:6. https://doi.org/10.3390/BLSF2024040006
59. Matloob A et al (2023) A review on edible coatings and films: advances, composition, production methods, and safety concerns. https://doi.org/10.1021/acsomega.3c03459
60. Janowicz M, Galus S, Ciurzyńska A, Nowacka M (2023) The potential of edible films, sheets, and coatings based on fruits and vegetables in the context of sustainable food packaging development. Polymers 15(21):4231. https://doi.org/10.3390/POLYM15214231
61. Wu S et al (2018) Electrochemical writing on edible polysaccharide films for intelligent food packaging. Carbohydr Polym 186:236–242. https://doi.org/10.1016/J.CARBPOL.2018.01.058
62. Abdolmaleki K, Rezaei F, Mohammadi R, Zare L, Shahmoradi S (2024) The application of film based on gelatin/hydroxymethyl cellulose and red beetroot betalain in smart food packaging. Food Sci Technol Int 18:112. https://doi.org/10.1177/10820132241266112/ASSET/IMAGES/10.1177_10820132241266112-IMG8.PNG
63. Hamann D et al (2021) Active edible films for application in meat products. Res Soc Develop 10(7):e13610716379. https://doi.org/10.33448/rsd-v10i7.16379
64. Tavassoli-Kafrani E, Shekarchizadeh H, Masoudpour-Behabadi M (2016) Development of edible films and coatings from alginates and carrageenans. Carbohydr Polym 137:360–374. https://doi.org/10.1016/J.CARBPOL.2015.10.074
65. Rosseto M et al (2023) Whey-based polymeric films for food packaging applications: a review of recent trends. J Sci Food Agric 103(7):3217–3229. https://doi.org/10.1002/JSFA.12310

66. Abdolmaleki K, Rezaei F, Mohammadi R, Zare L, Shahmoradi S (2024) The application of film based on gelatin/hydroxymethyl cellulose and red beetroot betalain in smart food packaging. Food Sci Technol Int 17:66112. https://doi.org/10.1177/10820132241266112/ASSET/IMAGES/10.1177_10820132241266112-IMG8.PNG
67. Olawade DB, Wada OZ, Ige AO (2024) Advances and recent trends in plant-based materials and edible films: a mini-review. Front Media SA. https://doi.org/10.3389/fchem.2024.1441650
68. Odetayo T, Tesfay S, Ngobese NZ (2022) Nanotechnology-enhanced edible coating application on climacteric fruits. Food Sci Nutr 10(7):2149–2167. https://doi.org/10.1002/FSN3.2557
69. Edible Packaging Market Trends, Size, Share and Insights 2031. https://www.coherentmarketinsights.com/market-insight/edible-packaging-market-819. Accessed on 06 March 2025

Chapter 2
Edible Films for Food-Grade Packaging

Dheeraj Gunwant, Mahima, Yusuf Parvez, Anjali Bora, and Manuraj

1 Introduction

The tremendous growth in world population seriously threatens ecosystems and climate stability, leading to irreversible environmental degradation. As per the Food and Agriculture Organization (FAO), approximately 132 million individuals globally face food insecurity, and 811 million people do not have access to food [1]. Despite this, approximately 931 million tons of food items and other crucial resources like water, soil, manufacturing, logistical, and human resources are discarded annually [2]. As the global population reaches 9.7 billion in 2050, the future is grim, and adverse environmental changes are inevitable [1]. The global demand for fresh, nutritious, healthy food products with shorter shelf life has resulted in tremendous food waste. The shelf life of edibles is the period for which it is safe while maintaining their physical, chemical, functional, microbiological, and sensory characteristics in situ. It depends on structure, ingredients, composition, manufacturing conditions, and packaging systems [3]. Enhancing the shelf life of edibles reduces food and packaging wastage as it can retain its original characteristics for longer. Food-grade packaging ensures packaged food protection by averting external spoiling agents and maintaining the supply chain. Their primary functions include separating edibles from

D. Gunwant (✉) · A. Bora · Manuraj
Department of Mechanical Engineering, Apex Institute of Technology, Rampur, Uttar Pradesh, India
e-mail: mailto.dheerajgunwant@gmail.com

Mahima
Department of Electronics and Electrical Engineering, Apex Institute of Technology, Rampur, Uttar Pradesh, India

Y. Parvez
Department of Computer Science and Engineering, Apex Institute of Technology, Rampur, Uttar Pradesh, India

A. Verma et al. (eds.), *Polymers and Composite Materials for Packaging: Smart Food Packaging and Solutions*, Materials Horizons: From Nature to Nanomaterials,
https://doi.org/10.1007/978-981-95-0677-4_2

the surrounding environment, reducing exposure to spoiling agents, and mitigating the loss of desirable compounds. Retrospectively, fruit waxing and meat larding were used in China and England in the twelfth and sixteenth centuries to delay their moisture loss. In the US, applying paraffin wax in the 1930s on citrus fruits marked the start of waxing. The global food industry is shifting toward greener, biodegradable, sustainable, and edible packaging options. These coatings are designed to imitate the desirable qualities of conventional waxes, viz., organic barrier toward moisture and gas [4]. In recent decades, plant-derived materials, casein, wheat starch, gluten, and soy protein-based materials offer protective properties identical to traditional materials. While highly efficient, these biopolymers are aligned with the global sustainability goals. These materials can protect the food quality, enhancing their aesthetic appeal by adding gloss and color [5]. Figure 1 illustrates an overview of the plant-derived materials and their applications in EFs.

Adequate packaging enhances food safety, shelf life, and quality while attracting consumers through an aesthetic appeal [7]. Packaging acts as a barrier against external factors, circumventing edibles from moisture, heat, light, and microorganisms, effectively augmenting their usable life by avoiding spoilage. These packaging materials should possess adequate thermomechanical properties to sustain shipping,

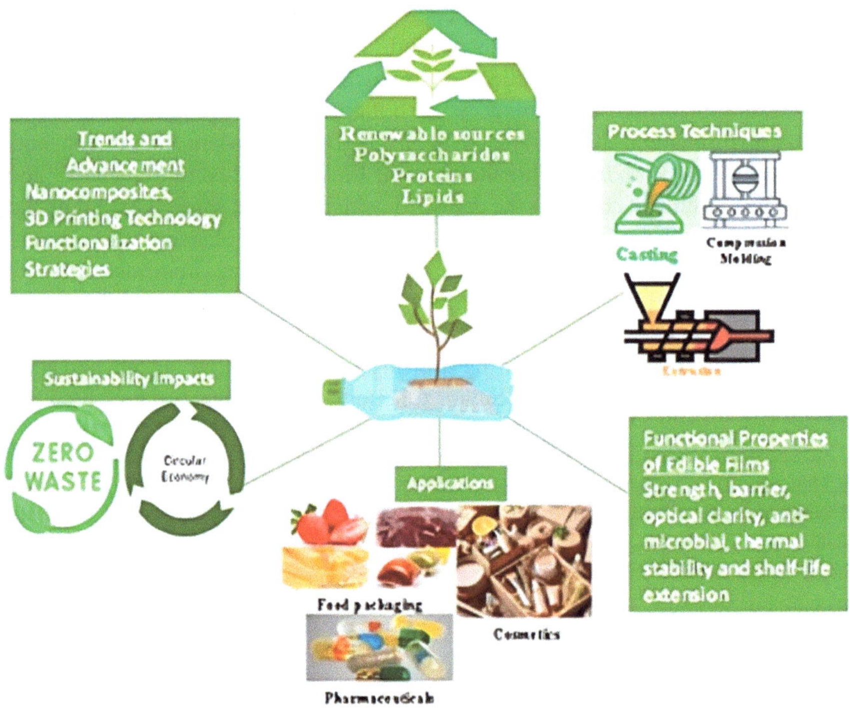

Fig. 1 Overview of plant-derive materials and their use in EFs [6]

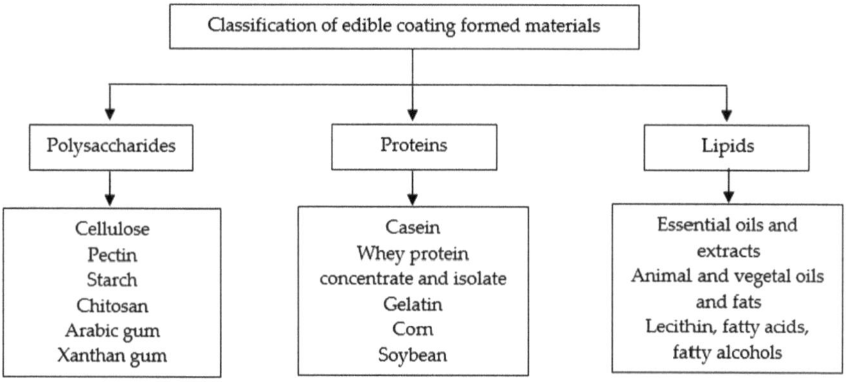

Fig. 2 General categorization of edible coating materials for applications in edibles [12]

handling, and thermal distress [8]. In circumstances where glass and metal are inappropriate, these conditions are sufficed by utilizing synthetic plastics. However, the non-renewability and non-biodegradability of plastics owing to their petroleum origin raises concerns over their environmental impact [9]. Moreover, conventional packaging materials pose a severe environmental threat due to variations in the CO_2 cycle, high toxicity, and decomposing problems [10]. Packaging waste contributes a tremendous amount of solid waste due to ignorant disposal and non-biodegradability, pushing the demand for biodegradable, novel, eco-friendly packaging materials [11].

Given the above, considerable attention is being given to developing biodegradable polymers from renewable sources [9]. Several biopolymers, viz., polypeptides and polysaccharides, are edible, unleashing many possibilities and application areas [9]. EFs are biomacromolecules comprising matrix, binder, and binding agents capable of forming incongruous cohesive layers. Plasticizers such as sugars and polyols are usually utilized to alleviate the brittleness of these films and enhance their usability. Figure 2 depicts the different types of coatings for improving the shelf life of the edibles.

These materials depicted in the figure are biodegradable and possess sensory values [13]. Moreover, incorporating nanomaterials can augment the physicomechanical properties, engendering antioxidative, and antimicrobial properties critical for food preservation. The antimicrobial agents are mixed with a polymer matrix to obtain packaging films with antimicrobial characteristics [14].

Effective implementation of these packaging materials can alleviate the environmental impact of petroleum-derived plastic materials. EFs offer several benefits, such as reducing waste disposal, eliminating the risk of accidental ingestion, and being sustainable. One can expect a drastic reduction in waste accumulation due to zero disposal of packing. It can be easily ingested without any danger of choking or health implications. The self-reliant EFs support the sustainability of packaging options by reducing dependence on fossil-based plastics. This packaging can incorporate antimicrobial agents to restrict bacterial and fungal growth, enhancing the shelf life of edibles and maintaining food quality [15, 16]. On the contrary, conventional packaging serves only as a barrier against contamination without any antimicrobial properties [17].

1.1 Need and Significance of EFs

The extensive utilization of petroleum-derived packaging materials is an imminent environmental danger. Fueled by the industrial revolution, within the last 70 years, the production of these plastic materials has increased exponentially by up to 225 times [18]. These materials have widespread applications in industries and households due to their cost-effectiveness and versatility. Recent reports indicate that the lifecycle of plastic materials adds approximately 2 billion tons of carbon emissions annually, with barely 1/10th part being recycled [19]. Shockingly, in developed countries, ~ 85% of municipal plastic waste is in landfill sites [20]. With such substantial manufacturing, usage, and improper disposal of plastics, the population is exposed to microplastic pollution and degradation of aquatic and terrestrial ecosystems. Furthermore, large-scale exposure to plastic waste-induced pollution is one of the key reasons behind diseases such as infertility, hormonal imbalance, and even cancer [21]. As far as the beverage and alcohol industry is concerned, glass bottles are the first choice due to their ability to inhibit moisture and gas motion [22]. However, glasses pose a severe environmental threat due to their non-biodegradability, high recycling cost, and landfilling.

Adequate handling of packaging residue, reaching ~ 173 kg/capita in 2017, is imperative in the circular economy action plan to ensure a cleaner and more competitive nation [23]. The strategy emphasizes the reduction and reutilization of the materials employed in developing packaging solutions. The aesthetic and antimicrobial requirements in the food packaging industry are the main driving forces behind cutting-edge innovations. Effective implementation of these innovations can contribute toward drastically reducing food wastage and helping countries achieve sustainable development goals. The EU has introduced guidelines to promote the circular economy to achieve carbon neutrality by 2050 via the "Green Deal Initiative" [23]. These guidelines impose stringent rules regarding the packaging, including clear labeling aimed at their reuse and recycling. The food industry faces the daunting challenge of addressing a demanding market and paving ways for achieving novel solutions to ensure quality with sustainability and safety. Conventional non-biodegradable petroleum-based packaging materials are a serious environmental threat due to the excessively high toxicity emissions. These toxic emissions are directly related to the unsolicited alterations in the CO_2 cycle, compounding the problem [10]. As a result, the packaging industry is being coerced to adopt biological sources for manufacturing biodegradable packaging materials. This issue has motivated researchers to search for eco-friendly packaging solutions that amalgamate food safety, quality, and sustainability.

Contrary to petroleum-derived plastics, plant-derived biofilms offer a significant advantage in food packaging, particularly regarding eco-friendliness. Life cycle assessments have revealed that these biofilms consume significantly lower non-renewable resources with smaller greenhouse gas emissions [24]. Polysaccharide and chitosan biofilms have been shown to augment the shelf life of meat products via dehydration prevention, CO_2 transfer, and adaptive permeability [25, 26]. Similarly, chitosan-based film-coated cheese exhibited excellent antimicrobial properties against various organisms [27]. Interestingly, the cheese retained its stability and sensory attributes throughout the duration. The retention is mainly attributed to the increased oxidation stability due to the antimicrobial microcrystalline cellulose embedded in the chitosan films used for coating. Edible packaging is a thin, consumable coating applied directly on the perishable food item's surface. These coatings augment the shelf life of such food items by mitigating oxidative damage, contamination, and moisture loss. The synthesis of these coatings involves several critical parameters, such as transparency, CO_2 permeability, and sensory neutrality. These coatings must be odorless, tasteless, and not affect the food's aesthetics [28]. Figure 3 depicts the overview of the main techno-functional properties and different applications of EFs in the food packaging industry. An efficient packaging system must ensure the minimization of food wastage and extend its shelf life by effectively controlling moisture levels. Perishable edibles such as vegetables and fruits undergo biological post-harvest processes, resulting in moisture loss. Even a minor reduction in moisture can result in a drastic loss in the quality of these edibles within a short span after harvest.

Edible coatings can provide a barrier layer that inhibits moisture transfer and tapers off their degradation. The adaptive absorption and release of moisture from polysaccharide-based EFs depends on ambient conditions such as humidity and temperature. Their utilization in EFs can offer a highly responsive technique for regulating moisture in food items [22]. While the ubiquitous conventional plastic-based packaging materials are the primary choice for packing edibles, approximately 93% of consumers are personally motivated to reduce their consumption [29]. These

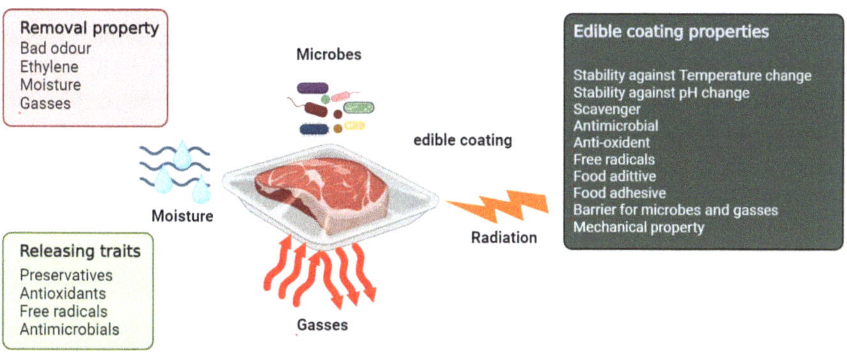

Fig. 3 Overview of the diverse applications and the properties of the edible coatings [29]

packaging can be manufactured in the form of films and shaped as pouches, liners, or wraps, depending on the shape and size of the food item. For example, mango kernel starch pouches outperformed the conventional low-density polyethylene (LDPE) pouches as packaging for red chili powder. Not only this, these pouches successfully preserved the color of the chili powder, simultaneously inhibiting capsaicinoid proportion [30]. Such advancements in packaging exhibit a high potential for edible protection, simultaneously paving the way for eco-friendly and efficient food packaging options.

However, EFs do not entirely replace archaic petroleum-derived packaging materials. These macromolecules suffer from various demerits, such as lower degradation points inhibiting their thermal processing via compression molding, extrusion, and injection molding [31]. They have significantly lower elongation, impact, and tensile strengths, limiting their utilization to low-strength applications only [32]. The tensile strength, Young's modulus, and elongation of polypropylene are 31–43 MPa, 1100–1500 MPa, and 500–600%, respectively. In comparison, gelatine has just 9 MPa, 22 MPa, and 45% values. Additionally, the water vapor transmission rate of cellulose acetate is as high as 257.8 MPa compared to just 0.1–0.3 for polypropylene [33]. Plant-derived biofilms have higher water absorption and permeation characteristics, drastically limiting their applications. Amylase and chitosan-based films have relatively higher gas barrier and tensile properties, making them interesting candidates for future advancements in the field. Moreover, the intensive agriculture required for obtaining these plant-derived biofilms involves over-utilizing herbicides, pesticides, and fertilizers, negatively impacting the environment. Widespread utilization of these chemicals for high yield may lead to eutrophication and acidification over time [34]. These caveats require the researchers to focus on conducting detailed physicomechanical and lifecycle assessments on these biofilms to widen their applications.

1.2 Market Trends and Consumer Preferences

Edible food packaging is made of consumable materials that can be cast into different shapes. Due to the pressing demand, the food packaging industry employs bio-derived by-products to achieve carbon neutrality and a circular bioeconomy. The current state-of-the-art in this relatively newer technology encompasses the fabrication of edible coatings with exceptional thermal sealability and moisture aversion [35]. However, the consumers lack the conceptual knowledge and terminologies to use these eco-friendly coatings.

They are mainly concerned with the type of coating material, sensory attributes, shelf life, and packaged food stability. The cost of coating material is also vital in deciding the consumers' acceptability. Despite the increasing interest in edible packaging due to its health and environmental benefits, it faces challenges due to stakeholders' obliviousness [36]. The key parameters the consumers consider while choosing packaging materials are the "texture," "freshness," and "nutritive value" of the edibles. A study on the acceptability of coated apples in Australia revealed their rejection by consumers due to the absence of information regarding the coating's benefits [37]. It implies that without adequate information, it is difficult for the consumers to perceive their benefits, ultimately hindering their acceptance. Taylor et al. have shown that food items coated with EFs exhibited up to an 8.06 rating for the edible fruit leather wraps on the nine-point hedonic scale [38]. The participants were offered three food items coated with eco-friendly edible coatings and asked to describe their sensory characteristics, qualities, and purchase intent. Up to 64–68% of participants reacted positively, distinctly understanding that these films are a potential solution to decrease the use of single-use plastic.

A similar study by Dey et al. emphasized the importance of developing completely edible coatings [39]. As per them, up to 40% of participants asserted that they will buy edible coatings in the future, provided the cost does not increase much. Present-day consumers are relatively well aware of the concerns about sustainable packaging solutions. This awareness motivates the researchers to develop new plant-derived packaging materials to address consumer demand. Manufacturers continuously create EFs with attributes comparable to petroleum-derived materials. As a result, the market share of the edible packaging industry has achieved a compound annual growth rate (CAGR) of 6.81% between 2017 and 2023. From being valued at USD 697 million in 2016, it has grown to USD 1097 million by 2023 [40]. In 2020, the CAGR of edible packaging was 7.64%, with the projected global market value at 2.06 billion between 2021 and 2027 [41].

The increase in market trends is attributed to the emergence of EFs vis-à-vis conventional food packaging materials. The stringent regulations of food packaging laws ensure the global sustainability goals pushing countries toward the circular economy. However, the high manufacturing cost of these films due to additives and biopolymers makes them costlier than conventional plastic-based packaging. Both traditional and edible packaging require approval from regulatory authorities, and EFs must undergo thorough scrutiny due to their ingest ability. In particular, edible packaging materials and all their components must be considered food that may be eaten without any potential health concerns [42]. The packaging industry's future is an amalgamation of conventional and edible materials to leverage the merits of each to develop efficient packaging solutions. Since the last decade, consumers have shown an increased interest in purchasing healthier edibles, such as fresh fruits with zero or minimal additives. Additionally, increased environmental awareness has reduced plastic consumption and waste disposal. The recent consumer demands are driving the food and food packaging industry to venture into alternatives, viz., EFs and coatings. These films can effectively replace conventional packaging solutions when fortified with innovative techniques. The development of these coatings aligns with the market trends toward eco-friendly, sustainable solutions.

2 Active, Intelligent, and Nanotechnology-Based EFs

2.1 Active EFs

Active EFs improve food quality and preservation using active components such as enzymes, antioxidants, and antimicrobials. These agents reduce oxidation and microbial growth, enhancing food's shelf-life extension by slowing down or inhibiting oxidation and microbial growth [43]. These active biofilms offer significant health benefits to the consumer in addition to the primary function of preserving perishable edibles. It withholds bioactive agents, which are eventually released into the edibles, fortifying them with nutrients [44].

Adding active antimicrobial and antioxidant compounds into the EFs offers several benefits in improving their physicomechanical properties. The key reason behind this is the microbial activity and oxidation occurring in the perishable items. Considering the growing consumer demand for fewer chemicals and food processing, the researchers are searching for newer techniques for greener food preservation methods. As a result, active additives which can be directly added to the biofilms have gained attention recently. These active additives can be directly applied to the food surface or incorporated within the packaging [45]. Food-grade natural polymer additive-based EFs have recently gained attraction in the packaging industry. These films are manufactured using various natural materials, viz., proteins, polysaccharides, resins, and lipids with (or without) additives. The antimicrobial and antioxidant films contain, for example, essential oils, plant extracts (that include polyphenols), or other synthetic or natural agents. Active films are pivotal in changing the consumer perception of conventional plastic-based packaging films. Figure 4 illustrates the applications of plant- and food-derived bioactive compounds used in manufacturing active food packaging solutions.

2.1.1 Polysaccharides

Polysaccharides are the prominent renewable biomass component and are identified as the most abundant natural resource. Complex carbohydrates comprise long monosaccharide units attached via glycosidic linkages [47]. The natural abundance and cost-effectiveness of these materials are gaining attraction in novel food packaging systems. Due to their relatively low cost compared to biopolymers, they are extensively used as matrices and coatings. Polysaccharides have tremendous potential as food packaging coatings due to environmental friendliness and biodegradability. Moreover, the ability of the human body to metabolize them makes them sound candidates for manufacturing EFs. Pectin, an amorphous, colloidal, white-colored, high molecular weight carbohydrate, is present in ripe fruits such as currants, apples. It is substantially utilized in manufacturing pharmaceuticals, fruit jellies, and cosmetics as an emulsifier and gel-forming ability [48]. Pectin forms a class of complex polysaccharides composed primarily of high molecular weight acidic

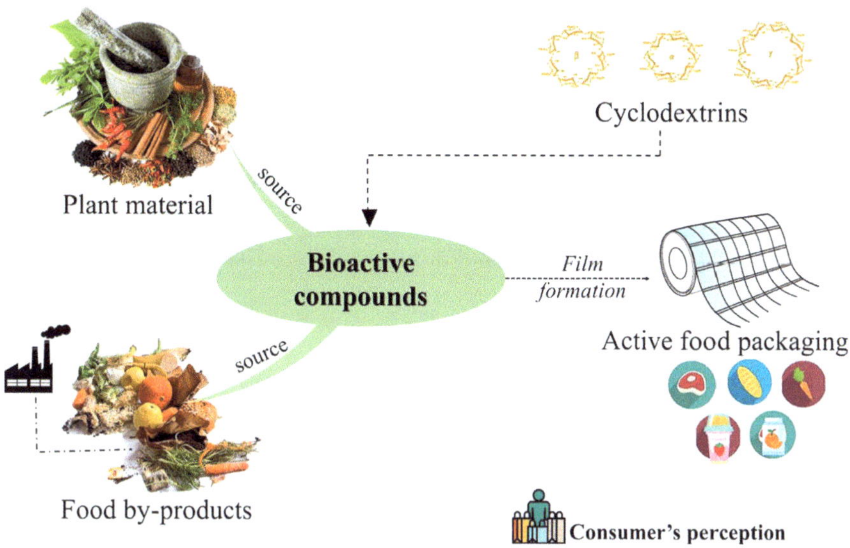

Fig. 4 Application of bioactive compounds from vegetable residues and food by-products in active food packaging [46]

structural polysaccharides and glycan galacturonans with diverse structures. It is one of the most desired materials for manufacturing biopolymers, and it has excellent possibilities for applications in developing EFs.

Polysaccharide-based EFs possess mediocre moisture resistance properties along with selective permeability for CO_2 and O_2 [48]. However, polysaccharide-based films exhibit better water resistance vis-à-vis protein-based films, which are better than the former regarding mechanical and barrier properties. Due to their excellent gel-forming ability, these films are typically made from seaweed extracts, pectin, starch, cellulose, gums, chitosan, and pullulan [49]. These precursor compounds possess crispness, compactness, adhesiveness, and viscosity and are mainly derived from microalgae, bacteria, and seaweed [50]. Protein-based biofilms are primarily derived from animal or plant derivatives such as egg albumin, gelatin, whey protein, casein and soybean, corn, cottonseed, rice, and peanuts. The flexibility of these biofilms can be improved by manipulating the protein network by adding plasticizers. Similarly, hydrophobicity, molecular size, and crystallinity can be augmented by adding hydrophobic additives, such as oils and bee wax [48]. The development of transparent and translucent films from *Pinus densiflora* hemicellulose has been reported [51]. These films displayed high potential for use as EFs with properties close to Xylan. Composite films prepared via acid hydrolysis of cellulose fibrils showed improved mechanical, optical, barrier, and thermal properties [52]. Polysaccharides have been employed extensively for EFs fabrication for cheese preservation and shelf-life extension [53]. These authors utilized galactomannan, chitosan, and agar to prepare cheese coatings with selective moisture transfer, lipid oxidation,

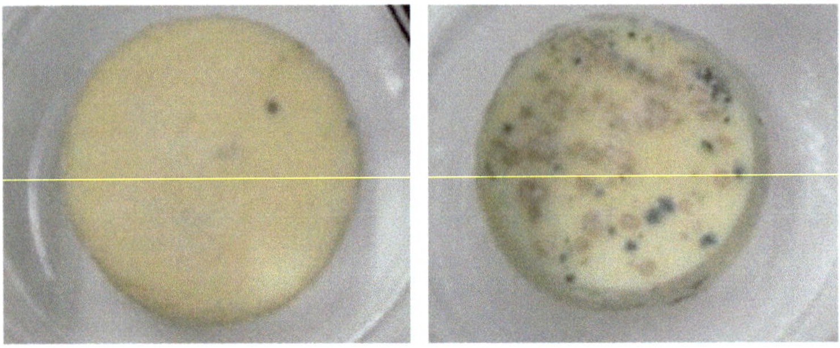

Fig. 5 Visual comparison of **a** coated and **b** uncoated cheese after 48 h [54]

and oxygen uptake. The coatings formed a specific environment around the cheese, preventing it during storage and transportation. The optimum galactomannan formulation with 0.5% oil, 2.0% glycerol, and 1.5% galactomannan displayed excellent water permeability, wettability, and opacity. An opacity of just 5.27% was instrumental in restricting the incident light on the cheese. The optimal mix showed reduced mold growth and respiration rates on the cheese surface. The better moisture retention was manifested by a smaller relative weight loss of 0.11% in coated cheese compared to 0.84% in uncoated cheese. Figure 5 depicts the condition of coated and uncoated cheese after 48 h. Extensive mold growth on the uncoated cheese can be discerned from the figure over the entire surface.

2.1.2 Probiotics

Probiotics, defined as living microorganisms with health benefits for the host, have been added to biopolymer matrices for use in active food packaging. The term "probiotic" implies "for life" and includes bacteria with beneficial effects on human and animal health [54]. These biopolymers can develop active food packaging with excellent safety and health benefits. Probiotics are assimilated in dairy products such as ice cream, desserts, and yogurt from the bovine, caprine, and ovine [55, 56]. These are also extensively utilized in non-dairy products such as fruit juices, fermented vegetables, and processed fruits for developing cholesterol and lactose-free products [57]. Probiotics have recently found use cases in edible polymeric matrices that are safe for food packaging due to the inherent antimicrobial characteristics that simultaneously benefit health. For them to have the desired effect on human health, maintaining their viability and metabolic activities throughout food processing and supply chains plays a pivotal role. In addition, probiotics must retain their key properties within the gastrointestinal tract of human beings.

Starches derived from various sources such as tapioca, potato, and corn have been used to develop EFs by assimilating probiotic cultures *Lactobacillus reuteri*, *Lactobacillus rhamnosus*, and *Lactobacillus acidophilus* [58]. Pullulan exhibited a probiotic viability of 80% after ten days of storage at 25 °C. The cell viability decreased to just 35% after twenty days. However, the starch content was inversely proportional to the cell viability. Another potential probiotic microorganism, *Carnobacterium maltaromaticum*, found in the normal intestine of several fish species, exhibited noticeable activity against gram-positive and negative pathogens along with antibiotic characteristics [59]. Biofilms developed by incorporating *C. maltaromaticum* into alginate have been utilized to preserve smoked salmon [60]. These films, applied on the salmon's surface, showed bacterial resistance, inhibiting infection under predetermined conditions. Sayed et al. evaluated the efficacy of edible coatings developed from sodium alginate, chitosan, and carboxymethyl cellulose with probiotic strains and antimicrobial microcrystalline cellulose [27]. Successful film formation was discerned through the FTIR analysis, showing a strong interaction between biopolymers and the probiotic strains. The developed probiotic films were employed to preserve soft cheese for 45 days with better sensory, microbiological, and chemical performance. The addition of antimicrobial microcrystalline cellulose enhances the mechanical properties of the films due to active biopolymer/crystal interaction, resulting in stress transfer. These films efficiently preserve the cheese from spoilage and maintain its stability even after the stipulated duration. Using lactic acid bacteria as a probiotic is another medium for food preservation by fungal contamination inhibition in dairy food items. Their abundance of several food items, especially dairy products, has earned them the status of GRAS [61]. EFs can effectively carry lactic acid bacteria and improve food quality by controlling gas exchange and lipid oxidation. Due to its semi-permeable nature, it can successfully prevent aroma, flavor, and freshness loss [62]. Whey protein-based EFs were developed by incorporating *Lactobacillus buchneri* probiotic against the spoiling effects of *Penicillium nordicum* bacteria in cheese. The developed films were thicker than the control films but had identical water permeability, moisture content, mechanical properties, and chemical structure. Interestingly, the *L. buchneri* cells were viable even after 30 days and highly effective against fungal contamination. Incorporating probiotics into EFs has a high potential as an alternative technique compared to the existing chemical techniques. Figure 6 compares the uncoated and *L. buchneri*-infused edible film-coated cheese. Extensive bacterial growth in the uncoated cheese sample can be seen in the uncoated sample.

2.1.3 Organic Compounds

Antimicrobial agents, viz., sodium benzoate, propionic acid, acetic acid, sorbic acid, potassium sorbate, and benzoic acid, are most commonly used in edibles. According to the United States Food and Drug Administration (USFDA), these acids are effectively used in food products due to GRAS rating. As a result, these chemicals have been used in EFs and coatings to suppress fungal and bacterial growth in food systems

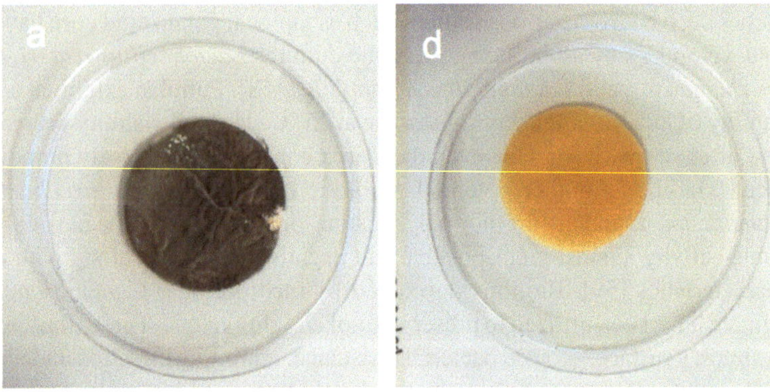

(a) (b)

Fig. 6 *P. nordicum* bacterial growth in **a** uncoated and **b** *L. buchneri*-infused edible film-coated cheese [63]

[63]. Additionally, salts of organic acids, commonly known as carboxylate salts, are extensively utilized as antimicrobial agents due to their high efficiency. The application of chitosan films loaded with propionic and acetic acids for meat preservation has been reported as early as 2000 [64]. These films successfully delayed the processed meat stored at 4 and 10 °C for 7 days by inhibiting *Serratia liquefaciens* and *Enterobacteriaceae* growth. Similarly, whey protein films incorporating p-aminobenzoic or sorbic acid inhibited the growth of *monocytogenes*, *Escherichia coli*, and *Salmonella typhimurium* [65]. These antimicrobials formed an inhibition zone of up to 26.7 mm around them, indicating a high lethality. Moreover, these antimicrobial compounds restricted the microorganism growth on the food surface.

The antimicrobial attributes of sorbic and benzoic acids and their salts against molds, fungi, and several bacteria have been reported in the literature [66, 67]. Similarly, potassium sorbate incorporated with wheat gluten has demonstrated antifungal properties [68]. *Aspergillus flavus* is a vital fungus that is particularly detrimental to peanuts and maize due to the aflatoxin synthesis in these plants. Another critical fungus is the *Rhizopus oryzae*. This fungus is predominantly found in spoiled food, vegetation, soil, and other substrates. Carboxymethyl cellulose-based films with varying potassium sorbate concentrations displayed excellent antifungal properties against *A. parasiticus* and *A. flavus* fungi [69]. Incorporating benzoic and sorbic acid into Argentine anchovy protein films showed significant antifungal properties against *R. oryzae* and *A. flavus* [66]. Films with just 1.5% sorbic or benzoic acid exhibited noticeable effectiveness against *R. oryzae*. A remarkable interaction was observed between the benzoic acid and Argentine protein isolate in the FTIR spectroscopy. The results indicated the potential utilization of these films in food safety applications to inhibit fungal growth, improving its shelf life. Additionally, at all concentrations of sorbic acid (0.5–1.5%), remarkable improvements were observed in the elongation of water solubility with a simultaneous reduction in the tensile strength.

2.2 *Intelligent EFs*

Intelligent packaging is a special type that keeps track of quality, safety, location, storage, retail sales, and consumption. As per the European Food Safety Authority, intelligent packaging implies packaging systems that monitor the packaged food's status and environment [70]. Next-generation packaging systems are set to disrupt the current trend of conventional packaging systems. It comprises two systems: one comes directly in contact with the edible, whereas another is concerned with monitoring the outside factors. The first system must comply with the regulatory guidelines concerning food quality and safety [71]. Intelligent packaging systems consist of three components, namely, indicators, data carriers, and sensors. These technologies are actively employed to collect and analyze food safety, storage, transportation, consumption, retail sales data, etc. Smart packaging systems provide valuable information about food items to manufacturers, shopkeepers, and consumers regarding the quality and freshness of edibles. Figure 7 depicts the concept of smart and intelligent packaging in the food packaging industry. Smart packaging maintains the stability of edibles by actively preventing or delaying their spoilage. It enhances the quality characteristics of the food, such as sensory feel, texture, and flavor, and monitors any deviations from the baseline.

Smart packaging informs end users about the food's history and present status and ensures its authenticity by preventing counterfeiting. The smart packaging systems

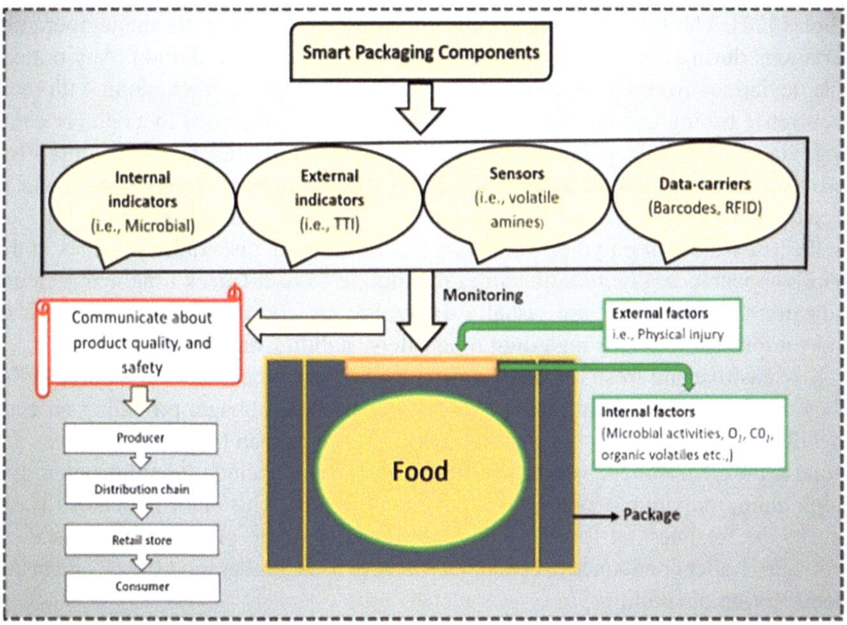

Fig. 7 Smart food packaging concept [72]

depend on indicators, data carriers, and sensors to gather and convey this information to the intended entity. Indicators indicate the alterations in the food item due to the environmental changes by generating a visual signal. Sensors are specialized devices comprising electronic circuitry integrated with software for quantifying the edibles' biological, chemical, and physical changes. Data carriers such as radio-frequency identification (RFID) tags and barcodes store and transmit data for identification and tracking. These electronic devices must be non-toxic, sensitive, cost-effective, and easy to use with packaging systems. These devices must possess high accuracy and precision and be able to monitor the food's condition with high fidelity [72].

2.2.1 Indicators

The indicators play the primary role of detecting an edible item's presence (or absence). It also detects any specific agent's concentration and the interaction between different components. These indicators, placed outside the packaging, provide information about the status of packaged items through various parameters, such as changing color intensities [73]. Leakage, freshness, and time–temperature indicators are the most common indicators used in intelligent packaging. Temperature plays a critical role in deciding the shelf life of food items. It is well known that significant temperature alterations may lead to premature decaying of the edibles due to microbial growth and product deterioration. Moreover, various chemical, enzymatic, and physical changes occur due to temperature variations in the ambiance of edibles [74]. The time–temperature indicators are used to monitor the temperature variations during food items' storage, transportation, and distribution. Any noticeable deviations from the prescribed temperature are duly recorded along with their occurrence timing and magnitude. These deviations are recorded in a belt (or color ring) via suitable pre-decided colors to indicate the food's status. The food items often lose their freshness due to unfavorable ambient conditions. A freshness indicator is used to monitor the food quality during storage, logistics, and distribution. It is based on the assessment of polymerization reaction resulting in discernible changes in the label's appearance. Figure 8 illustrates the lifeline's Fresh-Check time–temperature indicator. Color changes are visually discernible by consumers, enabling them to make informed decisions regarding food safety, stability, and quality.

pH-sensitive and fresh indicator films are generally prepared from Anthocyanin. They can be isolated from plants like berries and red cabbage, providing an eco-friendly replacement for conventional dyes [75]. More than 600 different types are found in the environment, while six different plant forms include cyanidin, petunidin, delphinidin, malvidin, peonidin, and pelargonidin [76]. This color indicator for pH detects the freshness of food through color change due to pH variation caused by spoilage. Higher confidence in consumer safety of food, higher food safety, and faster identification of spoiling.

Fig. 8 Color fading in colorimetric time-temperature indicator after predefined time at 25 °C [74]

2.2.2 Sensors

Sensors detect, locate, and quantify any quantity by measuring physical–chemical properties [77]. It consists of two main components: a receptor and a sensor. The physical and chemical alterations in the edibles can be detected by measuring deviations in the baseline activity, composition, and concentrations via the sensors. The detected signal is then converted into an energy form, which can be manipulated as required through another device known as a "transducer" [78]. Various gas, chemical, colorimetric, and biosensors are generally used in edible packaging systems. The electronic nose imitates the olfactory system to identify the odors emanating from food items [78]. These noses employ chemical sensors to detect the volatile organic compounds evolving from the food, informing the consumer about its freshness. Similarly, biosensors detect particular biological reactions occurring in the food [79]. These sensors generally capture the biogenic amines and other spoilage indicators. The colorimetric sensors are specialized chemical dyes that fade after a predefined duration. This fading directly indicates the freshness of food products (especially meat and fish) [74]. For example, the quality of meat products is monitored by determining the color change through specially designed cameras or other image-capturing instruments [80]. Figure 9 illustrates the colorimetric sensor for freshness detection of the food items. Gas sensors are specialized devices for detecting CO_2 and O_2 evolving from food items. These gases indicate the microbial activity in the edibles due to food spoilage and damaged packaging.

The beetroot waste extracts EFs developed are highly effective in monitoring food items' freshness [81]. These films exhibit pH sensitivity, undergoing alterations in color from red to yellow within a pH ranging from 1 to 13, rendering them useful in intelligent packaging systems. Encapsulated beetroot films with just 2% concentration could effectively monitor the freshness of hake medallions via color changes correlated to the spoilage indicators by the fourth day. This ability to visually offer

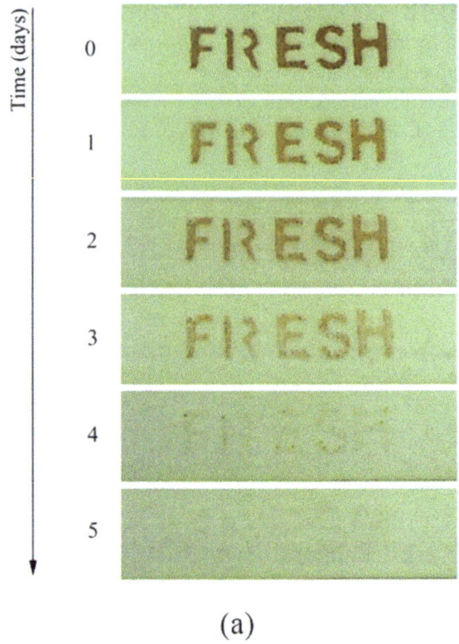

Fig. 9 Color fading in colorimetric sensor after a predefined time at 25 °C [74]

changes in the edibles' freshness and quality is a game-changer for consumer health and decision-making.

2.2.3 Data Carriers

Data carriers help communicate helpful information regarding food all over the supply chain. These ensure anti-theft, automation, traceability, and imitation protection throughout the food product supply chain [73]. The data carriers effectively store and transmit information on food product storage and logistics, which are applied to tertiary packaging. RFID tags and bar codes are the most common types of data carriers used in food packaging systems. The use of RFID technology in agri-food products has grown significantly in the last few years for automated traceability and identification. These use the RF electromagnetic field and promise high fidelity, accuracy, and precision at rapid speed vis-à-vis manual systems [82]. The RFID tags employ remote-controlled non-line-of-sight type techniques for collecting and transmitting information, viz., origin, file-tracking, and real-time locations of the food items [83]. Barcodes are devices based on the optical machine-readable picture principle attached to the product for identification [83]. Since their inception in the 1970s, they have been used in almost every product for inventory control, stock recording, and check-in and out presently [84]. It comprises spaced bars symbolizing twelve-digit numeric data with information like manufacturer information and item numbering. The barcodes employ reduced-space symbology, composite, and

Fig. 10 **a** One, **b** two-dimensional barcodes, and **c** QR codes used in the food industry as data carriers [72]

two-dimensional GS1 data bar families to acknowledge the demand for coding extensive data in minimal space. Figure 10 depicts the barcodes used for meat product packaging.

2.3 Enhancements Through Nanotechnology

Despite their usability, edible coatings pose significant challenges, such as weak water vapor barriers and low mechanical properties, primarily when employed with fresh produce. These challenges have attracted researchers to explore alternative packaging techniques, especially for perishable items. Nanotechnology is one of the techniques with numerous applications in biology, sports, medicine, energy production, chemistry, and the food industry [84]. It is based on the premise that the devices at the nanoscale (1–100 nm) compete with biomaterials found in the biomolecules, proteins, and membranes [85]. Nanotechnology applies to the ability to work with matters at the nanoscale. The applications of nanotechnology as a novel approach have resulted in effective post-harvest management of fresh produce. It is based on the premise that interesting properties are obtained in the edible coating at the nanoscale [86]. Nanoparticles possess a large surface area to volume ratio, morphology culminating in novel physicomechanical properties. Only a few nanoparticles are required to form robust interfacial bonding with the edible coatings. These particles have

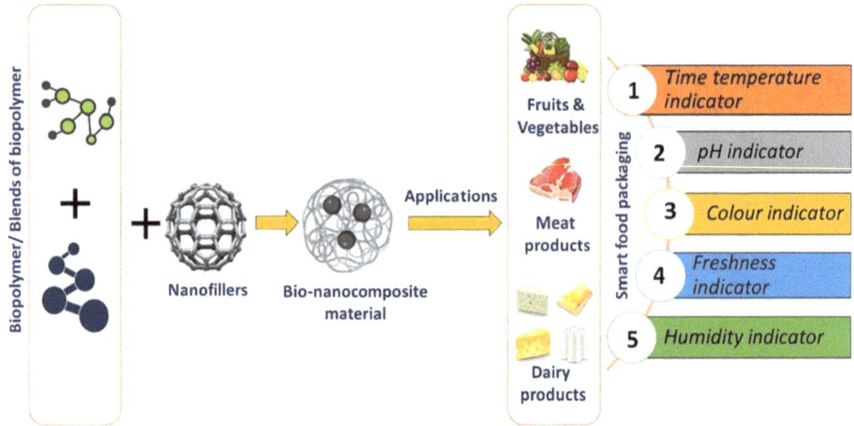

Fig. 11 Development of bio-nanocomposites using biopolymers and nanofillers [89]

been shown to enhance the shelf life of fruits compared to conventional polymer-based packaging materials. Adding nanoparticles to edible coating can improve their barrier and mechanical properties while augmenting thermal stability [87]. Figure 11 shows the process of making bio-nanocomposites using biopolymers and nanofillers to create innovative food packages for various edibles [88].

Compared to conventional packaging, nanotechnology-based packaging offers several advantages, like processability, thermal resistance, durability, and mechanical durability. Some nanofillers, such as chitosan, silver, ZnO, possess antimicrobial characteristics, engendering microbe resistance in food packaging [89]. Some metal and non-metals (Zn, Mg, C, Ti, Si) and their oxide nanoparticles have been shown to possess excellent antimicrobial characteristics and can be applied to food packaging [90]. As a result, incorporating these nanofillers into food packaging systems ensures food safety and shelf-life enhancement by vanquishing the deleterious microorganisms [13]. Nanoparticle-based EFs and coatings are highly effective in delaying food spoilage by providing sustained release of antimicrobial agents over an extended duration. The resulting enhancement in food's shelf life and quality leads to overall customer satisfaction and brand building [91]. Nanotechnology-based packaging solutions offer ecological advantage vis-à-vis their conventional counterparts by facilitating spoilage indications. These indications, such as off-flavors, harmful food toxins, and color, may be provided in any form. When applied to EFs, such a system boosts food security by sensing and reporting the nutraceutical food quality via nano-based techniques [92]. Essential oils are one of the critical materials for extending the shelf life of fruits by behaving as antimicrobials [93]. These oils have been shown to improve the longevity of perishable items for export to distant countries. Researchers have nano-encapsulated essential oils from cinnamon bark through whey protein concentrate [94]. The technique was employed to develop food-grade wax with excellent wettability on the produce surface. These coatings demonstrated a static contact angle of $85 \pm 1.6°$ and receding and advancing contact angles of 53.0

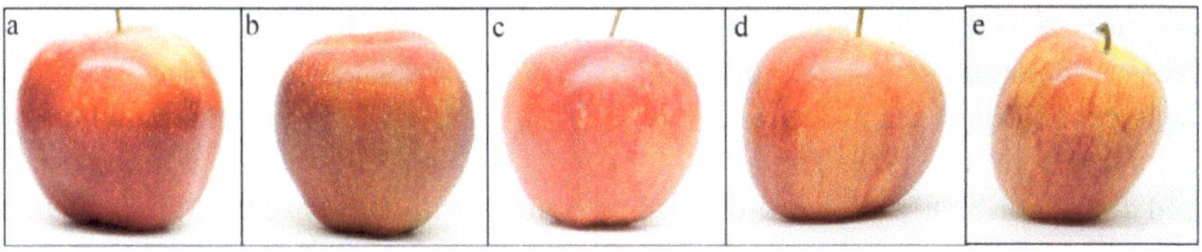

Fig. 12 **a** Freshly procured apples with a glossy appearance due to the applied wax, **b** apples washed with chloroform, **c** coated with paraffin, **d** apples coated with cinnamon essential oils, and **e** apples coated with freeze-dried cinnamon essential oil nanoparticles [94]

$\pm$ 1.6° and 90 $\pm$ 1.1° on the apple surface. These nanoencapsulations improved the shelf life by enhancing the half-life of produce by 61 h vis-à-vis bare produce. The bacterial population was reduced drastically upon using the proposed encapsulations, which delayed the bacterial and fungal activities by 72 h. Figure 12 depicts the appearance of red apples at various treatment stages.

In contrast to the merits mentioned above, nanotechnology is associated with undesirable consequences, risks, and adverse effects on the human and environment. Their persistent nature and non-biodegradability are observed to have a toxicological impact on the health and well-being of animals and humans. The nano-sized particles exhibit high reactivity and active interaction with the biological components, rendering them perilous for widespread applications in the food industry. These tiny particles can transgress across even the delicate cell membranes, causing severe damage [94]. Alarmingly, the ZnO nanoparticles show high toxicity levels compared to the bulk ZnO, which is non-toxic [89]. The toxicity of nanoparticles increases with a decrease in their size, causing severe damage to the lung tissues and even cancer. This danger increases when nanoparticles are used in the EFs as their toxicity does not depend on particle migration. The ecotoxicity of nanoparticles also harms the sustainability of ecological balance [36]. Nanomaterials can affect the food's sensory qualities (taste, appearance, and texture), significantly altering the mouthfeel. Nanoparticle-based coatings have been shown to manipulate the flavor profile of vegetables and fruits under long-term applications [95]. However, long-term consumption of nanotechnology-based coatings may cause effects that are yet to be understood completely, raising grave suspicion over their perilous effects. Due to this, rigorous regulatory guidelines and strict implementation are needed to balance their benefits with environmental sustainability.

3 Fabrication of EFs

Biodegradable active packaging solutions have the potential to drastically reduce greenhouse gas emissions by circumventing plastic usage. Using biopolymers to fabricate edible biodegradable films has enormous potential for cutting the carbon footprints of packaging industries. These biopolymers should be derived from renewable raw materials to maintain the circularity. Since their inception 100 years ago as a wax coating on fruits and vegetables, edible packaging forms an essential class of packaging solutions [96]. The development of these coatings aligns with the urgent need for traditional petroleum-derived polymers with bio-based polymers. Other innovations, such as "intelligent" and "smart" packaging, monitor the condition of the edibles, providing valuable feedback regarding the product's freshness and stability. Liquid coating materials are applied to food items via spraying, painting, dipping, layer-by-layer coating, cooling, and drying [97]. For example, edible coating fabricated from aloe vera gel is applied to the food via dipping [98]. In the case of powdered solid coating materials, the fluidized bed coating technique is widely accepted [99]. This section describes the main edible film manufacturing processes used prominently in the food industry, such as dipping, spraying, vacuum impregnation, and 3D printing.

3.1 Dipping

One of the most common, easy, straightforward coating techniques is "dipping." It comprises four steps: soaking, immersion, removal, and drying [100]. In this technique, the food item is immersed into the coating solution and held there for a stipulated duration. It is dried once the edible has made complete contact with the solution for the predetermined time. This procedure develops a protective coating on the food item's surface. Its main advantage lies in the ability to coat the surface of irregular fruits and vegetables such as durian and dragon fruit. The coated packaging restricts the moisture loss from the stored, offering extended protection from spoilage. The success of the dipping technique depends upon various factors, viz., coating solvent, viscosity, surface tension, and temperature. Dipping frequency and immersion duration are also critical in determining the coating quality. Viscosity affects the coating thickness of both dry and liquid coatings, with high viscosity being particularly beneficial for the dipping coating method. Contrarily, low-viscosity coating materials are a better choice in the spraying technique [101]. The appearance of fresh-cut fruits is another crucial factor affecting the applications of the dipping method. Dipping is easy and favors mass production but needs accurate thickness control, environmental management, and solvent volatilization. Moreover, poor coating/food surface adhesion is another pivotal factor affecting its widespread applicability, especially on smooth and (or) hydrophobic surfaces.

3.2 Vacuum Impregnation

An extension to the conventional dipping technique, "vacuum impregnation" is a hydrodynamic mechanism-based non-destructive technique [102]. In this technique, the food surface is coated by introducing a predetermined solution concentration into the intracellular spaces of the food matrix. The coating also engenders desired nutritional value and functional requirements into the food item simultaneously coating it. As soon as the edible comes into contact with the solution, the air is sucked via an air pump, creating a vacuum. Once the stipulated hold time has elapsed, the pressure is restored to normal pressure, creating a thicker and uniform coating [100]. The excess coating (if any) is removed from the fruit surface, resulting in the final optimally coated unit.

Vacuum impregnation cannot be applied to food items with high porosity. When subjected to vacuum pressure, the coating solution may penetrate them up to a depth more than desirable. Additionally, the food may undergo irreversible deformation under vacuum pressure, leading to poor aesthetics post-coating [102]. This procedure requires strict control over the pressure to minimize the physical damage and undesirable alterations in the physical properties of the edibles. It may be noticed that both dipping and vacuum impregnation cannot be applied to protein-based products such as dairy and meat. These items' poor firmness renders them a candidate due to their inability to sustain vacuum coating pressures.

3.3 *Spraying*

In this technique, a pressurized coating solution is forced on the food surface through a low-diameter nozzle with rapidly flowing air. This atomized coating solution attaches itself to the food surface, which is to be protected. A protective coating is formed on the food surface after the coating solution has evaporated. This technique has a higher material utilization than dipping, making it widely accepted for mass production. Hole diameter, temperature, pressure, surface tension, rheology, and nozzle shape are the critical parameters affecting the coating quality [103]. Other factors, such as spray distance, coating material flow rate, angle, and overlap rate, affect the coating quality [104]. Spraying has widespread applications due to the absence of direct contact with the food items. The spatial separation between the sprayer and edibles allows large food items to be coated relatively quickly. The separation reduces the contamination risk, drastically improving food's shelf life [104]. Dipping, on the other hand, has fewer industrial applications due to the excessive residual solution on the food surface, which requires more time for drying [105]. Figure 13 elaborates on the steps involved in dipping, spraying, and vacuum impregnation.

3.4 *3D Printing*

3D printing has gained the interest of researchers due to its ability to generate intricate and complex geometry along with economic and environmental benefits [106]. The 3D printing technology can transform material composition and structure by

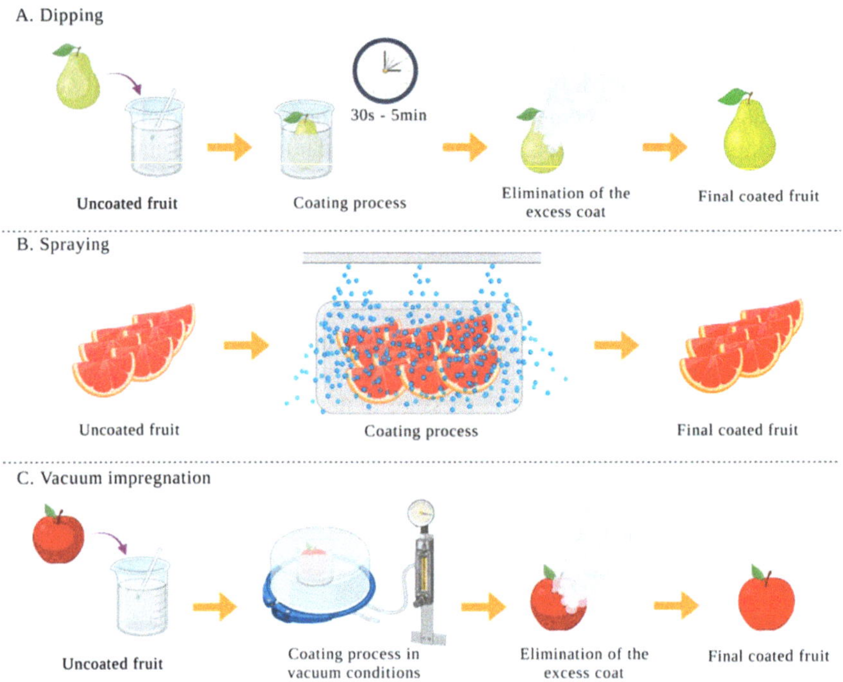

Fig. 13 Steps involved in dipping, spraying, and vacuum impregnation [17]

generating a printed object. It can generate regular gratings, thin films, variable depth patterns, lines and dots, and multiple heterogeneous designs. It allows adjustments of various parameters, such as shape optimization, surface roughness, bubble thickness, to develop attractive designs. This emerging technology has found extensive avenues in the food industry. It prepares customized meals based on specific calorie intake and nutritional requirements for pregnant women, older people, growing children, and athletes [107]. 3D printing offers a novel opportunity for patients and elderly persons with problems swallowing food to extract maximum nutritional benefits from food. Regarding food packaging solutions, 3D printing technology offers rapid manufacturing to print edible packaging systems using three-dimension models. It generates objects via sequential stacking and is extensively used in the construction, medical, biological, and aerospace industries [107]. 3D printing technology comprises various printing techniques, viz., stereolithography, powder bed fusion, binder jetting laminated, and object manufacturing. However, fused deposition additive manufacturing remains the most prominent manufacturing technique for edible coatings. The 3D printing of EFs is a three-step technique to fabricate products via additive manufacturing. The printing contours are decided by the computer-aided design developed on a computer. The edible material is then heated in the printer's heating chamber and melted into liquid to ensure flowability. This step is followed by moving the nozzle along the two-dimensional geometric trajectory and extruding the material.

The nozzle is traversed vertically, depositing the edible material layer-by-layer onto the base. The edible products are obtained from these depositions after adequate curing.

Leaw et al. used 3D printing to develop starch-gelatin-based EFs incorporating hawthorn berries extract [108]. The film-forming solutions were obtained using bovine gelatine and corn starch, and hawthorn berries extract was added by glycerol as plasticized. The obtained 3D-printed films were highly effective in containing the growth of *Pseudomonas aeruginosa*, showing high antibacterial activity. Moreover, the elongation at break and tensile strengths were comparable to the conventional standard for films developed with 4% glycerol. These films showed significant water solubility improvement from 30.02 to 49.75%, implying a noticeable enhancement in their edibility. Vegetable protein, chocolate, candy, and pancakes are ideal for fused deposition modeling [109]. The 3D printing technology has noticeable advantages vis-à-vis conventional film fabrication techniques. It results in significantly lower material wastage, fewer post-processing steps, and the ability to produce customized film shapes.

4 Applications and Challenges of EFs and Coatings

4.1 Applications

Over the last few years, EFs have been studied extensively in formulation and application strategy to improve the performance of coatings. However, edible film materials generally possess some limitations. For example, lipids-based EFs are not mechanically strong and provide unacceptable organoleptic properties through lipid oxidation. Similarly, protein-based EFs are brittle and susceptible to cracking, with their application restricted by possible allergens and ethical issues [110]. Meanwhile, polysaccharide-based EFs possess poor water barrier properties and suboptimal adhesion on the fruit and vegetable surface. The mechanical performance of these films can be improved by adding crosslinking agents, viz., organic acids, plasticizers, metallic ions, and phenolics [111].

Despite the novelty of the concept of EFs, numerous applications have been reported in the literature with limited success [112]. EFs are extensively employed for various purposes, from fruits and vegetables to meat and dairy products. Improved product aesthetics provide an advantage to the coated edible items. As discussed, these proteins, polysaccharides, or lipid-based coatings offer antimicrobial and antioxidative properties, enhancing shelf life. Figure 14 illustrates the application and action mechanism of EFs on various food products.

Biopolymer-derived EFs offer an innovative and sustainable approach to maintaining the freshness of edibles. These films reduce microbial contamination, lipid oxidation, and moisture loss from the food, extending their shelf life [114]. Biopolymer edible coatings are applied to grapes [115], bananas [116], mango [117],

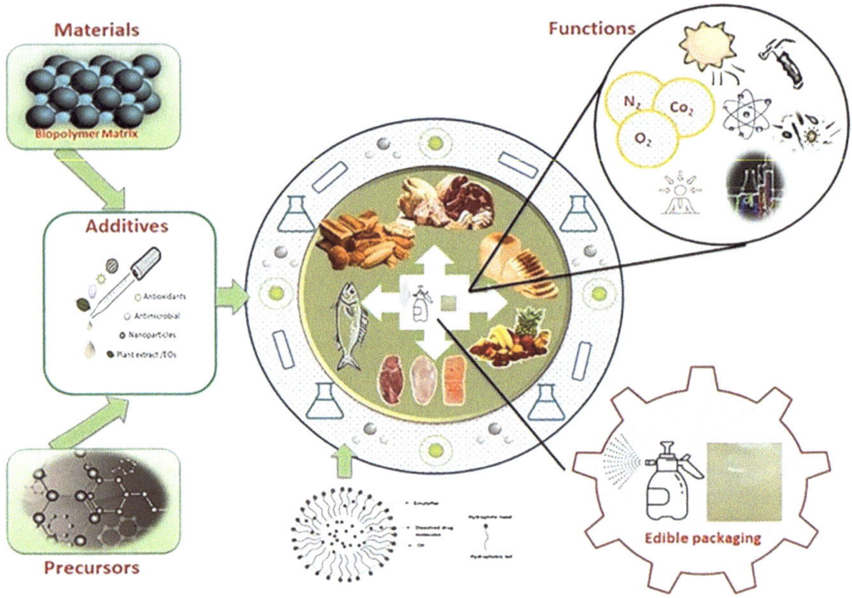

Fig. 14 Application and mechanism of edible coating on different types of food products [113]

and cucumber preservation [118]. Meat products, viz., chicken breast [119], trout fillets [120], and pork patties [121], have also been effectively preserved using these films.

Various biopolymers, such as polysaccharides, chitosan, alginate, pectin, and starch, have been used to develop edible coatings [122]. Polysaccharides and lipid's excellent gas and water barrier properties make them attractive precursors for EFs [122]. Commercially available lipid-derived coatings are highly efficient in reducing the respiration rates in cucumbers, presenting themselves as an option to conventional plastic-based packaging [123].

However, the coating performance depends upon the solvent type, deposition method, crystallinity, and polymer composition [124]. Their applications can be broadened by obtaining active, intelligent composite films with antioxidant and antimicrobial additives [36]. Such biodegradable, eco-friendly films act as a barrier against oil, oxygen, and moisture, making them sustainable packaging solutions. The following sections discuss the applications and challenges of edible coating and film on fruits and vegetables, dairy, and meat products, are discussed below.

4.1.1 Meat and Meat Products

Meat is a critical source of essential nutrients such as proteins, minerals, fatty acids, B-complex vitamins, and micronutrients [125]. However, it is influenced in terms of quality by biological, chemical, and physical properties like temperature, pH, light, and microbial contamination [126]. Spoilage is caused by lipid oxidation, microbial growth, and enzymatic autolysis, which influence meat's texture, color, flavor, and nutrients [49]. Due to its perishable nature, meat preservation demands the proper packaging. Since, consumers increasingly seek sustainable alternatives, edible packaging is a good solution for extending shelf life and enhancing food safety. Implementing EFs for meat and meat product packaging started with gelatine films and collagen casing [127]. Currently, regenerated and reconstituted collagen films are used extensively in the meat packaging industry to pack hams and sausages [128]. Noticeably, these films can be fabricated into the desired shape to fit the meat product. Antimicrobial coatings with augmented mechanical properties are manufactured by adding starch and cellulose derivatives or extruded collagen fibers [129]. Implementing these edible coatings on meat products has drastically reduced moisture loss, lipid oxidization, fading, and water drippage [130]. However, the quantity of vegetable oil and biopolymer affects the edibles' water loss, lipidation, and sensory properties [131]. Collagen is the most significant protein found in the bodies of animals, comprising 20–25% of the total proteins available in their body. Noticeably, collagen or its hydrolyzed derivative gelatin is substantially used for manufacturing these EFs. Gelatin films display substandard barrier properties due to their high moisture affinity [132]. Natural casings obtained from pigs, goats, and sheep are used to produce sausages. The collagen-rich submucosal layer of the intestine provides ample strength and flexibility and is one of the oldest EFs. These films show high moisture permeability in the high-humidity environment. However, at approximately 15% concentration, it exhibits a tensile strength of up to 100 MPa [133].

Different studies have indicated the ability of EFs and coatings to improve meat preservation during storage. Improved meat consumption, however, is linked to serious health and environmental degradation [132]. Meat alternatives derived from pulses, grains, and oils can replace conventional meat by mimicking taste and texture [134]. Their manufacture also promotes environmental sustainability through greenhouse gas emission reduction and climate change mitigation [135].

4.1.2 Fruits and Vegetables

Fruits and vegetables contain vitamins, proteins, minerals, fiber, and antioxidants essential in the human diet due to their rich nutrient and photochemical content [136]. Their consumption has been linked to low risks of cardiovascular diseases, diabetes, and stroke [137]. The WHO suggests a daily intake of 400 g of fruits and vegetables for optimal well-being [138]. However, their short lifespan is a problem for post-harvest management and causes humongous losses and environmental damage by physical, mechanical, chemical, and biological means [139].

Edible packaging provides a sustainable and eco-friendly way of minimizing post-harvest losses through shelf-life extension, moisture loss prevention, respiration rate, and enzymatic browning. The semi-permeable EFs provide a barrier against moisture and gas transpiration, which helps maintain aroma, freshness, and color [113]. Studies

Table 1 Overview of various EF formulations for fruits and vegetables

Type of EFs	Key components and additives	Functional benefits	References
Nanoparticles-based EFs	Chitosan, cellulose, lignocellulose, TiO_2, Ag, ZnO, CuO, SiO_2, selenium nanoparticles	Improved gas barrier, antimicrobial, reduced microbial spoilage, enhanced storage stability	[140]
Nano-emulsion-based EFs	Nano-emulsions of essential oils (clove, carvacrol, rosemary, oregano, cardamom), chitosan, alginate, starch	Smaller droplet size, enhanced bioavailability, antimicrobial, preserved texture, and sensory qualities	[141]
Bio-nanocomposite-based EFs	Chitosan, cellulose nanoparticles, carnauba wax, egg albumin, red roselle extract, ZnO nanoparticles, essential oils	Improved coating strength, reduced oxygen and water permeability, prolonged shelf life	[142]
Multilayer methodology EFs	Sodium alginate, citric acid, oxalic acid, chitosan	Enhanced barrier properties, controlled release of additives, prolonged storage	[143]

have shown that polysaccharide, protein, and lipids-based edible packaging prevents fruits and vegetables from mechanical, chemical, and biological losses during transportation and storage [114]. Various additives, viz., essential oils, chitosan, nanoemulsions, and nanoparticles, are added to these films to engender several functional benefits. Table 1 depicts the key components and their functional benefits in different EFs.

4.1.3 Dairy, Bakery, and Other Milk Products

EFs can preserve dairy products such as milk, yogurt, cheese by mitigating microbial activity-induced contamination [144]. These highly perishable food items are prone to mold and yeast spoilage due to the high moisture content [145]. Desrizal et al. have demonstrated the exceptional capability of biopolymer-based coatings vis-à-vis carrageenan in enhancing the shelf life of brown seaweed by inhibiting mold growth [146]. Similar studies on cheese preservation by sweet whey [147], chitosan [148], galactomannan [149], and agar [150] have also shown encouraging results. The potential of active EFs containing natural essential oils for soft cheese and butter shelf-life extension has been reported [151]. The high moisture and fat content in dairy products make them susceptible to mold growth due to lipid oxidation, causing premature spoilage. The developed films were efficacious in inhibiting spoilage by lowering the values of peroxide and thiobarbituric acid reactive substances, confirming their better preservation capability.

Moreover, essential oil-fortified films were also highly effective in preserving sensory attributes such as butter and cheese's appearance, flavor, and texture. Moisture loss was substantially reduced while maintaining high water content in the edibles, significantly decreasing the production of free fatty acids. This finding indicates slower lipolysis, culminating in better preservation of the dairy items by the proposed films. Nanotechnology-based EFs have enhanced the bioavailability and health benefits of yogurt's probiotics and bioactive compounds [152]. Nanomaterials viz., nano-emulsions, liposomes, and micelles are employed to encapsulate these ingredients within the yogurt, transforming it into a functional dairy product. In this regard, xanthum gum, alginate, and gum Arabic combined with maltodextrin have colossal potential as coating materials. These materials have been shown to promote good bacteria growth in yogurt, augmenting its health benefits significantly. They ensure the targeted delivery of nutrients to the intended location within the human digestive system, ensuring maximum benefits. In totality, edible coatings have tremendous potential in preserving (and enhancing) dairy products' stability, longevity, nutritional profile, and health. These films maintain color, acidity, pH, taste, firmness, and other sensory attributes while preventing weight loss due to moisture bereavement.

4.2 Challenges and Limitations of EFs

Food coatings present a green and sustainable solution to traditional plastic-based packaging, with a smaller environmental footprint and higher sustainability. Nevertheless, due to their widespread utilization, the problems related to their production, storage, and industrial applications and assuring food safety and consumer acceptability need to be overcome [153]. One of the significant drawbacks of biopolymer-based edible coatings and films, including polysaccharide- and protein-based ones, is their weak gas and water barrier characteristics. For this purpose, composite packaging solutions with emulsifiers, plasticizers, and other additives are suggested for augmenting mechanical strength, thermal resistance, and barrier efficiency [154]. In addition, excess levels of biopolymers and actives, including plant extracts and essential oils, will interfere with flavor and pose possible toxic threats, leading to unacceptable consumer concentrations. Governmental regulatory limitations of active agent concentration levels in food coating still exist. Still, the regulatory agency requires higher consumer acceptance and control [155]. To make edible coatings a viable commercial alternative, regulatory agencies need to encourage their environmental and consumer advantages. Overcoming these hurdles using research and innovation will be the most significant factor in making edible coatings a commonly used packaging option in the future.

4.2.1 Mechanical Strength

Food products' shelf life can be extended through edible coatings that are promising but sometimes lack stability in their mechanical strength and endurance. This could pose challenges, especially when juxtaposed with conventional petroleum-based plastic strength. Researchers are deeply engaged in bridging the present limitation by studying the overall assessment of methodologies utilized to enhance the mechanical properties of edible coatings. These efforts include systematically incorporating diversified techniques, the prominent being mixing multi-biopolymers with reinforcing additives such as nanomaterials. The objective is to considerably augment such coatings' overall structural tensile strength and long-term service life, eventually facilitating their extensive practical utilization and implementation across multiple sectors using such revolutionary methodologies.

4.2.2 Regulatory Requirements

EFs for food packaging and the genesis of nanomaterials mandate regulation with direction from regulatory agencies like the FDA and the EU [156]. The FDA must approve edible coatings, making them GRAS before use in commerce. Some essential oils and plant extracts employed in such coatings are allergenic based on the dosage. They, thus, must be constantly tested for toxicity and allergenicity [157]. For safe usage, edible coating materials must be food-grade and made within sanitary conditions.

Edible coatings were controlled by FDA regulation in 2006 and the European Directive of 1998 as food coating ingredients, additives, and packaging material [158]. The application of synthetic and modified resins as coating for vegetables and fruits is controlled by the US FDA and European law. Moreover, food coating ingredients include nanoparticles and antifungal, antioxidant, and antibacterial compounds. In contrast, food contact articles and food contact substances are said to describe packaging coatings and materials [124]. Nanomaterial-based food packaging has drawn significant attention since it can be employed to extend the shelf life of food. Even though they have the potential to benefit, there must be global regulation for their application to ensure that they are used safely. Regulation 10/2011 in the EU states that only approved and listed nanoparticles must be used behind a functional barrier. Regulation 450/2009/EC states that active ingredients in intelligent packaging must be incorporated directly into the packaging material [156].

4.2.3 Ensuring Coating Shelf Life

Edible coatings help extend the shelf life and improve appearance. Still, they also have a finite life, detracting from the merit of perpetuating food stability and quality over time. This consideration emphasizes the need to carefully formulate coatings to provide both the initial benefits and long-term stability. Researchers and business stakeholders are working to prevent moisture and oxygen entry by developing innovative preservation technology and optimizing packaging methods to overcome this obstacle. The efforts include combining new technology and constructing protective packaging to ensure that edible coatings remain functional for extended periods and have higher quality. The technologies allow coatings to endure storage and transportation and, as a result, influence industries that depend on such technologies.

5 Prospects and Conclusion

Biodegradable EFs replace plastic packaging because they are focused on sustainability. After all, they enhance food safety while bettering the natural on-market characteristics. With incorporated bioactive units, antioxidants, antimicrobials, and probiotics, these films enhance the fight against development by microbes, oxidants, and spoilage. As a result, they are highly beneficial in preserving dairy, meat, and fresh produce [159, 160]. For example, dairy products can inhibit mold growth; in meats, they can lessen lipid oxidation; and in fresh produce, they can reduce moisture loss and prolong freshness. Nevertheless, scalability poses a significant challenge, as issues related to cost, mechanical properties, and regulatory approval need to be resolved for these films to move from niche uses to widespread acceptance [161]. Despite these challenges, bioactive EFs are apt to meet the ever-increasing consumer demand for functional and sustainable packaging, making them a likely favorite among eco-conscious shoppers [162]. Also, regarding its potential to change food safety monitoring, smart packaging technologies like color-changing indicators that signal spoilage are at a nascent stage of development [163]. From an economic point of view, it may require higher initial investments. On the other hand, these options would become more attractive in the long term: smaller plastic waste and carbon footprints [164]. Collaborative research combining material science, food technology, and biotechnology is essential for addressing current obstacles and achieving film functionality and production efficiency [165]. Insights gained from existing applications, such as the effective use of chitosan-based films in preserving seafood, may pave the way to finding the best forms and processes for edible film production [166]. In the future, bioactive EFs may imply a whole new concept for packaging that will transform the food industry in the following years regarding sustainability, functionality, and innovation [167].

5.1 Emerging Trends in Bioactive and Functional Packaging: Bioactive EFs

A significant progression has been made in bioactive EFs, especially in antioxidant, antimicrobial, and nutrient-rich structures. Numerous research studies have focused on inculcating essential oils, probiotics, plant extracts, and nanostructures to enhance film characteristics and food preservative quality. The research has suggested that the antimicrobial properties of bioactive EFs with phenolic compounds and essential oils drastically reduce microorganism growth. Films containing nanofibers and nanoparticles act in a manner that encompasses and offers the partial release of antimicrobial agents [168]. The films enriched by aloe vera extract phenolic compounds showed increased oxidative stability and reduced lipid oxidation in dairy products and preserved foods [169]. The prebiotics and probiotics in the EFs have improved functional characteristics, positively affected gut health, and enhanced mechanical and barrier properties [170]. The properties of EFs can be improved using bioactive compounds, with additional value beyond food preservation.

The durability can be increased using the extracts of phenolic compounds, improving antioxidant features [171]. Chitosan and aloe vera strengthen the film's mechanical strength and protective capabilities, notably suppressing microbial growth in stored foods [172]. Innovations in nanotech have addressed better-controlled delivery and enhanced stability of bioactive substances. Incorporating nano-emulsions and nanofibers supports long stability and extended release of antimicrobial and antioxidative chemicals [173]. Polymeric nanocarriers enable the controlled release of bioactive ingredients, increasing the film's efficiency in reducing spoilage [174]. Bioactive EFs represent an upcoming alternative to synthetic packaging, possessing some merits but entirely with a few limitations. Extension of shelf life and control over spoilage through these films, when incorporated with antimicrobial and antioxidant agents, are comparable to conventional plastic films [175]. They greatly benefit the environment since they are biodegradable and sustainable, thus reducing plastic debris and pollution [176]. Unfortunately, prohibitive production costs constitute a significant disadvantage, as bioactive films are still pricier than the synthesized types due to few scalability opportunities in manufacturing processes; technological progress in production shall, however, reduce costs to achieve more economic viability and broader commercial use of edible bioactive films [177].

5.2 Innovations in Biodegradable and Zero-Waste Packaging

Modern EFs consist predominantly of fibrous proteins, polysaccharides, and lipids, which are then combined with new materials like plant residues, flours, and gum, constituting a strong protective film of certain food products [178]. The composite materials included sodium alginate and chitosan, which were assumed to enhance their strength and durability [179]. The algal polysaccharides ulvan have similar food packaging non-toxic properties, biodegradable and bioactive [180]. EFs replace traditional plastics by reducing non-biodegradable solid waste. These films extend food shelf life, lower spoilage, and maintain food quality [181]. The dry thermoplastic process enables the production of dissolvable, food-grade coatings, promoting circular economy principles [182].

EFs face many problems, such as poor mechanical strength and low water vapor resistivity. In their work, [183] indicated that gelatin-, cornstarch-, and carrageenan-based films differ in mechanical and chemical properties. Still, they are less stable and degrade more quickly than plastic. Films based on lipids provide better water resistance but do not have good strength, while those based on proteins are very strong and provide feeble moisture barriers [184]. Wheat straw, rice peel, and corn stalks are emerging agricultural by-products for EFs [125]. Waste generated during food processing, mainly fruit and vegetable residues, increases the toughness and biodegradability of these films [178]. It further conforms to circular economy principles, turning waste into valuable packaging. Biodegradable films replace petroleum-based plastics with renewable and bio-based materials, thus diminishing carbon footprints [182]. Further, these films counter landfill wastes while providing a food preservation platform, reducing food losses and enhancing economic value in their wake [185]. Circular economy mechanisms attain a sustainable production, consumption, and waste regeneration cycle. The prospects of active packaging systems include the development of encapsulations allowing controlled release of the active ingredients. These future EFs must possess desirable properties such as mechanical strength, stability, and impermeability to maintain the sensory attributes of the edibles. Figure 15 depicts the biodegradable biopolymer used in active packaging.

5.3 Commercialization Potential and Regulatory Aspects

The three notable factors involved in the commercial extent of EFs include their cost, the scale of production, and their acceptance by the consumer. High production costs and drying times discourage large-scale optional manufacture [186]. Market acceptance is also restrained by consumers' concerns regarding safety, taste, and appearance [187]. The future of research must enhance mechanical strengths and sensory qualities to enable the driving force to commercial success. The market for edible packaging is proliferating with the demand for developing sustainable alternatives. Valued at $697 million in 2016, the market is poised to cross $1.09 billion by 2023, registering a 6.81% annual growth rate [188]. The meat, poultry, dairy, and fruit industries are playing a dominant role in this respect. Rigid laws govern EFs' safety: in the US, the FDA controls them through the Food Contact Notification program. In Europe, the EFSA oversees food additives and packaging safety, while in India, the Food Safety and Standards Authority of India (FSSAI) monitors edible packaging safety regarding food. Each framework focuses on toxicity, allergenic status management, and ingredient transparency [189]. The consumption of these packages should be done with extra caution since they involve food-grade biopolymers and toxicity assessments to be done on them by the producers. Films must also avoid microbial growth, oxidation, and moisture loss to remain edible. Ingredients, especially allergens, must be carefully listed on the labels for consumer protection [190]. However, due to the lack of elongation, slow drying, and uneven thickness of films, scaling-up production is still a hard nut to crack [186]. Industrial viability

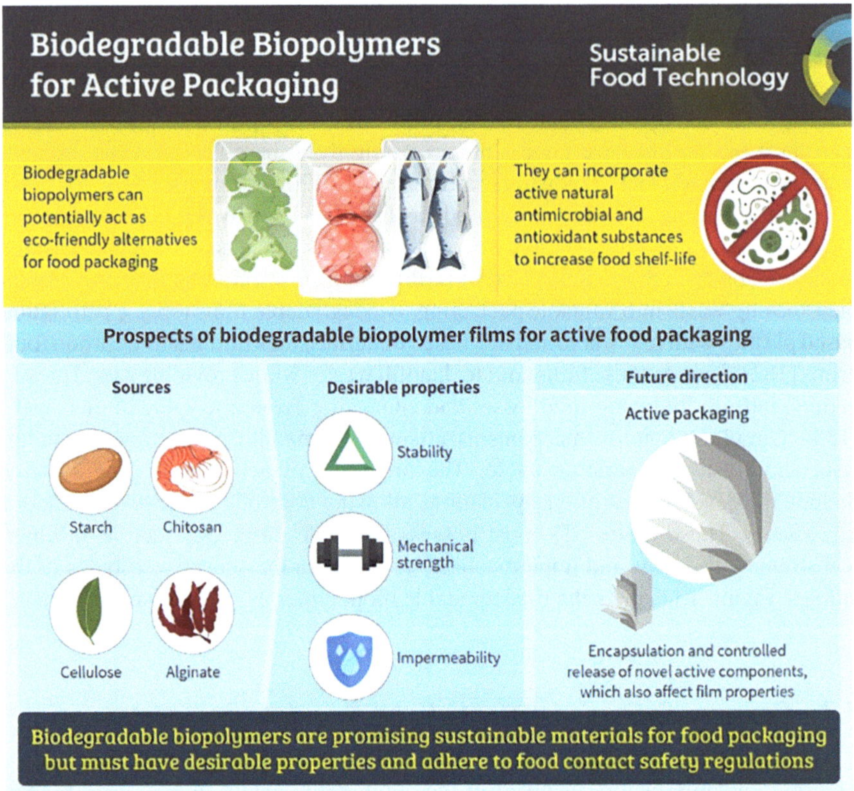

Fig. 15 Biodegradable biopolymers for active packaging demand, development, and directions [96]

is limited by fluctuating mechanical/barring properties. Although innovations like nanocomposite and multilayer films have been introduced to improve performance, cost implications and consumer acceptance remain barriers [191].

5.4 *Summary and Future Aspects*

EFs signal a potential advancement in food packaging that provides barrier properties against moisture, gas, and oil, which could extend shelf life. While relatively newer to the food packaging market, materials science and food technologies have allowed functional film compositions to utilize polysaccharides, proteins, and composite blends. These materials enhance the strength, flexibility, and transparency of films, broadening their applicability. Eco-friendly designs, such as water-soluble packaging, promote sustainability by lessening environmental waste. Nevertheless,

Table 2 EFs for food packaging: comprehensive research summary

Category	Key points	Details	References
Key findings	Barrier properties	EFs act as barriers to moisture, gas, and oil, extending shelf life	[192]
	Improved composition	Polysaccharides, proteins, and composites enhance strength, flexibility, and transparency	[193]
	Eco-friendly design	Water-soluble films dissolve by 90% within 24 h, reducing waste	[194]
Major gaps	Scalability issues	High production costs, slow drying times, and poor mechanical performance limit industrial use	[186]
	Safety and acceptance	Limited data on long-term toxicity, biodegradability, and consumer acceptance across regions	[187]
Future directions	Smart packaging	EFs are integrated with sensors to monitor freshness and safety	[195]
	3D printing	Customizable, nutrient-enriched EFs tailored to specific foods	[196]
	Active films	Incorporating natural antimicrobials and antioxidants to extend shelf life	[193]
Interdisciplinary approaches	Material science	Enhances film strength, transparency, and flexibility	[192]
	Food technology	Develops bio-based polymers from proteins, polysaccharides, and lipids	[192]
	Biotechnology	Engineers' films with antimicrobial and antioxidant properties	[196]
Consumer behavior	Consumer perception	Concerns about taste, safety, and cultural acceptance slow adoption	[38]
	Education campaigns	Promoting environmental and health benefits increases trust and acceptance	[187]

the ability to scale to industrial production has been difficult due to high costs, slow drying times, and weak mechanical properties. However, emerging technologies such as 3D printing allow EFs to be customized and infused with food-grade nutrients depending on existing food products. The introduction of smart packaging can extend this concept by providing real-time monitoring of food freshness and safety with the introduction of sensors embedded in the packaging. Active films made with natural antimicrobials and antioxidants can give prolonged shelf stability.

Furthermore, new biotechnologies are engineered, providing EFs with antioxidative and antimicrobial properties. Safety concerns regarding the materials, long-term toxicity, biodegradability, and conflicts with local consumer acceptance perpetuate barriers to adopting EFs. Consumer perceptions regarding taste, safety, and cultural factors continue to challenge the issue of local acceptance, underscoring the need for educational campaigns that build awareness and trust in edible film technology. Table 2 summarizes the key points and details about EFs as food packaging.

Conflicts of Interest There are no conflicts of interest to declare by the authors.

References

1. FAO warns that 930 million tonnes of food will end up in the trash. Accessed 27 Feb 2023. https://news.un.org/pt/story/2021/09/1764812
2. Peralta HC. Um Terço Dos Alimentos Produzidos No Mundo é Desperdiçado. Accessed 27 Feb 2025. Available online: https://www.dn.pt/sociedade/um-terco-dos-alimentos-produzidos-no-mundo-e-desperdicado-14049871.html
3. Kilcast D, Subramaniam P (eds) (2011) Food and beverage stability and shelf life. Elsevier
4. de Castro e Silva P, de Oliveira AC, Pereira LA, Valquíria M, Carvalho GR, Miranda KW, Marconcini JM, Oliveira JE (2020) Development of bionanocomposites of pectin and nanoemulsions of carnauba wax and neem oil pectin/carnauba wax/neem oil composites. Polym Compos 41(3):858–870
5. Choeybundit W, Karbowiak T, Lagorce A, Ngiwngam K, Auras R, Rachtanapun P, Noiwan D, Tongdeesoontorn W (2024) Eco-friendly straws: a fusion of soy protein isolate and cassava starch coated with beeswax and shellac wax. Polymers 16(13):1887
6. Olawade DB, Wada OZ, Ige AO (2024) Advances and recent trends in plant-based materials and edible films: a mini-review. Front Chem 12:1441650
7. Elkhattat D, Medhat M (2022) Creativity in packaging design as a competitive promotional tool. Inf Sci Lett 11(1):135–145
8. Ambaw A, Fadiji T, Opara UL (2021) Thermo-mechanical analysis in the fresh fruit cold chain: a review on recent advances. Foods 10(6):1357
9. Otoni CG, Avena-Bustillos RJ, Azeredo HM, Lorevice MV, Moura MR, Mattoso LH, McHugh TH (2017) Recent advances on edible films based on fruits and vegetables—a review. Comp Rev Food Sci Food Saf 16(5):1151–1169
10. Ezeoha SL, Ezenwanne JN (2013) Production of biodegradable plastic packaging film from cassava starch. IOSR J Eng 3(10):14–20
11. Zhong Y, Godwin P, Jin Y, Xiao H (2020) Biodegradable polymers and green-based antimicrobial packaging materials: a mini-review. Adv Ind Eng Polym Res 3(1):27–35
12. Hanani ZN, Roos YH, Kerry JP (2014) Use and application of gelatin as potential biodegradable packaging materials for food products. Int J Biol Macromol 71:94–102

13. de Azeredo HC, Mattoso LC, McHugh TH (2011) Nanocomposites in food packaging—a review. In: Advances in diverse industrial applications of nanocomposites, pp 57–78
14. Sung SY, Sin LT, Tee TT, Bee ST, Rahmat AR, Rahman WAWA, Tan AC, Vikhraman M (2013) Antimicrobial agents for food packaging applications. Trends Food Sci Technol 33(2):110–123
15. Oyom W, Zhang Z, Bi Y, Tahergorabi R (2022) Application of starch-based coatings incorporated with antimicrobial agents for preservation of fruits and vegetables: a review. Prog Org Coat 166:106800
16. Malhotra B, Keshwani A, Kharkwal H (2015) Antimicrobial food packaging: potential and pitfalls. Front Microbiol 6:611
17. Perez-Vazquez A, Barciela P, Carpena M, Prieto MA (2023) Edible coatings as a natural packaging system to improve fruit and vegetable shelf life and quality. Foods 12(19):3570
18. Ritchie H, Samborska V, Roser M (2023) Plastic pollution. Accessed 28 Feb 2025. Available at: https://ourworldindata.org/plastic-pollution
19. Waste and Resources Action Programme (2024) Redesigning the plastics system. Available at: https://www.wrap.ngo/taking-action/plastic-packaging#:%7E:text=The%20world%20produces%20141%20million,tonnes%20of%20carbon%20emissions%20annually
20. Nanda S, Berruti F (2021) Municipal solid waste management and landfilling technologies: a review. Environ Chem Lett 19(2):1433–1456
21. Goswami S, Adhikary S, Bhattacharya S, Agarwal R, Ganguly A, Nanda S, Rajak P (2024) The alarming link between environmental microplastics and health hazards with special emphasis on cancer. Life Sci 122937
22. Dharhshini A, Sistla YS (2023) Beeswax and castor oil to improve the moisture barrier and tensile properties of pectin-based edible films for food packaging applications. Eng Proc 37(1):33
23. Waste is still a taboo but change is on the way! Accessed 28 Feb 2025. https://www.wearephenix.com/pt-pt/o-desperdicio-alimentar-em-dados/
24. Gupta P, Toksha B, Rahaman M (2022) A review on biodegradable packaging films from vegetative and food waste. Chem Rec 22(7):e202100326
25. Kim S, Baek SK, Song KB (2018) Physical and antioxidant properties of alginate films prepared from *Sargassum fulvellum* with black chokeberry extract. Food Packag Shelf Life 18:157–163
26. Casariego ABWS, Souza BWS, Cerqueira MA, Teixeira JA, Cruz L, Díaz R, Vicente AA (2009) Chitosan/clay films' properties as affected by biopolymer and clay micro/nanoparticles' concentrations. Food Hydrocolloids 23(7):1895–1902
27. El-Sayed HS, El-Sayed SM, Mabrouk AM, Nawwar GA, Youssef AM (2021) Development of eco-friendly probiotic edible coatings based on chitosan, alginate and carboxymethyl cellulose for improving the shelf life of UF soft cheese. J Polym Environ 29:1941–1953
28. Abedi A, Lakzadeh L, Amouheydari M (2021) Effect of an edible coating composed of whey protein concentrate and rosemary essential oil on the shelf life of fresh spinach. J Food Process Preserv 45(4):e15284
29. Walker TR, McGuinty E, Charlebois S, Music J (2021) Single-use plastic packaging in the Canadian food industry: consumer behavior and perceptions. Humanit Soc Sci Commun 8(1):1–11
30. Yadav A, Kumar N, Upadhyay A, Singh A, Anurag RK, Pandiselvam R (2022) Effect of mango kernel seed starch-based active edible coating functionalized with lemongrass essential oil on the shelf-life of guava fruit. Qual Assur Saf Crops Foods 14(3):103–115
31. Zhao X, Cornish K, Vodovotz Y (2020) Narrowing the gap for bioplastic use in food packaging: an update. Environ Sci Technol 54(8):4712–4732
32. Filho WL, Barbir J, Abubakar IR, Paço A, Stasiskiene Z, Hornbogen M, Christin Fendt MT, Voronova V, Klõga M (2022) Consumer attitudes and concerns with bioplastics use: an international study. PLoS ONE 17(4):e0266918

33. Nilsen-Nygaard J, Fernández EN, Radusin T, Rotabakk BT, Sarfraz J, Sharmin N, Sivertsvik M, Sone I, Pettersen MK (2021) Current status of biobased and biodegradable food packaging materials: impact on food quality and effect of innovative processing technologies. Comp Rev Food Sci Food Saf 20(2):1333–1380
34. Gironi F, Piemonte V (2011) Bioplastics and petroleum-based plastics: strengths and weaknesses. Energy Sources Part A Recov Util Environ Effects 33(21):1949–1959
35. Cristofoli NL, Lima AR, Tchonkouang RD, Quintino AC, Vieira MC (2023) Advances in the food packaging production from agri-food waste and by-products: market trends for a sustainable development. Sustainability 15(7):6153
36. Chawla R, Sivakumar S, Kaur H (2021) Antimicrobial edible films in food packaging: current scenario and recent nanotechnological advancements—a review. Carbohydr Polym Technol Appl 2:100024
37. Bucher T, Malcolm J, Mukhopadhyay SP, Vuong Q, Beckett E (2023) Consumer acceptance of edible coatings on apples: the role of food technology neophobia and information about purpose. Food Qual Prefer 112:105024
38. Taylor S, Colonna A, Jung J, Gutierrez J, Zhao Y (2024) Consumer perception and acceptance of edible packaging for various food products. J Food Sci 89(4):2423–2437
39. Dey A, Dhumal CV, Sengupta P, Kumar A, Pramanik NK, Alam T (2021) Challenges and possible solutions to mitigate the problems of single-use plastics used for packaging food items: a review. J Food Sci Technol 58(9):3251–3269
40. Sharma P, Shehin VP, Kaur N, Vyas P (2019) Application of edible coatings on fresh and minimally processed vegetables: a review. Int J Veg Sci 25(3):295–314
41. Galus S, Arik Kibar EA, Gniewosz M, Kraśniewska K (2020) Novel materials in the preparation of edible films and coatings—a review. Coatings 10(7):674
42. Azeredo HM, Otoni CG, Mattoso LHC (2022) Edible films and coatings—not just packaging materials. Curr Res Food Sci 5:1590–1595
43. Flórez M, Guerra-Rodríguez E, Cazón P, Vázquez M (2022) Chitosan for food packaging: recent advances in active and intelligent films. Food Hydrocolloids 124:107328
44. Lopez-Rubio A, Gavara R, Lagaron JM (2006) Bioactive packaging: turning foods into healthier foods through biomaterials. Trends Food Sci Technol 17(10):567–575
45. Echeverría I, López-Caballero ME, Gómez-Guillén MC, Mauri AN, Montero MP (2018) Active nanocomposite films based on soy proteins-montmorillonite—clove essential oil for the preservation of refrigerated bluefin tuna (*Thunnus thynnus*) fillets. Int J Food Microbiol 266:142–149
46. Arruda TR, Bernardes PC, e Moraes ARF, Soares NDFF (2022) Natural bioactives in perspective: the future of active packaging based on essential oils and plant extracts themselves and those complexed by cyclodextrins. Food Res Int 156:111160
47. Valdés A, Burgos N, Jiménez A, Garrigós MC (2015) Natural pectin polysaccharides as edible coatings. Coatings 5(4):865–886
48. Costa MJ, Cerqueira MA, Ruiz HA, Fougnies C, Richel A, Vicente AA, Teixeira JA, Aguedo M (2015) Use of wheat bran arabinoxylans in chitosan-based films: effect on physicochemical properties. Ind Crops Prod 66:305–311
49. Sánchez-Ortega I, García-Almendárez BE, Santos-López EM, Amaro-Reyes A, Barboza-Corona JE, Regalado C (2014) Antimicrobial edible films and coatings for meat and meat products preservation. Sci World J 2014(1):248935
50. Paula GA, Benevides NM, Cunha AP, de Oliveira AV, Pinto AM, Morais JPS, Azeredo HM (2015) Development and characterization of edible films from mixtures of κ-carrageenan, ι-carrageenan, and alginate. Food Hydrocolloids 47:140–145
51. Shimokawa T, Togawa E, Kakegawa K, Kato A, Hayashi N (2015) Film formation and some structural features of hemicellulose fractions from *Pinus densiflora* leaves. J Wood Sci 61:53–59
52. Oun AA, Rhim JW (2015) Preparation and characterization of sodium carboxymethyl cellulose/cotton linter cellulose nanofibril composite films. Carbohydr Polym 127:101–109

53. Cerqueira MA, Lima AM, Souza BW, Teixeira JA, Moreira RA, Vicente AA (2009) Functional polysaccharides as edible coatings for cheese. J Agric Food Chem 57(4):1456–1462
54. Probiotics in food. Accessed 10 Mar 2025. https://openknowledge.fao.org/server/api/core/bitstreams/382476b3-4d54-4175-803f-2f26f3526256/content
55. Batista AL, Silva R, Cappato LP, Almada CN, Garcia RK, Silva MC, Raices RS, Arellano DB, Sant'Ana AS, Junior CAC, Freitas MQ (2015) Quality parameters of probiotic yogurt added to glucose oxidase compared to commercial products through microbiological, physical–chemical and metabolic activity analyses. Food Res Int 77:627–635
56. Ranadheera CS, Evans CA, Adams M, Baines SK (2016) Co-culturing of probiotics influences the microbial and physico-chemical properties but not sensory quality of fermented dairy drink made from goats' milk. Small Rumin Res 136:104–108
57. Martins EMF, Ramos AM, Vanzela ESL, Stringheta PC, de Oliveira Pinto CL, Martins JM (2013) Products of vegetable origin: a new alternative for the consumption of probiotic bacteria. Food Res Int 51(2):764–770
58. Kanmani P, Lim ST (2013) Development and characterization of novel probiotic-residing pullulan/starch edible films. Food Chem 141(2):1041–1049
59. Kim DH, Austin B (2008) Characterization of probiotic carnobacteria isolated from rainbow trout (*Oncorhynchus mykiss*) intestine. Lett Appl Microbiol 47(3):141–147
60. Concha-Meyer A, Schöbitz R, Brito C, Fuentes R (2011) Lactic acid bacteria in an alginate film inhibit Listeria monocytogenes growth on smoked salmon. Food Control 22(3–4):485–489
61. Varsha KK, Nampoothiri KM (2016) Appraisal of lactic acid bacteria as protective cultures. Food Control 69:61–64
62. Guimaraes A, Abrunhosa L, Pastrana LM, Cerqueira MA (2018) Edible films and coatings as carriers of living microorganisms: a new strategy towards biopreservation and healthier foods. Comp Rev Food Sci Food Saf 17(3):594–614
63. Cha DS, Chinnan MS (2004) Biopolymer-based antimicrobial packaging: a review. Crit Rev Food Sci Nutr 44(4):223–237
64. Ouattara B, Simard RE, Piette G, Bégin A, Holley RA (2000) Inhibition of surface spoilage bacteria in processed meats by application of antimicrobial films prepared with chitosan. Int J Food Microbiol 62(1–2):139–148
65. Cagri A, Ustunol Z, Ryser ET (2001) Antimicrobial, mechanical, and moisture barrier properties of low pH whey protein-based edible films containing p-aminobenzoic or sorbic acids. J Food Sci 66(6):865–870
66. da Rocha M, Loiko MR, Tondo EC, Prentice C (2014) Physical, mechanical and antimicrobial properties of Argentine anchovy (*Engraulis anchoita*) protein films incorporated with organic acids. Food Hydrocolloids 37:213–220
67. Su QZ, Lin QB, Chen CF, Wu LB, Wang ZW (2017) Effect of organic additives on silver release from nanosilver–polyethylene composite films to acidic food simulant. Food Chem 228:560–566
68. Türe H, Gällstedt M, Hedenqvist MS (2012) Antimicrobial compression-moulded wheat gluten films containing potassium sorbate. Food Res Int 45(1):109–115
69. Sayanjali S, Ghanbarzadeh B, Ghiassifar S (2011) Evaluation of antimicrobial and physical properties of edible film based on carboxymethyl cellulose containing potassium sorbate on some mycotoxigenic Aspergillus species in fresh pistachios. LWT Food Sci Technol 44(4):1133–1138
70. European Food Safety Authority (EFSA) (2009) Guidelines on submission of a dossier for safety evaluation by the EFSA of active or intelligent substances present in active and intelligent materials and articles intended to come into contact with food. EFSA J 7(8):1208
71. Dainelli D, Gontard N, Spyropoulos D, Zondervan-van den Beuken E, Tobback P (2008) Active and intelligent food packaging: legal aspects and safety concerns. Trends Food Sci Technol 19:S103–S112
72. Ahmed I, Lin H, Zou L, Li Z, Brody AL, Qazi IM, Lv L, Pavase TR, Khan MU, Khan S, Sun L (2018) An overview of smart packaging technologies for monitoring safety and quality of meat and meat products. Packag Technol Sci 31(7):449–471

73. Müller P, Schmid M (2019) Intelligent packaging in the food sector: a brief overview. Foods 8(1):16
74. Galagan Y, Su WF (2008) Fadable ink for time–temperature control of food freshness: novel new time–temperature indicator. Food Res Int 41(6):653–657
75. Yin W, Qiu C, Ji H, Li X, Sang S, McClements DJ, Jiao A, Wang J, Jin Z (2023) Recent advances in biomolecule-based films and coatings for active and smart food packaging applications. Food Biosci 52:102378
76. Smeriglio A, Barreca D, Bellocco E, Trombetta D (2016) Chemistry, pharmacology and health benefits of anthocyanins. Phytother Res 30(8):1265–1286
77. McGrath MJ, Scanaill CN, McGrath MJ, Scanaill CN (2013) Sensing and sensor fundamentals. In: Sensor technologies: healthcare, wellness, and environmental applications, pp 15–50
78. Rabehi A, Helal H, Zappa D, Comini E (2024) Advancements and prospects of electronic nose in various applications: a comprehensive review. Appl Sci 14(11):4506
79. Park YW, Kim SM, Lee JY, Jang W (2015) Application of biosensors in smart packaging. Mol Cell Toxicol 11:277–285
80. Ghaani M, Cozzolino CA, Castelli G, Farris S (2016) An overview of the intelligent packaging technologies in the food sector. Trends Food Sci Technol 51:1–11
81. Tshamisane M, Adeyemi JO, Fawole OA (2025) Fabrication of active and intelligent bio-based edible films using encapsulated beetroot powders for smart packaging. Mater Adv 6(3):1051–1066
82. Bibi F, Guillaume C, Gontard N, Sorli B (2017) A review: RFID technology having sensing aptitudes for food industry and their contribution to tracking and monitoring of food products. Trends Food Sci Technol 62:91–103
83. Yam KL, Takhistov PT, Miltz J (2005) Intelligent packaging: concepts and applications. J Food Sci 70(1):R1–R10
84. Tomar R, Singh NB (2023) Overview of applications of nanotechnology. In: Applications of emerging nanomaterials and nanotechnology, pp 1–26
85. Sanguansri P, Augustin MA (2006) Nanoscale materials development—a food industry perspective. Trends Food Sci Technol 17(10):547–556
86. Parisi C, Vigani M, Rodríguez-Cerezo E (2015) Agricultural nanotechnologies: what are the current possibilities? Nano Today 10(2):124–127
87. Shankar S, Jaiswal L, Selvakannan PR, Ham KS, Rhim JW (2016) Gelatin-based dissolvable antibacterial films reinforced with metallic nanoparticles. RSC Adv 6(71):67340–67352
88. Mishra B, Panda J, Mishra AK, Nath PC, Nayak PK, Mahapatra U, Sharma M, Chopra H, Mohanta YK, Sridhar K (2024) Recent advances in sustainable biopolymer-based nanocomposites for smart food packaging: a review. Int J Biol Macromol 135583
89. Sharma C, Dhiman R, Rokana N, Panwar H (2017) Nanotechnology: an untapped resource for food packaging. Front Microbiol 8:1735
90. He X, Deng H, Hwang HM (2019) The current application of nanotechnology in food and agriculture. J Food Drug Anal 27(1):1–21
91. Boz Z, Korhonen V, Koelsch Sand C (2020) Consumer considerations for the implementation of sustainable packaging: a review. Sustainability 12(6):2192
92. Nile SH, Baskar V, Selvaraj D, Nile A, Xiao J, Kai G (2020) Nanotechnologies in food science: applications, recent trends, and future perspectives. Nano-Micro Lett 12:1–34
93. Arcot Y, Mu M, Lin YT, DeFlorio W, Jebrini H, Kunadu APH, Yegin Y, Min Y, Castillo A, Cisneros-Zevallos L, Taylor TM (2024) Edible nano-encapsulated cinnamon essential oil hybrid wax coatings for enhancing apple safety against food borne pathogens. Curr Res Food Sci 8:100667
94. Nienhaus K, Nienhaus GU (2014) Fluorescent proteins for live-cell imaging with super-resolution. Chem Soc Rev 43(4):1088–1106
95. Onyeaka H, Passaretti P, Miri T, Al-Sharify ZT (2022) The safety of nanomaterials in food production and packaging. Curr Res Food Sci 5:763–774

96. Westlake JR, Tran MW, Jiang Y, Zhang X, Burrows AD, Xie M (2023) Biodegradable biopolymers for active packaging: demand, development and directions. Sustain Food Technol 1(1):50–72
97. Weng S, Marcet I, Rendueles M, Díaz M (2024) Edible films from the laboratory to industry: a review of the different production methods. Food Bioprocess Technol 1–27
98. Sarker A, Grift TE (2021) Bioactive properties and potential applications of aloe vera gel edible coating on fresh and minimally processed fruits and vegetables: a review. J Food Meas Charact 15:2119–2134
99. Dehghani F, Farhadian N (2023) Encapsulation: fluidized bed coating technology. In: Principles of biomaterials encapsulation: volume one. Woodhead Publishing, pp 143–156
100. Senturk Parreidt T, Schmid M, Müller K (2018) Effect of dipping and vacuum impregnation coating techniques with alginate based coating on physical quality parameters of cantaloupe melon. J Food Sci 83(4):929–936
101. Valentino M, Volpe S, Di Giuseppe FA, Cavella S, Torrieri E (2020) Active biopolymer coating based on sodium caseinate: physical characterization and antioxidant activity. Coatings 10(8):706
102. Soares ADS, Ramos AM, Vieira ÉNR, Vanzela ESL, de Oliveira PM, Paula DDA (2018) Vacuum impregnation of chitosan-based edible coating in minimally processed pumpkin. Int J Food Sci Technol 53(9):2229–2238
103. Lara GR, Uemura K, Khalid N, Kobayashi I, Takahashi C, Nakajima M, Neves MA (2020) Layer-by-layer electrostatic deposition of edible coatings for enhancing the storage stability of fresh-cut lotus root (*Nelumbo nucifera*). Food Bioprocess Technol 13:722–726
104. Aayush K, McClements DJ, Sharma S, Sharma R, Singh GP, Sharma K, Oberoi K (2022) Innovations in the development and application of edible coatings for fresh and minimally processed apple. Food Control 141:109188
105. Pham TT, Nguyen LLP, Dam MS, Baranyai L (2023) Application of edible coating in extension of fruit shelf life. AgriEngineering 5(1):520–536
106. Leontiou A, Georgopoulos S, Karabagias VK, Kehayias G, Karakassides A, Salmas CE, Giannakas AE (2023) Three-dimensional printing applications in food industry. Nanomanufacturing 3(1):91–112
107. Nida S, Moses JA, Anandharamakrishnan C (2025) 3D printed food package casings from sugarcane bagasse: a waste valorization study. Biomass Convers Bioref 15(2):1835–1845
108. Leaw ZE, Kong I, Pui LP (2021) 3D printed corn starch–gelatin film with glycerol and hawthorn berry (*Crataegus pinnatifida*) extract. J Food Process Preserv 45(9):e15752
109. Nachal N, Moses JA, Karthik P, Anandharamakrishnan C (2019) Applications of 3D printing in food processing. Food Eng Rev 11(3):123–141
110. Aaqil M, Peng C, Kamal A, Nawaz T, Gong J (2024) Recent approaches to the formulation, uses, and impact of edible coatings on fresh peach fruit. Foods 13(2):267
111. Kong P, Rosnan SM, Enomae T (2024) Carboxymethyl cellulose–chitosan edible films for food packaging: a review of recent advances. Carbohydr Polym 122612
112. McHugh TH, de Jesús Avena-Bustillos R (2011) Applications of edible films and coatings to processed foods. In: Edible coatings and films to improve food quality. CRC Press, pp 304–331
113. Kumar N, Pratibha, Prasad J, Yadav A, Upadhyay A, Neeraj, Shukla S, Petkoska AT, Heena, Suri S, Gniewosz M (2023) Recent trends in edible packaging for food applications—perspective for the future. Food Eng Rev 15(4):718–747
114. Riseh RS, Vatankhah M, Hassanisaadi M, Kennedy JF (2023) Chitosan-based nanocomposites as coatings and packaging materials for the postharvest improvement of agricultural product: a review. Carbohydr Polym 309:120666
115. Silva-Vera W, Zamorano-Riquelme M, Rocco-Orellana C, Vega-Viveros R, Gimenez-Castillo B, Silva-Weiss A, Osorio-Lira F (2018) Study of spray system applications of edible coating suspensions based on hydrocolloids containing cellulose nanofibers on grape surface (*Vitis vinifera* L.). Food Bioprocess Technol 11:1575–1585

116. Maqbool M, Ali A, Ramachandran S, Smith DR, Alderson PG (2010) Control of postharvest anthracnose of banana using a new edible composite coating. Crop Prot 29(10):1136–1141
117. Ranjith FH, Adhikari B, Muhialdin BJ, Yusof NL, Mohammed NK, Ariffin SH, Hussin ASM (2022) Peptide-based edible coatings to control postharvest fungal spoilage of mango (*Mangifera indica* L.) fruit. Food Control 135:108789
118. Adetunji CO, Ogundare MO, Ogunkunle ATJ, Kolawole OM, Adetunji JB (2014) Effects of edible coatings from xanthum gum produced from *Xanthomonas campestris* pammel on the shelf life of *Carica papaya* Linn fruits. Asian J Agric Biol 2(1):8–13
119. Garavito J, Moncayo-Martínez D, Castellanos DA (2020) Evaluation of antimicrobial coatings on preservation and shelf life of fresh chicken breast fillets under cold storage. Foods 9(9):1203
120. Ojagh SM, Rezaei M, Razavi SH, Hosseini SMH (2010) Development and evaluation of a novel biodegradable film made from chitosan and cinnamon essential oil with low affinity toward water. Food Chem 122(1):161–166
121. Gagaoua M, Bhattacharya T, Lamri M, Oz F, Dib AL, Oz E, Uysal-Unalan I, Tomasevic I (2021) Green coating polymers in meat preservation. Coatings 11(11):1379
122. Kumar N, Neeraj, Pratibha Kishore A, Mithul Aravind S, Singh A (2023) Bionanocomposites for active and smart food packaging: a review on its application, safety, and health aspects. J Food Process Eng 46(5):e14320
123. Rux G, Labude C, Herppich WB, Geyer M (2023) Investigation on the potential of applying bio-based edible coatings for horticultural products exemplified with cucumbers. Curr Res Food Sci 6:100407
124. Hamed I, Jakobsen AN, Lerfall J (2022) Sustainable edible packaging systems based on active compounds from food processing byproducts: a review. Comp Rev Food Sci Food Saf 21(1):198–226
125. Gupta S, Kaur A, Ghoshal G (2024) Sustainable packaging solution from agriculture waste: production of bioplastic, biocomposite, biopolymer. In: Transforming agriculture residues for sustainable development: from waste to wealth. Springer Nature Switzerland, Cham, pp 245–272
126. Shah MA, Bosco SJD, Mir SA (2014) Plant extracts as natural antioxidants in meat and meat products. Meat Sci 98(1):21–33
127. Trajkovska Petkoska A, Daniloski D, Kumar N, Pratibha, Broach AT (2021) Biobased materials as a sustainable potential for edible packaging. Sustain Packag 111–135
128. Gil M, Rudy M (2023) Innovations in the packaging of meat and meat products—a review. Coatings 13(2):333
129. Mukherjee C, Varghese D, Krishna JS, Boominathan T, Rakeshkumar R, Dineshkumar S, Rao CB, Sivaramakrishna A (2023) Recent advances in biodegradable polymers—properties, applications and future prospects. Eur Polym J 192:112068
130. Sun XD, Holley RA (2012) Antimicrobial and antioxidative strategies to reduce pathogens and extend the shelf life of fresh red meats. Comp Rev Food Sci Food Saf 11(4):340–354
131. Umaraw P, Verma AK (2017) Comprehensive review on application of edible film on meat and meat products: an eco-friendly approach. Crit Rev Food Sci Nutr 57(6):1270–1279
132. Rubio NR, Xiang N, Kaplan DL (2020) Plant-based and cell-based approaches to meat production. Nat Commun 11(1):1–11
133. Fadini AL, Rocha FS, Alvim ID, Sadahira MS, Queiroz MB, Alves RMV, Silva LB (2013) Mechanical properties and water vapour permeability of hydrolysed collagen–cocoa butter edible films plasticised with sucrose. Food Hydrocolloids 30(2):625–631
134. Santo RE, Kim BF, Goldman SE, Dutkiewicz J, Biehl E, Bloem MW, Neff RA, Nachman KE (2020) Considering plant-based meat substitutes and cell-based meats: a public health and food systems perspective. Front Sustain Food Syst 4:569383
135. Poore J, Nemecek T (2018) Reducing food's environmental impacts through producers and consumers. Science 360(6392):987–992
136. Oz AT, Kafkas E (2017) Phytochemicals in fruits and vegetables. In: Waisundara V (ed) Superfood and functional food. IntechOpen, London, pp 175–184

137. Lourenço SC, Moldão-Martins M, Alves VD (2019) Antioxidants of natural plant origins: from sources to food industry applications. Molecules 24(22):4132
138. Appleton KM, McGrath AJ, McKinley MC, Draffin CR, Hamill LL, Young IS, Woodside JV (2018) The value of facial attractiveness for encouraging fruit and vegetable consumption: analyses from a randomized controlled trial. BMC Public Health 18:1–12
139. Miteluţ AC, Popa EE, Drăghici MC, Popescu PA, Popa VI, Bujor OC, Ion VA, Popa ME (2021) Latest developments in edible coatings on minimally processed fruits and vegetables: a review. Foods 10(11):2821
140. de Menezes FLG, de Lima Leite RH, Dos Santos FKG, Aria AI, Aroucha EMM (2024) TiO_2 incorporated into a blend of biopolymeric matrices improves film properties and affects the postharvest conservation of papaya fruits under UV light. Food Chem 433:137387
141. Iqbal SZ, Haider A, ur Rehman F, Cui G, Waseem M, Iqbal M, Khaneghah AM (2024) Application of carboxymethylcellulose in combination with essential oils nano-emulsions edible coating for the preservation of kiwifruit. Int J Biol Macromol 261:129947
142. Li C, Li F, Wang K, Xie D (2024) Green and facile fabrication of multifunctional cellulose nanocrystal and carvacrol together reinforced chitosan bio-nanocomposite coatings for fruit preservation. Int J Biol Macromol 265:130651
143. Ali S, Hassan M, Ejaz S, Arshad MN, Sabir IA (2024) Edible coatings for postharvest fruit and vegetable storage. In: Sustainable postharvest technologies for fruits and vegetables. CRC Press, pp 370–385
144. Díaz-Montes E, Castro-Muñoz R (2021) Edible films and coatings as food-quality preservers: an overview. Foods 10(2):249
145. Garnier L, Valence F, Mounier J (2017) Diversity and control of spoilage fungi in dairy products: an update. Microorganisms 5(3):42
146. Desrizal KI, Mardiah A (2020) Microbiological quality of brown seaweed (*Sargassum* sp.) dodol packed with carrageenan and chitosan coatings. Pak J Nutr 19:309–316
147. Siriwardana J, Wijesekara I (2021) Analysis of the effectiveness of an antimicrobial edible coating prepared from sweet whey base to improve the physicochemical, microbiological, and sensory attributes of Swiss cheese. Adv Agric 2021(1):5096574
148. Nottagh S, Hesari J, Peighambardoust SH, Rezaei-Mokarram R, Jafarizadeh-Malmiri H (2020) Effectiveness of edible coating based on chitosan and Natamycin on biological, physico-chemical and organoleptic attributes of Iranian ultra-filtrated cheese. Biologia 75:605–611
149. Cerqueira MA, Sousa-Gallagher MJ, Macedo I, Rodriguez-Aguilera R, Souza BW, Teixeira JA, Vicente AA (2010) Use of galactomannan edible coating application and storage temperature for prolonging shelf-life of "regional" cheese. J Food Eng 97(1):87–94
150. Guitián MV, Ibarguren C, Soria MC, Hovanyecz P, Banchio C, Audisio MC (2019) Anti-Listeria monocytogenes effect of bacteriocin-incorporated agar edible coatings applied on cheese. Int Dairy J 97:92–98
151. Arshad R, Sameen A, Huma N, Zia MA (2020) Exploring the potential of active edible coating on the shelf stability of dairy products. Pak J Agric Sci 57(1)
152. Salama HH, Trif M, Rusu AV, Bhattacharya S (2022) Application of functional and edible coatings and films as promising strategies for developing dairy functional products—a review on yoghurt case. Coatings 12(6):838
153. Shlush E, Davidovich-Pinhas M (2022) Bioplastics for food packaging. Trends Food Sci Technol 125:66–80
154. Abdullah, Cai J, Hafeez MA, Wang Q, Farooq S, Huang Q, Tian W, Xiao J (2022) Biopolymer-based functional films for packaging applications: a review. Front Nutr 9:1000116
155. Duguma HT (2022) Potential applications and limitations of edible coatings for maintaining tomato quality and shelf life. Int J Food Sci Technol 57(3):1353–1366
156. Trajkovska Petkoska A (2021) Improved shelf life and quality of tomato (*Solanum lycopersicum* L.) by using chitosan-pullulan composite edible coating enriched with pomegranate peel extract. ACS Food Sci Technol 1(4):500–510

157. Priya K, Thirunavookarasu N, Chidanand DV (2023) Recent advances in edible coating of food products and its legislations: a review. J Agric Food Res 12:100623
158. Cloete L, Picot-Allain C, Ramasawmy B, Neetoo H, Ramful-Baboolall D, Emmambux MN (2023) Drivers and barriers for commercial uptake of edible coatings for fresh fruits and vegetables industry—a review. Food Rev Int 39(6):3481–3514
159. Hassan B, Chatha SAS, Hussain AI, Zia KM, Akhtar N (2018) Recent advances on polysaccharides, lipids and protein based edible films and coatings: a review. Int J Biol Macromol 109:1095–1107
160. Youssef AM, El-Sayed SM (2018) Bionanocomposites materials for food packaging applications: concepts and future outlook. Carbohydr Polym 193:19–27
161. Galus S, Kadzińska J (2015) Food applications of emulsion-based edible films and coatings. Trends Food Sci Technol 45(2):273–283
162. Ramos ÓL, Pereira RN, Martins A, Rodrigues R, Fucinos C, Teixeira JA (2016) Design of bioactive edible films and coatings for food applications. Food Eng Rev 8(3):351–366
163. Realini CE, Marcos B (2014) Active and intelligent packaging systems for a modern society. Meat Sci 98(3):404–419
164. Ribeiro-Santos R, Andrade M, de Melo NR, Sanches-Silva A (2017) Use of essential oils in active food packaging: recent advances and future trends. Trends Food Sci Technol 61:132–140
165. Falguera V, Quintero JP, Jiménez A, Muñoz JA, Ibarz A (2011) Edible films and coatings: structures, active functions and trends in their use. Trends Food Sci Technol 22(6):292–303
166. Elsabee MZ, Abdou ES (2013) Chitosan based edible films and coatings: a review. Mater Sci Eng C 33(4):1819–1841
167. Han JH (2014) Innovations in food packaging. Elsevier
168. Gürdal A, Çetinkaya T (2023) Advancements in edible films for aquatic product preservation and packaging. Rev Aquac. https://doi.org/10.1111/raq.12880
169. Mahajan K, Kumar S, Bhat Z, Naqvi Z, Jayawardena R (2021) Development of bioactive edible film using phytochemicals from aloe vera for improved microbial and lipid oxidative stability of frozen dairy products. Food Bioprocess Technol. https://doi.org/10.1007/s11947-021-02699-8
170. Hashemi S, Khorshidian N, Mohammadi M (2022) An insight to potential application of synbiotic edible films and coatings in food products. Front Nutr 9. https://doi.org/10.3389/fnut.2022.875368
171. Pedreiro S, Figueirinha A, Silva A, Ramos F (2021) Bioactive edible films and coatings based in gums and starch: phenolic enrichment and foods application. Coatings. https://doi.org/10.3390/coatings11111393
172. Kaur N, Somasundram C, Razali Z, Mourad A, Hamed F, Ahmed Z (2024) Aloe vera/chitosan-based edible film with enhanced antioxidant, antimicrobial, thermal, and barrier properties for sustainable food preservation. Polymers 16. https://doi.org/10.3390/polym16020242
173. Rangaraj V, Rambabu K, Banat F, Mittal V (2021) Natural antioxidants-based edible active food packaging: an overview of current advancements. Food Biosci 43:101251. https://doi.org/10.1016/J.FBIO.2021.101251
174. Zareie Z, Yazdi T, Mortazavi S (2020) Development and characterization of antioxidant and antimicrobial edible films based on chitosan and gamma-aminobutyric acid-rich fermented soy protein. Carbohydr Polym 244:116491. https://doi.org/10.1016/j.carbpol.2020.116491
175. Singh A, Kim J, Lee Y (2022) Phenolic compounds in active packaging and edible films/coatings: natural bioactive molecules and novel packaging ingredients. Molecules 27. https://doi.org/10.3390/molecules27217513
176. Olawade D, Wada O, Ige A (2024) Advances and recent trends in plant-based materials and edible films: a mini-review. Front Chem 12. https://doi.org/10.3389/fchem.2024.1441650
177. Janowicz M, Galus S, Ciurzyńska A, Nowacka M (2023) The potential of edible films, sheets, and coatings based on fruits and vegetables in the context of sustainable food packaging development. Polymers 15. https://doi.org/10.3390/polym15214231

178. Galus S, Kibar E, Gniewosz M, Kraśniewska K (2020) 4 novel materials in the preparation of edible films and coatings—a review. Coatings. https://doi.org/10.3390/coatings10070674
179. Kuprina E, Yakkola A, Manuylov A, Kabysh O, Zhirkova M, Zhernyakova A, Shmelkova M, Vitalyevich T, Stepanov A (2020) Development of biodegradable packaging films and edible coatings for food products based on composite materials. Prog Chem Appl Chitin Deriv. https://doi.org/10.15259/pcacd.25.009
180. Wang H, Cao Z, Yao L, Feng T, Song S, Sun M (2023) Insights into the edible and biodegradable ulvan-based films and coatings for food packaging. Foods 12. https://doi.org/10.3390/foods12081622
181. Choudhary U, Bhinchhar B, Paswan V, Kharkwal S, Yadav S, Singh P (2021) Utilization of agro-industrial wastes as edible coating and films for food packaging materials. In: Food processing—new insights [working title]. https://doi.org/10.5772/intechopen.99786
182. Prakash G, Mishra R (2023) Edible and bio-based food packaging: a review. Pharm Innov. https://doi.org/10.22271/tpi.2023.v12.i7sr.21623
183. Dyshlyuk L, Babich O, Belova D, Prosekov A (2017) Comparative analysis of physical and chemical properties of biodegradable edible films of various compositions. J Food Process Eng 40. https://doi.org/10.1111/JFPE.12331
184. Shenju Y (2014) Research progress 11 on application of edible films in animal foodstuff. China Brewing
185. Durango A, Soares N, Arteaga M (2011) Edible films and coatings as biodegradable active packaging in the preservation of food products. Biotecnol Sector Agropec Agroind 9:112–118
186. Jeevahan J, Chandrasekaran M, Venkatesan S, Sriram V, Joseph G, Mageshwaran G, Durairaj R (2020) Scaling up difficulties and commercial aspects of edible films for food packaging: a review. Trends Food Sci Technol 100:210–222. https://doi.org/10.1016/j.tifs.2020.04.014
187. Wan VCH, Lee CM, Lee SY (2007) Understanding consumer attitudes on edible films and coatings: focus group findings. J Sens Stud 22(3):353–366
188. Suhag R, Kumar N, Petkoska A, Upadhyay A (2020) Film formation and deposition methods of edible coating on food products: a review. Food Res Int 136:109582. https://doi.org/10.1016/j.foodres.2020.109582
189. Aguirre-Joya J, León-Zapata M, Álvarez-Pérez O, Torres-León C, Nieto-Oropeza D, Ventura-Sobrevilla J, Aguilar M, Ruelas-Chacón X, Rojas R, Ramos-Aguiñaga M, Aguilar C (2018) Basic and applied concepts of edible packaging for foods, pp 1–61. https://doi.org/10.1016/B978-0-12-811516-9.00001-4
190. Fonte P, Machado B (2021) Embalagens comestíveis: aplicação, efeito na segurança alimentar e aceitabilidade. Acta Port Nutr. https://doi.org/10.21011/apn.2020.2514
191. Maftoonazad N, Badii F, Shahamirian M (2014) Recent innovations in the area of edible films and coatings. Recent Pat Food Nutr Agric 5(3):201–213. https://doi.org/10.2174/2212798405666131129151640
192. Bozkurt S, Altay Ö, Koç M, Ertekin FK (2023) Edible films and coatings in food systems
193. Hamann D, Puton B, Colet R, Steffens J, Ceni G, Cansian R, Backes G (2021) Active edible films for application in meat products. Res Soc Dev. https://doi.org/10.33448/rsd-v10i7.16379
194. Marinovich A, Aluru R, Zyryanov G, Fomin A (2021) Scientists created edible films for food packaging
195. Mellinas C, Valdés A, Ramos M, Burgos N, Garrigós M, Jiménez A (2016) Active edible films: current state and future trends. J Appl Polym Sci. https://doi.org/10.1002/APP.42631
196. Avramescu S, Butean C, Popa C, Orțan A, Moraru I, Temocico G (2020) Edible and functionalized films/coatings—performances and perspectives. THE Coat 10:687. https://doi.org/10.3390/coatings10070687

Chapter 3
Enhancing Composites for Food Packaging: Exploring the Impact of Additives and Fillers

Esha Trivedi and Ritam Patra

1 Introduction

In the food industry, packaging plays a very important role in preserving the shelf life and maintaining the quality of a food product. Packaging offers new approaches for maintaining food quality, enhancing shelf life, and minimizing food losses, leading to an increase in food security [1]. Traditional packaging materials derived from plastic have excellent barrier properties, durability, and mechanical strength, but they cannot be recycled. This limitation has led to an increasing demand for innovation in materials that are sustainable and can be recycled easily [2]. The demand for biodegradable materials has resulted in the development of materials in which additives and fillers are added to enhance performance and lower environmental impact.

Composites are made up of multiple components that collectively form a multiphase, multicomponent system having distinct identities and characteristics for each part [3]. Composite packaging materials create better packaging options by combining the distinctive qualities of different substrates. By increasing the shelf life and visual appeal of a food product, we can address issues in the food industry such as distribution and preservation. Innovative food packaging relies heavily on composite materials. Since they give the packages greater usefulness and flexibility, they have several advantages over traditional materials. These modifications make it possible for packaging materials to meet customer needs and requirements pertaining to sustainability, food safety, and extended shelf life.

E. Trivedi (✉)
Department of Paper Technology, Indian Institute of Technology Roorkee, Saharanpur, Uttar Pradesh, India
e-mail: esha_t@pt.iitr.ac.in

R. Patra
Department of Materials Science and Engineering, Swansea University, Swansea, Wales, UK

A. Verma et al. (eds.), *Polymers and Composite Materials for Packaging: Smart Food Packaging and Solutions*, Materials Horizons: From Nature to Nanomaterials,
https://doi.org/10.1007/978-981-95-0677-4_3

The benefits of using composites include improved mechanical strength, enhanced barrier properties, innovation, and sustainability. Mechanical strength is significantly enhanced by combining polymers with reinforcing additives and fillers [4]. Blended cellulose acetate propionate with starch and biodegradable polycaprolactone will produce promising polymer composites for food packaging applications. It was observed that the elongation at the break, tensile strength, and yield of blends were enhanced. Incorporating additives and fillers like graphene oxide and nanoclay improves the barrier properties as these reduce the permeability to gases, moisture, and light. Organomodified montmorillonite was incorporated into the zein matrix, which improved the barrier properties (water vapor and oxygen barrier) of coated polypropylene films. It was seen that oxygen permeability was decreased to 25% by the coating, and water vapor permeability decreased by 30% [5]. Composites drive sustainability and innovation by enabling the development of materials that are high-performance materials, have less environmental impact, and are eco-friendly [6].

Packaging materials prepared from biodegradable polymers have some limitations because of their poor properties, but these can be improved by the incorporation of some additives and fillers. Additives like antioxidants, stabilizers, antimicrobial agents, and plasticizers play an important role in enhancing polymer properties, extending the lifespan of materials and durability, and improving food preservation [7]. Fillers that enhance barrier qualities, structural integrity, and biodegradability include nanoclays, graphene derivatives, nanocellulose, and starch-based compounds [8]. The next-generation packaging industry's advancement depends mainly on the conjunction between them.

This book chapter reviews the function of fillers and additives in packaging composites and how they impact mechanical properties, barrier performance, and sustainability. It also highlights challenges in integrating them, including cost, dispersion efficiency, and recyclability, while discussing market trends and future research targets. Using bio-based and nanofiller-based packaging materials promotes innovations. Using them also helps maintain the balance between functionality and environmental concerns. Hence, we have more efficient and sustainable packaging alternatives.

2 Role of Additives and Fillers in Packaging Composites

Additives and fillers are critical in improving packaging composites' physical, chemical, and functional qualities. They improve the performance, durability, and cost-effectiveness of composite packaging materials. They modify polymer matrix to achieve desired properties such as greater mechanical strength, higher barrier characteristics, extended shelf life, and increased biodegradability. To overcome the disadvantages of existing packaging materials and develop high-performing and green replacements, additives and fillers are critical [9, 10].

2.1 Additives

Additives play a key role in making packaging materials stronger, more durable, and better looking. They help improve the quality of packaging so that it can be used for various purposes across many industries. For example, antioxidants, thermal stabilizers, and UV inhibitors protect the packaging from damage caused by oxygen, heat, and sunlight, thus helping to increase shelf life and keep packaging intact. Plasticizers are another common additive that makes the material soft and easy to shape, which is especially useful for flexible packaging like food wrap. In addition, additives such as antimicrobial agents and antifogging chemicals are often added to food packaging to keep products fresher, safer, and visible, making them more appealing to consumers [11].

2.1.1 Plasticizers

Plasticizers are added to polymers to make them more flexible and easier to work with. Their main purpose is to lower the temperature at which the material hardens, allowing it to remain softer and more adaptable. They can also increase fracture and impact resistance [12]. Plasticizers increase the flexibility and processing ability of polymers by weakening the forces holding them together and by interfering with their chains [13]. In bio-based packaging materials, these additives are necessary because they can enhance the films' maintain their integrity, avoid pores and openings in the polymeric matrix, and maintain their handling and flexibility [14]. The most widely used plasticizers in biodegradable and/or edible films are glycerols, ethylene glycol, sorbitol, lecithin waxes, vegetable oils, and amino acids. In the absence of a plasticizing agent, biodegradable films have a tendency to become brittle due to high intermolecular forces involving polymer chain-to-chain contact [15]. The right choice of plasticizers is crucial in maintaining the harmony between mechanical strength, permeability, and elasticity. Overuse-induced migration issues may affect food safety and product integrity.

2.1.2 Stabilizers

In packaging materials, stabilizers are essential because they prevent the polymer matrix from degrading due to a variety of factors, including heat, UV rays, and humidity [16]. Additives known as stabilizers improve food texture, conserve color and flavor, and boost the overall quality standards and acceptance of food products by facilitating in the uniform dispersion of two or more immiscible elements [17]. The primary stabilizer types used in packaging composites include antioxidants, UV stabilizers, heat stabilizers, and light stabilizers. Samper et al. utilized flavonoids for improving properties of polypropylene. They found that these natural derivatives can act as UV stabilizers and antioxidants; hence, the material properties are improved [18].

2.1.3 Antioxidants

For a few years now in packaging industry antioxidants are being extensively used to preserve and protect food to enhance their longevity. Environmental oxygen causes oxidation which is one of the primary reasons for food deterioration and to cater this problem antioxidants are used to safeguard their quality by preventing the oxidation of proteins and lipids. The packaging industry uses a number of approved antioxidants, including artificial tert-butylhydroquinone, propyl gallate, butylated hydroxytoluene, and butylated hydroxyanisole. Further, from natural resources there are some natural antioxidants that are derived. For example, catechin, caffeic acid, and α-tocopherol are some prominent natural antioxidants [19]. Antioxidants are also added to biodegradable polymer matrix to create coatings and films to introduce excellent antioxidant qualities and easy recycling. Siddiqui et al. [20] studied that naturally sourced antioxidants, including thymol, carvacrol, limonene, or cinnamaldehyde added to the polylactic acid polymer matrix, creates active food packaging materials.

2.1.4 Antimicrobial Agents

Harmful bacteria that causes food spoiling can be controlled and resisted by the introduction of antimicrobial agents into the polymer films. This can as well extend the shelf life and maintain food quality properly [21]. Essential oils, bacteriocins, and plant extracts are the commonly used natural antibacterial agents. Another promising antibacterial agent is chitosan, and its derivatives are also being studied widely by the researchers [22]. Metal oxide nanoparticles such as $ZnAl_2O_4$ and $CuAl_2O_4$ are incorporated with chitosan and shellac polymers for producing active packaging materials which are prominent example of using synthetic antibacterial agents into packaging composites [23]. Packaging manufacturers thus enjoy a diverse array of antimicrobial agents with a range of alternatives to modify packaging into unique and custom options [24].

2.1.5 Oxygen and Moisture Scavengers

For packaged food products, the quality and shelf life are maintained by adding oxygen and moisture scavengers. Food oxidation results in bacterial infection, color changes, and an unappealing flavor. By protecting the food from oxidative damage within the packaging, oxygen scavengers improve nutritional value and provide better storage. This is accomplished by making the packaging materials more active; they can be utilized for sachet or incorporated into the packaging materials [25]. An inorganic–organic hybrid polymer-based transparent oxygen scavenger that is also

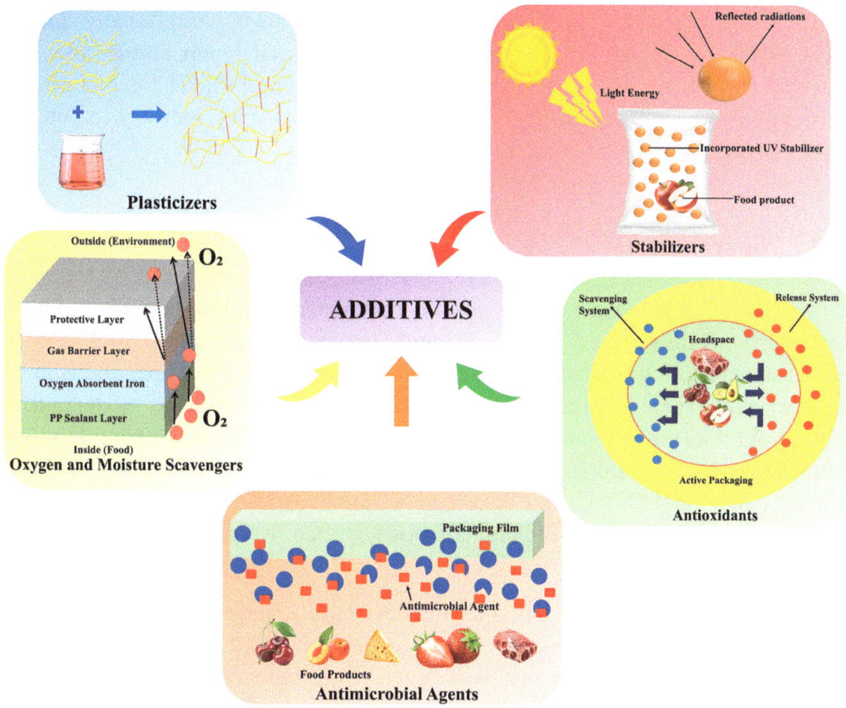

Fig. 1 Types of additives used in packaging composites

UV-activated was created. After developing and applied on aluminum foil, a scavenger demonstrated a high capacity for absorbing oxygen when triggered by ultraviolet light [26]. Particularly for high-water activity products like raw meat and fresh fruit, too much moisture in the package alters both the product's freshness and the integrity of the packaging. Moisture absorbers like cellulose-based biopolymers, silica gel, polyvinyl alcohol (PVA), activated charcoal, chitosan biopolymers, and starch copolymers are incorporated into the packaging and the container to promote moisture-absorbing [27] (Fig. 1).

2.2 *Fillers*

In order to boost the mechanical, thermal, and barrier qualities of packaging composites and hence improve material efficiency, fillers are crucial [28]. Natural fillers offer several benefits such as corrosion tolerance, resistance to chemicals and heat, high strength, and stiffness making them an eco-friendly option for packaging composites [29]. Various fillers are employed in packaging composites to make them more functional and lower production cost. Common fillers used in the packaging industry include calcium carbonate, talc, kaolin, and carbon black. Additionally, natural fibers like lignocellulose, which is sourced from jute, sisal, hemp, cotton, and other materials, are also used as fillers in composite packaging [30]. The exploration of their application in composites continues with the packaging industry to enhance performance and sustainability.

2.2.1 Natural Fillers

Natural agricultural wastes can be used as raw materials for sustainable products, particularly for the plastics industry. Natural fibers are gaining interest as bio-fillers due to their appealing attributes, which include low cost, low density per unit volume, and acceptable specific strength, in addition to their renewable and biodegradable features. A broad range of natural fibers can be utilized as reinforcement or fillers [31]. Natural fibers are categorized into three domains: plant (cellulose/lignocellulose), mineral, and animal [32]. Except for silk, most animal fibers often have low mechanical strength. Among natural fibers, composites often use plant-derived cellulose/lignocellulosic fibers [30].

- Starch-based fillers: Starch-based polymers have emerged as one of the most promising alternatives for conventional petroleum-based polymers in the food packaging industry. Waste materials found from ordinary sources are upcycled into high-value resources through the incorporation of natural biopolymer like starch which can be obtained from conventional sources such as rice, potatoes, and corn as well as from non-traditional feedstocks including peels and seeds of fruit and vegetable [10]. To optimize the functional performance of starch-based composite, recent investigations have focused on tailoring filler characteristics, including type, size, and composition [33]. For instance, banana midribs have been employed to enhance Young's modulus, tensile strength, and vapor barrier properties of starch-based films [34]. Furthermore, the reinforcement of the thermoplastic starch matrices with nanoclays has been reported to increase the barrier and mechanical functionality of the composites [35].
- Cellulose fibers: Cellulose fibers are widely explored as reinforcing filler and gas barrier in food packaging composites because of their differentiable properties, such as lightweight nature, environmental safety, eco-friendliness, and cost efficiency. However, by altering cellulose fiber, certain drawbacks can be addressed [36]. The integration of cellulose fibers into polylactic acid was examined by Elena Popa et al. The outcome proved that the addition enhances processing, reduces melt viscosity, and makes it less transparent but more permeable [37].
- Lignin-based fillers: In order to create solutions with improved biodegradability and possibly lower production costs, lignin can be used as a filler in bio-composites and bioplastics [38]. To develop eco-friendly antimicrobial composite films, lignin has been incorporated into chitosan in varying proportions. Post-characterization, these composite films demonstrated enhanced crystallinity and notable structural modifications, contributing to controlled film thickness and reduced moisture

absorption and solubility. The films' exhibited pronounced antibacterial properties suggesting their applicability in active food packaging systems [39]. Techniques like chemical modification and the integration of stabilizing additives have been explored to improve the mechanical robustness and functional longevity of lignin-based packaging materials. Using lignin's resistance to environmental degradation, this functional enhancement extends the shelf life of packaging solutions [40].

2.2.2 Synthetic Fillers

Synthetic fillers have become essential components in the development of high-performance packaging composites, due to their significant role in enhancing the functional properties and cost efficiency in materials Some examples of synthetic fillers used in packaging composites are: calcium carbonate, titanium dioxide, silicon carbide, boron carbide, silicon dioxide, zirconium dioxide, zeolite, aluminum oxide, zinc oxide, and tungsten carbide [41]. In polymer matrices, inorganic fillers have gained more attention compared to natural fillers due to their superior mechanical properties. These properties include enhanced wear resistance, thermal stability, and improved interfacial bonding between fiber and matrix. These fillers not only contribute to optimize the inert properties of composites but also support eco-friendly design when used in combination with natural fibers [42]. Typically synthetic fillers are incorporated at concentration ranges from 4 to 5% by weight, depending on the nature of the filler and polymer matrix system. Fillers are chosen based on the final packaging product's desired qualities [28]. Silicon carbide was used as a filler in sisal and E-glass fabrics with an epoxy matrix. The results showed an improvement in flexural strength and physical properties. Fiber pull-out is reduced via hybridization, and reduced voids were seen [43].

2.2.3 Nanomaterial-Based Fillers

A variety of nanomaterials are used as a filler to modify the functional properties of polymer material because they lack high mechanical qualities, thermal conductivity, and thermal stability [44, 45]. The mechanical strength, gas barrier, and UV barrier properties of nanocomposites are higher than those of original biopolymers. Commonly used nanofillers include carbon nanotubes, starch nanocrystals, cellulose-based nanofibers or nanowhiskers, silicon dioxide nanoparticles, chitosan nanoparticles, and nanoflakes [46]. Nano-zinc oxide was used as a filler, and neem essential oil was added to a chitosan polymer to create enhanced nanocomposite films, using different concentrations of zinc oxide. The testing indicated improvements in the films' thermal stability, morphology, and mechanical properties [47]. The antibacterial qualities and mechanical barrier can both be enhanced by nanotechnology. Most of the nano-powered food packaging is made of silver nanoparticles, nanoclay, titanium, and zinc oxide. Nanomaterials' physical and chemical characteristics make them widely applicable in numerous sectors [48] (Table 1).

3 Impact on Mechanical Properties

Packaging materials need to be durable, and physically strong, so we evaluate their mechanical properties [67]. This includes how this material would bend, resist impact, stretch, and break when we add various fillers and additives [68]. Studies have shown that the type and amount of additives affect the overall stability and strength of the polymer in packaging composites [69]. Important factors include size of the particle, inherent strength, flexibility, and impact resistance. Proper dispersion and adhesion of the fillers and additives are also crucial for the material.

3.1 Impact of Particle Size and Distribution

The performance of polymer composites is significantly influenced size, concentration, and adhesion of the particles within the matrix. Particle size in particular plays a crucial role in deriving final material's mechanical behavior [70]. For example, because of their extremely small size, carbon nanotubes could be used as reinforcements in various materials which are often lightweight and soft to make them strong [71]. Nanofiller's high surface area and high aspect ratio contribute in improving the mechanical properties of fiber-reinforced polymer composites [28]. Higher friction bonding can be achieved by increasing matrix compactness with the inclusion of smaller particles with smooth, spherical surfaces. The addition of small-sized particles can increase matrix compactness. However, a reduced space between particles can also cause agglomeration, which can make the material more brittle. The toughness and elastic modulus, for instance, increased significantly from 0 to 20 wt.% with a 25 μm filler size [68]. Overall, smaller particle sizes typically have better advantages, especially at the micro and nanoscales because large particles can create more stress concentrations within the matrix.

3.2 Impact on Tensile Strength and Elastic Modulus

Filler addition generally increases Young's modulus. A study was conducted in which lignocellulosic fillers were added to polycaprolactone composites; results indicate that Young's modulus of composites increased [72]. The elastic modulus of composites can be readily increased by adding micro- or nanoscale particles. Efficient stress transfer from the polymer matrix to the particle is made feasible by the particles'

Table 1 Properties of additives and fillers

Additive/filler	Type	Function	Advantages	References
Plasticizers (glyceryl tributyrate and triethyl citrate)	Natural	Improving thermal and physical properties Enhancing flexibility and processability	Lowering thermal transition temperatures Non-toxic and environmentally friendly	[49]
Stabilizers	Natural (Tannins)	Improve the films' thermal and UV stability	Improved thermo-oxidative stability Enhanced UV stability Radical scavenging Improved compatibility	[50]
	Synthetic	Prevents degradation from UV, heat, and oxygen	Increases shelf life and material stability	[51]
Antioxidants	Natural (α-tocopherol)	Delays undesired oxidation processes in food Radical scavenging	Food quality preservation Improved material properties Migration control	[52]
	Synthetic (selenium nanoparticles)	Free radical scavenging Antioxidant action	Extended shelf life Maintained food quality Non-migratory behavior	[53]
Antimicrobial agents	Synthetic (silver nanoparticles)	Strong antibacterial activity Cellular disruption	Enhanced food safety and shelf life Improved film properties	[54]
	Natural	Inhibit the growth of microorganisms Enhanced barrier properties	Synergistic effects Enhanced mechanical strength and thermal stability Suitable for various foods	[55]
Oxygen scavengers	Natural/synthetic	Absorbs residual oxygen	Prevents oxidation and spoilage	[56]
Enzyme-based additives	Natural	Antimicrobial activity Structural configuration	Enhanced structural and barrier properties Food preservation	[57]

(continued)

Table 1 (continued)

Additive/filler	Type	Function	Advantages	References
Nanocellulose	Natural	Reinforcement Enhances barrier properties	High stiffness Low oxygen permeability High specific surface area	[58]
Starch-based fillers	Natural	Reinforcing filler Thermal stability	Increased biodegradability Low cost Modified properties	[59]
Lignin-based fillers	Natural	Surface properties modification Compatibilizer	Renewability Low density in composites and low abrasion Improved strength and thermal properties	[60]
Calcium carbonate	Synthetic	Reducing surface energy Functional properties Enhancement	Cost-effective Increased impact and electric resistance Waste reduction Control over the manufacture of products	[61]
Silica-based fillers	Synthetic	Enhances thermal and barrier properties Active packaging	Improved antibacterial and antioxidant activity Improves durability and mechanical strength	[62]
Titanium dioxide	Synthetic	Imparts antimicrobial properties Crosslinking agent Improves film properties	Enhances mechanical properties Low toxicity and economical Extended shelf life	[63]
Nanoclays	Nanomaterial	Improves physical, chemical, mechanical, and barrier properties Active and intelligent packaging	It can act as an ethylene scavenger Thermal stability Improve transparency of the film Cost-effectiveness	[64]

(continued)

Table 1 (continued)

Additive/filler	Type	Function	Advantages	References
Silver nanoparticles	Nanomaterial	Provides antimicrobial properties Biosensing properties	Improved shelf life Prevents visible and ultraviolet light penetration Physicochemical properties enhancement	[65]
Graphene oxide	Nanomaterial	Improves barrier properties Antimicrobial activity	Improved mechanical properties Enhances durability and functionality Enhanced thermal stability	[66]

higher degree of rigidity compared to the polymeric matrices [68]. Young's modulus is greatly influenced by the amount of filler in amorphous materials, but crystalline materials are less susceptible. Fillers consisting of silica or graphene can increase Young's modulus. Elastic behavior is greatly affected by minimum filler loading [73]. One essential mechanical property that ensures the structural integrity of a composite is its tensile strength. Fillers and additives can be employed to boost tensile strength, as numerous studies have shown. The tensile characteristics of kenaf/polyester composites could be enhanced by the addition of alumina trihydrate [74]. Adding tamarind seed powder to flax/elium composites led in a 6% improvement in ultimate tensile strength and a 10% increase in Young's modulus [75]. When titanium dioxide particles were introduced to poly (ethylene-co-1-octadene) composites, their tensile modulus increased [76].

3.3 *Effect on Impact Resistance and Toughness*

The inclusion of fillers and additives has a significant impact on the toughness and impact resistance of composites, especially in packaging applications. In order to improve their mechanical properties, polylactic acid-based composite materials that lack hardness should be reinforced with inorganic, high-strength reinforcements such as basalt microfibers, talc nano-platelets, and clay nanoparticles [77]. The findings of fish scale powder and wood dust as fillers demonstrated improved mechanical characteristics with a 20.48% improvement in impact resistance [78].

3.4 *Concentration of Filler Content*

The concentration of fillers determines the mechanical performance, agglomeration, and particle distribution in composites. The size and concentration of the filler have been shown to impact the agglomeration and spatial distribution of the filler particles. The distribution of filler particles may be uniform or non-uniform, depending on the concentration [79]. There is often an appropriate limit for filler material. Too much filler can cause some beneficial features to deteriorate. More filler, for instance, has been shown to have a less detrimental effect on impact resistance than on strain at break [80]. Polylactic acid-based composite films were created in which micro-boron nitride and nano-boron nitride were incorporated to improve the barrier and mechanical properties. Low concentrations (1–3%) of fillers improved mechanical and barrier properties, greater concentrations frequently result in agglomeration, non-uniform distribution, lower flexibility, and decreased mechanical strength. Therefore, to achieve the appropriate qualities in composite films for food packaging, it is imperative to optimize the filler content [81]. Therefore, especially for food packaging applications, optimal filler concentration is essential.

3.5 *Effects of Temperature*

The functionality of the additives and fillers in the composite might be affected by the temperature at which it is utilized or generated. Temperature can affect how fillers and additives work in a composite matrix like higher temperature can cause low viscosity and decreased packing stress of composites. This impacts handling and mechanical performance of the composites [82]. High temperatures can cause weakening of the resin; on the other hand, low temperature can increase the flexural strength of carbon fiber composites. The glass transition and heat deflection temperature are greatly affected by composite's behavior at various temperatures [83].

4 Enhancement of Barrier Properties

To ensure overall food safety including extended shelf life and reduced spoilage, it is paramount to resist light, heat, moisture, and gases from the contact of food items. In this sector, composite material's barrier properties are of paramount importance in the protection of food products against external environmental factors. The barrier properties of packaging materials are enhanced by the addition of different additives and fillers, increasing their effectiveness. The developments in nanotechnology and bio-based materials have contributed to barrier enhancement, which has produced innovative solutions including biodegradable polymer blends, smart packaging systems with active barriers, and coatings [84]. These technological advances not only improve protection but also reduce the impact of packaging waste on the environment by the inclusion of bio-based materials [85]. Researchers are concentrating on creating multilayer barrier systems that maximize performance by combining several materials with different qualities and enhancing sustainability [13].

- Oxygen Barrier: Upon exposure to higher O_2 concentrations, food items constituting elevated amounts of lipids or oils might lead to physicochemical changes affecting the food quality. This exposure induces negative chemical transformations that compromise food quality, including discoloration, lipid oxidation, and the development of undesirable flavors. Subsequently, oxygen permeation can facilitate microbial advancements, thereby accelerating food spoilage. The integrity of food items can also be compromised during transportation and storage. So utilization of high oxygen barrier composite packaging materials helps in safeguarding of food quality [86, 87].
- Moisture Barrier: The stability of perishable goods often depends on moisture and humidity, frequently resulting in the proliferation of mold, subsequent spoilage, and further economic losses. Elevated relative humidity conditions create a suitable environment to microbial proliferation, thus compromising quality and safety of food [88]. Through the incorporation of additives and fillers, the water vapor transmission rate (WVTR) can be controlled. These additives and fillers include nanoclays, silica nanoparticles, and hydrophobic coatings. Wet and humid weather's adverse effect could be resisted and controlled by introducing multilayer coatings and hydrophobic fillers in the polymer matrix [89].
- Microbial Barrier: Packaging materials are important to resist microbial contamination. Various methods enhance antimicrobial properties. The simplest way to reduce food contamination and deterioration is to use antibacterial agents. Several modifications have been introduced in the field of food packaging, one of the demanding methods for food preservation, to support the characteristics and functions of antibacterial materials [90]. Silver nanoparticles, chitosan coatings, and antimicrobial compounds all prevent bacterial and fungal growth, which lowers spoilage and increases shelf life. These can be added to active packaging solutions [91].
- UV Light Barrier: Photochemical reactions caused by ultraviolet (UV) radiation can deteriorate package materials and the contents within. In the packaging industry, new developments have been made in the creation of several kinds of synthetic, inorganic, and organic UV absorbers to stop food goods from deteriorating [92]. UV-absorbing coatings, lignin-based fillers, and titanium dioxide are all helpful in preventing product photodegradation. This type of barrier property is important when focusing on the packaging of pharmaceuticals, dairy products, and drinks that are light-sensitive [93].

4.1 Mechanisms of Barrier Enhancement

The mechanisms of improving barrier properties of packaging materials are explained in this section. Improvement relies upon the type of additives and fillers used, the polymer matrix, and the intended application of the packaging.

4.1.1 Tortuosity Effect

Using the tortuosity effect, fillers and additives enhance barrier properties. Adding nanofillers (various clays, layered silicates, graphene oxide, cellulose nanocrystals, and spherical silica) to a polymer matrix improves gas barrier properties. These fillers create a complex path for gas and moisture molecules. This extended and restricted path limits diffusion, which forces the gases to travel around the fillers, slowing down their movement [94]. The tortuosity effect can be applied in packaging applications where gas and moisture resistance are necessary.

4.1.2 Crystallinity Modification

In this mechanism, the crystalline structure of the polymer matrix is modified with the incorporation of additives and fillers acting as nucleating agents; this addition improves the barrier properties of the material. Additives and fillers enhance final material properties and processability as they can accelerate the crystallization process and increase the temperature of crystallization. For instance, nanoparticles such as carbon nanotubes, kaolinite, and montmorillonites may increase the crystallinity degree of poly(m-xylylene adipamide) to 42% from 20% [95]. The type of polymers used and their properties are important factors that can affect the degree of modification of crystallinity. For example, when nanofibrillated cellulose was added to the polylactic acid composites, it was found that the addition improved the crystallinity and mechanical properties of the composites [96].

4.1.3 Hydrophobic Surface Modification

Additives and fillers like lignin, titanium dioxide, and silica nanoparticles are added to composites to enhance their water-repellent property. Titanium dioxide and silicon dioxide are also added because they give dual-scale roughness being micrometer-sized [97]. In a research study, graphene oxide was modified to turn it into hydrophilic graphene oxide to produce thin films that have a polystyrene matrix. This modification enhanced the hydrophobicity of the nanocomposite films [98].

4.1.4 Cross-Linking Mechanism

Additives, like citric acid, epoxidized soybean oil, etc., are used that strengthen intermolecular interactions because they promote cross-linking between polymer chains. The interaction helps in lowering the moisture and gas permeability because they inhibit the movement of polymer chains. For example, citric acid used with hydroxyl groups like starch or cellulose resulted in increasing the cohesion of the polymer chain because of citric acid's crosslinking behavior [99]. For improvising the barrier properties of edible and biodegradable packaging materials, manufacturers can use suitable crosslinking agents and can control the degree of crosslinking to extend the shelf life and food security [100].

4.1.5 Multi-layered Packaging System

High-performing packaging materials are created through these systems as they combine different materials together, resulting in highly improved barrier properties. Polyvinylidene chloride, ethylene vinyl alcohol, and nanoclay-reinforced layers are commonly used in these systems to offer a better resistance to moisture, oxygen, and aromas [101, 102]. To extend the shelf life of packaged foods, for example, zein nanocapsules containing carvacrol and sodium alginate have been used to create nano-laminated films [103] (Fig. 2).

5 Sustainability and Environmental Considerations

Industries are being forced to provide environmentally friendly packaging solutions that strike a compromise between performance and environmental impact due to the growing worldwide concern over plastic waste and environmental sustainability. Even while they work well, conventional packaging materials made of petroleum greatly increase the amount of non-biodegradable waste. One promising strategy for lowering carbon footprints and improving recycling is the use of sustainable fillers and additives in packaging composites. Key factors that should be considered for sustainable packaging materials are listed below:

- Biodegradability and Composability: By adding bio-based fillers like cellulose nanofibers, lignin, and starch, packaging materials become more biodegradable and can break down organically in the environment. Because of their biodegradable qualities, polymers like polylactic acid, polyhydroxyalkanoates, and polybutylene adipate terephthalate are extensively researched as sustainable substitutes for conventional plastics [104].
- Reduction of Carbon Footprint: To reduce greenhouse gas emissions, sustainable packaging solutions emphasize the utilization of waste-derived materials and renewable resources. Compared to petroleum-based plastics, the production

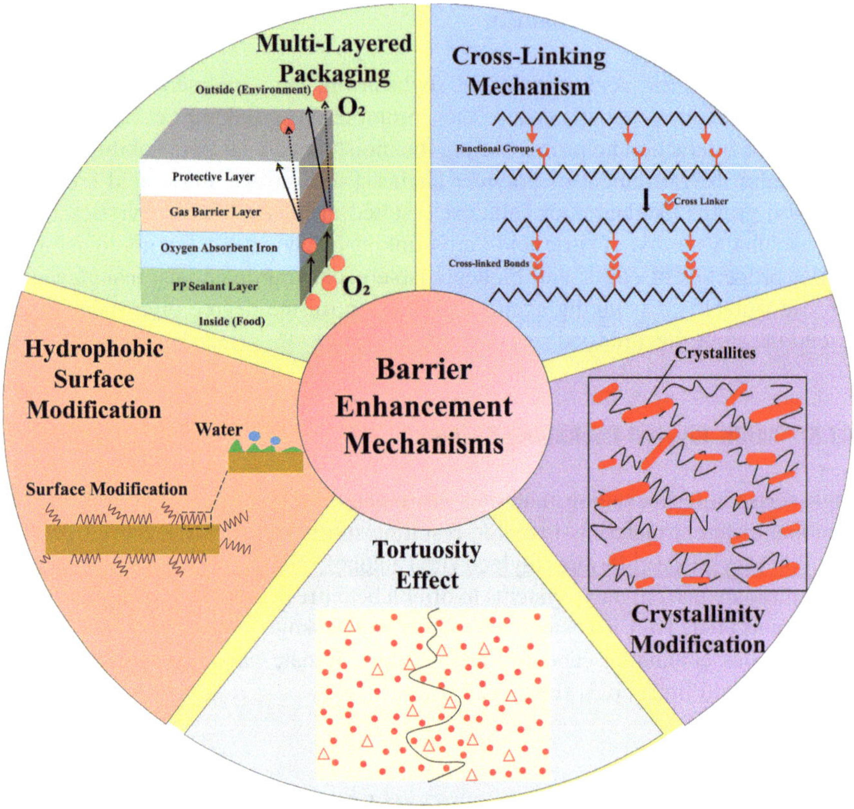

Fig. 2 Mechanism of barrier enhancement in packaging composites

of bioplastics and composites based on nanocellulose uses fewer fossil fuels, lowering carbon footprints. Fermentation-derived biopolymers like bio-based polyethylene and polyhydroxyalkanoates can also be used in the development of sustainable materials. These biopolymers use carbon dioxide present in the atmosphere when they are manufactured, helping to achieve carbon neutrality [105, 106].

- The Circular Economy and Recyclability: While designing a packaging material, recyclability is an important factor that is considered for promoting a circular economy. Using recyclable additives and environmentally friendly design processes, high structural integrity is maintained across several applications and recycling cycles. Mechanical and thermal strength are enhanced by calcium carbonate, silica nanoparticles, and graphene oxide, leading to a stabilized polymer matrix and thus reducing deterioration of material when reprocessing. Recycling processes using chemicals like pyrolysis, enzymatic depolymerization, and solvolysis break down the plastic matrix into monomers that can be reused. Mono-materials and biopolymers can be used to improve the recyclability [105, 107].

- Renewable and Bio-based Fillers: Recent research on packaging materials is focusing on replacing synthetic fillers by using biopolymers like chitosan, lignin, nanocellulose, and agricultural waste fibers. Bio-based polymers reinforced in high-performance packaging are more convenient to use because they are sustainable and have better barrier, thermal, and mechanical properties. They also help in sustainability projects because they reduce dependency on fossil fuels. Natural fibers combined with nanomaterials make them biopolymer composites, leading to a shift to fully bio-based materials, which enhance materials durability and recyclability [108, 109].
- Water and Energy Efficiency in Production: Sustainable packaging manufacturing methods are designed to use less energy and water, which lessens their impact on the environment. High-performance barrier qualities are maintained while environmentally friendly production is further enhanced by advancements in solvent-free and green synthesis methods [110].
- Smart and Intelligent Packaging for Waste Reduction: Oxygen scavengers, freshness indicators, and antimicrobial coatings are examples of active and intelligent packaging solutions that help increase product shelf life and decrease food waste. By guaranteeing that packaged items stay fresh for extended periods and reducing the environmental effect of wasted products, these advances support sustainability goals [111].

Although the use of recyclable and bio-based materials is expanding quickly, there are still several obstacles to overcome, especially when it comes to striking a balance between long-term material performance, cost-effectiveness, and the viability of large-scale production. To promote industry adoption, sustainable packaging solutions must not only satisfy legal and environmental requirements but also retain competitive cost.

6 Conclusion and Future Perspective

The incorporation of additives and fillers in packaging composites has transformed the food packaging composites by providing a sustainable, high-performance alternative to traditional packaging solutions. The integration of additives into biodegradable polymers has particularly enhanced the food safety, flexibility, and durability. The bio-based reinforcement and nanofillers enhanced the barrier and mechanical properties. The functionality of packaging materials was enhanced by bio-based coatings, multilayered, and active packaging. Despite these innovations, there are still some challenges in developing completely scalable and sustainable packaging solutions. More research into polymer modifications, hybrid composite constructions, and advanced reinforcement techniques is required because many bio-based polymers still have poor mechanical and barrier properties compared to petroleum-based packaging.

To overcome these challenges, an extensive approach is required, one that combines innovations in materials, effective production techniques, and the ideas of a circular economy to create packaging solutions that are both economically viable and environmentally friendly. Innovations in smart and active packaging technologies, as well as the growing use of renewable and biodegradable materials, will greatly influence the future of food packaging. Furthermore, active packaging that includes freshness indicators, oxygen scavengers, and antimicrobials will reduce food spoiling while maintaining consumer safety. Innovations in recycling, strict rules for managing recyclable plastics, along with ongoing research into biopolymer development, and growing consumer demand for low-waste and environmentally friendly packaging will encourage the formulation of new material combinations and effective recycling techniques.

Conflicts of Interest There are no conflicts of interest to declare by the authors.

References

1. Opara UL, Mditshwa A (2013) A review on the role of packaging in securing food system: adding value to food products and reducing losses and waste. Afr J Agricult Res 8(22):2621–2630. https://doi.org/10.5897/AJAR2013.6931
2. Ibrahim ID, Hamam Y, Sadiku ER, Ndambuki JM, Kupolati WK, Jamiru T, Eze AA, Snyman J (2022) Need for sustainable packaging: an overview. Polymers 14(20):1–16. https://doi.org/10.3390/polym14204430
3. Ambaye TG, Vaccari M, Prasad S, van Hullebusch ED, Rtimi S (2022) Preparation and applications of chitosan and cellulose composite materials. J Environ Manag 301:113850. https://doi.org/10.1016/j.jenvman.2021.113850
4. Abdellah Ali SF (2016) Mechanical and thermal properties of promising polymer composites for food packaging applications. IOP Conf Ser Mater Sci Eng 137(1):35. https://doi.org/10.1088/1757-899X/137/1/012035
5. Ozcalik O, Tihminlioglu F (2013) Barrier properties of corn zein nanocomposite coated polypropylene films for food packaging applications. J Food Eng 114(4):505–513. https://doi.org/10.1016/j.jfoodeng.2012.09.005
6. Mohanty AK, Vivekanandhan S, Pin JM, Misra M (2018) Composites from renewable and sustainable resources: challenges and innovations. Science 362(6414):536–542. https://doi.org/10.1126/science.aat9072
7. Fornaro M, Liguori B, Ambrogi V, Caputo D (2024) Zeolite additives for flexible packaging polymers: current status review and future perspectives. Polymers 16(23):1–17. https://doi.org/10.3390/polym16233399
8. Siddiqui SA, Yang X, Deshmukh RK, Gaikwad KK, Bahmid NA, Castro-Muñoz R (2024) Recent advances in reinforced bioplastics for food packaging: a critical review. Int J Biol Macromol 263:399. https://doi.org/10.1016/j.ijbiomac.2024.130399
9. Helanto K, Matikainen L, Talj R, Rojas OJ (2019) Bio-based polymers for sustainable packaging and biobarriers: a critical review. BioResources 14(2):4902–4951

10. Muñoz-Gimena PF, Oliver-Cuenca V, Peponi L, López D (2023) A review on reinforcements and additives in starch-based composites for food packaging. Polymers 15(13):972. https://doi.org/10.3390/polym15132972
11. Singh P, Saengerlaub S, Wani AA, Langowski HC (2012) Role of plastics additives for food packaging. Pigm Resin Technol 41(6):368–379. https://doi.org/10.1108/03699421211274306
12. Bocqué M, Voirin C, Lapinte V, Caillol S, Robin JJ (2016) Petro-based and bio-based plasticizers: chemical structures to plasticizing properties. J Polym Sci A Polym Chem 54(1):11–33. https://doi.org/10.1002/pola.27917
13. Alhanish A, Abu Ghalia M (2021) Developments of biobased plasticizers for compostable polymers in the green packaging applications: a review. Biotechnol Progr 37(6):1–17. https://doi.org/10.1002/btpr.3210
14. Vieira MGA, Da Silva MA, Dos Santos LO, Beppu MM (2011) Natural-based plasticizers and biopolymer films: a review. Eur Polym J 47(3):254–263. https://doi.org/10.1016/j.eurpolymj.2010.12.011
15. Nur Hanani ZA, McNamara J, Roos YH, Kerry JP (2013) Effect of plasticizer content on the functional properties of extruded gelatin-based composite films. Food Hydrocoll 31(2):264–269. https://doi.org/10.1016/j.foodhyd.2012.10.009
16. Plota A, Masek A (2021) Plant-origin stabilizer as an alternative of natural additive to polymers used in packaging materials. Int J Mol Sci 22(8):12. https://doi.org/10.3390/ijms22084012
17. Kumar A, Kumari P, Gupta K, Singh M, Tomer V (2023) Recent advances in extraction, techno-functional properties, food and therapeutic applications as well as safety aspects of natural and modified stabilizers. Food Rev Int 39(4):2233–2276. https://doi.org/10.1080/87559129.2021.1950174
18. Samper MD, Fages E, Fenollar O, Boronat T, Balart R (2013) The potential of flavonoids as natural antioxidants and UV light stabilizers for polypropylene. J Appl Polym Sci 129(4):1707–1716. https://doi.org/10.1002/app.38871
19. Sanches-Silva A, Costa D, Albuquerque TG, Buonocore GG, Ramos F, Castilho MC, Machado AV, Costa HS (2014) Trends in the use of natural antioxidants in active food packaging: a review. Food Add Contam A 31(3):374–395. https://doi.org/10.1080/19440049.2013.879215
20. Siddiqui MN, Redhwi HH, Tsagkalias I, Vouvoudi EC, Achilias DS (2021) Development of bio-composites with enhanced antioxidant activity based on poly(Lactic acid) with thymol, carvacrol, limonene, or cinnamaldehyde for active food packaging. Polymers 13(21):652. https://doi.org/10.3390/polym13213652
21. Sung SY, Sin LT, Tee TT, Bee ST, Rahmat AR, Rahman WAWA, Tan AC, Vikhraman M (2013) Antimicrobial agents for food packaging applications. Trends Food Sci Technol 33(2):110–123. https://doi.org/10.1016/j.tifs.2013.08.001
22. Mousavi Khaneghah A, Hashemi SMB, Limbo S (2018) Antimicrobial agents and packaging systems in antimicrobial active food packaging: an overview of approaches and interactions. Food Bioprod Process 111:1–19. https://doi.org/10.1016/j.fbp.2018.05.001
23. Mohamed SAA, Mekkey SD, Othman AM, El-Sakhawy M (2024) Novel antimicrobial biodegradable composite films as packaging materials based on shellac/chitosan, and $ZnAl_2O_4$ or $CuAl_2O_4$ spinel nanoparticles. Sci Rep 14(1):1–22. https://doi.org/10.1038/s41598-024-78261-1
24. Fadiji T, Rashvand M, Daramola MO, Iwarere SA (2023) A review on antimicrobial packaging for extending the shelf life of food. Processes 11(2):1–30. https://doi.org/10.3390/pr11020590
25. Gupta P (2023) Role of oxygen absorbers in food as packaging material, their characterization and applications. J Food Sci Technol 61(2):242–252. https://doi.org/10.1007/s13197-023-05681-8
26. Amberg-Schwab S, Müller K, Somorowsky F, Sängerlaub S (2023) UV-activated, transparent oxygen scavenger coating based on inorganic–organic hybrid polymer (ORMOCER®) with high oxygen absorption capacity. Coatings 13(2):473. https://doi.org/10.3390/coatings13020473
27. Gaikwad KK, Singh S, Ajji A (2019) Moisture absorbers for food packaging applications. Environ Chem Lett 17(2):609–628. https://doi.org/10.1007/s10311-018-0810-z

28. Ramesh M, Rajeshkumar LN, Srinivasan N, Kumar DV, Balaji D (2022) Influence of filler material on properties of fiber-reinforced polymer composites: a review. E-Polymers 22(1):898–916. https://doi.org/10.1515/epoly-2022-0080
29. Moutinho LG (2025) Bio-based thermoplastic composites reinforced with natural fillers for sustainable packaging: a comprehensive review, pp 1–25. https://doi.org/10.1002/pc.29589
30. Sydow Z, Bieńczak K (2019) The overview on the use of natural fibers reinforced composites for food packaging. J Nat Fibers 16(8):1189–1200. https://doi.org/10.1080/15440478.2018.1455621
31. Majeed K, Jawaid M, Hassan A, Abu Bakar A, Abdul Khalil HPS, Salema AA, Inuwa I (2013) Potential materials for food packaging from nanoclay/natural fibres filled hybrid composites. Mater Des 46:391–410. https://doi.org/10.1016/j.matdes.2012.10.044
32. Prathipa R, Sivakumar C, Shanmugasundaram B (2018) Biodegradable polymers for sustainable packaging applications. Int J Mech Eng Technol 9(6):293–303
33. Pérez HF, Galicia García M (2020) J Appl Res Technol 18:245–258
34. Wahab A (2024) Mechanical, structural and barrier properties of starch—based film reinforced 8:117–123
35. Adewumi FD, Ogona DF, Ogundolie FA, Ogunmodede O, Peters O, Idowu O (2024) Starch-based composites: an eco-friendly alternative for food packaging. In: International conference on science, engineering and business for driving sustainable development goals, SEB4SDG 2024, pp 1–6. https://doi.org/10.1109/SEB4SDG60871.2024.10629865
36. Saedi S, Garcia CV, Kim JT, Shin GH (2021) Physical and chemical modifications of cellulose fibers for food packaging applications. Cellulose 28(14):8877–8897. https://doi.org/10.1007/s10570-021-04086-0
37. Elena Popa E, Rapa M, Popa O, Mustatea G, Ioan Popa V, Carmen Mitelut A, Elena Popa M (2017) Polylactic acid/cellulo se fibres based composites for food packaging applications 4:673–677
38. Tribot A, Amer G, Abdou Alio M, de Baynast H, Delattre C, Pons A, Mathias JD, Callois JM, Vial C, Michaud P, Dussap CG (2019) Wood-lignin: Supply, extraction processes and use as bio-based material. Eur Polym J 112:228–240. https://doi.org/10.1016/j.eurpolymj.2019.01.007
39. Rai S, Dutta PK, Mehrotra GK (2017) Lignin incorporated antimicrobial chitosan film for food packaging application. J Polym Mater 34(1):171–183
40. Priyadarshi R, Ghosh T, Purohit SD, Prasannavenkadesan V, Rhim JW (2024) Lignin as a sustainable and functional material for active food packaging applications: a review. J Clean Product 469:143151. https://doi.org/10.1016/j.jclepro.2024.143151
41. Kiran MD, Govindaraju HK, Jayaraju T, Kumar N (2018) Review-effect of fillers on mechanical properties of polymer matrix composites. Mater Today Proceed 5(10):22421–22424. https://doi.org/10.1016/j.matpr.2018.06.611
42. Praveenkumara J, Madhu P, Yashas Gowda TG, Sanjay MR, Siengchin S (2022) A comprehensive review on the effect of synthetic filler materials on fiber-reinforced hybrid polymer composites. J Text Instit 113(7):1231–1239. https://doi.org/10.1080/00405000.2021.1920151
43. Arpitha GR, Sanjay MR, Senthamaraikannan P, Barile C, Yogesha B (2017) Hybridization effect of sisal/glass/epoxy/filler based woven fabric reinforced composites. Exper Techn 41(6):577–584. https://doi.org/10.1007/s40799-017-0203-4
44. Domun N, Hadavinia H, Zhang T, Sainsbury T, Liaghat GH, Vahid S (2015) Improving the fracture toughness and the strength of epoxy using nanomaterials-a review of the current status. Nanoscale 7(23):10294–10329. https://doi.org/10.1039/c5nr01354b
45. Huang C, Qian X, Yang R (2018) Thermal conductivity of polymers and polymer nanocomposites. Mater Sci Eng R Rep 132:1–22. https://doi.org/10.1016/j.mser.2018.06.002
46. Zhou X, Zhou X, Zhou L, Jia M, Xiong Y (2024) Nanofillers in novel food packaging systems and their toxicity issues. Foods 13(13):1–34. https://doi.org/10.3390/foods13132014
47. Sanuja S, Agalya A, Umapathy MJ (2015) Synthesis and characterization of zinc oxide-neem oil-chitosan bionanocomposite for food packaging application. Int J Biol Macromol 74:76–84. https://doi.org/10.1016/j.ijbiomac.2014.11.036

48. Chadha U, Bhardwaj P, Selvaraj SK, Arasu K, Praveena S, Pavan A, Khanna M, Singh P, Singh S, Chakravorty A, Badoni B, Banavoth M, Sonar P, Paramasivam V (2022) Current trends and future perspectives of nanomaterials in food packaging application. J Nanomater 2022:416. https://doi.org/10.1155/2022/2745416
49. Malbos LB, Iglesias-Montes ML, Seoane IT, Cyras VP, Manfredi LB (2024) Biobased ternary composites for food packaging: influence of natural plasticizers and starch on polylactic acid performance. J Mater Sci 59(43):20304–20324. https://doi.org/10.1007/s10853-024-10375-3
50. Bridson JH, Kaur J, Zhang Z, Donaldson L, Fernyhough A (2015) Polymeric flavonoids processed with co-polymers as UV and thermal stabilisers for polyethylene films. Polym Degrad Stabil 122:18–24. https://doi.org/10.1016/j.polymdegradstab.2015.10.002
51. Sönmez HE, Bodur MS, Baysan AA, Bakkal M, Serhatli E (2021) Photostabilizers performance on the surface analysis of green composites. Polym Polym Comp 29(1):57–64. https://doi.org/10.1177/0967391120901620
52. Dumitriu RP, Stoleru E, Mitchell GR, Vasile C (2021) Incorporating α-tocopherol for food packaging applications
53. Vera P, Echegoyen Y, Canellas E, Nerín C, Palomo M, Madrid Y, Cámara C (2016) Nano selenium as antioxidant agent in a multilayer food packaging material. Anal Bioanal Chem 408(24):6659–6670. https://doi.org/10.1007/s00216-016-9780-9
54. Roy S, Shankar S, Rhim JW (2019) Melanin-mediated synthesis of silver nanoparticle and its use for the preparation of carrageenan-based antibacterial films. Food Hydrocoll 88:237–246. https://doi.org/10.1016/j.foodhyd.2018.10.013
55. Shamene B, Tee SM, Rosli NA (2024) Strategic addition of rosemary essential oil in natural rubber-toughened poly(lactic acid)/cinnamon oil blends for antimicrobial packaging. Fut Foods 10:100506. https://doi.org/10.1016/j.fufo.2024.100506
56. Ahn BJ, Gaikwad KK, Lee YS (2016) Characterization and properties of LDPE film with gallic-acid-based oxygen scavenging system useful as a functional packaging material. J Appl Polym Sci 133(43):1–8. https://doi.org/10.1002/app.44138
57. Alkhair AY, Melesse EY, Anatol I, Filinskaya A, Tveritnikova IS, Mihryachev OI (2025) Effect of lysozyme and glucose oxidase on the physical-mechanical and barrier properties of linear low-density polyethylene, pp 1–14. https://doi.org/10.22063/POJ.2025.35579.1336
58. Muthulakshmi L, Mohan S, Rajam R (2023) Bio-based composites for food packaging. emerging techniques for food processing and preservation, pp 121–142. https://doi.org/10.1201/9781003217138-5
59. Jayarathna S, Andersson M, Andersson R (2022) Recent advances in starch-based blends and composites for bioplastics applications. Polymers 14(21):1–24. https://doi.org/10.3390/polym14214557
60. Toriz G, Denes F, Young RA (2002) Lignin-polypropylene composites. Part 1: composites from unmodified lignin and polypropylene. Polym Comp 23(5):806–813. https://doi.org/10.1002/pc.10478
61. Thenepalli T, Jun AY, Han C, Ramakrishna C, Ahn JW (2015) A strategy of precipitated calcium carbonate ($CaCO_3$) fillers for enhancing the mechanical properties of polypropylene polymers. Korean J Chem Eng 32(6):1009–1022. https://doi.org/10.1007/s11814-015-0057-3
62. Lawal U, Kumar N, Samyuktha R, Gopi A, Robert V, Pugazhenthi G, Loganathan S, Valapa RB (2024) Poly (lactic acid)/amine grafted mesoporous silica-based composite for food packaging application. Int J Biol Macromol 277:67. https://doi.org/10.1016/j.ijbiomac.2024.134567
63. Zhang W, Rhim JW (2022) Titanium dioxide (TiO_2) for the manufacture of multifunctional active food packaging films. Food Packag Shelf Life 31:100806. https://doi.org/10.1016/j.fpsl.2021.100806
64. Deshmukh RK, Hakim L, Akhila K, Ramakanth D, Gaikwad KK (2023) Nano clays and its composites for food packaging applications. Int Nano Lett 13(2):131–153. https://doi.org/10.1007/s40089-022-00388-8
65. Ediyilyam S, George B, Shankar SS, Dennise TT, Wacławek S, Cerník M, Padil VVT (2021) Chitosan/gelatin/silver nanoparticles composites films for biodegradable food packaging applications. Polymers 13(11):680. https://doi.org/10.3390/polym13111680

66. Silva-Leyton R, Quijada R, Bastías R, Zamora N, Olate-Moya F, Palza H (2019) Polyethylene/graphene oxide composites toward multifunctional active packaging films. Compos Sci Technolo 184:107888. https://doi.org/10.1016/j.compscitech.2019.107888
67. Briassoulis D, Giannoulis A (2018) Evaluation of the functionality of bio-based food packaging films. Polym Test 69:39–51. https://doi.org/10.1016/j.polymertesting.2018.05.003
68. Siraj S, Al-Marzouqi AH, Iqbal MZ, Ahmed W (2022) Impact of micro silica filler particle size on mechanical properties of polymeric based composite material. Polymers 14(22):830. https://doi.org/10.3390/polym14224830
69. Kościuszko A, Czyżewski P, Wajer Ł, Ościak A, Bieliński M (2020) Properties of polypropylene composites filled with microsilica waste. Polim Polym 65(2):99–104
70. Fu SY, Feng XQ, Lauke B, Mai YW (2008) Effects of particle size, particle/matrix interface adhesion and particle loading on mechanical properties of particulate-polymer composites. Comp B Eng 39(6):933–961. https://doi.org/10.1016/j.compositesb.2008.01.002
71. Lau AKT, Hui D (2002) The revolutionary creation of new advanced materials: carbon nanotube composites. Comp B Eng 33(4):263–277. https://doi.org/10.1016/S1359-8368(02)00012-4
72. Ludueña L, Vázquez A, Alvarez V (2012) Effect of lignocellulosic filler type and content on the behavior of polycaprolactone based eco-composites for packaging applications. Carbohydr Polym 87(1):411–421. https://doi.org/10.1016/j.carbpol.2011.07.064
73. Zhu J, Abeykoon C, Karim N (2021) Investigation into the effects of fillers in polymer processing. Int J Lightweight Mater Manufact 4(3):370–382. https://doi.org/10.1016/j.ijlmm.2021.04.003
74. Atiqah A, Ansari MNM, Kamal MSS, Jalar A, Afeefah NN, Ismail N (2020) Effect of alumina trihydrate as additive on the mechanical properties of kenaf/polyester composite for plastic encapsulated electronic packaging application. J Mater Res Technol 9(6):12899–12906. https://doi.org/10.1016/j.jmrt.2020.08.116
75. Selvaraj DK, Silva FJG, Campilho RDSG, Baptista A, Pinto GFL (2019) Influence of the natural additive on natural fiber reinforced thermoplastic composite. Proced Manufact 38(2019):1121–1129. https://doi.org/10.1016/j.promfg.2020.01.200
76. Sit M, Dashatan S, Zhang Z, Dhakal HN, Khalfallah M, Gamer N, Ling J (2023) Inorganic fillers and their effects on the properties of flax/PLA composites after UV degradation. Polymers 15(15):3221
77. Al-Maharma AY, Sendur P (2019) Review of the main factors controlling the fracture toughness and impact strength properties of natural composites. Mater Res Exp 6(2):28. https://doi.org/10.1088/2053-1591/aaec28
78. Tripathy C, Mohapatra DK, Deo CR, Mishra C (2025) In fluence of layering variation with addition of bio-fillers on mechanical and thermal performances of hybrid polyester composites. https://doi.org/10.1177/00952443241306701
79. Pavlyuchkova EA, Malkin AY, Kornev YV, Simonov-Emelyanov ID (2024) Distribution of filler in polymer composites role of particle size and concentration. Polym Sci Ser A 66(1):113–120
80. Post W, Kuijpers LJ, Zijlstra M, van der Zee M, Molenveld K (2021) Effect of mineral fillers on the mechanical properties of commercially available biodegradable polymers. Polymers 13(3):1–14. https://doi.org/10.3390/polym13030394
81. Ara I, Haskaraca G, Yildiz S, Ayhan Z (2024) Composite PLA films incorporated with micro/nano boron nitride for sustainable food packaging. Green Mater. https://doi.org/10.1680/jgrma.24.00109
82. Elbishari H, Satterthwaite J, Silikas N (2011) Effect of filler size and temperature on packing stress and viscosity of resin-composites. Int J Mol Sci 12(8):5330–5338. https://doi.org/10.3390/ijms12085330
83. Muna II, Mieloszyk M (2021) Temperature influence on additive manufactured carbon fiber reinforced polymer composites. Materials 14(21):413. https://doi.org/10.3390/ma14216413
84. Panou A, Karabagias IK (2024) Migration and safety aspects of plastic food packaging materials: need for reconsideration? Coatings 14(2):1–17. https://doi.org/10.3390/coatings14020168

85. Kaur R, Pathak L, Vyas P (2024) Biobased polymers of plant and microbial origin and their applications: a review. Biotechnol Sustain Mater 12:1–21. https://doi.org/10.1186/s44316-024-00014-x
86. Ghaani M, Cozzolino CA, Castelli G, Farris S (2016) An overview of the intelligent packaging technologies in the food sector. Trends Food Sci Technol 51:1–11. https://doi.org/10.1016/j.tifs.2016.02.008
87. Zabihzadeh Khajavi M, Ebrahimi A, Yousefi M, Ahmadi S, Farhoodi M, Mirza Alizadeh A, Taslikh M (2020) Strategies for producing improved oxygen barrier materials appropriate for the food packaging sector. Food Eng Rev 12(3):346–363. https://doi.org/10.1007/s12393-020-09235-y
88. Chandrababu V, Parameswaranpillai J, Gopi JA, Pathak C, Midhun Dominic CD, Feng NL, Krishnasamy S, Muthukumar C, Hameed N, Ganguly S (2024) Progress in food packaging applications of biopolymer-nanometal composites: a comprehensive review. Biomater Adv 162:213921. https://doi.org/10.1016/j.bioadv.2024.213921
89. Kathuria A, Zhang S (2022) Sustainable and repulpable barrier coatings for fiber-based materials for food packaging: a review. Front Mater 9:1–21. https://doi.org/10.3389/fmats.2022.929501
90. Ahmed S, Sameen DE, Lu R, Li R, Dai J, Qin W, Liu Y (2022) Research progress on antimicrobial materials for food packaging. Crit Rev Food Sci Nutr 62(11):3088–3102. https://doi.org/10.1080/10408398.2020.1863327
91. Shankar S, Khodaei D, Lacroix M (2021) Effect of chitosan/essential oils/silver nanoparticles composite films packaging and gamma irradiation on shelf life of strawberries. Food Hydrocoll 117:106750. https://doi.org/10.1016/j.foodhyd.2021.106750
92. Tripathi S, Kumar L, Deshmukh RK, Gaikwad KK (2024) Ultraviolet blocking films for food packaging applications. Food Bioprocess Technol 17(6):1563–1582. https://doi.org/10.1007/s11947-023-03221-y
93. Muzata TS, Gebrekrstos A, Orasugh JT, Ray SS (2023) An overview of recent advances in polymer composites with improved UV-shielding properties. J Appl Polym Sci 140(14):1–31. https://doi.org/10.1002/app.53693
94. Idris A, Muntean A, Mesic B (2022) A review on predictive tortuosity models for composite films in gas barrier applications. J Coat Technol Res 19(3):699–716. https://doi.org/10.1007/s11998-021-00579-6
95. Zhao Z, Tan Y, Guo S, Ni X (2022) An efficient composite modifier prepared for enhancing the crystallization and flame-retardancy of poly(m-xylylene adipamide). Polymers 14(17):626. https://doi.org/10.3390/polym14173626
96. Chu H, Chen Z, Chen Y, Wei D, Liu Y, Zhao H (2024) Mechanical properties and crystallinity of specific PLA/cellulose composites by surface modification of nanofibrillated cellulose. Polymers 16(17):474. https://doi.org/10.3390/polym16172474
97. Pathreeker S, Chando P, Chen FH, Biria S, Li H, Finkelstein EB, Hosein ID (2021) Superhydrophobic polymer composite surfaces developed via photopolymerization. ACS Appl Polym Mater 3(9):4661–4672. https://doi.org/10.1021/acsapm.1c00744
98. Ghanem AF, Youssef AM, Abdel Rehim MH (2020) Hydrophobically modified graphene oxide as a barrier and antibacterial agent for polystyrene packaging. J Mater Sci 55(11):4685–4700. https://doi.org/10.1007/s10853-019-04333-7
99. Zhang W, Roy S, Assadpour E, Cong X, Jafari SM (2023) Cross-linked biopolymeric films by citric acid for food packaging and preservation. Adv Colloid Interf Sci 314:102886. https://doi.org/10.1016/j.cis.2023.102886
100. Grabska-Zielińska S (2024) Cross-linking agents in three-component materials dedicated to biomedical applications: a review. Polymers 16(18):679. https://doi.org/10.3390/polym16182679
101. Gupta RK, Pipliya S, Karunanithi S, Eswaran UGM, Kumar S, Mandliya S, Srivastav PP, Suthar T, Shaikh AM, Harsányi E, Kovács B (2024) Migration of chemical compounds from packaging materials into packaged foods: interaction, mechanism, assessment, and regulations. Foods 13(19):125. https://doi.org/10.3390/foods13193125

102. Tamizhdurai P, Mangesh VL, Santhosh S, Vedavalli R, Kavitha C, Bhutto JK, Alreshidi MA, Yadav KK, Kumaran R (2024) A state-of-the-art review of multilayer packaging recycling: challenges, alternatives, and outlook. J Clean Product 447:141403. https://doi.org/10.1016/j.jclepro.2024.141403
103. Wang Q, Chen W, Zhu W, McClements DJ, Liu X, Liu F (2022) A review of multilayer and composite films and coatings for active biodegradable packaging. NPJ Sci Food 6(1):32. https://doi.org/10.1038/s41538-022-00132-8
104. Naser AZ, Deiab I, Darras BM (2021) Poly(lactic acid) (PLA) and polyhydroxyalkanoates (PHAs), green alternatives to petroleum-based plastics: a review. RSC Adv 11(28):17151–17196. https://doi.org/10.1039/d1ra02390j
105. Mori R (2023) Replacing all petroleum-based chemical products with natural biomass-based chemical products: a tutorial review. RSC Sustain 1(2):179–212. https://doi.org/10.1039/d2su00014h
106. Ranakoti L, Gangil B, Misha SK, Singh T, Sharma SAIR, El-khatib S (2022) Critical review on polylactic acid: properties, structure. Materials 15(4312):1–29
107. Zhu Z, Liu W, Ye S, Batista L (2022) Packaging design for the circular economy: a systematic review. Sustain Product Consump 32:817–832. https://doi.org/10.1016/j.spc.2022.06.005
108. Ilyas RA, Aisyah HA, Nordin AH, Ngadi N, Yusoff M, Zuhri M, Rizal M, Asyraf M, Sapuan SM, Zainudin ES, Sharma S, Abral H, Asrofi M, Syafri E (2022) Nanocomposites for various advanced applications. Polymers 14(874):1–36
109. Punia Bangar S, Ilyas RA, Chaudhary N, Dhull SB, Chowdhury A, Lorenzo JM (2023) Plant-based natural fibers for food packaging: a green approach to the reinforcement of biopolymers. J Polym Environ 31(12):5029–5049. https://doi.org/10.1007/s10924-023-02849-3
110. Wang L (2014) Energy efficiency technologies for sustainable food processing. Energy Effic 7(5):791–810. https://doi.org/10.1007/s12053-014-9256-8
111. Realini CE, Marcos B (2014) Active and intelligent packaging systems for a modern society. Meat Sci 98(3):404–419. https://doi.org/10.1016/j.meatsci.2014.06.031

Chapter 4
Application of Nanocomposites Materials for Packaging

Ramgulam Mahloniya, Abhilasha Mishra, Adeeba Mirza, and Rekha Goswami

1 Introduction

The excellent packaging materials have the intrinsic property to increase the shelf life of many products of food by retarding spoiling processes along with hindering the growth and population of pathogens [1]. Nanomaterials exhibit unique characteristics because of their particle size, which is in the nonorange [2]. Materials known as nanomaterials are applicable in diverse sectors, including research and development, electronics, energy, and pharmaceuticals. The intrinsic properties, along with the size, shape, interfacial area, and interfacial interactions of nanocomposites, have a significant role in their applications in the health and manufacturing sectors [1]. Additionally, the nanomaterials have diverse properties and behaviours from bulk materials [3]. In recent times, nanomaterials with diverse functions have been particularly studied for their antibacterial and antioxidant characteristics. The industries, including automotive, aerospace, electronics, packaging, and the biomedical sector, have found use for nanomaterials [4]. Nanomaterials are becoming increasingly important because of their many applications, especially in the packaging industries [5].

Biodegradable, active, and intelligent packaging materials have made substantial use of a wide variety of nanocomposites. Particularly, based on polymeric materials, improved packaging with super-barrier qualities that compete with those of glass or

R. Mahloniya
Engineering Chemistry Department, Lakshmi Narain College of Technology and Excellence, Bhopal, Madhya Pradesh, India

A. Mishra (✉) · A. Mirza
Department of Chemistry, Graphic Era Deemed to be University, Dehradun, Uttarakhand, India
e-mail: abhi1680@geu.ac.in

R. Goswami
Department of Environmental Science, Graphic Era Hill University, Dehradun, Uttarakhand, India

A. Verma et al. (eds.), *Polymers and Composite Materials for Packaging: Smart Food Packaging and Solutions*, Materials Horizons: From Nature to Nanomaterials,
https://doi.org/10.1007/978-981-95-0677-4_4

metal can be developed. In addition, they have improved biodegradability and can offer bacteriostatic and bactericidal applications at the nanoscale to lower microbial activity and protect the packaging materials [6]. As a result, polymeric nanocomposites offer an improved packaging with advantage of lowering the necessary filler loading levels while retaining the essential qualities as compared to conventional fillers. Additionally, this decreases the weight of the nanocomposite film [7]. Thus, there is a great opportunity to fabricate packaging materials with advanced properties and using intelligent and active packaging [8]. The beneficial effect of the packaging film's barrier properties, mechanical strength, and structural integrity not only enhances the food shelf life but also makes it easier to communicate information about the quality and safety of the enclosed food product. Providing a detailed investigation of the wide range of innovative packaging techniques created especially for food product preservation and containment is the main aim of this study. The biodegradability and antibacterial qualities of the packaging materials were investigated. Furthermore, a comprehensive examination has been conducted on the characteristics and advantages of the nanocomposites used as packaging materials.

Nanotechnology, or the development of a material on a length scale of <1 mm in one dimension, is a key factor in innovation and industrial progress in recent decades. The multidisciplinary field of development and characterization of nanomaterials, which show distinct behavioural pattern and properties than materials at higher length scales. These distinct characteristics of nanoscale materials have a significant impact on the interaction of the materials at molecular level [9]. The distinct property of nanomaterials includes material morphology, nanoparticle form and aspect ratio, size distribution, surface-to-volume ratio, surface charge, aggregation state, and so on. As the material size shrinks to the nanoscale, the surface area to volume ratio grows rapidly, and the roles of surface/interface become more essential, whereas gravitational effects are minimal. Nanotechnology uses the unique features of materials at nanolevel to restructure materials related to their majority qualities. The surface plasmon resonance (SPR) observed by metal nanoparticles at the nanoscale, which is used to fabricate biosensors [10].

The polymer nanocomposites which are hybrid material can also be created by incorporating nanoparticles into a polymer matrix. A polymer nanocomposite is built using various fundamental nanotechnology principles. Choosing a nanomaterial with the right surface chemistry will influence the dispersion and distribution of nanoparticle in the polymeric material, affecting the electrical conduction, optical characteristics, and mechanical strength of the nanocomposite of polymer. Polymeric nanocomposites play an important role in several industries, including engineering, consumer goods, biomedicine, and food. Various nanomaterials are used for making nanocomposite packaging which provide non-reinforcement, active packaging, and smart packaging as shown in Fig. 1 [11]. The manufacture and ingestion of these polymer nanocomposites frequently increase major environmental concerns, such as plastic pollution in the natural environment, CO_2 emissions, health and safety risks, and depletion of resources [12]. The United Nations (UN) General Assembly sectioned 17 interrelated objectives to resolve global challenges by 2030 [13].

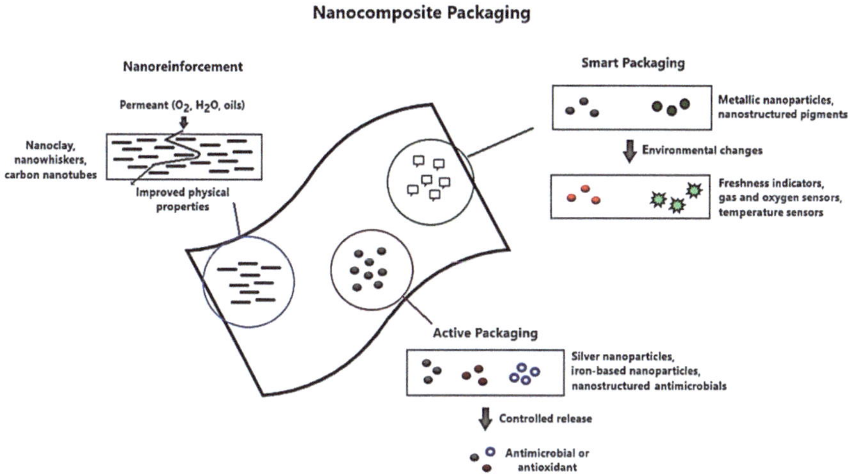

Fig. 1 Different properties incorporated in nanocomposite packaging in the food packaging industries (reproduced with permission from ref. [11])

2 Characterization of the Packaging Materials

All the possible properties for packaging films are barrier qualities, mechanical strength, optical properties, sealability, printability, and eco-friendliness [1]. Many packaging films feature product information, logos, and other graphics [14]. The film compatibility for ink or other printing methods is a key feature in the packaging sector. With a growing emphasis on film which are sustainable and environmentally friendly, films are now engineered to exhibit antibacterial properties, hydrophilicity, biodegradability, biocompatibility, and non-toxicity [15]. Antimicrobial and biological compatibility are especially important to protect the product, and packaging engineers must evaluate the properties of film, when choosing a substantial for a given particular application, as detailed below.

2.1 Antibacterial Activity

A major and growing issue in daily life, bacterial infection and spread do significant harm to a variety of sectors and aspects of daily life. Textile, packaging, medical, water and land transportation, and water purification facilities are the industries most impacted [16]. Among the nosocomial infections linked to it, *Escherichia coli*, *Staphylococcus aureus,* and *Listeria monocytogenes* are regarded as extremely deadly bacterial strains and pathogens [17]. The scenario necessitates quick action to develop strategies for limiting the spread of microbes, which is crucial for everyday tasks such as packaging as well as industry. Packaging materials affect directly the contents of materials, particularly when it comes to food goods. The contents within could deteriorate, become contaminated with microorganisms, or become unsafe for human consumption if the packaging materials are not protected [1]. There is a chance of microbial infection if the packaging material does not block the passage of oxygen or moisture. In these circumstances, microbes that are innocuous in the absence of oxygen and moisture can grow and endanger the consumer [18]. Therefore, the main characteristics of a material used in packaging to prevent undesired infections is its antimicrobial potential. To test the antibacterial properties of different microorganisms, such as Gram-positive and Gram-negative bacteria on the material's surface (contact technique), particular protocols of different methodologies must be followed for the antimicrobial characterization of materials [19]. Theoretical and commercial interest in nanomaterials produced in various ways has grown over time because of their likely antibacterial properties [20]. The metals especially silver, gold, and zero-valent iron metal nanoparticles have been shown to weaken and damage bacterial membranes, preventing them from growing and evolution of reactive oxygen, which causes cell lysis and death (Fig. 2).

A variety of technical methods, nanomaterials, and bacterial strains have been used to demonstrate the antibacterial effectiveness of various nanomaterials used in packaging. Ag, ZnO, and CuO nanoparticles were used to create films of starch-based nanocomposite materials [22]. These films demonstrated improved mechanical qualities at a lower dosage, which increased their antimicrobial effectiveness against *S.*

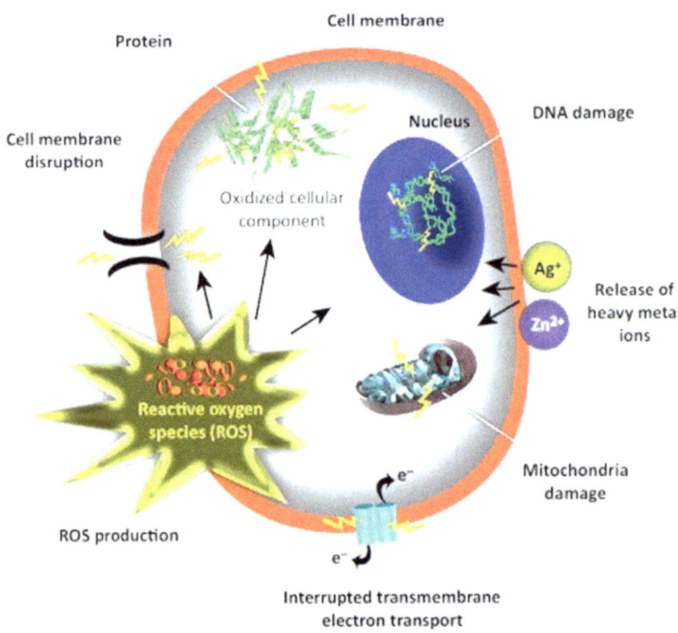

Fig. 2 Antibacterial action of nanoparticles (reproduced with permission from ref. [21])

aureus and *E. coli*. The antibacterial potential of chitosan/gelatin films containing green-synthesized zinc oxide nanoparticles against *S. aureus* and *E. coli* was investigated recently by Kumar et al. [23]. The addition of ZnO nanoparticles made with Cassia fistula fruit extract was responsible for the biodegradable film's improved thermal stability and elongation at break. The chitosan–gelatin–ZnO nanocomposite films showed noteworthy antibacterial activity, indicating their potential in packaging of food and medical materials, other applications due to both biodegradability and antimicrobial qualities. The antibacterial effectiveness of cellulose acetate films boosted with green-synthesized Ag-NPs demonstrates improved activity for food shelf life [24]. The strong antibacterial activity these films demonstrated against foodborne pathogens indicated that they could be used for food packaging with improved antimicrobial qualities.

2.2 *Biodegradability*

One of the biggest effects on the environment is the accumulation of trash from packing materials, and this problem is becoming more and more of a concern. Recently, experts are now concentrating on creating packaging materials that are biologically decomposable, that is capable of breaking down naturally into innocuous, non-toxic simple compounds—in order to address this [8]. This reduces the pollution that these materials contribute to the environment by addressing the issue of packaging trash accumulation. Nanomaterials that can degrade quickly and totally mineralize the environment are being employed in packaging as the need for packaging rises because of industrialization and consumerism [25]. Various international laws of composability and biodegradability have been proposed. Adopting and following certain rules and norms is necessary to evaluate the degradability of packaging material [6]. The choice of a suitable standard should be made based on the kinds of tests that must be used and the inferences that must be made from them. These tests can be further separated into laboratory, simulation, and field tests. From an application perspective, field testing is the best way to evaluate the biodegradability of the plastics used in packaging. We are unable to control several aspects of the environment, including pH, humidity, and temperature, due to its complexity [26]. Innovative applications in a variety of industries have been spurred by the development of nanotechnology, particularly in those of packaging using composite nanomaterials, or nano-packaging [27]. For instance, many polymeric materials have been investigated to create packaging-related nano-encapsulates. Because of their superior biodegradability, nanoclays have potential to develop biodegradable nanocomposite materials [18]. Figure 3 shows the proven effectiveness of nanocomposites in offering sufficient protection against harmful UV radiation, unfavourable climatic conditions, and microbiological pathogens.

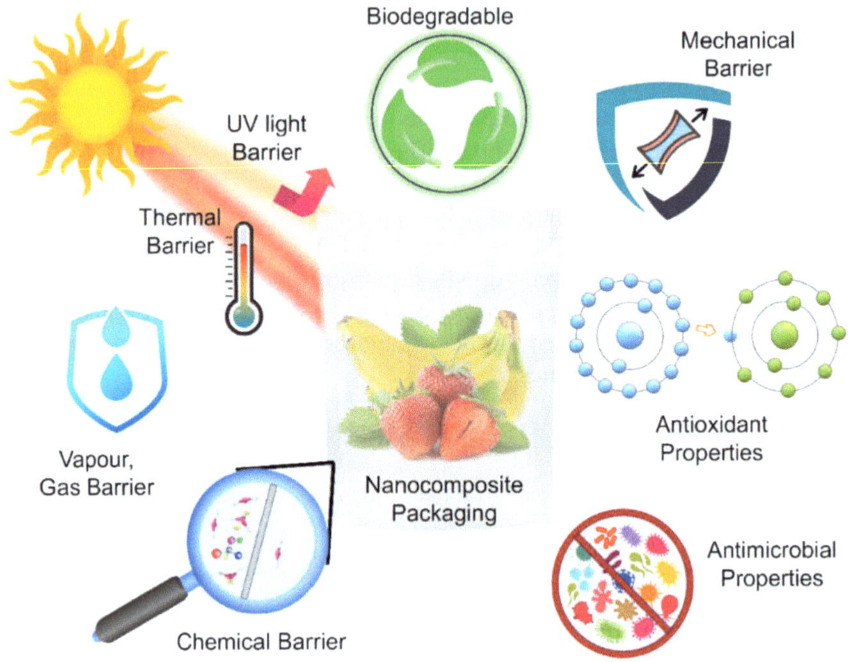

Fig. 3 Function of nanomaterials in food packaging (reproduced with permission from ref. [27]

3 Types of Nanocomposite Packaging

Conventional food packaging maximizes shelf life, maintains product integrity, and ensures maximum food safety. Food security is a critical issue in biological chemistry, and to develop antimicrobial packaging material is a focused effort to manage microorganisms that negatively affect the achievement of these goals. The increasing awareness for natural, locally sourced, and organic products is necessary to develop the best approach for protecting food integrity against harmful microbial intrusion [28]. To ensure optimal food quality, more effective regulatory systems must be developed because the current conventional methods used to preserve food quality have fallen short of customer expectations [29]. The applications of active, biodegradable, and smart/intelligent packaging have developed in this context as a remedy for these issues, as will be covered in more detail below.

3.1 Biodegradable Packaging

Waste builds up because of the traditional plastic-based packaging material's limited ability to degrade physically, chemically, and biologically. This accumulation, in turn, produces a multitude of significant environmental and health-related challenges [30]. The accumulated waste materials on urban streets and roads have been investigated to have adverse effects on the drainage system, resulting in obstructions and subsequent overflow [31]. The aquatic ecosystem is seriously threatened by the plastic waste that is deposited into freshwater and marine water. It has been investigated that the incineration process releases harmful gases, which deteriorate the quality of air. In addition, it makes the environmental issue of global warming worse, and the emission of hazardous gases presents a serious threat to the ecosystem [32]. Furthermore, incineration has been linked to a number of health aspects, underscoring the possible drawbacks of this practice. There is intense research over the garbage disposal issues, and the adverse impacts on the different ecosystem and human health caused by the non-biodegradable nature of different human-made polymers have been performed throughout the world [33]. Therefore, the search for alternate materials with eco-friendly abilities is therefore becoming more important. Applications of biodegradable packaging materials have become a feasible substitute approach in a number of industries with the goal of reducing the possible hazards materials related to non-biodegradable synthetic plastic materials [34]. The biodegradable plastics are a kind of plastic that decomposes by the action of naturally occurring microorganisms, such as bacteria, fungi, and algae as per American Society for Testing and Materials [35]. In terms of tensile strength, thermal qualities, and oxygen and moisture transmission, the materials in issue are similar to those of traditional plastics and come from renewable resources. The breakdown of biodegradable plastics produces important products such as water, carbon dioxide, or biomass [36]. The environment benefits from the lack of garbage accumulation. The agriculture and food packaging sectors are where biodegradable polymers are most commonly used. The hydrolysis, catalysed by microorganisms, which is often divided into three steps: fragmentation, hydrolysis, and assimilation is a biodegradation (Fig. 4). In the first step, weathering, UV irradiation, mechanical forces, microorganisms are the factors which fragments the polymers into small pieces or microplastics. In the second step, the ester bond of the polymer hydrolyzes at eventually leading to the reduced molar mass and form soluble oligomers, dimers, and monomers. Finally, in bio-assimilation and mineralization process, the intracellular enzymes use these degraded products as carbon sources and energy to produce increased cell biomass and simple end products like CO_2 and water.

3.2 *Active Packaging*

In order to protect the contents from the outside environment by film of packaging material. This packaging film protects contents from oxygen, light, moisture, and other environmental factors that can cause spoilage or deteriorate the property of the product. Films of packaging material with advanced barrier properties can improve the shelf life of the food [38]. Packaging films with the good strength and flexibility

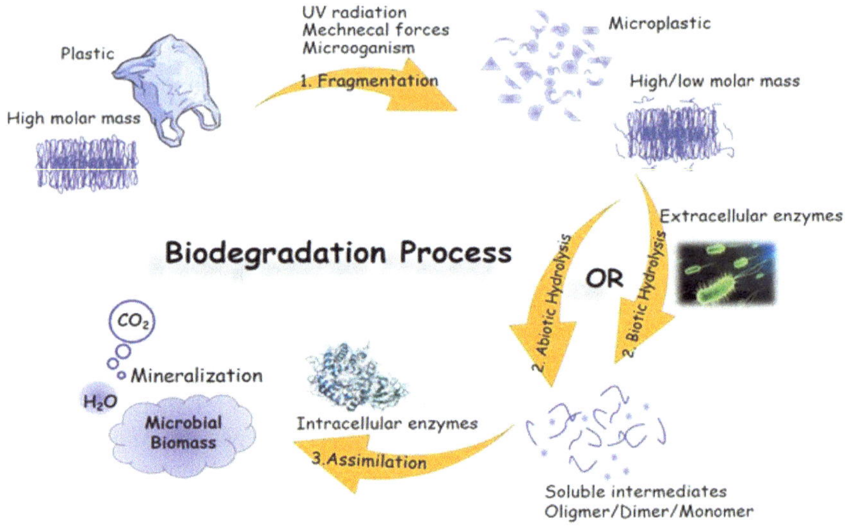

Fig. 4 Representation of the various steps involved in biodegradation (reproduced with permission with ref. [37])

to withstand the stresses of manufacturing, transportation, and handling are desirable. The mechanical properties which are significant for packaging film include tensile strength, tear resistance, and puncture resistance. Transparent or translucent is also one of the other properties of the film which allow the product to be visible to the consumer [39]. The preservatives in food are frequently added to inhibit the population of germs, prevent undesirable chemical reactions, or control the activity of enzymes. The objective of active packaging is to preserve and enhance the food's shelf life of the packaged product by incorporating particular additives into the packaging film or into the containers themselves [28]. To alter the normal barrier, mechanical, and optical properties [40] displayed by conventional packaging materials, these additives are especially functionalized and developed. These functions include a different element such as macronutrients [41], phytochemicals, and plant extracts . phytochemicals [43], plant extracts, essential oils [42], and inorganic and organic nanoparticles. Currently, packaging materials contain nanoparticles that have the potential to be antibacterial and/or antioxidant. Instead of being integrated directly into the food product, special attention has been paid to the use of biogenic nanomaterials that are localized within the film itself [1]. Because of this localization, there may be less of these nanoparticles required to protect the food during storage and transportation, which could lead to less of them being used and eventually consumed [44]. To extend the shelf life of food in a packaging material using various processes that depend on the particular is the ultimate goal.

3.3 Smart and Intelligent Packaging

Certain materials used in smart packaging have been developed to react in particular ways as the system changes. These modifications could involve shifts in the food's maturity, safety, or quality. Although they do not directly affect the product they are enclosing, the packaging materials can communicate information about the condition of the wrapped item. The main goal of packaging is to provide a signal of such changes, improving the overall functionality of the packaging [45], as seen in Fig. 5.

4 Applications of Nanomaterials in Packaging

As a potential remedy for the conventional shortcomings of technologies that cannot meet the ongoing demand for food items, nanotechnology has emerged [46, 47]. The success of food production, as determined by product safety and quality retention, depends on sustainable food packaging [48]. The applications of nanomaterials such as nanocomposites (NCSs), nanoparticles (NPs), and NCS films for diagnostics, coating, and packaging, nanotechnology has produced creative solutions [1]. To relate the expanding tendency towards the developing of nanomaterials in food matrices led the numerous qualities that revealed at the scale of the nanometric [49].

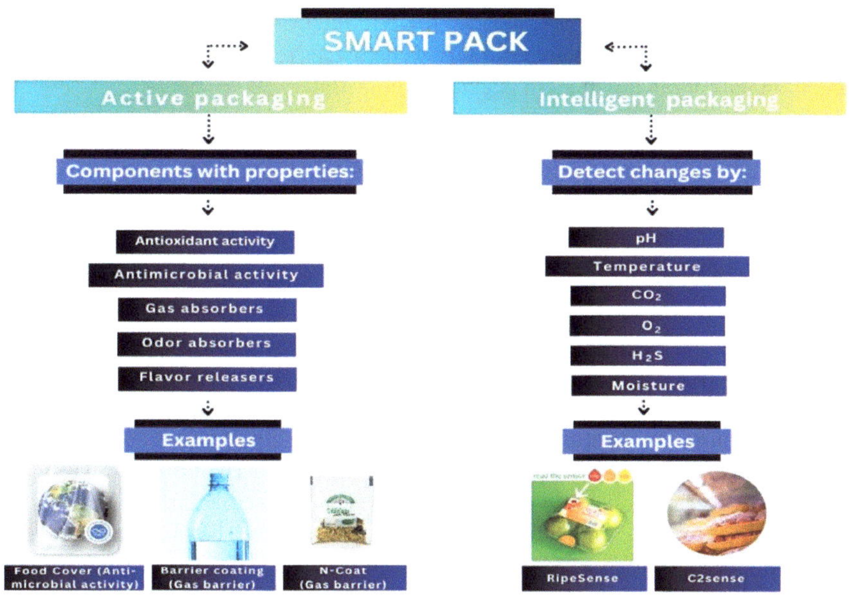

Fig. 5 Different types of packaging and their functions with examples (reproduced with permission with ref. [45])

The discussed exceptional qualities are significantly applicable in different industries, especially in the food business [50]. In food packaging, physical contamination and microbiological damage are prevented. For extension of the shelf life of food and reduce spoiling, food is packaged and kept in a hygienic atmosphere [51–53]. The technological advancements, such as their chemical composition, crystal structure, physicochemical stability, huge surface and area size shape, nanoparticles have become more well-known [54]. NCS demonstrates good mechanical strength and physical qualities than polymeric films [55], and other qualities like moisture, gas, or barrier property for antimicrobial were substantially optimized when mixed with NPs [56–58]. The surface-to-volume rate excessively high indicates as a sign of NCS's exceptional qualities [59]. The interface between biomolecules and nanostructures arises unique characteristics which are physically, chemically, and mechanically advanced compared to biomolecule; it is vital to ascertain how to generate and describe these nanomaterials [49]. The application of two-dimensional (2D) materials in food packaging material can improve safety and shelf life of food [60, 61]. Smart packaging materials with sensing abilities can be developed using materials with semiconducting properties, including tungsten diselenide (WSe2) and molybdenum disulfide (MoS2). On changing environment of packaging, such as temperature change or variation in gas composition, these transition-metal dichalcogenides (TMDs) can provide real-time information regarding the safety and freshness of food [62, 63]. Retailers and consumers can learn much more from this advanced packaging about the variables that could impact the quality of food material. Additionally, a further layer of research is performed on food packaging through the material like hexagonal boron nitride (h-BN) which is 2D materials [64]. These innovations in packaging materials address challenges related to food quality, safety, and sustainability in the modern packaging industry.

4.1 Organic Nanocomposites (NCSs)

Recently, academicians and industrial experts are interested in polymers having organic nature. These materials are made from organic biopolymers, which are biodegradable and environmentally benign and are derived from living things like plants and animals that are widely present in our daily life. Examples of these are chitosan, starch, cellulose alginate, and others, for applications in agriculture and food production. The hydrogels or NCS films based on biopolymers may be used as direct coatings for the extension of shelf life of food [58]. Therefore, availability is a benefit of organic polymeric nanoparticles over inorganic nanoparticles. Biocompatibility high-specific energy, cheap cost, biodegradability, flexibility, and ease of processing [65]. However, Priyadarshi et al. [66] noted that organic nanomaterials like nanonis and natural nanoemulsions such as essential oils (including chitosan, cellulose, curcumin, and melanin) have been used extensively lately. However, when NPs or other antimicrobials present short-term health hazards, their migration into food-contact materials or films can be difficult. Promoting the NCS biopolymer could assist in addressing global environmental issues, such as the generation of municipal garbage and extending food shelf life [66].

4.2 Polysaccharide-Based Nanoparticles (P-NPs)

Polysaccharides are the most prevalent forms of carbohydrates that found naturally and present in living things, microbes, plants body, and animals [67]. The natural biomaterials used to make P-NP include pectin-cellulose-alginate gum or chitosan-starch-chitin. The stability of pectin is a key factor because of its outstanding structural qualities, ease of modification, and promising future are observed. These biomaterials for screening are inexpensive. The important characteristics of polysaccharides are their non-toxicity and biological suitability, making them ideal for use in food packaging [68, 69]. As a food packaging material, polymeric nanocomposites have qualities that can be tested in various applications [70]. The biodegradation property of food packaging materials [71–73] and the gelling agents introduced [61] that form an intracellular matrix including metal ions such as calcium, sodium, and magnesium are two examples of such crucial characteristics. Many of these are extensively used as renewable and sustainable packaging materials. There are several that are commonly used as renewable and sustainable packaging materials. These characteristics make P-NPs excellent candidates for nano-reinforcement in organic nanoclusters. Chitosan, chitin, and alginate-starch are non-toxic and antibacterial in nano-forms. The cellulose, pectin, chitosan, and starch are polysaccharides, numerous researchers have created food packaging materials with the help of P-NCSs [70]. The resources, molecular structure, its application in composite reinforcement, antibacterial property, and as intelligent packaging material are shown in Fig. 6. The entire food packaging business is anticipated to grow as a result of relying on this type of lipopolysaccharide nanocomposites enhance the physical characteristics of biopolymers by using these P-NPs as fillers [68]. However, P-NPs are known to be environmentally friendly with few concerns about their toxic biodegradability or physiological stability [74].

4.3 Starch Nanoparticles (S-NPs)

The glucose molecules polymerize to form starch with glycosidic connections [76, 77]. Typically, fruits, cereals, and vegetables contain this type of polysaccharide. S-NPs, which contain amorphous particles, can induce controlled nanoprecipitation of gelatin and starch. S-NPs and S-NCs have both been employed as blending agents in polymer base to enhance mechanical strength and barrier property, especially moisture resistance [68]. Adding S-NCs to starch membranes could be a way to make them more applicable [78]. Additionally, the permeability extent of water vapour (WV) by the addition of rice S-NCs-starch membranes. But solubility extent

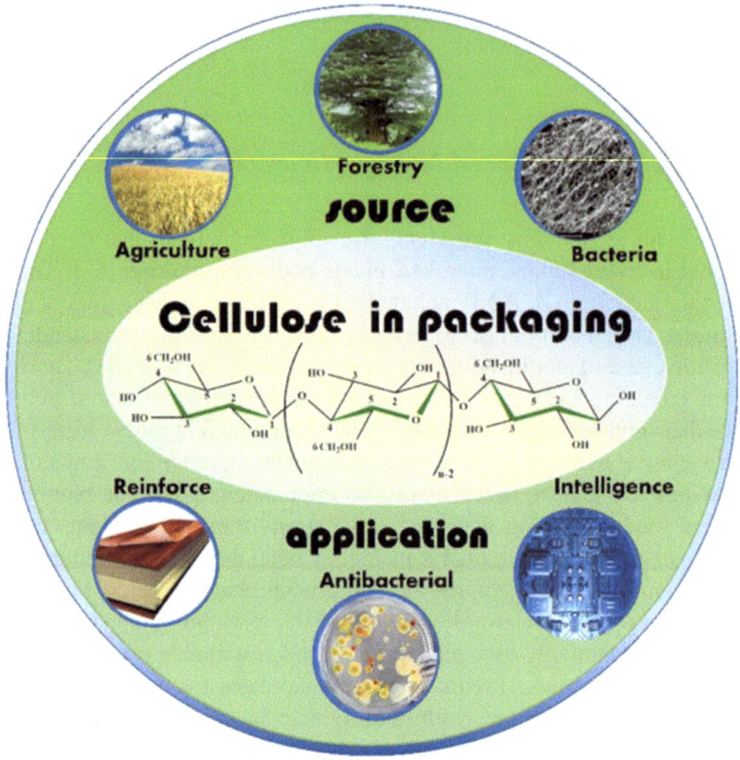

Fig. 6 Resources, molecular structure, and packaging application of cellulose (reproduced with permission with ref. [75])

was decreased by adding S-NCs to the potato starch films. Rice S-NCs and potato S-NCs were applied to the membranes, which increased the cohesiveness and reduced surface roughness and fragility. To improve the mechanical qualities, packaging based on S-PNs is currently created with a combination of natural polymers like starch, cellulose, chitosan, gelatin, whey protein. S-NCs thus serve as structural reinforcing agents that improve a variety of film properties. Thus, the manufacturing of S-NCs-starch packaging materials has greatly improved quality of food, extension of shelf life, tensile strength, modulus of elasticity, and barrier qualities [79]. The preparation, properties, and application of S-NPs are summarized in Fig. 7.

4.4 Cellulose Nanoparticles

It is easy to extract cellulose from the plant cell walls. Since polymers are widely available, cellulose, which may be made from a variety of agricultural and forestry

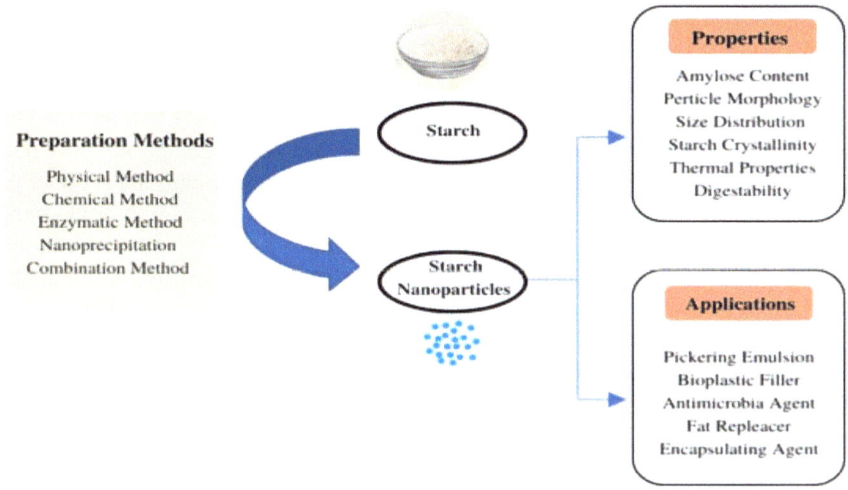

Fig. 7 Schematic presentation of synthesis, characterization, and applications packaging materials (reproduced with permission with ref. [80])

wastes, including fruits, vegetables, rice, and cotton stalks, is relatively inexpensive to buy as a raw material. Cellulose can also be extracted from food industry waste, including bagasse and mango peel [70]. In general, three types of Ce-NPs are discussed here: bacterial Ce-NPs (BNC), cellulose nanofibrils (Ce-NF), and cellulose nanocrystals (Ce-NCs). The variety of Ce-NPs can be obtained using various methods as per the size morphology. [81]. The parameters before or after the treatment decide the origin the cellulose [69]. To fabricate biodegradable membranes with the help of Ce-NPs, processed from agricultural waste, non-wood biomass, and various plant extracts. The biodegradable membrane of Ce-NPs can replace plastic packaging materials of composites. There are more advantages in using Ce-NPs film in terms of physical property and health issue. Producing biodegradable films from Ce-NPs can significantly solve this problem by replacing composite plastic packaging. Ce-NC enhanced the physical characteristics of the carboxymethyl-cellulose/agar combination films in study, leading to improved mechanical and water resistance [82].

4.5 Chitosan Nanoparticles (Ch-NPs)

Nature has a lot of chitins in the form of amino-polysaccharide polymers that are abstracted from the cell membrane of insects, crustaceans, and fungi [83]. Marine life, particularly crustaceans, is the primary sources of chitin. Two processes (chemical and enzymatic deacetylation) can convert chitin to chitosan [65]. According to Roy and Rhim [84], functionalized Ch-NPs that have been synthesized with ideal functional and physical characteristics hold promise for functional packaging applications. Chitosan is a viable choice for the creation of eco-friendly materials because it is a bio-decomposable natural polymer obtained from natural resources. Additionally, excellent biocompatibility and antibacterial qualities of chitosan make it a useful material for food packaging. Thus, chitosan-based nanocomposites have been employed as a food packaging material, particularly to notify consumers about the quality of the food being packaged or to build intelligent packaging product that extends the shelf life of packed food [85]. The antibacterial approach of chitosan is against various microorganisms like bacteria, fungi, and viruses. Additionally, the various factors which affect this antibacterial activity of chitosan, for example the nature of chitosan or its derivatives, molecular weight, chemical composition, degree of deamination, and the microorganism [86]. The various researches on Ch-NPs have done which are concerned many interesting causes of their attending physical and chemical properties [74]. To manage and maintain shrimp freshness, a multipurpose, pH-sensitive, and antibacterial film has been developed by combining quercetin-loaded Ch-NPs with synthetic purple sweet potato juice rich in anthocyanins in a polymer agar/sodium alginate matrix [87]. Specifically, this film exhibits excellent mechanical qualities, antibacterial and WV barrier qualities, and high UV-blocking capabilities. To create active film by mixing pullulan as biopolymers and chitosan, employed Ch-NPs encapsulating curcumin [88]. The film's mechanical and hydrophobic qualities, as well as its UV protection, were greatly enhanced by the addition of functionalized nanofillers without sacrificing its transparency. The qualities of the WV barrier were somewhat enhanced. The potential of maize starch, Ch-NPs, and thymol bio-NCS films as efficient packing materials was demonstrated by Othman et al. [89] and demonstrated the potential to increase the shelf life of packed cherry tomato slices while exhibiting less weight loss and no discernible changes in firmness. In addition, during the 7 days of storage, no mould growth was seen on those who were directly exposed to the film.

4.6 Protein Nanoparticles (NPs)

The colloidal nanoparticles of protein are ubiquitous and act as building blocks with a variety of uses in food formulations and biological processes [90, 91]. By adjusting environmental factors to trigger protein self-assembly behaviour, food scientists have created a chance to create distinctive and predictable nanoparticles that are inspired by natural proteins. The two most nanostructured food proteins which are intensively studied in recent years are casein and zein, which are derived from milk and corn, respectively. Additionally, to serve as a standard to maintain food stability, the protein casein and zein maintain the food impacts and benefits [84]. To create functional nanoparticles (NPs) to enhance the freshness or consistency of food compositions and packaging is the keen interest of scientist [91]. Natural antimicrobial ingredients are frequently applied for increasing the shelf life of food and have been developed as food additives. The binary structure of protein shields the binding sites more effectively and stops trophic degradation in unfavourable environmental circumstances [92]. The pathogen's surface receives adequate amounts of essential oils from this conjugation, which also alters the taste of the particular essential oils. Additionally, P. aeruginosa-treated zein-NaCas NPs loaded with ginger oil and produced in nanoemulsions, which react with the cell membrane and harm the surface of the cell, resulting in shrinkage and wrinkles of the cell. To extend the shelf life of meat and prevent it from spoiling at cold chicken breasts, these nanoemulsions also prevent microbiological development [93]. An effective technique to increase the biological availability and activity of stabilizers for poorly soluble or hydrophobic nutrients is the creation of nanostructures, which are based on dietary proteins as nanocarriers [94].

4.7 *Inorganic Nanoparticles (NPs)*

Bacterial growth in packaging may be inhibited by inorganic nanoparticles [95]. Packaging waste material output can be decreased by combining nanoenhancers with biopolymer-based materials. The majority of inorganic nanoparticles used in packaging are inorganic materials, such as Au–NPs CuO–NPs, TiO_2–NPs, Ag–NPs, and ZnO–NPs. The physicochemical and functional qualities of these materials can be improved, and food systems can benefit from their use in packaging [54]. To preserve and extend the shelf life of food, inorganic nanoparticles (NPs) are used in food packaging films. In the realm of research, inorganic nanofillers fall into four major categories: firstly, carbon-based nanostructures (CNTs, fullerenes, graphite, graphene, and carbon fibres). Metal or metallic oxide nanostructures like Ag, Cu, CuO, ZnO, TiO_2, Au belong to second category. While montmorillonite, bentonite, halloysite are nanoclays, the last category is non-metallic nanostructures (sulphur and silica NPs). Although the majority of the fillers under the category of nanoscale used in this field have been approved by GRAS for a thorough examination of their physical and chemical characteristics. The improved procedures require to evaluate the transfer of nanoparticles from packaging material to food content and the relationship between particle toxicity and characteristics. Considering food distribution at the nanoscale, there is a noteworthy necessity [66]. The coated layers of nanofillers on polymer films, create an excellent oxygen barrier and UV protection that can prolong shelf life, preserve or enhance food quality and safety, and slow down food decomposition [96]. Blending additives and fillers into film or coatings guarantees food quality by successfully preventing lipid oxidation and microbiological growth [91]. Metal/metal oxide NPs can be functionalized by polymeric materials based on protein and fillers like plasticizers, antibacterial compounds. The physiochemical characteristics of metallic nanoparticles including large surface area and antibacterial activity are useful as standard packaging materials and as functional materials when added to biopolymeric-based matrices and utilized in a variety of nanomaterials [95]. The P-NCSs can be strengthened and given antimicrobial properties by the addition fillers like TiO_2–NPs, ZnO–NPs, Ag–NPs, Au–NPs, and CuO–NPs [71–73]. Silver or zinc oxide, zeolites, and nanoclay are among the nanomaterials that now make up the majority of NCS packaging [97]. Inorganic filters such as zinc oxide (ZnO) and titanium dioxide (TiO_2) are often known as mineral filters, and they have been approved by the US Food and Drug Administration (FDA) [98]. The following are some inorganic nanoparticles that we shall discuss in this study.

4.7.1 Silver Nanoparticles (Ag-NPs)

Silver-based nanoparticles have application in packaging sector because of their superior antimicrobial property, least toxicity, and good thermal stability [99, 100]. Previously using a variety of biopolymers such as chitosan, cellulose, and starch, Ag-NPs have been used to develop antimicrobial packaging materials [101, 102]. Prior studies have concentrated on creating functionalized special films based on chitosan to pack food items. These NPs enhanced a number of characteristics as compared to traditional film [100, 103, 104]. The Ag-NP-coated film showed spherical clusters, increased roughness, flexibility, and transparency. Additionally, its thickness, solubility, and WV permeability all decreased. Additionally, it demonstrated antibacterial efficacy against S. aureus and E. coli. Additionally, the film preserved the quality and longevity and postponed ripening when applied to fruits, but untreated fruits showed increased wrinkles and wilting [105]. According to Ceballos et al. [101], adding NPs increases the thickness and opacity of starch films. They showed that Ag-NP had a reinforcing effect. The material had the lowest moisture content and the highest toughness and elongation at break. Furthermore, after being buried for four weeks, all the materials broke down. Ag-NPs have been combined with various materials in many investigations. An antibacterial layer was created when Ag-NPs were combined with chitosan/carbon spheres [106]. By the electrostatic interaction and filling behaviour of micro-nanometres, chitosan-based films were enhanced swelling, solubility in water thermal properties [107]. The function of the biological film improved with the help of inclusion of silver nanoparticles. For instance, how long do the carrot slices that are encased in this film last? According to Pandian et al. [108], Ag-NPs were applied on microcrystalline cellulose to create a film. This layer decreased food deterioration from microorganisms, extending the shelf's vegetables life. PVA-starch (PSt) films used in bioactive packaging enhanced tensile strength and tensile modulus while also improving water resistance. PSt-Ag-NPs found in various food products were measured for their ability to release silver ions [109].

4.7.2 Titanium Dioxide Nanoparticles (NPs of TiO_2)

Titanium dioxide nanoparticles are frequently utilized in biopolymeric nanocomposites matrix films which can extend the shelf life of consumable goods because of their low toxicity, gas-barrier qualities, antibacterial activity, biocompatibility, and photocatalytic property [110, 111]. Titanium dioxide particles are work as semiconductor that absorbs light in particular electromagnetic spectrum regions to demonstrate its exceptional photocatalytic activity. Oxygen species, which are reactive, are formed as a result of this process, and they have strong antibacterial qualities. Importantly, TiO_2 in bulk and TiO_2 nanostructures were shown the remarkable antibacterial properties, which completely depend on photocatalytic activity [112]. TiO_2 biodegradable films were efficient to extend the shelf life of bananas due to their C_2H_4 scavenging property, which slowed the rate of deterioration of the food quality [113]. TiO_2 gelatin was combined with high methoxyl pectin and curcumin to create environmentally favourable NCS hybrid membranes. Higher TiO_2-NP concentrations affected the physical properties of film, such as its surface, roughness, and compactness. The moisture content dropped sharply, transmission of light and solubility in water decreased, and the thermal properties of the film increased as the nanocomposites based on TiO_2 content increased. In the film of nanocomposites, biodegradation tests showed a notable reduction in weight [114].

4.7.3 Zinc Oxide Nanoparticles (ZnO-NPs)

The antibacterial qualities are also shown by zinc oxide which is semiconductor material as TiO_2. When Zn^{2+} ions are produced in microbial environment, ZnO nanostructures have antibacterial properties and absorb UV radiation. The microbial enzyme system is disrupted, and amino acid metabolism is greatly impacted [115]. However, two factors primarily influence the release of Zn^{2+}. (1) the physical and chemical characteristics of the nanoparticles, including their concentration, size, porosity, and shape; and (2) the properties of chemical, such as its pH, exposure duration, and other variables [112]. ZnO-NPs have unique chemical and physical characteristics that decide its suitability for a variety of uses; other applications are also photovoltaic testing, biomaterials, and packaging materials. These characteristics include a broad absorption spectrum of radiation, UV barrier, antibacterial and antifungal activity, big excitonic binding energy, and large energy band gap, high electrochemical coupling coefficient [116]. ZnO-NPs were synthesized using various commercial techniques, including hydrothermal approaches [117]. Because of their UV-blocking properties, reduced water permeability, and sustained oxygen transmission rate, PVA film based on ZnO nanoparticles is prepared for packaging applications of fresh food products and to extent shelf life of food [102]. Through the inhibition of microbial development and the prevention of lipid/protein oxidation, ZnO-NPs effectively preserve the quality of meat. Whether or if ZnO-NPs are produced by plants, their usage in active packaging is a viable [118].

4.8 Carbon Nanotubes (C-NPSs)

Numerous new structures with distinctive characteristics have been found as a result of the capacity of carbon atoms to arrange and produce various allotropes and phases.

Carbon nanoparticles (C-NPs) show special physiochemical and biological characteristics and are very attractive. These materials are frequently blended with metals or polymers to increase their inherent properties that use in a variety of settings and domains [119]. C-NPs, acid poly(adipic acid-butylene-co-terephthalate) PBAT and tannic acid were combined using a solution casting technique. Tannic acid cross-linked the C-NP and PBAT composite. The nanocomposites demonstrated strong antimicrobial properties against E. coli and S. aureus and two foodborne pathogens. Additionally, food quality investigations revealed that the chemical could increase the shelf life of peeled carrots by up to 12 days. The outcomes imply that the materials are a good choice for food packaging [120]. C-NPs with PBAT alone and got a similar outcome [121]. Recently, metal nanoparticles known as C-NPs have been used to make a variety of PBAT composites because of their eco-friendliness, affordability, non-toxicity, and biocompatibility [119, 122].

4.9 Nanoclay

Natural nanoclays are abundant and feature soft structures, low specific gravity, high aspect ratios, and plate-like forms. Incorporating nanoclay into a polymer matrix (as shown in Fig. 8) boosts the physiochemical barrier activity and mechanical strength. The desired properties in film packaging are made possible by a variety of nanoclay modification techniques [97]. Hydrated silicate and alumina combine to form nanoclay, a layered clay. It is a very sought-after material for food packaging design because of its qualities, which include its capacity to expand and plasticize as well as reinforce food packaging [123]. Oil-resistant paper with a chitosan/montmorillonite composite covering was prepared for use in packaging [124]. The Nanotechnology Product Database (NPD) examined more than 8694 items based on nanotechnology that was offered by 2194 businesses across more than 60 nations. The NPD has 222 nanoclay-based items as of August 2020, from 125 businesses across 25 countries, categorized into several classifications such as their packaging and food products [125].

The NCS films based on montmorillonite showed enhanced stiffness, moldability, WV barrier, and UV resistance. When montmorillonite, chitosan, and essential oil are combined to make films, the addition of chitosan to essential oil, the film with high mechanical properties is formed. This is explained by the initial stage's montmorillonite exfoliation, which strengthens the compact NCS structure. An intercalated structure is produced when montmorillonite is added essential oil [127].

4.10 Combined NPs

The produced nanocomposite films were regarded as promising antibacterial biomaterials for food packaging [128]. NCS films have superior heat resistance, barrier

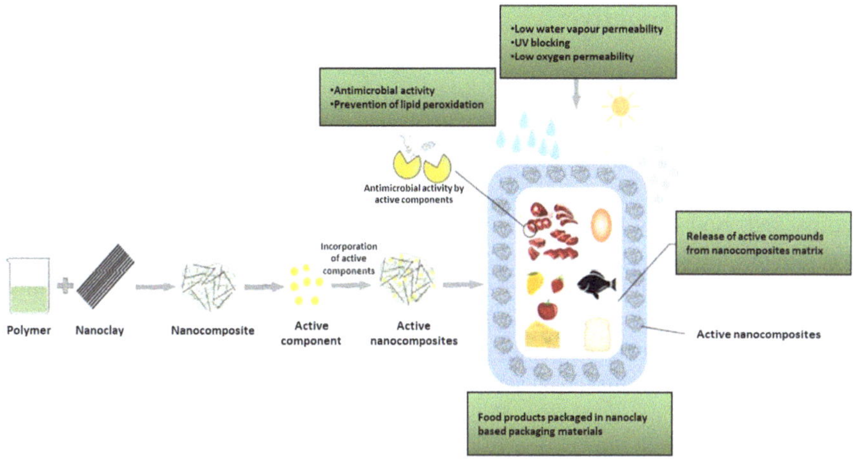

Fig. 8 Formation of nanoclay active composites for food packaging (reproduced with permission with ref. [126])

qualities, and the capacity to tolerate high external tensile stresses due to their microstructure which is dense and robust network [129]. Nanoparticles of zinc oxide and carbon atoms work together as active additives and nanofillers to enhance the bacteriostatic physiochemical characteristics of film based on gelatin [130]. Similarly, on addition of nanofillers like NPs can greatly improve the qualities of packaging (such as mechanical strength, gas and water resistance, and microbiological barriers) [58]. The combination of Ag-NPs and ZnO-NPs at starch/PBAT improves the films' material qualities and antibacterial effectiveness. Compared to LDPE packaging, antibacterial films prolong shelf life and prevent spoiling in peaches and nectarines, according to preliminary packaging experiments [28]. Likewise, carbon nanotubes and zinc oxide nanocomposites-based PVA film with different concentrations of zinc oxide-doped multiwalled (MW) carbon nanotubes offer significant potential for packaging and outstanding transparency, according to Wen et al. [131]. Silica oxide and zinc oxide nanoparticles-based PVA film were found to be highly effective [132]. These results suggest that nanocomposites films may extend the shelf life of chilled shrimp by using less chemicals and providing a novel, economical, secure, and efficient packaging method. The developing efficient food packaging to control both qualitatively and quantitatively food losses brought on by the combination of NPs. They also work as a packaging, which extend the shelf life of food and improve mechanical strength and barrier property.

5 Future Prospective

The development and advanced research in the applications of nanocomposites in packaging industries are highly promising. The quality of barrier activity of the packaging film can be enhanced by nanocomposites like graphene and carbon nanotubes, increasing their resistance to gas and moisture penetration. As a result, the packed food may be safe by increasing the shelf life [133]. Biodegradability and composability packaging material are the examples of more environmentally compatible options that can be created using nanocomposites. This can also minimize the adverse effects of waste packaging materials on the ecosystem. Furthermore, advanced research needs to maintain the freshness and safety of packed products, nanomaterials with antibacterial activity can be used to packaging films [1]. An intelligent material, that is used to packaging sector, provides real-time data related to product quality, including humidity, temperature, and freshness, can be produced using nanocomposites. By minimizing the chance of contamination, oxidation, and spoiling, nanomaterials can enhance the safety of food materials [134]. Advanced packaging, flexible, and lightweight films that develop enhanced functionality and usability can be created by nanocomposites. Thus, the performance and functionality of packaging content might be changed by using nanoparticles in packaging applications, which could also reduce the adverse impacts of materials on the environment and increase food safety. To anticipate in advancements, improvements, and research in this field keep progressing [135]. However, there are concerns about the possible toxicological and environmental adverse impacts of using nanomaterials in packaging applications for both customers and employees [4]. Nanocomposites also can penetrate the lungs and may cause inflammation and damage. The person involved in the production of nanocomposites are especially concerned about inhalation toxicity. Certain nanocomposites have the high potential to irritate skin and have other dermatological impacts. This is a concern for both customers who come into contact with the packaging materials and employees who handle nanocomposites. Nanomaterials, part of packaging material, may enter into the product itself, when food or medications come into direct touch with the packaging [43]. There are some nanocomposites that can build up in specific body tissues or organs and have long-term negative impacts on health. Therefore, consumers worry about possible toxicity of nanocomposites. It is particularly concerning that the applications, disposal, and production of packaging materials may release of nanomaterials into the surrounding environment [136]. The adverse effects of nanocomposites are not investigated completely, and advanced research is required to determine their possible impacts on environment. To work on these challenges, producers and authorities must carefully evaluate the safety and possible adverse impacts of nanocomposites prior to their usage in packaging. To aware customer related to any possible risks by using of nanocomposites in packaging, these demands carrying out comprehensive toxicological investigation, creating suitable safety guidelines and regulations, and putting in place suitable labelling and communication measures.

6 Conclusion

The applications of nanocomposites in the packaging sector and their different properties are discussed in this chapter. Sustainable packaging materials should be recyclable, renewable, biodegradable, environmentally friendly, and beneficial to businesses. Nanocomposites with antimicrobial activity can aid in the fight against the growing threat of microbial infection and spread. Environmental pollution and packaging waste can be reduced with the use of nanocomposites-based biodegradable packaging materials. This is also possible to modify the packaging material to have the required antibacterial and biodegradable properties by blending suitable nanocomposites. Investigating the infrastructure to produce environmentally friendly and biodegradable packaging materials and essential to minimize the detrimental adverse impacts of packaging waste on the environment. Prioritizing sustainable solutions calls for more work from citizens, corporations, and governments. Additionally, it is imperative that authorities take the initiative and put policies into place that seek to ban non-biodegradable plastic and control the use of cutting-edge packaging materials. By implementing mentioned parameters, the large amount of non-biodegradable plastic waste that ends up in landfilling and the ocean can be greatly reduced, hence lowering the adverse ecological impacts on plastic pollution. To anticipate more advanced research and improvements in the nanotechnological applications for environmentally friendly packaging.

Conflicts of Interest There are no conflicts of interest to declare by the authors.

References

1. Bhardwaj S, Lata S, Garg R (2023) Application of nanotechnology for preventing postharvest losses of agriproducts. J Horticult Sci Biotechnol 98(1):31–44
2. Sharma R, Garg R, Kumari A (2020) A review on biogenic synthesis, aspplications and toxicity aspects of zinc oxide nanoparticles. EXCLI J 19:1325–1340
3. Rai M, Ingle AP, Gupta I, Pandit R, Paralikar P, Gade A, Chaud MV, dos Santos CA (2019) Smart nanopackaging for the enhancement of food shelf life. Environ Chem Lett 17(1):277–290
4. Ghanem AF, Youssef AM, Abdel Rehim MH (2020) Hydrophobically modified graphene oxide as a barrier and antibacterial agent for polystyrene packaging. J Mater Sci 55(11):4685–4700
5. Zafar A, Khosa MK, Noor A, Qayyum S, Saif MJ (2022) Carboxymethyl cellulose/gelatin hydrogel films loaded with zinc oxide nanoparticles for sustainable food packaging applications. Polymers 14(23):5485
6. Suput D, Popovic S, Hromis N, Ugarkovic J (2021) Degradable packaging materials: sources, application and decomposition routes. J Process Energy Agricult 25(2):37–42
7. Yadav R, Singh M, Shekhawat D, Lee SY, Park SJ (2023) The role of fillers to enhance the mechanical, thermal, and wear characteristics of polymer composite materials: a review. Compos A Appl Sci Manuf 175:107775

8. Filip N, Radu I, Veliceasa B, Filip C, Pertea M, Clim A, Pinzariu AC, Drochioi IC, Hilitanu RL, Serban IL (2022) Biomaterials in orthopedic devices: current issues and future perspectives. Coatings 12(10):1–12
9. Jin R, Higaki T (2021) Commun Chem 4:28
10. Wu Q, Wu W, Chen F, Ren P (2022) Analyst 147:2809–2818
11. Brandelli A (2024) Nanocomposites and their application in antimicrobial packaging. Front Chem 12:1356304
12. Gioia C, Giacobazzi G, Vannini M, Totaro G, Sisti L, Colonna M, Marchese P, Celli A (2021) Chemsuschem 14:4167–4175
13. Nations United (2015) Transforming our world: the 2030 agenda for sustainable development. United Nations
14. De Moura MR, Mattoso LHC, Zucolotto V (2012) Development of cellulose-based bactericidal nanocomposites containing silver nanoparticles and their use as active food packaging. J Food Eng 109(3):520–524
15. Mittal V (2009) Polymer layered silicate nanocomposites: a review. Materials 2(3):992–1057
16. Anvar AA, Ahari H, Ataee M (2021) Antimicrobial properties of food nanopackaging: a new focus on foodborne pathogens. Front Microbiol 12:6585
17. Orsuwan A, Shankar S, Wang LF, Sothornvit R, Rhim JW (2016) Preparation of antimicrobial agar/banana powder blend films reinforced with silver nanoparticles. Food Hydrocoll 60:476–485
18. Iamareerat B, Singh M, Sadiq MB, Anal AK (2018) Reinforced cassava starch based edible film incorporated with essential oil and sodium bentonite nanoclay as food packaging material. J Food Sci Technol 55(5):1953–1959
19. Martí M, Frígols B, Serrano-Aroca A (2018) (2018) Antimicrobial characterization of advanced materials for bioengineering applications. J Vis Exper 138:1–10
20. Garg R, Garg R, Okon Eddy N, Ibrahim Almohana A, Fahad Almojil S, Amir Khan M, Ho Hong S (2022) Biosynthesized silica-based zinc oxide nanocomposites for the sequestration of heavy metal ions from aqueous solutions. J King Saud Univ Sci 34(4):101996
21. Diez-Pascual AM (2019) Antibacterial nanocomposites based on thermosetting polymers derived from vegetable oils and metal oxide nanoparticles. Polymers 11:1790
22. Peighambardoust SJ, Peighambardoust SH, Mohammadzadeh Pournasir N, Pakdel P (2019) Properties of active starch-based films incorporating a combination of Ag, ZnO and CuO nanoparticles for potential use in food packaging applications. Food Pack Shelf Life 22:100420
23. Kumar S, Mudai A, Roy B, Basumatary IB, Mukherjee A, Dutta J (2020) Biodegradable hybrid nanocomposite of chitosan/gelatin and green synthesized zinc oxide nanoparticles for food packaging. Foods 9(9):1–13
24. Marrez DA, Abdelhamid AE, Darwesh OM (2019) Eco-friendly cellulose acetate green synthesized silver nano-composite as antibacterial packaging system for food safety. Food Pack Shelf Life 20:100302
25. Nurfatihah Z, Siddiquee S (2019) Nanotechnology: recent trends in food safety, quality and market analysis. In: Siddiquee S, Melvin GJH, Rahman MM (eds) Nanotechnology: applications in energy, drug and food. Springer, New York, pp 283–293
26. Mistretta MC, Botta L, La Mantia FP, Di Fiore A, Cascone M (2021) Film blowing of biodegradable polymer nanocomposites for agricultural applications. Macromol Mater Eng 306(9):1–8
27. Ghosh S, Mandal RK, Mukherjee A, Roy S (2025) Nanotechnology in the manufacturing of sustainable food packaging: a review. Discover Nano 20(1):36
28. Zhai X, Zhou S, Zhang R, Wang W, Hou H (2022) Antimicrobial starch/poly (butylene adipate-co-terephthalate) nanocomposite films loaded with a combination of silver and zinc oxide nanoparticles for food packaging. Int J Biol Macromol 206:298–305
29. Lopez OV, Castillo LA, García MA, Villar MA, Barbosa SE (2015) Food packaging bags based on thermoplastic corn starch reinforced with talc nanoparticles. Food Hydrocoll 43:18–24

30. Adeyeye SAO, Surendra Babu A, Guruprasath N, Sankar Ganesh P (2023) Starch nanocrystal and its food packaging applications. Curr Res Nutr Food Sci 11(1):1–21
31. Bansal M, Garg R, Garg VK, Garg R, Singh D (2022) Sequestration of heavy metal ions from multi-metal simulated wastewater systems using processed agricultural biomass. Chemosphere 296:133966
32. Garg R, Garg R, Okon-Eddy N, Ibrahim-Almohana A, Fahad-Almojil S, Amir-Khan M, Ho-Hong S (2022) Biosynthesized silica-based zinc oxide nanocomposites for the sequestration of heavy metal ions from aqueous solutions. J King Saud Univ Sci 34(4):101996
33. Rani S, Sharma S, Bansal M, Garg R, Garg R (2022) Enhanced Zn(II) adsorption by chemically modified sawdust based biosorbents. Environ Sci Pollut Res 30:99046–99061
34. Odoemelam SA, Oji EO, Eddy NO, Garg R, Garg R, Islam S, Khan MA, Khan NA, Zahmatkesh S (2023) Zinc oxide nanoparticles adsorb emerging pollutants (glyphosate pesticide) from aqueous solutions. Environ Monit Assess 195(6):78195
35. Filiciotto L, Rothenberg G (2021) Biodegradable plastics: standards, policies, and impacts. Chemsuschem 14(1):56–72
36. Cao W, Yan J, Liu C, Zhang J, Wang H, Gao X, Yan H, Niu B, Li W (2020) Preparation and characterization of catechol-grafted chitosan/gelatin/modified chitosan-AgNP blend films. Carbohydr Polym 247:4
37. Wang GX, Huang D, Ji JH, Völker C, Wurm FR (2021) Seawater-degradable polymers—fighting the marine plastic pollution. Adv Sci 8(1):2001121
38. Alizadeh-Sani M, Maleki M, Eghbaljoo-Gharehgheshlaghi H, Khezerlou A, Mohammadian E, Liu Q, Jafari SM (2022) Titanium dioxide nanoparticles as multifunctional surface-active materials for smart/active nanocomposite packaging films. Adv Colloid Interf Sci 300:102593
39. Miao F, Shao C, Li X, Lu N, Wang K, Zhang X, Liu Y (2016) Polyaniline-coated electrospun carbon nanofibers with high mass loading and enhanced capacitive performance as freestanding electrodes for flexible solid-state supercapacitors. Energy 95:233–241
40. Cao W, Yan J, Liu C, Zhang J, Wang H, Gao X, Yan H, Niu B, Li W (2020) Preparation and characterization of catechol-grafted chitosan/gelatin/modified chitosan-AgNP blend films. Carbohydr Polym 247:68
41. Mellinas AC, Jimenez A, Garrigos MC (2020) Pectin-based films with cocoa bean shell waste extract and ZnO/Zn-NPs with enhanced oxygen barrier, ultraviolet screen and photocatalytic properties. Foods 9(11):875
42. Arabpoor B, Yousefi S, Weisany W, Ghasemlou M (2021) Multifunctional coating composed of *Eryngium campestre* L. essential oil encapsulated in nano-chitosan to prolong the shelf-life of fresh cherry fruits. Food Hydrocoll 111:106394
43. Jayakumar AKV, Heera TS, Joseph M, Mathew SG, Nair IC, Radhakrishnan EK (2019) Starch-PVA composite films with zinc-oxide nanoparticles and phytochemicals as intelligent pH sensing wraps for food packaging application. Int J Biol Macromol 136:395–403
44. Valipour Motlagh N, Aghazamani J, Gholami R (2021) Investigating the effect of nano-silver contained packaging on the olivier salad shelf-life. BioNanoScience 11(3):838–847
45. de Sousa MS, Schlogl AE, Estanislau FR, Souza VGL, dos Reis Coimbra JS, Santos IJB (2023) Nanotechnology in packaging for food industry: past, present, and future. Coatings 13(8):1411
46. Tonezzer M, Bazzanella N, Gasperi F, Biasioli F (2022) Nanosensor based on thermal gradient and machine learning for the detection of methanol adulteration in alcoholic beverages and methanol poisoning. Sensors 22(15):6881
47. Malik S, Muhammad K, Waheed Y (2023) Nanotechnology: a revolution in modern industry. Molecules 28(2):661
48. Leyva-Jimenez FJ, Oliver-Simancas R, Castangia I, Rodríguez-García AM, Alanon ME (2023) Comprehensive review of natural based hydrogels as an upcoming trend for food packing. Food Hydrocoll 135:108124
49. Gomez-Gomez AL, Martínez-Ayala AL, Moguel-Concha DDR, Borges-Martínez JE, Perea-Flores MDJ, Davila-Ortiz G (2023) Relationship of nanomaterials' structure based on their application in the food industry: physicochemical and techno-functional characteristics. Appl Sci 13(12):7167

50. Joudeh N, Linke D (2022) Nanoparticle classification, physicochemical properties, characterization, and applications: a comprehensive review for biologists. J Nanobiotechnol 20(1):262
51. Channa IA, Ashfaq J, Siddiqui MA, Chandio AD, Shar MA, Alhazaa A (2022) Multi-shaded edible films based on gelatin and starch for the packaging applications. Polymers 14(22):5020
52. Khodaei D, Alvarez C, Mullen AM (2021) Biodegradable packaging materials from animal processing co-products and wastes: an overview. Polymers 13(15):2561
53. Shaikh S, Yaqoob M, Aggarwal P (2021) An overview of biodegradable packaging in food industry. Elsevier 4:503–520
54. Dash KK, Deka P, Bangar SP, Chaudhary V, Trif M, Rusu A (2022) Applications of inorganic nanoparticles in food packaging: a comprehensive review. Polymers 14(3):521
55. Xing W, Tang Y (2022) On mechanical properties of nanocomposite hydrogels: searching for superior properties. Nano Mater Sci 4(2):83–96
56. Gupta RK, El-Gawad FA, Ali EAE, Karunanithi S, Yugiani P, Srivastav PP (2024) Measurement: food nanotechnology—current applications and future scope in food packaging systems. Mearsur Food 13:100131
57. Aththanayaka S, Thiripuranathar G, Ekanayake S (2022) Emerging advances in biomimetic synthesis of nanocomposites and potential applications. Mater Today Sustain 20:100206
58. Wypij M, Trzcinska-Wencel J, Golinska P, Avila-Quezada GD, Ingle AP, Rai M (2023) The strategic applications of natural polymer nanocomposites in food packaging and agriculture: chances, challenges, and consumers' perception. Front Chem 10:1106230
59. Hallad SA, Banapurmath NR, Khan TY, Umarfarooq MA, Soudagar MEM, Hunashyal AM, Ali MA (2021) Statistical analysis of polymer nanocomposites for mechanical properties. Molecules 26(14):4135
60. Wawrzyniak J (2023) Advancements in improving selectivity of metal oxide semiconductor gas sensors opening new perspectives for their application in food industry. Sensors 23(23):9548
61. Yadav A, Kumar H, Sharma R, Kumari R (2023) Synthesis, processing, and applications of 2D (nano) materials: a sustainable approach. Surf Interf 39:102925
62. Bandyopadhyay J, Ray SS (2023) Nanotechnology in innovative food preservation and packaging. Emerg Technol Food Preserv 349(350):349
63. Joshi N, Pransu G, Conte-Junior CA (2023) Critical review and recent advances of 2D materials-based gas sensors for food spoilage detection. Crit Rev Food Sci Nutr 63(30):10536–10559
64. Wang K, Li F, Sun X, Wang F, Xie D, Wei Y (2023) Transparent chitosan/hexagonal boron nitride nanosheets composite films with enhanced UV shielding and gas barrier properties. Int J Biol Macromol 251:126308
65. Ahankari SS, Mohanty AK, Misra M (2023) Nanomaterials from renewable resources for emerging applications. CRC Press, New York
66. Priyadarshi R, Roy S, Ghosh T, Biswas D, Rhim JW (2022) Antimicrobial nanofillers reinforced biopolymer composite films for active food packaging applications-a review. Sustain Mater Technol 32:e00353
67. Silva FCO, Malaisamy A, Cahú TB, de Araújo MIF, Soares PAG, Vieira AT, dos Santos Correia MT (2023) Polysaccharides from exudate gums of plants and interactions with the intestinal microbiota: a review of vegetal biopolymers and prediction of their prebiotic potential. Int J Biol Macromol 254(2):127715
68. Altaf A, Usmani Z, Dar AH, Dash KK (2022) A comprehensive review of polysaccharide-based bionanocomposites for food packaging applications. Discov Food 2(1):10
69. Nasrollahzadeh M, Sajjadi M, Iravani S, Varma RS (2021) Starch, cellulose, pectin, gum, alginate, chitin and chitosan derived (nano) materials for sustainable water treatment: a review. Carbohydr Polym 251:116986
70. Pal K, Sarkar P, Anis A, Wiszumirska K, Jarzębski M (2021) Polysaccharide-based nanocomposites for food packaging applications. Materials 14(19):5549

71. Brandelli A, Lopes NA (2021) Nanocomposite antimicrobial films based on biopolymers. Biopolym Based Nano Films 15:149–170
72. Lichade KM, Jiang Y, Pan Y (2021) Hierarchical nano/micro-structured surfaces with high surface area/volume ratios. J Manufact Sci Eng 143(8):081002
73. Xia GX, Wu YM, Bi YF, Chen K, Zhang WW, Liu SQ, Liu RH (2021) Antimicrobial properties and application of polysaccharides and their derivatives. Chin J Polym Sci 39:133–146
74. Yanat M, Schroen K (2021) Preparation methods and applications of chitosan nanoparticles; with an outlook toward reinforcement of biodegradable packaging. React Funct Polym 161:104849
75. Wang J, Han X, Zhang C, Liu K, Duan G (2022) Source of nanocellulose and its application in nanocomposite packaging material: a review. Nanomaterials 12(18):3158
76. Vianna TC, Marinho CO, Júnior LM, Ibrahim SA, Vieira RP (2021) Essential oils as additives in active starch-based food packaging films: a review. Int J Biol Macromol 182:1803–1819
77. Gupta RK, Guha P, Srivastav PP (2023) Effect of high voltage dielectric barrier discharge (DBD) atmospheric cold plasma treatment on physicochemical and functional properties of taro (*Colocasia esculenta*) starch. Int J Biol Macromol 253:126772
78. Martins PC, Latorres JM, Martins VG (2022) Impact of starch nanocrystals on the physicochemical, thermal and structural characteristics of starch-based films. LWT 156:113041
79. Adeyeye SAO, Babu AS, Guruprasath N, Ganesh PS (2023) Starch nanocrystal and its food packaging applications. Curr Res Nutr Food Sci J 11(1):01–21
80. Marta H, Rizki DI, Mardawati E, Djali M, Mohammad M, Cahyana Y (2023) Starch nanoparticles: preparation, properties and applications. Polymers 15(5):1167
81. Asha V, Sivaranjana P (2023) Development of biodegradable films using nanocellulose for food packaging application: review. Azerbaijan Chem J 1:30–54
82. Roy S, Kim HJ, Rhim JW (2021) Synthesis of carboxymethyl cellulose and agar-based multifunctional films reinforced with cellulose nanocrystals and shikonin. ACS Appl Polym Mater 3(2):1060–1069
83. Hisham F, Akmal MM, Ahmad F, Ahmad K, Samat N (2023) Biopolymer chitosan: potential sources, extraction methods, and emerging applications. Ain Shams Eng J 15(2):102424
84. Roy S, Rhim JW (2021) Fabrication of chitosan-based functional nanocomposite films: effect of quercetin-loaded chitosan nanoparticles. Food Hydrocoll 121:107065
85. Irastorza A, Zarandona I, Andonegi M, Guerrero P, de la Caba K (2021) The versatility of collagen and chitosan: from food to biomedical applications. Food Hydrocoll 116:106633
86. Mondal A, Dhar AK, Banerjee S, Hasnain MS, Nayak AK (2022) Antimicrobial uses of chitosan. Chitosan Biomed Appl 61:13–36
87. Dong S, Zhang Y, Lu D, Gao W, Zhao Q, Shi X (2023) Multifunctional intelligent film integrated with purple sweet potato anthocyanin and quercetin-loaded chitosan nanoparticles for monitoring and maintaining freshness of shrimp. Food Pack Shelf Life 35:101022
88. Roy S, Min SJ, Biswas D, Rhim JW (2023) Pullulan/chitosan-based functional film incorporated with curcumin-integrated chitosan nanoparticles. Coll Surf A Physicochem Eng Aspects 660:130898
89. Othman SH, Wane BM, Nordin N, Noor Hasnan NZ, Talib RA, Karyadi JNW (2021) Physical, mechanical, and water vapor barrier properties of starch/cellulose nanofiber/thymol bionanocomposite films. Polymers 13(23):4060
90. Lu W, Nishinari K, Phillips GO, Fang Y (2021) Colloidal nutrition science to understand food-body interaction. Trends Food Sci Technol 109:352–364
91. Wang Y, Luo Y (2023) Colloidal nanoparticles prepared from zein and casein: interactions, characterizations and emerging food applications. Food Sci Hum Wellness 12(2):337–350
92. Bagheri Darvish H, Bahrami A, Jafari SM, Williams L (2021) Micro/nanoencapsulation strategy to improve the efficiency of natural antimicrobials against Listeria monocytogenes in food products. Crit Rev Food Sci Nutr 61(8):1241–1259
93. Tang M, Liu F, Wang Q, Wang D, Wang D, Zhu Y, Xu W (2022) Physicochemical characteristics of ginger essential oil nanoemulsion encapsulated by zein/NaCas and antimicrobial control on chilled chicken. Food Chem 374:131624

94. Tang CH (2021) Assembled milk protein nano-architectures as potential nanovehicles for nutraceuticals. Adv Colloid Interf Sci 292:102432
95. Adeyemi JO, Fawole OA (2023) Metal-based nanoparticles in food packaging and coating technologies: a review. Biomolecules 13(7):1092
96. Channa IA, Ashfaq J, Gilani SJ, Shah AA, Chandio AD, Jumah MNB (2022) UV blocking and oxygen barrier coatings based on polyvinyl alcohol and zinc oxide nanoparticles for packaging applications. Coatings 12(7):897
97. Deshmukh RK, Hakim L, Akhila K, Ramakanth D, Gaikwad KK (2023) Nano clays and its composites for food packaging applications. Int Nano Lett 13(2):131–153
98. Ujjan ZA, Bhatti MA, Shah AA, Tahira A, Shaikh NM, Kumar S, Ibupoto ZH (2022) Simultaneous doping of sulfur and chloride ions into ZnO nanorods for improved photocatalytic properties towards degradation of methylene blue. Ceram Int 48(4):5535–5545
99. Jaswal T, Gupta J (2023) A review on the toxicity of silver nanoparticles on human health. Mater Today Proceed 81:859–863
100. Neupane NP, Kushwaha AK, Karn AK, Khalilullah H, Khan MMU, Kaushik A, Verma A (2022) Anti-bacterial efficacy of bio-fabricated silver nanoparticles of aerial part of *Moringa oleifera* lam: rapid green synthesis, in-vitro and in-silico screening. Biocatal Agricult Biotechnol 39:102229
101. Ceballos RL, Von Bilderling C, Guz L, Bernal C, Famá L (2021) Effect of greenly synthetized silver nanoparticles on the properties of active starch films obtained by extrusion and compression molding. Carbohydr Polym 261:117871
102. Shankar S, Khodaei D, Lacroix M (2021) Effect of chitosan/essential oils/silver nanoparticles composite films packaging and gamma irradiation on shelf life of strawberries. Food Hydrocoll 117:106750
103. Yang D, Liu Q, Gao Y, Wan S, Meng F, Weng W, Zhang Y (2023) Characterization of silver nanoparticles loaded chitosan/polyvinyl alcohol antibacterial films for food packaging. Food Hydrocoll 136:108305
104. Zhao R, Guan W, Zhou X, Lao M, Cai L (2022) The physiochemical and preservation properties of anthocyanidin/chitosan nanocomposite-based edible films containing cinnamon-perilla essential oil pickering nanoemulsions. LWT 153:112506
105. Silva LSC, Martim SR, Gomes DMD, Prado FB, Marinho NMV, Silva TDA (2021) Amazonian tuber starch based films incorporated with silver nanoparticles for preservation of fruits. Res Soc Develop 10(6):e23510615304
106. Wu Z, Tang S, Deng W, Luo J, Wang X (2021) Antibacterial chitosan composite films with food-inspired carbon spheres immobilized AgNPs. Food Chem 363:130342
107. Ediyilyam S, George B, Shankar SS, Dennis TT, Wacławek S, Cerník M, Padil VV (2021) Chitosan/gelatin/silver nanoparticles composites films for biodegradable food packaging applications. Polymers 13(11):1680
108. Pandian H, Senthilkumar K, Naveenkumar M, Samraj S (2023) Azadirachta indica leaf extract mediated silver nanoparticles impregnated nano composite film (AgNP/MCC/starch/whey protein) for food packaging applications. Environ Res 216:114641
109. Srikhao N, Kasemsiri P, Ounkaew A, Lorwanishpaisarn N, Okhawilai M, Pongsa U, Chindaprasirt P (2021) Bioactive nanocomposite film based on cassava starch/polyvinyl alcohol containing green synthesized silver nanoparticles. J Polym Environ 29:672–684
110. Mesgari M, Aalami AH, Sahebkar A (2021) Antimicrobial activities of chitosan/titanium dioxide composites as a biological nanolayer for food preservation: a review. Int J Biol Macromol 176:530–539
111. El-Kadib A, Wronska N, Lisowska K, Anouar A, Katir N, Miłowska K, Bryszewska M (2023) Functional bio-based chitosan films: from material design to biological properties. Funct Mater Biomed Appl 59:1–50
112. de Dicastillo CL, Correa MG, Martínez FB, Streitt C, Galotto MJ (2020) Antimicrobial effect of titanium dioxide nanoparticles. Antimicrob Resist One Health Perspect 1:95–112
113. Phothisarattana D, Wongphan P, Promhuad K, Promsorn J, Harnkarnsujarit N (2021) Biodegradable poly (butylene adipate-co-terephthalate) and thermoplastic starch-blended

TiO2 nanocomposite blown films as functional active packaging of fresh fruit. Polymers 13(23):4192
114. Candra A, Tsai HC, Saragi IR, Hu CC, Yu WT, Krishnamoorthi R, Lai JY (2023) Fabrication and characterization of hybrid eco-friendly high methoxyl pectin/gelatin/TiO_2/curcumin (PGTC) nanocomposite biofilms for salmon fillet packaging. Int J Biol Macromol 232:123423
115. Stałanowska K, Szablinska-Piernik A, Okorski A, Lahuta LB (2023) Zinc oxide nanoparticles affect early seedlings' growth and polar metabolite profiles of pea (*Pisum sativum* L.) and wheat (*Triticum aestivum* L.). Int J Mol Sci 24(19):14992
116. Channa IA, Shah AA, Rizwan M, Makhdoom MA, Chandio AD, Shar MA, Mahmood A (2021) Process parameter optimization of a polymer derived ceramic coatings for producing ultra-high gas barrier. Materials 14(22):7000
117. Alrajhi AH, Ahmed NM (2023) Green synthesis of zinc oxide nanoparticles using salvia officinalis extract. Handbook of green and sustainable nanotechnology: fundamentals, developments and applications. Springer, New York, pp 1–21
118. Smaoui S, Cherif I, Hlima HB, Khan MU, Rebezov M, Thiruvengadam M, Lorenzo JM (2023) Zinc oxide nanoparticles in meat packaging: a systematic review of recent literature. Food Pack Shelf Life 36:101045
119. Slepickova Kasalkova N, Slepicka P, Svorcík V (2021) Carbon nanostructures, nanolayers, and their composites. Nanomaterials 11(9):2368
120. Venkatesan R, Sivaprakash P, Kim I, Eldesoky GE, Kim SC (2023) Tannic acid as a crosslinking agent in poly (butylene adipate-co-terephthalate) composite films enhanced with carbon nanoparticles: processing, characterization, and antimicrobial activities for food packaging. J Environ Chem Eng 11(4):110194
121. Venkatesan R, Surya S, Suganthi S, Muthuramamoorthy M, Pandiaraj S, Kim SC (2023) Biodegradable composites from poly (butylene adipate-co-terephthalate) with carbon nanoparticles: preparation, characterization and performances. Environ Res 235:116634
122. Holmannova D, Borsky P, Svadlakova T, Borska L, Fiala Z (2022) Carbon nanoparticles and their biomedical applications. Appl Sci 12(15):7865
123. Mohammadi Sadati SM, Shahgholian-Ghahfarrokhi N, Shahrousvand E, Mohammadi-Rovshandeh J, Shahrousvand M (2022) Edible chitosan/cellulose nanofiber nanocomposite films for potential use as food packaging. Mater Technol 37(10):1276–1288
124. Pires J, Paula CDD, Souza VGL, Fernando AL, Coelhoso I (2021) Understanding the barrier and mechanical behavior of different nanofillers in chitosan films for food packaging. Polymers 13(5):721
125. NPD (2021) Nanotechnology in food industry
126. Nath D, Santhosh R, Pal K, Sarkar P (2022) Nanoclay-based active food packaging systems: a review. Food Pack Shelf Life 31:100803
127. Cui R, Yan J, Cao J, Qin Y, Yuan M, Li L (2021) Release properties of cinnamaldehyde loaded by montmorillonite in chitosan-based antibacterial food packaging. Int J Food Sci Technol 56(8):3670–3681
128. Shahvalizadeh R, Ahmadi R, Davandeh I, Pezeshki A, Moslemi SAS, Karimi S, Mohammadi M (2021) Antimicrobial bio-nanocomposite films based on gelatin, tragacanth, and zinc oxide nanoparticles–microstructural, mechanical, thermo-physical, and barrier properties. Food Chem 354:129492
129. Yuan T, Zeng J, Wang B, Cheng Z, Gao W, Xu J, Chen K (2021) Silver nanoparticles immobilized on cellulose nanofibrils for starch-based nanocomposites with high antibacterial, biocompatible, and mechanical properties. Cellulose 28:855–869
130. Mo X, Peng X, Liang X, Fang S, Xie H, Chen J, Meng Y (2021) Development of antifungal gelatin-based nanocomposite films functionalized with natamycin-loaded zein/casein nanoparticles. Food Hydrocoll 113:106506
131. Wen YH, Tsou CH, de Guzman MR, Huang D, Yu YQ, Gao C, Wang ZH (2022) Antibacterial nanocomposite films of poly (vinyl alcohol) modified with zinc oxide-doped multiwalled carbon nanotubes as food packaging. Polym Bull 79:3847–3866

132. Shao J, Wang L, Wang X, Ma J (2021) Enhancing microbial management and shelf life of shrimp *Penaeus vannamei* by using nanoparticles of metallic oxides as an alternate active packaging tool to synthetic chemicals. Food Pack Shelf Life 28:100652
133. Lata S, Bhardwaj S, Garg R (2022) Nanomaterials for sensing and biosensing: applications in agri-food diagnostics. Int J Environ Anal Chem 59:1–18
134. Priyadarshi R, Kim HJ, Rhim JW (2021) Effect of sulfur nanoparticles on properties of alginate-based films for active food packaging applications. Food Hydrocoll 110:106155
135. Alghdeir M, Mayya K, Dib M (2019) Characterization of nanosilica/low-density polyethylene nanocomposite materials. J Nanomater 2019:1–9
136. Castro-Mayorga JL, Fabra MJ, Lagaron JM (2016) Stabilized nanosilver based antimicrobial poly(3-hydroxybutyrate-co-3-hydroxyvalerate) nanocomposites of interest in active food packaging. Innov Food Sci Emerg Technol 33:524–533

Chapter 5
Polymer Composites Used for Biomedical Packaging: An Overview

Piyush Kumar

1 Introduction

A composite is a material construction composed of two or more in detail different components that work together to create superior properties. There are usually two phases in composites: a matrix phase and a dispersion phase [1]. Crucially, they provide novel attributes that distinguish them from their component elements. Investigations have taken note Polymer composites' adaptability, lightweight nature, and capacity to be customized for particular uses have made them more important in biomedical packaging [2]. These materials are perfect for protecting biological samples, medications, and medical devices because they combine mechanical strength, biocompatibility, and barrier qualities [2–4] They are made of a polymer matrix reinforced with fibers or particles (such as glass, carbon, or natural fibers).

In biomedical packaging applications, polymer composites are utilized due to their cost-effectiveness, superior strength and protection, thermal and chemical resistance, conformability, biocompatibility, excellent barrier characteristics, and lightweight biological form [4, 5]. This is especially helpful when it comes to shipping medications and medical equipment. Polymer composites can be designed to offer superior resistance to light, oxygen, and moisture—all of which are critical for preserving the stability of biologics, medications, and medical equipment.

Polymer composites' characteristics can be adjusted to meet particular needs. For instance, mechanical strength, thermal stability, and permeability can all be tailored for specific purposes by changing the kind or quantity of reinforcement [1]. It's

P. Kumar (✉)
Department of Polymer and Process Engineering, Indian Institute of Technology Roorkee, Saharanpur, Uttar Pradesh, India
e-mail: pkumar@pe.iitr.ac.in

Divisional Manager: R&D-2W: Division, Rico Auto Industries Ltd Gurugram, Gurugram, Haryana, India

A. Verma et al. (eds.), *Polymers and Composite Materials for Packaging: Smart Food Packaging and Solutions*, Materials Horizons: From Nature to Nanomaterials,
https://doi.org/10.1007/978-981-95-0677-4_5

critical that materials used in medical packaging don't negatively interact with the products being packaged or endanger public health. It is possible to design certain polymer composites to be biocompatible, which guarantees that no toxic materials will leak into the final product [4–7]. Additionally, polymer composites can be made to resist harsh chemical or temperature conditions. These materials offer further protection for biomedical products that need to be sterilized or that are susceptible to temperature changes [2, 3, 6]. Compared to metals or ceramics, which frequently need more intricate manufacturing procedures, advanced composite materials can be created in more economical methods.

2 Types of Polymer Composites

A polymer composite material consists of a continuous polymeric matrix phase reinforced by the dispersed reinforcement embedded in it, and the reinforcement may be particles or fibers. Matrix material for a polymer composite is either a thermoplastic or a thermoset, although, in some cases an elastomer may also be used as matrix [8, 9]. The polymeric matrix holds and binds the reinforcement, transfers applied stress to the reinforcement and protects the composite from the external environmental effects. Additionally, the polymer matrix governs the physical, thermal and also the temperature-dependent properties and high-temperature applications of a polymeric composite [10]. Reinforcement, may consist of particles or fibers. These may be macro or nano particles or fibers. The reinforcement, embedded in the matrix, carries the applied stress, and thus, governs the mechanical properties of the composite [10]. The efficiency of a reinforcement depends on its aspect ratio, [length to diameter (l/d) ratio]. Continuous fibers have a higher aspect ratio than a short fiber and therefore are more efficient reinforcements. Fiber/matrix interfacial adhesion also plays an important role in determining the properties of a polymer composite [10]. Thus, a polymer composite offers synergistic properties resulting from the properties of the matrix and of the reinforcement and their interfacial adhesion [8–10]. Polymer matrix composites have been classified into two categories depending on (i) the nature of the polymer matrix and (ii) the type of the reinforcement.

2.1 Polymer as Matrix for the Composites

On the basis of the type of the polymer matrix, polymer composites are divided into three categories [10]: (i) Thermoplastic matrix composite (ii) Thermoset matrix composites and (iii) Elastomeric matrix composites. (i) A thermoplastic matrix composite is based on a matrix that is a thermoplastic, that may be amorphous or crystalline. Polyethylene (PE), polypropylene (PP), acrylonitrile–butadiene–styrene (ABS), polystyrene (PS), and many other thermoplastics have been used as matrix for composites. (ii) A thermoset matrix composite uses a thermoset polymer as the matrix [10]. Epoxy, polyester, phenolic, polyimide, and a few other thermosets have been used as matrix for composites. (iii) Elastomers have also been used as matrix for composites, including those based on natural fibers as reinforcements [11].

2.2 Reinforcements for Polymer Composites

Based on the type of reinforcement, polymer composites have been classified into three groups:(i) Particle Reinforced Polymer Composites, also termed as particulate composites; (ii) Fiber Reinforced Polymer Composites, that are further sub-divided as (a) short fiber reinforced polymer composites and (b) continuous fiber reinforced polymer composites; and (iii) Nano Polymer Composites that are further sub-divided as (a) nano particle reinforced polymer composites and (b) nano fiber reinforced polymer composites. In a particle reinforced polymer composite, reinforcement is in the form of a particle, that is non-fibrous and generally has no long dimension. Examples of particulate reinforcements include, silica, talc, calcium carbonate, etc. [12]. A short fiber reinforced polymer composite consists of a polymer matrix reinforced by short fibers, that may be inorganic, synthetic or natural [13]. Examples of inorganic short fibers used as reinforcements for polymer composites are: glass and carbon; of synthetic fibers are: aramid fibers; and of natural fibers are: mineral, animal, and plant-based fibers. A continuous fiber reinforced polymeric composite consists of a polymer matrix reinforced by continuous fibers, that have a higher aspect ratio, than a short fiber. Examples of continuous fibers, used as reinforcements for polymer composites, include, carbon, glass, and high silica fibers [12, 14, 15]. A nano polymer composite consists of either a nano particle or a nano fiber as reinforcement. Examples of nano materials used as reinforcements include, carbon nanotubes, nano clay, and nano silica [16].

2.3 Polymer Blend Matrix Composites and Polymer Hybrid Composites

Polymer composites have been extended to composites where the matrix is a blend of two polymers, and these are known as polymer blend matrix composites. Polymer composites have also been extended to polymer hybrid composites, wherein, a thermoplastic or a thermoset matrix is reinforced by more than one reinforcement. The two reinforcements may be, both particles, both fibers, or, one particle and one fiber, and the two reinforcements may be macro or nano particle or fiber.

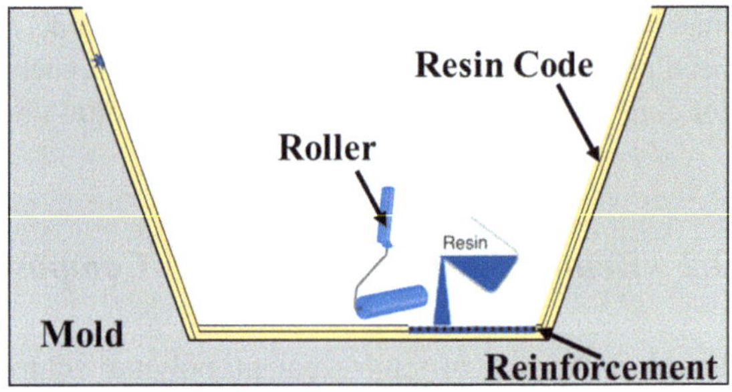

Fig. 1 Hand lay-up process

3 Technique for Fabricating Polymer Composite

Polymer composites, which enhance a material's mechanical, thermal, or electrical properties, are made by combining a polymer matrix with reinforcing elements (glass, carbon, aramid, etc.) or fillers (clay, nanoparticles, etc.). Depending on the type of polymer, the reinforcing material, the desired benefits of the finished composite, and the intended use, there are various ways of manufacturing polymer composites.

3.1 Hand Lay-Up Method

The most simple and traditional open molding technique for creating composites is hand lay-up. Initially, the mold is filled with dry fibers in the form of woven, knitted, stitched, or bond fabrics, and the resin matrix is applied to the reinforcing material using a brush [1, 9, 17]. The wet composite is then rolled using hand rollers to assist a consistent resin distribution, improve the interaction between the reinforcement and the matrix, and achieve the desired thickness. Lastly, the laminates are allowed to cure in a typical atmosphere. The four processes of this procedure are typically lay-up, curing, gel coating, and mold preparation. Heat is not necessary during the curing process, which builds up the fiber reinforced resin composite. Prior to coating the mold surface with a colored gel coat, a superior product surface gets created. An overview of the hand lay-up method used to manufacture resins is provided in Fig. 1.

3.2 Injection Molding

Injection molding is a popular technique for creating biomedical packaging. In order to create the required shape, polymer composites are melted, injected under pressure into a mold cavity, cooled, and then solidified [1, 18]. This technique's capacity to produce intricate designs, great precision, and effective mass manufacturing make it especially well-suited for biomedical applications. The injection molding machine

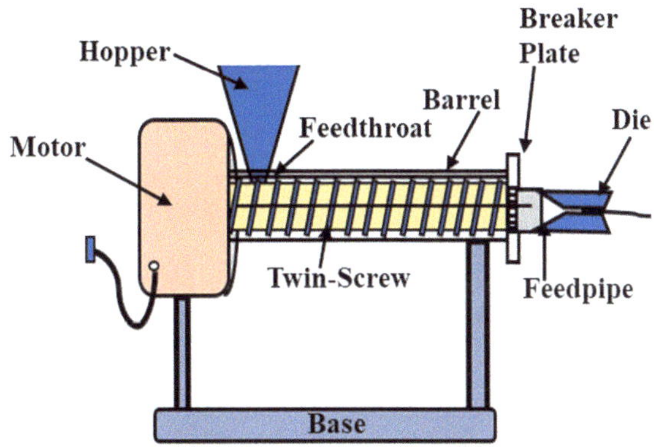

Fig. 2 Overview of the injection molding machine

entails softening a thermoplastic material in a heated cylinder before injecting it under pressure into the cavity of the mold, where it hardens as a result of cooling. The injection unit, clamping unit and mold pressure are crucial parameters for operating an injection molding machine. Melting the polymer before injecting it into the mold is the primary function of the injection unit. The clamping unit ensures that the opening and closing actions of mold are reliable, so that the mold closes and that the molded product is removed conveniently from the machine. The molten material is pushed through the cylinder and into a mold that is attached to the ends of the cylinder by applying pressure.

Clamping unit ensures a reliable opening and closing action, mold closing, and works to take out the molded product. Mold gives shape to the product and governs the quality of the specimen. Cycle times for the injection molding process typically range from 10 to 100 s. Specimen plates are generally loaded, one by one, for molding the samples. Mold gives shape to the product and controls its quality. An overview of the injection molding machine used to manufacture of polymers composites provided in Fig. 2.

3.3 Compression Molding

The fabrication of lighter components has historically made extensive use of compression molding [18]. This is an outdated method of sandwiching a mold between the hydraulic press's two plates. Composite components, such as resins or composite granules, can be charged between the cavities of a preheated mold. Once the mold cavity reaches the desired temperature, pressure can be applied to the mold cavity to give the composite components the form they need. For thermoset materials in particular, the machine's molds are maintained closed until the resin matrix has undergone the appropriate curing. Heat and pressure are applied to the mold cavity, causing the polymeric ingredients to melt. Nevertheless, the temperature must be lowered for thermoplastic composites before the composite parts

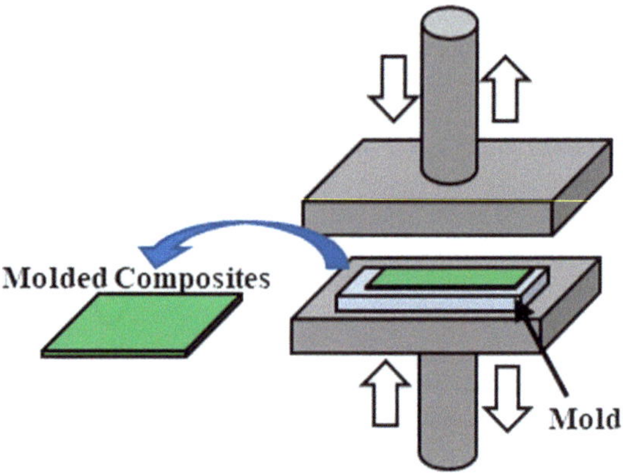

Fig. 3 Overview of the compression molding

may be ejected. Furthermore, since thermoset plastic composites must have strong crosslinking during solidification under pressure, this cooling step is not required. The density, dimensions, and thickness of the composites are determined by the heating/cooling process as well as the compression molding machines' pressure [18]. An overview of the compression molding used to manufacture of polymers composites is provided in Fig. 3.

3.4 Spray-Up Method

The spray-up process is another widely used technique for producing composite materials [19]. This technology deposits resins and composite materials onto a mold by means of controlled spraying, which concurrently projects fibers and resins onto the mold surface to generate the desired part. The layers of reinforcement and resin are applied to the mold uniformly using spray guns and other specialized tools. This process saves a lot of time and is especially effective when creating large numbers of basic biomedical packaging components.

3.5 Resin Transfer Molding

A combined resin and catalyst are injected into a closed mold that contains a fiber pack or preform in the low-pressure molding process known as resin transfer molding [20]. After the resin has dried, the mold can be opened to extract the completed part.

3.6 Blow Molding

Blow molding is another important polymer processing method [21]. The process of blow molding uses air pressure to push the plastic against the mold walls, creating hollow components. Although the method is usually used to make containers, including milk jugs and soda bottles, blow molding is not just utilized to make containers for biomedical packaging.

3.7 3D Printing (Additive Manufacturing)

The design is possible to create geometrically complicated items, forms, and textures by additive manufacturing or 3D printing [22]. It enables the fabrication of goods that were just not economically feasible to produce using traditional manufacturing techniques, and it frequently consumes less material than those methods. Comparing 3D printing to conventional manufacturing techniques could be useful if you're new to the technology.

The two main conventional manufacturing techniques are: Machining is an example of Subtractive Manufacturing (SM), which is the controlled elimination of material. Formative Manufacturing (FM) is the process of reshaping a material, typically plastic, to produce a finished good. Contrast these with additive manufacturing or 3D printing, which creates the finished product by building on component.

4 Polymer Composite Materials Used for Biomedical Packaging

Recently, several biomedical packaging manufacturers has made use of polymer composite materials by altering them to be biocompatible. The outstanding mechanical and fire retardant capabilities of several innovative composite materials also lead to their employment in other diverse properties [2, 23].

4.1 Medical Device Packaging

Polymer composites used for packaging implants, diagnostic devices, or surgical tools that require sterile conditions.

4.2 Pharmaceutical Packaging

Use in packaging for sensitive drugs, vaccines, or biologics, where barrier properties are critical.

4.3 Controlled Drug Release

Introduction of packaging systems that allow for controlled drug release or active ingredients for prolonged efficacy.

4.4 Medical Device Packaging

Packaging for implants, surgical instruments, diagnostic tools, and other medical devices often requires high strength, sterility, and protection from external environmental factors. Composites can provide these benefits, while maintaining ease of handling and transport.

4.5 Pharmaceutical Packaging

For drugs and biologics, polymer composites can be used to produce blister packs, vials, syringes, and other forms of packaging that require moisture and oxygen barrier properties to ensure product stability.

4.6 Sterilization Packaging

Some medical packaging needs to withstand processes like autoclaving (high-temperature steam sterilization). Polymer composites can be engineered for high-temperature resistance, ensuring that the package retains its integrity during sterilization procedures.

4.7 Drug Delivery Systems

In specialized applications like drug delivery patches or implants, polymer composites can be tailored to release drugs over time, or to encapsulate sensitive compounds, providing protection and controlled release.

4.8 Environmental Packaging

The growing interest in sustainability has led to the development of bio-based and biodegradable polymer composites, which offer an environmentally-friendly alternative to traditional packaging materials. These materials can decompose without causing harm to the environment, reducing the overall environmental impact of medical packaging.

5 Challenges and Considerations

Despite their many advantages, there are several challenges and considerations when using polymer composites in biomedical packaging:

- **Regulatory Approval**: Medical packaging is subject to stringent regulatory requirements, such as those from the Food and Drug Administration (FDA) and European Medicines Agency (EMA). Polymer composites must undergo extensive testing to meet these standards, particularly regarding biocompatibility, leachability, and sterility.
- **Cost and Manufacturing Complexity**: While polymer composites can offer excellent performance, they may be more expensive to produce compared to traditional materials. Advanced composites (e.g., those involving nanomaterials) may require specialized manufacturing techniques.
- **Environmental Concerns**: Although bio-based and biodegradable composites are emerging as sustainable options, many polymer composites are still derived from petrochemicals, contributing to environmental concerns related to their disposal and recycling.
- **Long-Term Durability**: Over time, polymer composites may degrade due to exposure to UV radiation, temperature fluctuations, or mechanical stress, potentially compromising their protective properties.

6 Future Directions

The following trends indicate that polymer composites in biomedical packaging have a bright future:

- **Smart Packaging:** By adding sensors or indications to polymer composite materials, real-time knowledge of the product's condition (such as moisture, temperature, or contaminants) may be obtained, enhancing reliability as well as security.
- **Nanotechnology:** Because nanoparticles provide improved qualities such improved barrier performance, antibacterial activity, and self-healing capabilities, their use in polymer composites is anticipated to grow.
- **Green Composites:** As sustainability gains traction, renewable resources and recyclable polymer composites are being developed, along with recycling options for pre-existing composite materials.
- **3D Printing:** In the packaging of medical devices, additive manufacturing methods like 3D printing are becoming more popular. 3D printing makes it simple to create bespoke polymer composite structures, which makes it possible to create packaging methods that are specifically suited for particular medical items.

7 Conclusion

Composite materials are regarded as fortunate modern materials because of their exceptional superior qualities and broad range of multipurpose packaging uses. There are numerous benefits of using composite materials in packaging applications, including reduced costs, enhanced biocompatibility, and increased life cycle. The polymer composite materials work well in pharmaceutical packaging, medical device packaging, controlled drug release pharmaceutical packaging, drug delivery systems, and environmental packaging, among other applications. Bio-composite materials have recently found application in the production of pharmaceuticals, medical equipment, and packaging for medical products. Polymer matrix composites are manufactured using a variety of contemporary techniques, including melt-extrusion, solution mixing, electrospinning, latex technology, and in situ processes. It is anticipated that the use of composite materials in medical applications will be inspiring, and new surgical instruments made of polymer composite materials will also have improved prosthesis stability. While these composite materials have several advanced qualities, like low inflammatory, biodegradable, biocompatible, and greater cell adhesion, they also have a number of disadvantages, including the release of acidic byproducts and poor cell affinity. In order to commercialize polymer composite materials, these disadvantages must be eliminated, and additional research on the design and development of polymer composite materials must be conducted.

Declaration of Competing Interest The authors declare that they have no known competing financial interests or personal relationships that could have appeared to influence the work reported in this paper.

Acknowledgements To whom the author is grateful for the required resources have been provided by Rico Auto Industries Ltd., Gurugram, Haryana, India. The authors are also grateful for the monthly research fellowship offered by the Ministry of Education and an outstanding research environment offered by the Indian Institute of Technology, Roorkee, Saharanpur Campus, Saharanpur, Uttar Pradesh, India.

References

1. Kamal A, Ashmawy M, Shanmugan S, Algazzar AM, Elsheikh AH (2022) Fabrication techniques of polymeric nanocomposites: a comprehensive review. Proc Inst Mech Eng Part C J Mech Eng Sci 236(9):4843–4861. https://doi.org/10.1177/09544062211055662
2. Biswal T, BadJena SK, Pradhan D (2020) Synthesis of polymer composite materials and their biomedical applications. Mater Today Proc 30:305–315. https://doi.org/10.1016/J.MATPR.2020.01.567
3. Ramakrishna S, Mayer J, Wintermantel E, Leong KW (2001) Biomedical applications of polymer-composite materials: a review. Compos Sci Technol 61(9):1189–1224. https://doi.org/10.1016/S0266-3538(00)00241-4
4. Gopanna A, Rajan KP, Thomas SP, Chavali M (2019) Polyethylene and polypropylene matrix composites for biomedical applications. Mater Biomed Eng Thermoset Thermoplast Polym 175–216. https://doi.org/10.1016/B978-0-12-816874-5.00006-2
5. Udayakumar GP et al (2021) Biopolymers and composites: Properties, characterization and their applications in food, medical and pharmaceutical industries. J Environ Chem Eng 9(4):105322. https://doi.org/10.1016/J.JECE.2021.105322
6. Zhang M et al (2022) Recent advances in polymers and polymer composites for food packaging. Mater Today 53:134–161. https://doi.org/10.1016/J.MATTOD.2022.01.022
7. Omerović N et al (2021) Antimicrobial nanoparticles and biodegradable polymer composites for active food packaging applications. Compr Rev Food Sci Food Saf 20(3):2428–2454. https://doi.org/10.1111/1541-4337.12727
8. Valente R (2021) Design methods for composite materials: preface. Encycl Mater Compos 3:1. https://doi.org/10.1016/B978-0-12-819724-0.00111-7
9. Priyaranjan Samal MMM, Pandu RV, Arabinda M et al (2025) Progress in research on hybrid metal matrix composites. J Alloys Compd 838:131–152. https://doi.org/10.1016/j.jmapro.2020.09.010
10. Lechkov M, Prandzheva S (2010) Encyclopedia of polymer composites : properties, performance and applications. 1245
11. Starr TF, Starr M (1998) Thermoset resins for composites : directory & databook, p 339
12. Leblanc JL (2010) Filled polymers : science and industrial applications, p 428
13. Fu SY, Lauke B, Mai YW (2009) Science and engineering of short fibre reinforced polymer composites. Sci Eng Short Fibre Reinf Polym Compos 1–338 https://doi.org/10.1533/9781845696498
14. Rothon RN (2003) Particulate-filled polymer composites, p 544
15. Agarwal BD, Broutman LJ, Chandrashekhara K (2025) Analysis and performance of fiber composites, p 562. [Online]. Available: https://books.google.com/books/about/Analysis_and_Performance_of_Fiber_Compos.html?id=T1CtwL9uK5MC. Accessed on April 20, 2025

16. Nanocomposite Science and Technology (2003) Nanocomposite Sci Technol. https://doi.org/10.1002/3527602127
17. Sanjay MR et al (2016) Applications of natural fibers and its composites: an overview. Nat Resour 7(3):108–114. https://doi.org/10.4236/NR.2016.73011
18. Maurya AK, Manik G (2023) Advances towards development of industrially relevant short natural fiber reinforced and hybridized polypropylene composites for various industrial applications: a review. J Polym Res 30(1):1–20. https://doi.org/10.1007/S10965-022-03413-8
19. Gascons M, Blanco N, Matthys K (2012) Evolution of manufacturing processes for fiber-reinforced thermoset tanks, vessels, and silos: a review. https://doi.org/10.1080/0740817X.2011.590177
20. Shojaei A, Ghaffarian SR, Karimian SMH (2003) Modeling and simulation approaches in the resin transfer molding process: a review. John Wiley and Sons Ltd. https://doi.org/10.1002/pc.10050
21. Cantor KM, Watts P (2011) Plastics Processing. In: Applied plastics engineering handbook: processing and materials, Elsevier, pp 195–203. https://doi.org/10.1016/B978-1-4377-3514-7.10012-1
22. Jadhav A, Jadhav VS (2022) A review on 3D printing: an additive manufacturing technology. In: Materials today: proceedings. Elsevier Ltd, pp 2094–2099. https://doi.org/10.1016/j.matpr.2022.02.558
23. Sapuan SM, Nukman Y, Abu Osman NA, Ilyas RA (2020) Composites in biomedical applications, p 302

Chapter 6
Self-healing Composites in Packaging

Pawan Kumar Pandey

1 Introduction to Self-healing Packaging

1.1 Overview and Definition of Self-healing Materials

Self-healing materials can be defined as an emerging type of smart materials with extended maintaining functionality because of having ability to repair the damage autonomously. These materials are inspired by biological healing system, and they can trigger a repairing mechanism by responding against an internal or external stimuli, without any type of human interference [1].

In packaging context, these materials are helpful in addressing the common issues (e.g., barrier degradation, punctures, and cracks), which are reducing the packaging product integrity. On the basis of their healing mechanism, which is intrinsic (external healing agents are not required) or extrinsic (external healing agents are required) approach, these materials can be categorized [2]. Intrinsic self-healing materials depends on intrinsic molecular rearrangements as well as reversible chemical bonds, while extrinsic self-healing materials utilize several types of embedded healing agents (e.g., vascular networks or microcapsules) who release different types of repairing substances after get damage. The durability and efficiency are enhanced because of this self-healing ability which ultimately improves the sustainability and reduce the waste [3].

P. K. Pandey (✉)
Department of Polymer and Process Engineering, IIT Roorkee, Roorkee, India
e-mail: pk_pandey@pe.iitr.ac.in

A. Verma et al. (eds.), *Polymers and Composite Materials for Packaging: Smart Food Packaging and Solutions*, Materials Horizons: From Nature to Nanomaterials,
https://doi.org/10.1007/978-981-95-0677-4_6

1.2 Benefits, Importance, and Applications in Packaging

1.2.1 Benefits of Self-healing Packaging

The ability of self-healing materials to enhance the product protection, increase shelf life, and improve sustainability make them valuable in packaging stems. Packaging materials serves as the first line barrier between the product and external environment by preventing the mechanical damage, moisture ingress, and contamination. Although the conventional packaging materials are susceptible to the deterioration, which could result in product degradation, additional waste, and financial losses [4].

1.2.2 Importance of Self-healing Packaging

Self-healing materials maintain the product integrity, extend their self-life, and reduce the waste production; thus, offer significant advantages in packaging materials. They help in preventing contaminations and spoilage, ensure quality of the packaged items for a relatively longer duration. In addition, these materials minimize the need for replacements and lowers the packaging waste, ultimately contribute to the environmental sustainability. Although there is a higher initial investment, the self-healing materials emerge as a promising solution for efficient and sustainable materials due to their long-term cost-effectiveness [5].

1.2.3 Applications of Self-healing Packaging

The self-healing materials can be integrated with packaging industries across the various sectors to improve the functionality and durability. In addition, these materials not only incorporate several features like thermosensitive and antimicrobial features; but also makes a revolution in conventional packaging industry by improving safety, environmental sustainability, and resilience. In electronics sector, packaging with these materials protects the sensitive components from environmental factors and mechanical stress. In food packaging, these materials prevent the oxygen and moisture infiltration, preserve freshness and reduce spoilage. In medical and pharmaceutical packaging, these materials improve integrity and sterility by repairing the microdamage [1].

1.3 Objectives of the Chapter

In this chapter, the overview of packaging by self-healing materials along with fundamental concepts, mechanism, and applications are discussed. It also explains their importance in packaging, working principles, and advantages in improving packaging performance. In addition, the key barriers in their development and successful commercialization with current trends and future prospects are also discussed. Finally, reader will able to understand their role, benefits and challenges in packaging industry.

1.4 Current Challenges in Packaging Materials

Although there are several advantages and applications of packaging materials in various industries, but there are several challenges also need to be resolved for its commercial applications. It is important to explore these materials in various industries with full potential by industrial collaboration, material development, and innovative research to overcome these challenges, [6], like.

1.4.1 Packaging Material Degradation

In case of conventional packaging materials, the degradation rate is faster due to various factors like mechanical stress, environmental exposure, and chemical interactions, which leads to the lesser protective efficiency, higher replacement cost, and increased waste production.

1.4.2 Environmental Impact and Sustainability

The environmental pollution, caused by some conventional packaging materials like plastic is a global concern. The replacement of these pollutants with self-healing recyclable or biodegradable materials is necessary for not only promoting sustainability, but also reducing the waste production.

1.4.3 Barrier Performance and Integrity Maintenance

The maintenance of higher barrier properties to protect against contaminants, moisture, and oxygen is must for a packaging material. A small damage like cracks can reduce the barrier ability of the packaging material, leading to reduce the self-life and spoilage.

1.4.4 Cost and Scalability

The self-healing functionalities can be integrated and developed into packaging materials by using some advanced manufacturing techniques with significant initial cost.

Due to the complexity of these materials, scaling-up their production for commercial applications is still a challenge.

1.4.5 Regulatory and Safety Considerations

The safety and toxicity of the packaging materials are the important parameters that must be considered specially for the pharmaceutical and food industries. These parameters ensure that the developed and integrated self-healing materials are biocompatible and do not cause any adverse effect on healthy individuals.

2 Fundamental Mechanisms of Self-healing

The basic mechanisms of self-healing materials cover a wide range of healing applications, from autonomic to non-autonomic and from intrinsic to extrinsic. Several factors like material property, activation conditions, and application determine the selection of a particular self-healing strategy. The in-depth understanding of these mechanisms will help in designing advanced self-healing packaging that would elevate durability, sustainability, and functionality for cutting-edge industries. Self-healing materials have an inherent property to repair itself after any damage, thus they can not only restore their function, but also enhance the sustainability and durability of packaging materials autonomously. This self-healing mechanism can be categorized on the basis of their repair efficiency, energy requirement, and activation mode [7].

2.1 Intrinsic Versus Extrinsic Healing Mechanisms

Based on the requirement of any external healing agent, the self-healing mechanism can be divided into intrinsic and extrinsic healing mechanism [8].

2.1.1 Intrinsic Healing Mechanisms

In this self-healing mechanism, these materials do not require any type of external healing agent. Their inherent structural or molecular property enables the self-repairing by some reversible reactions (e.g., supramolecular interactions, covalent or hydrogen-bonding) to reform the broken bonds and restore the integrity of materials. Reversible polymer networks, shape memory polymer, and ionic interactions are some common intrinsic healing mechanisms.

2.1.2 Extrinsic Healing Mechanisms

Unlike the intrinsic one, extrinsic self-healing materials require some external healing agents (e.g., embedded microcapsules or vascular networks), to release a repairing material after damage. This mechanism includes microcapsule-based healing systems, vascular healing systems, and phase-separated healing systems.

2.2 Autonomic Versus Non-autonomic Healing

Self-healing system can also be classified into automatic and non-automatic system on the basis on requirement of an external trigger or intervention. Although, unlike the automatic one, external intervention is required in non-automatic healing system, and it offers a controlled healing facility, which is highly demanding in special type of packaging applications [9].

2.2.1 Autonomic Healing

These healing systems do not require any type of external trigger or intervention and they can detect the damage and repair it. These materials generally incorporate capsule-based systems, self-healing polymers, and vascular networks. This system ensures the improved lifespan of materials without any manual repair and considered as an ideal system for packaging of those materials who face repeated stress.

2.2.2 Non-autonomic Healing

Unlike the automatic one, this system requires an external activator to detect and repair the damage. The activator may be physical (e.g., Light, Heat, or Moisture) or chemical (e.g., solvents).

2.3 Key Factors Influencing Self-healing Efficiency

There are several key factors influencing the efficiency of self-healing materials [10] including.

2.3.1 Material Composition and Chemistry

The material composition (e.g., thermoplastic, thermoset, biodegradable or elastomeric etc.) and material chemistry influence the healing ability of self-healing materials. In case of intrinsic healing, the self-repair is facilitated by the dynamic covalent bonding or supramolecular interactions due to the presence of functional groups with the ability of reversible bonding. While, the effectiveness of extrinsic healing is tuned by compatibility between polymer matrix and encapsulated healing agents, which ensures the proper release and post-damage interaction to restore the material integrity.

2.3.2 Healing Activation Conditions

Various healing activation conditions like environmental factors (e.g., humidity, pH, and temperature) controls the healing efficiency of these materials. The activation of light-responsive self-healing materials need light of an appropriate wavelength, while the thermally responsive self-healing materials need a standard heat exposure to initiate and formation of cross-links.

2.3.3 Damage Size and Type

Self-healing materials are more effective for repairing microscopic cracks than large-scale fractures, because repairing of smaller cracks is more efficient by extrinsic or intrinsic mechanisms. Since vascular healing systems are able to continuous healing by interconnected channels, which release the healing agents after damage; they can be heal the larger-scale damage more efficiently than microcapsule-based systems.

2.3.4 Healing Kinetics and Repeatability

Some self-healing materials (e.g., microcapsule-based systems) are of single use because of the depletion of healing agent after one cycle, while others can perform multiple healing cycles. The healing speed of these materials is very important specially in food packaging, where quick restoration of barrier properties is necessary to ensure product safety and prevent food contamination.

2.4 Stimuli-Responsive Healing Systems

To initiate the repair system, the external triggers are utilized by stimuli-responsive self-healing systems and these materials are able to control the healing based on some specific environmental conditions [11] including.

2.4.1 Thermally Responsive Healing

At specific temperatures, some thermos-responsive self-healing polymers get activated, softened, and upon heating, reform bonds, while the shape memory polymers utilize heat to recover their original structure, making them suitable candidates in temperature-sensitive packaging applications.

2.4.2 Light-Activated Healing

Ultraviolet or visible light triggers healing in polymer networks that respond photochemically. They are therefore widely used for coatings and protective films in packaging materials. This enables an accurate mode of control over the healing process without mechanical intervention; hence, durability and function of these materials are enhanced.

2.4.3 Moisture-Activated Healing

Since some bio-based self-healing polymers reestablish their hydrogen bonds on exposure to humidity or moisture, they are particularly suitable for food packaging applications typically exposed to moisture in the environment.

2.4.4 pH and Chemical-Responsive Healing

Some self-healing materials can change their chemical structure responding to changes in pH or chemicals, and they are, therefore, very effective in applications dealing with smart packaging that reacts to indicators of spoilage or external contamination.

2.4.5 Electric and Magnetic Field-Responsive Healing

Conductive self-healing polymers get triggered under electric or magnetic fields, rendering them valuable for applications in electronic packaging to ensure connectivity and protection against damages.

3 Types of Self-healing Composites for Packaging

The packaging industry is being completely transformed by self-healing composites as they increase toughness, enhance shelf life, and reduce the material waste. These materials can autonomically or manually repair damage such as cracks and microfractures and continue to functionally protect the packaged goods. There are many different kinds of self-healing mechanisms, each developed with various approaches and materials. The main self-healing approaches relevant to packaging applications include microencapsulation-based healing, vascular network healing, reversible polymer networks, and shape memory/thermally responsive systems [1].

3.1 Microencapsulation-Based Healing

Microencapsulation is one of the most popular techniques for self-healing composites. It includes microencapsulation of healing agent-filled microcapsule structures like monomers, resins, or adhesives embedded within the composite matrix. The damage to the encapsulation material causes rupture of microspheres and consequent release of the healing agent into the crack or damaged zone. The healing agent then undergoes either polymerization or cross-linking by exposure to air, moisture, or a catalyst incorporated in the matrix, sealing the damaged region.

Advantages: There are several advantages of microencapsulation-based healing. These include self-healing without external intervention which is very efficient for self-repairing materials. Economical because it can easily be integrated into traditional polymer-based packaging without requiring much modification. Besides that, the approach is highly versatile, as it can be incorporated into a wide range of polymer matrices, both biodegradable and synthetic, thus enhanced applicability in varied packaging solutions.

Challenges: Microencapsulation-based healing systems face many challenges, one of which is limited healing cycles, as healing once stops when microcapsules deplete. It is also necessary to ensure dispersion control for uniformly healing properties throughout the material. Encapsulation stability further becomes important because microcapsules must remain intact and functional during manufacturing and storage processes to retain effectiveness.

3.2 Vascular Network Healing

As a mimic for the natural healing of the biological circulatory system, the vascular healing network integrates the microchannels or capillary networks into the composite material. These networks act by storing and transporting healing agents to damaged areas, enabling multiple healing cycles. When a crack begins to form, it intersects with the vascular network, thereby allowing the healing agent to flow into the damaged site for repairing its process.

Advantages: Healing with a vascular network has a lot of advantages over microencapsulation. E.g., healing agents can be continuously supplied for more than one healing cycle, thus allowing for repeated self-repair. This provides for the efficient distribution of healing on damaged areas and enhances the durability of the materials. Furthermore, the vascular networks are considered highly customizable and can be designed with various architectures including interconnected channels as well as parallel channels to maximize healing efficiency for specific applications.

Challenges: Due to the requirement of very advanced fabrication techniques such as 3D printing or sacrificial templating, the complex manufacturing of precise vascular networks within polymer matrices has become one of the challenges for the healing vascular networks. Other than that, leakage can be a risk. The network should otherwise remain sealed, with the healing agent released only when the matrix is damaged. Both the cost of material and production needs to be weighed against the advantages of superior shelf life by their incorporation.

3.3 Reversible Polymer Networks

Reversible polymer networks, which are characterized by dynamic covalent bonds or supramolecular interactions, can be broken and reformed in response to external stimuli such as heat, light, or pH changes, allowing healing of damage to be done autonomously or manually by restoring the original molecular structure. They include Diels–Alder polymers (operate through reversible Diels–Alder reactions, whereby the bond will be broken by heat and reformed upon cooling, allowing for repeated healing); hydrogen-bonding networks (that are based on non-covalent interactions that could dynamically form and break, thus allowing for reversible self-healing); ionomer-based polymers (containing ionic bonds that rearrange when damaged to afford self-repair when subjected to suitable environment conditions).

Advantages of Reversible Polymer Networks: Reversible polymer networks have several advantages, such as healing multiple times without external healing agents, making them much more effective from a long-term perspective. They often find compatibility with sustainable and biodegradable materials for packing purposes, which increases the scope for recyclability and environmental credit. For some reversible systems, healing occurs at an extremely fast rate, which is very suitable for applications that require fast interventions after damage.

Challenges: Reversible polymer networks face several challenges in terms of their dependency on external stimuli like heat, light, or changes in pH, which may limit their applicability. In certain cases, whereas the reversible network can restore self-healing properties, the mechanical strength and barrier properties of these materials may lower than conventional polymers. Hence, a trade-off exists between self-healing capability versus structural integrity. The other complications lie in maintaining a fine balance between healing efficiency and mechanical performance, which requires an accurate object-molecular design, making the entire process of design and implementation very complicated.

3.4 Shape Memory and Thermally Responsive Systems

Shape memory polymers (SMPs) and thermally responsive systems exhibit self-healing properties by recovering their original shape when exposed to specific thermal conditions. These materials are designed with a "programmed" shape that can be temporarily deformed and, upon activation by heat, return to their initial structure, effectively closing cracks and restoring functionality. The self-healing mechanisms of SMPs include thermal activation, where heat-induced phase transitions restore the polymer structure; light-activated healing, where certain SMPs respond to infrared or UV light to trigger shape recovery; and hydrogel-based memory systems, which utilize moisture and temperature changes to facilitate healing, particularly in biodegradable packaging applications.

Advantages: One of the many advantages of shape memory and thermally responsive systems is that they are simple: they do not require additional healing agents or complex microstructures. They also contribute toward enhanced sustainability because of their integration into bio-based and recyclable packaging materials. These systems promise a fast response, with some shape memory polymers (SMPs) demonstrating almost instantaneous healing at activation.

Challenges: There are various challenges that exist for technologies that exhibit both shape memory and thermal responsiveness that restrict the practical applications of these technologies. Their temperature sensitivity could be a drawback, significantly affecting the cost of use in packaging applications for heat-sensitive food products, in cases where no external heat activation can be applied. Furthermore, fatigue resistance also presents a challenge as they are subjected to repeated cycles of shape recovery, which result in performance degradation over time. Processing these materials is also complex since it would require specific conditions before effects of shape memory can be achieved, which makes scalability difficult.

4 Materials and Design Strategies for Self-healing Packaging

The preparation of self-healing packaging materials is likely to involve the careful choice of base materials, healing agents, and possible reinforcements. The efficiency, longevity, and sustainability of the self-healing mechanism will at any time hinge on these components. The contents of the subsequent sections will describe various materials and design strategies for self-healing packaging applications [12].

4.1 Polymer Matrices (Thermoplastic, Thermoset, Elastomeric, Biodegradable)

The polymer matrix is the primary structure on which the self-healing packaging is built. Different polymers have various advantages in flexibility, durability, and self-healing properties.

4.1.1 Thermoplastic-Based Self-healing Systems

Thermoplastics are the polymers which on heating soften and become fully rigid on cooling, thus providing a reversible self-healing mechanism. Some examples are polyethylene (PE), polypropylene (PP), and polyvinyl chloride (PVC). The healing of these polymers is generally enhanced with reversible bonds, heating-responsive healing, or embedded healing agents. Due to lightness and processing simplicity, these polymers find applications in food and pharmaceutical packaging.

4.1.2 Thermoset-Based Self-healing Systems

They are known as thermosets, resins like epoxy and polyurethane that cure irreversibly, providing mechanical strength and durability. Most self-healing thermosets utilize either encapsulated healing agents or dynamic covalent bonds as their mechanism. These materials are ideally suited for high-performance packaging application, which requires structural integrity.

4.1.3 Elastomeric Self-healing Materials

Elastomers such as silicone and natural rubber are demonstrated to be extremely elastic and flexible to a large extent, which fits into their applicability in self-healing applications with resilience requirements. The self-healing function is designed based on dynamic bonds, apart from the intrinsic reorientation of polymer chains. Such materials are used as flexible packaging where the characteristics of resistance to impact and elongation are required.

4.1.4 Bio-based and Biodegradable Self-healing Polymers

The environment-friendly biodegradable self-healing polymers, such as polylactic acid (PLA), polyhydroxyalkanoates (PHA), and starch-based materials, are gaining importance in research because of sustainability issues. These polymers naturally degrade and exhibit self-healing properties through either reversible hydrogen-bonding or microencapsulated healing agents. Such materials are expected to play a vital role in green packaging applications.

4.2 Healing Agents (Encapsulated, Microvascular, Dynamic Bonds)

From the moment of initiation, healing agents are integral to the self-healing process in enhancing the restoration of materials upon damage. These agents can be incorporated into the polymer matrix by means of different strategies.

4.2.1 Encapsulated Healing Agents

Microencapsulation is a method of inserting microcapsules containing healing agents such as polymerizable monomers or adhesives within the packaging material. When damage occurs, the capsules break open to release the healing agents to seal the cracks. This approach has applications where healing occurs only once and is mostly employed with thermoset polymers.

4.2.2 Microvascular Network Systems

Acting like biological vascular systems, microvascular networks permit the introduction of interconnected capillary-like channels containing healing agents. The microvascular networks assure ever-continuing healing by supplying fresh healing agents on demand during damage. Thus, this work strengthens durability and extends the service life of the material for use in long-term packaging applications.

4.2.3 Dynamic Bonding Mechanisms

Polymers have self-healing properties due to dynamic covalent bonds, hydrogen bonds, and reversible supramolecular interactions. These bonds enable the material to reform and regain its full structural integrity without the need for any external healing agents. A few examples include repeated Diels–Alder reactions and disulfide bonds for self-healing in smart packaging materials.

4.3 Nanocomposites and Smart Additives

Nanomaterials and smart additives contribute to self-healing effectiveness by improving mechanical resistance, enhancing barrier properties, and making the material responsive to stimuli from the environment.

4.3.1 Nanoparticle-Enhanced Self-healing

Various types of nanoparticles like glass, graphene oxides, and carbon nanotubes strengthen polymer matrices, resulting in improved mechanical strength and crack closure. They provide electrical and thermal responsiveness for advanced packaging applications as well.

4.3.2 Smart Additives for Stimuli-Responsive Healing

Smart additives such as shape memory polymers, phase change materials, and conductive fillers impart adaptive ability of self-healing to packaging materials. Thermo- responsive additives can activate the healing process when exposed to heat, while moisture-responsive additives can allow the repair to be self-healed under a humid environment.

4.4 Fiber-Reinforced and Multi-phase Self-healing Composites

Reinforcements by fibers and multi-phase composite designs improve their self-healing properties while ensuring the mechanical robustness of the materials.

4.4.1 Fiber-Reinforced Self-healing Composites

By incorporating fibers like aramid, glass, or carbon into the polymer matrix, the structure obtains additional integrity and load transfer is accomplished in a better way. These fibers may also be functionalized with healing agents so as to provide localized self-healing within the packaging materials.

4.4.2 Multi-phase Self-healing Composites

In order to consolidate different healing mechanisms, multi-phase composites may combine encapsulated healing agents with dynamic bonding systems. Such hybrid designs offer reliability and performance of self-healing materials at an enhanced level in diverse packaging environments.

5 Manufacturing Techniques for Self-healing Packaging

The manufacturing techniques like extrusion, injection molding, coating, and 3D printing help bring life to self-healing packaging. Each offers different advantages for the incorporation of healing agents, durability, and sustainability. When the technologies of self-healing packaging take steps toward advancement, it will be important to optimize these manufacturing methods for their commercial scale-up [13].

5.1 *Extrusion and Injection Molding*

Extrusion and injection molding are flexible methods used in the packaging industry for the manufacture of polymer-based packaging materials. Such techniques allow for embedding healing agents, microcapsules, or dynamic polymer networks for the self-healing functionality of packaging materials.

5.1.1 Extrusion Process

Extrusion refers to a continuous manufacturing method that melts a polymeric material and forces it through a shaped die to produce films, sheets, or tubes. Self-healing property imparts can enter through this avenue:

Blending Self-healing Agents with Polymers

Microcapsules with healing agents or reversible cross-linking polymer networks can be uniformly dispersed in a polymeric melt.

Layered Extrusion

A multilayer extrusion process can embed a self-healing layer within a conventional packaging film, maintaining the mechanical integrity of the film while allowing healing efficiency.

Co-extrusion

It is a process wherein different layers of polymers with different properties are extruded at the same time to give advanced packaging materials with enhanced healing properties.

5.1.2 Injection Molding Process

Injection molding entails a process of melting the polymer pellets and injecting the molten polymer into the cavity of the mold, where it cools and solidifies into the desired shape. This process is widely used for rigid packaging components such as bottles, containers, and caps. Self-healing characteristics may be introduced into injection-molded packaging.

Embedding Microcapsules or Vascular Networks

At the time of molding, healing microcapsules with monomers or catalysts are mixed uniformly within the polymer matrix. When the packaging is damaged, the microcapsules break open, permitting the healing agent to flow into the crack and actuate the repair.

Thermally Reversible Polymers

Injection molding can process the self-healing polymer network based on reversible covalent bonds or supramolecular interactions for heat-induced damage recovery.

Using Nanocomposites

Functional nanomaterials such as carbon nanotubes or graphene can enhance the mechanical strength of injection-molded composites and provide self-repair upon applied mechanical stress or environmental triggers.

5.2 *Coating and Layered Approaches*

Coating and layered strategies in self-healing systems have proved their effectiveness in providing packaging surfaces with extended life as well as plausible barrier functionalities.

5.2.1 Self-healing Coatings

Thinly applying self-healing coatings to standard packaging systems equips them with protection and restoring functions. Coating formulations include the following.

Microcapsule-Based Coatings

Microcapsules filled with healing agents (e.g., polymeric monomers or reactive fluids) are dispersed in a coating matrix. Upon mechanical damage, the capsules rupture, releasing their contents to seal the cracks.

Polymer-Based Dynamic Coatings

Coatings made from reversible covalent or supramolecular polymer networks can autonomously repair minor scratches and abrasions.

Nanostructured Self-healing Coatings

Incorporation of nanostructure with, for instance, silica or titanium dioxide nanoparticles improves the scratch resistance property of coatings, while adding to their self-healing performance.

5.2.2 Multilayer Packaging Structures

Layered approaches are transitions that integrate self-healing functionality components in a packaging structure. Main strategies are as follows.

Core–Shell Multilayer Films

The self-healing material is placed between layers of conventional polymers, hence providing the necessary protection while still allowing for repair.

Responsive Barrier Layers

Inner thin films could be engineered into the active layer restoration of moisture or gas barrier properties after damage events to stop subsequent contamination and spoilage.

Self-healing Sheets Laminated

Adhesive layers of self-healing polymer can be laminated onto the packaging films, which can also further increase the lifespan of the films.

5.3 3D Printing Self-healing Packaging Materials

An advanced technique for designing complex structures is 3D printing, which creates possibility of adapting self-healing packaging materials entirely to the preference of the user.

5.3.1 Types of 3D Printing for Self-healing Packaging

The types of 3D printing techniques used include:

Fused Deposition Modeling (FDM)

Self-healing materials contained in thermoplastic filaments are layered one by one to develop strong packaging components through the extrusion process.

Stereolithography (SLA)

Light-sensitive resins with self-healing properties undergo controlled photopolymerization to form detailed packaging structures.

Direct Ink Writing (DIW)

Specialized nozzles print self-healing hydrogels and elastomers, thus rendering them flexible and adaptive in their design as packaging.

5.3.2 Integration of Self-healing Mechanisms in 3D Printing

Using Self-healing Polymers

Various types of specific polymers like, shape memory polymers, hydrogen-bonded networks, or Diels–Alder-based reversible polymers can be used in 3D printing to make an autonomous repair packaging material.

Embedding Vascular Networks

Microfluidic vascular networks can be printed within packaging structures to transport healing agents to damaged regions.

Customizable Healing Properties

By adjusting print parameters, self-healing packaging can be tailored for specific applications, such as moisture-sensitive food containers or impact-resistant pharmaceutical packaging.

5.3.3 Benefits of 3D Print for Self-healing Packaging

Design Flexibility

It enables the creation of complex geometries with self-healing functions that are typically designed for a purpose.

Efficient in Materials Consumption

This is by eliminating the wastes since the amount of material used is exactly the needed amount.

Rapid Modeling

New designs of self-healing packages can be fast prototyped to facilitate testing and validation.

6 Functional and Intelligent Aspects of Self-healing Packaging

Self-healing packaging tends not merely to repair damage but seeks to improve the functioning and so-called "smartness" properties of the packaging material concerned under various conditions to prolong the life, improve quality, and enhance safety. Such features ensure that the packaging materials provide inside integrity, effective preservation of contents, and dynamic response to the condition of their environment. This chapter reviews major functionalities of self-healing packaging like: oxygen scavenging, moisture regulating, antimicrobial properties, and temperature-responsive cold chain applications [2].

6.1 Oxygen Scavenging and Barrier Properties

The primary sources of food spoilage, the oxidation of products-sensitive properties, and degradation are exposure to oxygen. Self-healing packaging, which has been produced to scavenge oxygen, maintains quality on the active removal or blockage of oxygen ingress. Active components incorporated in such systems include:

6.1.1 Oxygen-Absorbing Additives

Iron-based compounds as well as organic oxygen scavengers react with oxygen to reduce its presence in the packaging's atmosphere.

6.1.2 Nano-Composite Barrier Layers

They add graphene oxide, nanoclay, and other nano-fillers to improve the oxygen-blocking ability of the packaging while still maintaining mechanical flexibility.

6.1.3 Self-repairing Barrier Coatings

Polymeric films embedded with microcapsules containing reactive agents that seal cracks or punctures, thereby restoring the material's gas barrier properties.

6.2 Moisture Regulation and Humidity Control

It is the appropriate moisture levels that are kept during the storage of perishable materials to prevent growth or development of microbes in them along with the fact they brought about deterioration in the properties of materials. Self-healing moisture-regulating packaging materials are designed either to absorb or repel moisture depending upon the requirements of the packaged product. Some key approaches are.

6.2.1 Hydrophilic or Hydrophobic Polymer Matrices

Some self-healing polymers develop their affinity to water that would prevent over moisture accumulation or dehydration from packaged goods.

6.2.2 Hygroscopic Fillers and Coatings

Silica gel, zeolites, or biopolymer composites can act as functions for humidity regulation within the packaging ecosystem.

6.2.3 Self-repairing Water Barrier Films

Damage in the packaging that can leak into moisture is sealed with stimuli-responsive materials, thus maintaining the required concentration of humidity. Such advancements suitably serve food, pharmaceutical, and electronic packaging areas with precise moisture control to ensure stability and performance.

6.3 Antimicrobial and Food Preservation Capabilities

The major concern of food packaging is microbial and thus spoilage and associated health risks. But the incorporation of antimicrobial agents with self-healing packaging provides an updated line of defense against bacterial and fungal growth. These systems typically operate through.

6.3.1 Incorporation of Antimicrobial Nanoparticles

Antimicrobial properties inherent to silver (Ag), copper (Cu) and zinc oxide (ZnO) nanoparticles allow them to be implemented as part of the self-healing film to prevent from being contaminated.

6.3.2 Essential Oils and Others Bioactive Agents

Natural antimicrobials like thyme oil, cinnamon extract, and chitosan would be released when the pack is damaged to eliminate microbial growth.

6.3.3 pH- or Enzyme-Triggered Antimicrobial Release

Smart self-healing packaging could detect a change in pH due to microbial activity and respond by releasing antimicrobial agents.

6.4 Temperature-Responsive Healing for Cold Chain Applications

Cold chain logistics require damage-resistant packaging solutions that can endure fluctuations in temperature and mechanical stress. Temperature-responsive self-healing packaging is engineered within its microcracks or punctures that are likely to take place during transit; thus, an insulation and protection function is maintained. Basic mechanisms include.

6.4.1 Thermally Responsive Polymers

Shape memory polymers and phase-change materials (PCMs) can mend damage within certain temperature ranges, thus guaranteeing package integrity in frozen or chilled conditions.

6.4.2 Microencapsulated Healing Agents

These release self-healing chemicals when activated by temperature changes, thus sealing cracks and preventing exposure to external contaminants.

6.4.3 Flexible Bio-based Films

Biodegradable self-healing materials designed for cold chain applications, thus having value for both sustainability and performance, can respond to the environmental problem they tackle while being robust.

7 Performance Evaluation and Characterization

Performance evaluation of self-healing composites in packaging is very necessary for understanding their effectiveness, durability, and overall functioning. This section provides an in-depth analysis of the essential evaluation methods such as mechanical testing, healing efficiency evaluation, recovering barrier properties, environmental stability, and long-term aging studies. Each subsection describes the methods and importance of these tests to ensure the reliability of self-healing packaging materials [14].

7.1 Mechanical Properties and Durability Tests

Mechanical properties of self-healing composites are fundamental to the application in packaging. Various testing methods assess the strength, toughness, and ability to withstand external forces before and after healing. Tests include.

7.1.1 Tensile Testing

Measures strength, elasticity, and elongation at break of the material. Self-healing packaging should retain its tensile properties after multiple healing occurrences.

7.1.2 Flexural and Compression Testing

Measures the material's resistance to bending and compressive forces which provide an important factor for rigid and semi-rigid packaging applications.

7.1.3 Impact Resistance Testing

Evaluates the composite's ability to absorb energy from sudden impacts like drops or mechanical shocks during transportation.

7.1.4 Fatigue Testing

Determines long-term durability under cyclic loading conditions to test how far the material can maintain performance over repeated stress cycles.

7.2 Healing Efficiency and Recovery Rate

Healing efficiency refers to the extent to which a self-healing material can actually regain its original properties after being damaged, and healing efficiency is therein one yardstick for measuring self-healing materials. Typically, this efficiency is expressed as a percentage representing recovered mechanical performance. Major evaluation methodologies include.

7.2.1 Healing Time Evaluation

Evaluates the time vital for the material to restore mechanical integrity.

7.2.2 Stress–Strain Recovery

Involves comparing pre-and post-damage tensile strengths to evaluate the recovery efficiency.

7.2.3 Fracture Surface Analysis

Includes the application of techniques such as scanning electron microscopy (SEM) to observe crack closure and material reformation.

7.2.4 Repeated Healing Cycles

Tests how many healing cycles the material can undergo before degradation in performance occurs.

7.3 Barrier Property Recovery Evaluation

Barrier properties have their utmost importance in packaging, especially with regard to foods and pharmaceuticals where another layer of protection is provided against oxygen, moisture, and other contaminants. Recovery of these properties after some damage is evaluated by the following.

7.3.1 Oxygen Transmission Rate (OTR)

Measures the permeability of the material to oxygen as a function of time before and after healing.

7.3.2 Water Vapor Transmission Rate (WVTR)

Measures the moisture ingress resistance of the material, which is crucial for humidity-sensitive products.

7.3.3 Gas Permeability Testing

Diffusion rates of various gases such as carbon dioxide and nitrogen will be examined to check whether the protected capabilities still reside in the healed material.

7.3.4 Chemical Resistance Testing

A test to assess whether the healed composite resist degradation upon contact with oils, acids, or solvents usually encountered in packaging applications.

7.4 *Environmental and Chemical Stability*

Packaging of self-healing composites requires the ability to resist environmental conditions, including temperature variation, humidity, and chemical handling. In stability testing.

7.4.1 Thermal Stability Tests

Differentiated by thermal analysis based on DSC and TGA, the tests determine material ability to hold its properties under varying temperature limits.

7.4.2 UV Aging Tests

The tests are focusing upon the prolonged exposure of samples to UV radiation to evaluate photodegredation resistance.

7.4.3 Resistance to Chemical Exposure

Determining the effect of food acids, alkalis, and preservatives on the self-healing capabilities of the material.

7.4.4 Hygroscopic Behavior

Assessing water absorption and swelling effects that could deteriorate structural integrity in the long run.

7.5 *Long-Term Durability and Aging Studies*

The long-term viability of self-healing composites is fundamental for market demands. The aging studies are set to simulate long-term exposure to real-world conditions to assess.

7.5.1 The Accelerated Aging Tests

High temperature and high humidity are utilized to predict the lifespan.

7.5.2 Long-Term Recovery

Whether materials sustain healing efficiency under a large number of damage and repair cycles.

7.5.3 Storage and Shelf-Life Studies

Assesses self-healing performance upon long-term storage in diverse environmental settings.

7.5.4 Microstructural Changes

Through microscopic and spectroscopic techniques, it is anticipated that the analysis will give understandings on structural degradation and rearrangement of polymer chain over time.

8 Applications of Self-healing Packaging

The self-healing packaging concept is being developed as a type of smart packaging that will help to improve the durability, functioning, and sustainability across various industries. Integrating self-healing properties into packaging enhances product shelf life and protection while reducing material waste. A detailed look into the applications of self-healing packaging-food packaging, pharmaceutical packaging, medical packaging, and electronic packaging are the ones of interest [15].

8.1 Food Packaging (Fresh Produce, Meat, Dairy, Beverages)

Food packaging plays an important role in assuring the freshness of products, protecting them from contamination, and minimizing food waste. Self-healing materials improve food packaging performance by fixing small damages like microcracks and punctures that may interfere with the integrity of the packaging. By incorporating self-healing technologies into food packaging, the manufacturers can increase food preservation, decrease spoilage risks, and minimize packaging waste toward a sustainable supply chain.

8.1.1 Fresh Produce Packaging

Fruits and vegetables perish very quickly, and with packaging and materials that confer both mechanical strength and ability to develop a controlled atmosphere. Self-healing polymer films with oxygen-scavenging capabilities assist in swooshing gas exchange to not allow wastage of perishable goods. Additionally, these materials have the capability to self-repair small perforations, reducing the risk of external contamination and dehydration.

8.1.2 Meat and Dairy Product Packaging

Being highly prone to microbial contamination, meat and dairy foods require airy, tough packaging. Self-healing barrier films can maintain a hermetic seal despite minor damage and prevent oxygen from entering and microbial growth. Advanced antimicrobial self-healing coatings can further resist bacteria, making for food safety.

8.1.3 Beverage Packaging Systems

Packages that hold liquids such as cartons, pouches, and bottles are susceptible to leakage from mechanical stress. Self-healing polymers with shape memory will seal the minor leaks in the beverage packaging so that no liquids are spilled, keeping the product usable for a longer time.

8.2 Pharmaceutical and Medical Packaging

Sterile packaging of pharmaceutical and medical products is subject to rigorous safety and regulatory requirements to guarantee the efficacy of its products and the safety of its patients. Self-healing packaging systems afford higher protection against contamination, mechanical stress, and environmental agencies that can degrade sensitive medical and pharmaceutical products.

8.2.1 Packaging and Blister Packs for Drugs

The packaging of medications, especially for liquids or in capsules, is one that provides barriers to moisture and air. The self-healing barrier film acts to provide an airtight seal that holds even where microcracks or perforations develop during handling or shipping. This is an important feature to keep drugs potent for any extended period.

8.2.2 Sterile Medical Packaging

For surgical instruments, syringes, and any other implements, sterile packaging is necessary to keep out infections and guarantee safety. Within either flexible pouches or rigid containers, self-healing coatings may repair small damage, keeping sterility intact till the package is opened by controlled conditions.

8.2.3 Temperature-Sensitive Medical Products

Vaccines, biologics, and any other temperature-sensitive pharmaceutical items require special packaging that agrees with establishing temperature ranges. Self-healing insulation materials in combination with smart temperature-responsive healing mechanisms secure thermal integrity of such products to minimize spoilage probabilities en route and in storage.

8.3 Electronic and Smart Packaging

Smart and electronic packaging have created new demands for durable, flexible materials capable of self-repair. Self-healing packaging provides even more protection for the electronic components, sensors, and interactive packaging features, thereby ensuring long-term functionality and reliability.

8.3.1 The Protection of Electronic Goods

Self-healing films and coatings can be used to protect electronic devices such as smartphones and tablets from minor scratches and cracks. This has special relevance for flexible displays and wearable electronics, where integrity needs to be sustained.

8.3.2 Intelligent Labels and Sensors

An increasing number of packing solutions incorporate RFID tags and temperature items and smart sensors to monitor product conditions. These self-healing polymer substrates will secure the operation of the sensors if minor damage occurs, raising the level of reliability of the data and monitoring of the supply chain.

8.3.3 Energy Harvesting and Active Packaging

Some self-healing materials in electronic packaging can also contribute toward energy harvesting, such as flexible batteries and circuits embedded in packaging.

By autonomously repairing damage, these components maintain functionality, thus increasing the efficiency and sustainability of active packaging solutions in general.

9 Sustainability and Regulatory Considerations

Sustainability and regulation would form the lifeblood of development and adoption of self-healing packaging. These factors, including LCA for environmental reduction, checks for non-violation of international standards for safety, would in totality determine whether or not self-healing materials are viable for production and commercialization. Future improvements on this concept must primarily focus on these advancements toward improving recyclability, biodegradability, and consumer safety, thus spearheading the vast acceptance of this innovative technology [16].

9.1 Life Cycle Assessment and Environmental Impact

Life Cycle Assessment (LCA) is a structured approach to assessing the environmental impact of self-healing packaging materials, from the extraction of raw materials to disposal. The assessment considers a multitude of parameters, such as energy consumption, carbon footprint, water usage, and the final self-healing packaging system should minimize environmental burden through the increase in product lifecycle, i.e., reducing waste formation and concomitantly decreasing the consumption of raw material.

Though decreases in the rates of replacements become another benefit to applying self-healing packaging, degradation of physical product packaging is often due to mechanical damage, which requires further production and disposal, thereby contributing to energy pollution. However, new systems, if self-healing, would minimifically require new packaging anyway, resulting in minimized manufacturing emissions and landfill accumulation.

Nonetheless, the sustainability of self-healing composites depends on the selection of materials. Biodegradable self-healing polymers and bio-based materials like PLA and polyhydroxyalkanoates (PHA) are great alternatives to fossil-based plastics, as they allow for the circular economy while at the same time being fairly non-toxic. However, energy-intensive production processes and the possible toxicity of an agent used to heal the majority of these materials must be addressed when viewing the environmental value of such materials.

9.2 Recyclability and End-of-Life Considerations

The recyclability of self-healing materials depends primarily on the type of polymer matrices and healing mechanisms employed. The thermoplastic self-healing composition is generally more recyclable than thermosets due to the possibility of melting and reprocessing thermoplastics. Alternatively, thermosetting compositions are often more durable and heat resistant, making it less reversible due to the development of irreversible cross-linking.

End-of-life strategies for the self-healing package may include mechanical, chemical recycling, and composting. Mechanical recycling involves shredding used packaging into pellets for re-use but might compromise the healing function. The chemical recycling method could be depolymerization and solvolysis, which can recover monomers and other valuable materials to manufacture high-quality recycled materials with all self-healing capabilities intact.

A second consideration is biodegradability. These account for some self-healing materials, especially from natural polymer sources (chitosan and starch-based-composites), which can decompose in controlled environments and, thereby, aid in reducing microplastic pollution. However, a compromise should be sought between biodegradability and self-healing efficiency for the best self-healing performance.

9.3 Food Contact Compliance and Safety Assessments

Self-healing packaging for food applications must abide by strict food safety regulations in order to prevent contamination and health hazards. Food contact materials (FCMs) must prove to be free from harmful substances, thus ensuring that healing agents and polymer matrices do not supervene with food products.

Migration studies are crucial to assess the release of chemical compounds from the packaging into the food environment under varying conditions of temperature, humidity, and time of storage. Different authorities including U.S. FDA; EFSA, etc., have set specific migration limits for various substances. Thus, self-healing polymers and additives must be rigorously evaluated to satisfy such safety thresholds.

Further, some of the nanomaterials and microencapsulation techniques used for self-healing packaging require scrutiny because of the potential toxicological implications. Encapsulating materials, generally used for containing healing agents, need to be biocompatible and non-toxic. Other good alternatives are natural and food-grade healing agents, such as proteins and polysaccharides, since they improve food safety and regulatory approval from consumers too.

9.4 International Regulations on Novel Packaging

Regulatory frameworks concerning new packaging mostly differ across different regions. Such agencies as the FDA, EFSA, and the China National Food Safety Standard (GB) provide material safety and approval processes in the framework of the novel packaging regulations. Conformity to such laws guarantees market acceptance and legal authorization for commercial use.

The European Union: Regulation (EC) No 1935/2004 which comprehensively regulates all requirements for food contact materials including that food packaging does not release anything harmful into the food. Specific rules for plastics which are made from self-healing composites fall under the Regulation (EU) No 10/2011 which outlines the migration limit and testing.

In the United States, the FDA's Code of Federal Regulations (CFR) Title 21 prohibits the use of any substance in food packaging unless this provision makes an exception. The requirements for self-healing materials must also be analyzed for direct or indirect applicability for any food contact where material degradation products could be released.

The above-mentioned GB 4806 standards of China cover provisions on material safety for food applications mainly based on stringent migration tests and toxicological assessments as obligatory. Thus, standard compliance is needed by prospective companies introducing self-healing packaging into international markets.

Given the dynamic scene of packaging regulations, ongoing research, and contact with regulatory bodies is the lifeblood to ascertain that self-healing materials are at the very top when it comes to safety and quality standards.

9.5 Consumer Safety and Perception

Consumer adoption of self-healing packaging depends on their awareness about the same, perceived benefits regarding and their trust in its safety. Self-healing technology, while providing numerous benefits, such as extended runtime of packaging and better preservation of food, raises questions in minds of people on safety regarding the healing agents and nanomaterials.

Establishing transparency of material composition and regulatory approval would aid in building confidence among consumers. Premises and consequences thereof can assure customers that self-healing packaging is safe for use, alongside educational campaigns and industry partnerships. Governmental initiatives can further push for more benefits of self-healing technology while addressing common myths and stressing sustainability involved.

Studies among consumers show that, in general, they lean toward environmentally friendly and sustainable packaging solutions. Therefore, companies integrating self-healing packaging schemes into their sustainability initiatives would further add value and enhance brand loyalty and appeal among eco-conscious consumers.

Affordability and accessibility remain the main factors to be considered for the adoption in the market. Self-healing technology should be able to compete with conventional packaging solutions in terms of cost-effectiveness and performance to do so.

10 Case Studies and Commercial Applications

Self-healing composites have made the transition from the research laboratory to the actual world and have proved their worth in strengthening packaging, saving wastage, and ensuring safer products. This section presents case studies, performance metrics in reality, cost-effectiveness analysis, and consumer acceptance alongside market viability. Alongside lower production costs, progress in research onto self-healing packaging will thrive and begin to revolutionize packaging norms within the industries toward being a routine innovation [15].

10.1 Successful Implementation Examples

Many firms and research laboratories have been at the forefront of developing self-healing packaging. The few case examples of which bear testimony as follows.

10.1.1 Self-healing Film for Food Packaging

One multinational food packaging firm developed a polymeric self-healing film that heals minor scratches and punctures and thus is being incorporated into packaging to preserve perishable items with an oxygen-impermeant barrier even after small damages.

10.1.2 Beverage Cartons Using Microencapsulation

One of the leading beverage firms microencapsulated healing agents in the coating of its cartons. When the surface is scratched, the healing agents release and seal the damage area from contaminants.

10.1.3 Antimicrobial Self-healing Packaging

This is the medication for service providers that developed a healing blister pack for tablet packaging with antimicrobial properties accompanying healing. Such innovations would reduce bacterial contamination and ensure prolonged integrity of the product.

10.2 Performance in Reality

Laboratory studies point out the promise of self-healing composites, but real-life applications testify to their performance across different conditions.

10.2.1 Food Supply Chains

Self-healing packaging was studied and proved to seal even micro-tears when put to spoilage pretreatments, thus greatly reducing food waste during transportation and storage.

10.2.2 Cold Chain Logistics

For temperature-sensitive, self-healing films respond to temperature conditions in the environment and thus are reliable for frozen and refrigerated commodities.

10.2.3 Retail and Consumer Handling

Self-healing packaging can withstand repeated handling by consumers in retail environments. As a result, a lot fewer packages fail and products are not returned because of compromised package integrity.

10.3 Economic Analysis

The cost of self-healing packaging materials is generally higher compared to conventional options; however, most of the long-term benefits of using these materials are greater than the costs involved. From a cost–benefit analysis point of view, the benefits are as follows.

10.3.1 Reduced Packaging Waste

Self-healing composites prolong the use phase of their components and thereby contribute to waste reduction with regards to disposal and recycling costs.

10.3.2 Fewer Product Losses

Damage to the product is physical, and the product may also be exposed to environmental influences; thus, less product losses occur within the supply chain.

10.3.3 Reduces the Need for Secondary Packaging

Increased strength of packaging material would omit the need of superfluous protective layers, thus saving material and shipping costs.

10.3.4 Consumers Willing to Pay

Surveys reveal that such consumers are willing to spend more for increased safety or sustainability of product packaging.

10.4 Market Acceptance and Consumer Response

It appears from market studies that self-healing technologies are becoming popular among consumers. The following are some reasons that is believed to be behind or at least promoting other buyer acceptance.

10.4.1 Consumer Perception of Innovation

For people, self-healing packaging is a new technological perk that ratifies an instantly better level of safety and longevity for products.

10.4.2 Sustainability Factor

Earth-friendly consumers are sympathized with the possibility of saving waste destruction by self-healing materials and endorse it with the global trends of sustainability.

10.4.3 Government Backing

Trust and potential acceptance could be improved as endorsements for these developing innovations to consider the safety benefits for food and environmental regulators.

10.4.4 Trends in Industry Adoption

The industry's most powerful companies are those that have applied self-healing technology to their packaging and other uses by the industry.

11 Challenges, Future Perspectives, and Emerging Trends

There are several challenges associated with self-healing materials like Scalability and Cost Considerations, Integration with Smart and Active Packaging Technologies, Future Self-Healing Materials, and Prospective New Applications and Markets [15].

11.1 Scalability and Cost Considerations

The practical implementation of self-healing composites in packaging heads toward numerous challenges, namely, scalability and cost issues. While fully lab-scale studies have demonstrated promising self-healing properties, the translation and industrial-scale production of these materials are encountered with complex scenarios. These scenarios arise from several reasons.

11.1.1 Manufacturing Complexity

Many healing mechanisms, such as microcapsule and vascular network healing, require highly delicate fabrication techniques, which may not be compatible with conventional high-throughput packaging processes.

11.1.2 Material Costs

The attempt of incorporating specialized healing agents, nanomaterials, and engineered polymer matrices elevates the costs of raw materials, thereby discouraging commercialization on a larger scale.

11.1.3 Processing Requirements

Certain self-healing materials entail extra expenses for curing, activation, or thermal treatment, which amounts to energy and labor costs.

11.1.4 Economic Viability

For self-healing packaging to be commercially viable, it must show a clear profit versus loss compared with ordinary materials with respect to shelf life, waste reduction, or sustainability.

11.2 Integration with Smart and Active Packaging Technologies

Self-healing composites, by their interaction with smart and active packaging technologies, would boost this packing sector to new heights. These high-tech packaging systems have real-time monitoring capabilities, better protection features, and improved shelf life for perishables. The interface among self-healing materials and smart packaging innovations encompasses.

11.2.1 Sensor-Embedded Healing Systems

Integration of self-healing polymers with embedded sensors permits real-time inspection of package integrity, detects damages, and activates auto-repair.

11.2.2 Active Preservation Features

Self-healing packaging can incorporate antimicrobial agents, oxygen scavengers, and moisture regulators for food safety and spoilage protection.

11.2.3 Temperature-Responsive Healing

Packaging that repair itself autonomously against environmental stimuli, like heat or humidity fluctuations, would be particularly beneficial for real-time products such as pharmaceuticals and perishable foods.

11.2.4 Recyclability and Circular Economy Integration

These smart packaging solutions endowed with self-healing properties must be developed with sustainability considerations that allow them to be integrated into recycling and biodegradation pathways.

11.3 Future Self-healing Materials

It is likely that the future of self-healing materials in packaging will evolve into next-generation materials with better healing capability, reactivity, and sustainability. Emerging areas for future research include.

11.3.1 Bio and Biodegradable Self-healing Polymers

New and sustainable adhesives in packaging are sought for reducing environmental footprint impacts. Bio-based self-healing materials from naturally occurring polymers, including chitosan, cellulose, and polylactic acid (PLA), are promising substitutes for synthetic materials.

11.3.2 Multi-responsive Systems for Healing

Self-healing materials of the next generation will be designed to respond to different stimuli, which will include changes in temperature, pH, and stress, thus making the healing process more robust and adaptive.

11.3.3 Self-healing Hydrogels for Packaging

Hydrogels with dynamic covalent bonds or supramolecular interactions are of growing interest in applications requiring high flexibility and reusability, such as soft package films.

11.3.4 Graphene- and Nanomaterial-Based Composites

With respect to self-healing functionality, mechanical strength, conductivity, and barrier properties are multiplied with the use of nanomaterials, such as graphene oxide and carbon nanotubes.

11.4 Prospective New Applications and Markets

Outside of traditional food-packing, self-healing composites could potentially reach other applications and markets-usually projected to bring about paradigm shifts in nearly all industries.

11.4.1 Pharmaceutical and Medical Packaging

Self-healing materials would be applicable in protective packaging for sensitive pharmaceuticals, ensuring sterility and integrity, while prolonging the shelf life.

11.4.2 Electronics Packaging

Integration of self-healing composites in packaging of electronic devices may safeguard componentry from mechanical stress, moisture, and damage to environment, thereby enhancing reliability and longevity.

11.4.3 Automotive and Aerospace Packaging

Self-healing materials would serve to cushion damage to high-value automotive and aerospace components during transportation and storage.

11.4.4 Luxury and Sustainable Packaging

Self-healing materials promise to enhance the affordability of premium and permanent packaging for consumers of cosmetic and premium goods.

12 Conclusion and Future Prospectives

This self-healing composites packaging is a disruptive innovation able to alleviate challenges concerning material longevity, sustainability, and protection of products. This chapter has delved into the pathways and mechanisms, materials, processes, applications, and future strategies around self-healing packaging technologies. This section will summarize key findings, delimit current limitations and recommendations for further research and development, and finally provide an outlook on the future of self-healing packaging solutions.

12.1 Summary

Research and development of self-healing composites for packaging have shown great merit in terms of durability, waste management, and sustainability. This chapter highlights the following important findings.

12.1.1 Mechanisms of Self-healing

Healing in packaging materials may be instigated through intrinsic and extrinsic mechanisms including microencapsulation, vascular networks, reversible polymer networks, and shape memory induction.

12.1.2 Material Innovations

Within a polymer matrix (thermoplastics, thermosets, elastomers, and/or biodegradable polymers), the incorporation of healing agents (microcapsules, microvascular networks and dynamic bonds) has proved to enhance the longevity of the material.

12.1.3 Manufacturing Approaches

Packaging materials with healing functions are manufactured by using extrusion, injection molding, coating, and 3D printing.

12.1.4 Performance and Functionalities

Self-healing packaging restores not only the structural integrity but also smart properties such as oxygen scavenging, antimicrobial protection, and moisture regulation.

12.1.5 Sustainability and Regulations

The use of bio-based self-healing materials can reduce environmental impact, and regulatory frameworks are progressing toward the safety and compliance of food packaging.

12.1.6 Commercial Applications

Some self-healing composites in packaging have been tested in the real world to provide evidence toward the enhancement of product lifespan and reduction of material waste for food, pharmaceuticals, and electronics packaging.

12.2 Current Limitations and Challenges

The amazing potential for self-healing composites packaging notwithstanding, there are challenges that need to be addressed before such systems can achieve acceptability on a large scale.

12.2.1 Cost and Scalability

The most common methods of preparing self-healing materials require complex syntheses and high costs, economically inhibiting large-scale manufacture.

12.2.2 Healing Efficiency and Durability

Many self-healing systems have limitations in healing speed, efficiency, and long-term durability under repeated damage cycles.

12.2.3 Compatibility with Packaging Processes

Existing industrial packaging processes may require modifications to efficiently integrate self-healing materials, adding logistical complexity.

12.2.4 Environmental and Toxicological Concerns

Certain self-healing agents may have the potential of posing a risk due to migration into the packaged goods, particularly in food and pharmaceutical applications; thus, a comprehensive examination of safety is required.

12.2.5 Consumer Awareness and Acceptance

Better awareness and confidence of self-healing packaging technology by the general public and industry stakeholders is required among them.

12.3 Recommendations for Research and Development

In order to combat these horizons and accelerate the adoption of self-healing packaging, future research and development activities should focus on the following areas.

12.3.1 Cost-Effective Synthesis

Design and manufacture these self-healing materials at huge volumes in an economically viable manner by amicable production techniques including green chemistry and approaches driven by nanotechnology.

12.3.2 Improved Healing Mechanisms

Advance the self-healing materials through improvements in healing kinetics, repeatability, and responsiveness to the environment in increasing their self-healing capability.

12.3.3 Harmonization with Smart Packaging

Integrate these self-healing properties with other smart packaging functionalities, as sensors, indicators and active protection features.

12.3.4 Sustainability and Circular Economy

Research into bio-based, biodegradable and recyclable self-healing materials in keeping with the global sustainability goals.

12.3.5 Regulation Compliance and Safety

Acquisition of materials compatibility studies along with extensive studies on migration behavior, toxicology, and interactions to meet food safety and environmental regulations.

12.3.6 Consumer and Industry Engagement

Raise awareness, pilot studies, and even more, and encourage cooperation between academia and industry and policy-makers regarding innovation readiness and market penetration.

12.4 Future Outlook for Self-healing Packaging

The future is bright for self-healing packaging. More advances will come including material sciences, manufacturing technologies, and their commercial implementation tomorrow. Some indicative trends and resulting opportunities are likely to be as follows.

12.4.1 Stimuli-Responsive

Autonomous healing stimuli-responsive materials set to be developed for diverse external environments, including humidity and temperature changes and mechanical stresses.

12.4.2 Combined with Sustainable Packaging

Compostable and biodegradable self-healing materials in the advent of a circular economy to ensure reduced plastic waste and environmental footprint.

12.4.3 Recent Developments in Additive Manufacturing

The use of 3D printing techniques to produce intricately designed self-healing structures that function for a variety of applications in packaging.

12.4.4 Personalized and Adaptive Packaging

Intelligent packaging response to external stimuli will increase shelf life and be more convenient for consumers.

12.4.5 Going Mainstream Within Industry

With anticipated reductions in costs and performance improvements, self-healing packaging is forecast to become an everyday solution in the food, pharmaceutical, and electronic industries.

Declaration of Competing Interest The authors declare that they have no known competing financial interests or personal relationships that could have appeared to influence the work reported in this paper.

References

1. Kontiza A, Kartsonakis IA (2024) Smart composite materials with self-healing properties: a review on design and applications. Polymers 16:2115
2. Huang K, Wang Y (2023) Recent advances in self-healing materials for food packaging. Packag Technol Sci 36(3):157–169. https://doi.org/10.1002/pts.2701
3. Wang J, Gao Q, Zhao F, Ju J (2024) Repair mechanism and application of self-healing materials for food preservation. Crit Rev Food Sci Nutr 64(30):11113–11123. https://doi.org/10.1080/10408398.2023.2232877
4. Yang S, Zhou S, Yuan J (2025) Self-healing elastomers and coatings via metal coordination bonds. Chem Eur J 15:38. https://doi.org/10.1002/chem.202404038
5. Ismadi NFB, Zaini NAM, Samsudin D, Hashim F, Salim ZASBA (2025) Preparation and characterization of self-healing natural rubber (NR) based on zinc diacrylate networks. Eng Headway 14:3–9. https://doi.org/10.4028/p-1EG0Qf
6. Ekeocha J, Ellingford C, Pan M, Wemyss AM, Bowen C, Wan C (2021) Challenges and opportunities of self-healing polymers and devices for extreme and hostile environments. Adv Mater 33(33):52. https://doi.org/10.1002/adma.202008052
7. Gautam BR, Khan NI, Gosvami NN, Das S (2025) Recent advancements in self-healing materials and their application in coating industry. Proceed Instit Mech Eng L J Mater Des Appl 239(3):411–427. https://doi.org/10.1177/14644207241269558
8. Zibafar M, Eslami-Farsani R (2024) Reversible ionic bond mechanism in self-healing of natural rubber: a review. Polym Rev 15:1–30. https://doi.org/10.1080/15583724.2024.2392124
9. Wang C, Boulatov R (2025) Autonomic self-healing of polymers: mechanisms, applications, and challenges. Molecules 30(3):469. https://doi.org/10.3390/molecules30030469
10. Abdullah AM, Yu K (2024) Self-healing polymers and composites: a review of recent developments. In: Structural health monitoring/management (SHM) in aerospace structures. Elsevier, Amsterdam, pp 229–263. https://doi.org/10.1016/B978-0-443-15476-8.00015-0
11. Poornima Vijayan P, George JS, Revathy RV (2023) Self-healing epoxy resin with multi-stimuli-responsive behaviour, pp 161–174. https://doi.org/10.1007/978-981-19-6038-3_6
12. Lee YB, Chang KM, Wilson GO, Andersson HM, Moore JS, Sottos NR, Braun PV (2023) Self-healing polymers. In: Encyclopedia of polymer science and technology. Wiley, Hoboken, pp 1–47. https://doi.org/10.1002/0471440264.pst469.pub2
13. Hoque ME, Shafoyat MU (2024) Self-healing polymer nanocomposite films and coatings. In: Polymer nanocomposite films and coatings. Elsevier, Amsterdam, pp 181–217. https://doi.org/10.1016/B978-0-443-19139-8.00010-3
14. Panahi P, Khorasani SN, Mensah RA, Das O, Neisiany RE (2024) A review of the characterization methods for self-healing assessment in polymeric coatings. Prog Organ Coat 186:108055. https://doi.org/10.1016/j.porgcoat.2023.108055

15. Kim S, Jeon H, Koo JM, Oh DX, Park J (2024) Practical applications of self-healing polymers beyond mechanical and electrical recovery. Adv Sci 11(16):463. https://doi.org/10.1002/advs.202302463
16. Cerdan K, Thys M, Costa Cornellà A, Demir F, Norvez S, Vendamme R, Van den Brande N, Van Puyvelde P, Brancart J (2024) Sustainability of self-healing polymers: a holistic perspective towards circularity in polymer networks. Prog Polym Sci 152:101816. https://doi.org/10.1016/j.progpolymsci.2024.101816

Chapter 7
Self-healing Composite Materials in Packaging Applications

Ritam Patra and Esha Trivedi

1 Introduction

Ancient human civilizations used natural materials like animal skins, leaves, and containers made of wood to store and transport food and essential items, making these the earliest known packaging which reflects a fascinating journey from ancient history to modern times shaped by both innovation and necessity [1, 2]. Archaeologists unearthed pottery and glass items dating back to 7000 B.C., that indicate initial attempts to utilize more enduring materials. Amphoras and glass pottery that were used in the Mediterranean region from 1500 B.C. to 500 A.D. are now showcased by Museums and were widely used by the ancients peoples of Egypt, Roman Empire, and Greece to carry and trade grains, wine and oil [3]. Indigenous tribes of Latin-America used ceramic materials during religious occasions circa 1000 B.C., which demonstrates the diversity of indigenous packaging crafts [4]. Packaging technology evolved alongside civilizations. The invention of Glass around 1200 B.C. is one of the most important events in the history of packaging that utilizes basic elements such as silica, limestone, sand, and soda. The invention of the blowpipe around 300 B.C. revolutionized glass making by allowing the manufacture of spherical vessels of different shapes and sizes. The split molding procedure, originated around the seventeenth and eighteenth centuries allows the manufacturing of glass with irregular shapes and complex embellishments. Despite these developments, metal packaging was not widely used until around 1200 A.D., when stronger alloys and the tin plating process were developed [5, 6].

R. Patra (✉)
Department of Materials Science and Engineering, Swansea University, Swansea, Wales, UK
e-mail: ritam_p@pe.iitr.ac.in

E. Trivedi
Department of Paper Technology, Indian Institute of Technology Roorkee, Saharanpur, Uttar Pradesh, India

A. Verma et al. (eds.), *Polymers and Composite Materials for Packaging: Smart Food Packaging and Solutions*, Materials Horizons: From Nature to Nanomaterials,
https://doi.org/10.1007/978-981-95-0677-4_7

The Industrial Revolution marked a revolutionary moment in the history of packaging. Nicholas Appert invented the process canning as a technique of preserving food for the French Army in the early nineteenth century, stepping the first event in the Modern Packaging. During and after WWII introduction of materials such as paper, paperboard, and the first plastics that were made from cellulose and petroleum, which significantly altered the packing landscape, notably [7].

The sustainability factor is the most crucial drawback of traditional packaging material rather being historically important. Plastics and metals, while durable and cost-effective, are non-biodegradable hence questionable to sustainability due to their nature and energy-intensive manufacturing processes. These problems highlight the need for materials that can meet modern needs while maintaining environmental integrity [8].

Conventional materials often suffer limitations due to non-biodegradability and commercial-social shift in consumer requirement. This phenomenon fuels the need of innovative and novel packaging in the market by engineering durability and adaptability in these materials to meet higher performance standards while remaining sustainable. Self-healing composite technology is nothing but an innovative approach to cater this demand by increasing product life and reducing waste.

These advanced materials can self-repair when broken, replicating nature's regenerative capabilities. Adding these self-healing functionality into materials is a major solution in the field of packaging technology. It will make products to last long, cut down on waste and improve efficiency [9, 10]. This self-healing functionality not only saves resources but also provides protection to the packed items, which makes them advantageous in domains where packaging integrity is important.

This book chapter explores the persistent search for innovation while addressing the current obstacle: cost, scalability, and regulatory barriers by presenting both research prototypes and commercially available self-healing composite materials. Ultimately, self-healing composites mark a paradigm leap, bringing with them a new era of packaging that is more durable, produces less waste, and actively supports a regenerative, sustainable economy as mentioned in the abstract. This detailed illustration summarizes the important principles covered in the book chapter on self-healing composites in packaging. "Self-Healing Composites" serves as the central theme in the diagram, surrounded by a variety of interrelated pieces that illustrate the topic's extent and depth (Fig. 1).

2 Self-healing Materials

2.1 Lessons from Living Being

Living things have an amazing ability to heal themselves naturally when they detect any injury. Such complex process commences by first detecting the damage mostly through inflammatory chemical signals. Next, a series of events are initiated that

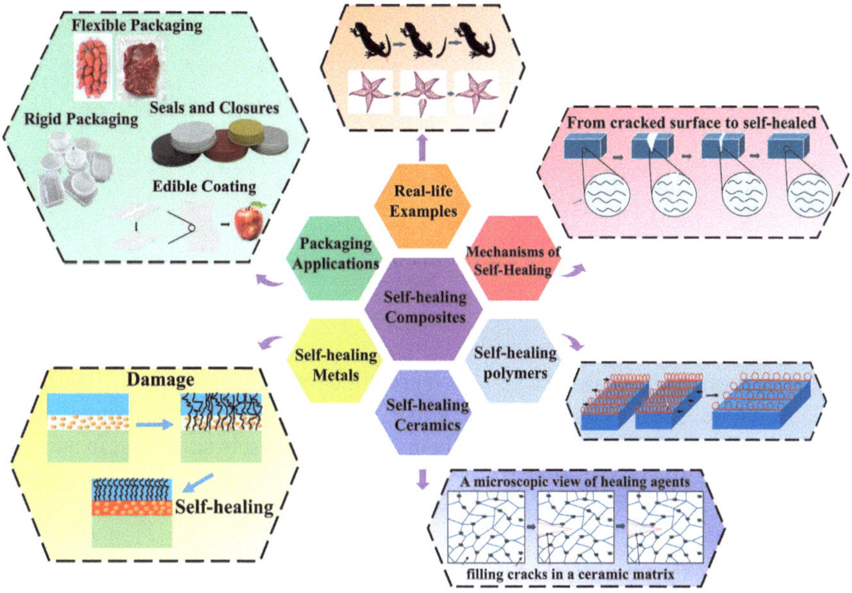

Fig. 1 Overview of packaging self-healing composites: examining uses, processes, and material types

lead to tissue remodeling and wound closure. This healing occurs at multiple scales, commencing at the molecular level and advancing to the tissue level. Researchers in polymer chemistry are drawing inspiration from these biological repair mechanisms to develop effective self-healing materials [9]. In biological systems, wound healing involves a multifaceted, well-coordinated response involving various cell types. The first stage is to quickly identify the wound, which can be done in a matter of seconds to minutes, even when hundreds of micrometers are involved. To identify this, the disruption caused by the injury must first be converted into a chemical or physical signal that is then sent throughout the tissue. This signal propagation synchronizes the protective responses of cells both near and distant from the wound site [10].

We'll now look at the amazing self-healing properties of humans, animals, and plants in nature and how these biological processes might serve as inspiration for the creation of self-healing materials for engineering uses. Animals are the main emphasis here since they have developed more complex methods than plants, which are capable of self-healing. Additionally, human healing both psychological and physical will be emphasized.

2.1.1 Examples of Self-healing in Nature

When cut, grass grows back quickly. This is a very common example of a self-healing in nature. For instance, the cut edges looks rough when grass is mowed but they

Fig. 2 Grass can heal after cutting

disappear soon and regrows back. This regrowth is helped by water, nutrients, and sunlight. Some grasses can regenerate either above-ground or underground stems, which develop new shoots and sustain them until they are self-sufficient, in addition to repairing their leaves [9] (Fig. 2).

2.1.2 Self-healing Examples in Animals

A variety of self-healing techniques have been developed by animals. Certain animals, such as cats and dogs, naturally clean and cure wounds using their saliva. Growth factors, defensins, and enzymes like lysozyme and peroxidase are among the antiviral and antibacterial substances found in saliva. Additionally, licking aids in clearing the wound's debris [11, 12].

When under danger, a lot of lizards can separate their tails. As the lizard flees, the detached tail wiggles to divert predators. Following that, the lizard grows a new tail, however it might not look exactly like the old one in terms of structure. This is a survival strategy that involves giving up a portion of the body in order to preserve the whole [13].

Some species of starfish can grow back from a single arm, especially if a portion of the central disk is still attached. Starfish can also grow back lost arms. Months or even years may pass before this regeneration occurs. Another asexual method of reproduction used by starfish is fragmentation, in which the center disk brakes and each piece grows back into a whole starfish [14].

Salamanders' "perfect" regeneration is well known. They can heal sliced spinal cords, eyes, injured lungs, whole limbs, and even parts of their brains without leaving scars. Eventually, the regenerated body parts work and look just like the original. Macrophages are immune cells that are essential to their scarless healing process [15]. Many earthworms may regrow missing segments from their head or tail, though research findings vary by species. If each half contains a portion of the clitellum, a sexual organ, many species can even regenerate a whole worm from two halves [16]. Sea cucumbers have a remarkable capacity for regeneration, which includes the ability to regenerate internal organs. They can deceive predators and flee by expelling their poisonous internal organs when confronted. After that, they can grow

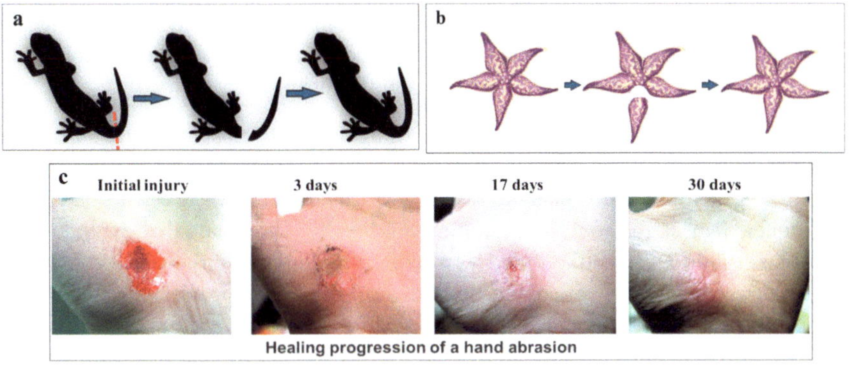

Fig. 3 Set of illustrations showing how animals can repair themselves: **a** a lizard growing back its tail; **b** a starfish growing back a lost arm; and **c** human skin healing a wound

new organs in roughly a month or two. Certain species of sea cucumbers are even capable of dividing their bodies in half, with each half growing back into a whole new individual [17].

Self-healing in humans is a complicated process that involves both mental and physical health. Humans have extraordinary healing powers, particularly when it comes to skin and bone, even though they are unable to regrow limbs [18]. The epidermis and dermis that make up human skin are constantly renewing and growing. Blood clotting, which involves platelets and a complicated series of chemical processes, happens when skin is wounded. To halt the bleeding, platelets collect at the wound site and form a plug. This blockage is strengthened by a fibrin mesh, which traps red and white blood cells to create a clot. After white blood cells combat infection, the wound goes through three stages: proliferation, during which new tissue grows, and maturation, during which the tissue is altered. Weeks or even years may pass during this entire process. For general health, psychological self-healing, which addresses mental and emotional well-being—is also essential. Physical health can be impacted by psychological issues such as stress and emotional tension. Stress management, mindfulness, and getting enough sleep are among strategies that can help people heal psychologically and physically [19–22] (Fig. 3).

2.2 *Mechanisms of Self-healing: Extrinsic Versus Intrinsic*

The self-healing system's inherent ability to detect and sense any damage enables the system to regain its original state. In particular, self-healing mechanisms seek to detect and fix small defects, such cracks, in order to restore the material's mechanical or functional performance [23].

Scientists have created a variety of self-healing methods for synthetic materials, drawing inspiration from nature's varied healing techniques [24]. These mechanisms fall into two general categories: extrinsic and intrinsic.

- Extrinsic Self-healing: Using specialized agents that release healing substances when harm occurs is known as extrinsic self-healing. Imagine it as a Band-Aid that automatically heals a wound. To fill in cracks or breaks, this method frequently uses microcapsules containing healing chemicals that explode when struck. Alternatively, like blood vessels in a biological body, vascular networks can direct healing chemicals to the location of harm. However, the intricacy of building these networks and the microcapsules' one-time usage constraint make them impractical [25, 26].
- Intrinsic Self-healing: Conversely, intrinsic self-healing is compared to the skin's innate restorative mechanisms. This approach doesn't depend on other parties. Rather, it makes use of the material's natural qualities. Because of their dynamic, reversible connections that can separate and reattach, polymers made for intrinsic healing are able to heal themselves. There are two primary types of these ties: noncovalent interactions like hydrogen bonds, which provide a balance between strength and flexibility to break and reconstruct under the correct circumstances, and dynamic covalent bonds, which can undergo chemical transformations to reform after injury [27, 28]. Because it may be repeated and doesn't require extra healing agents, intrinsic self-healing is a more effective and sustainable method. In situations where manual repair is not feasible, materials possessing inherent healing properties are especially well-suited. The applications are as diverse as they are significant, ranging from common electronics that gain from greater durability to aerospace applications where materials must endure harsh environments. Essentially, intrinsic healing is incorporated into the material itself, ready to react and restore integrity at the molecular level whenever necessary, whereas extrinsic healing mechanisms are like a repair kit that is assembled and ready to use. Because of this, intrinsic self-healing mechanisms are a key component of the next generation of intelligent, long-lasting materials that have the potential to completely transform industries by incorporating sustainability and longevity into their list of material attributes.

Self-healing materials exceptional capacity to self-heal damage including micro cracks, increases product lifecycles and improves performance enables them to act as a special category of compounds especially in the packaging sector. To categorize self-healing materials we can classify them by their matrix material, which includes metals, ceramics, and polymers (Fig. 4).

2.3 Self-healing Polymers

Polymers are among the most extensively researched and used self-healing materials due to their diversity and ability to mend themselves. Polymers' natural self-healing

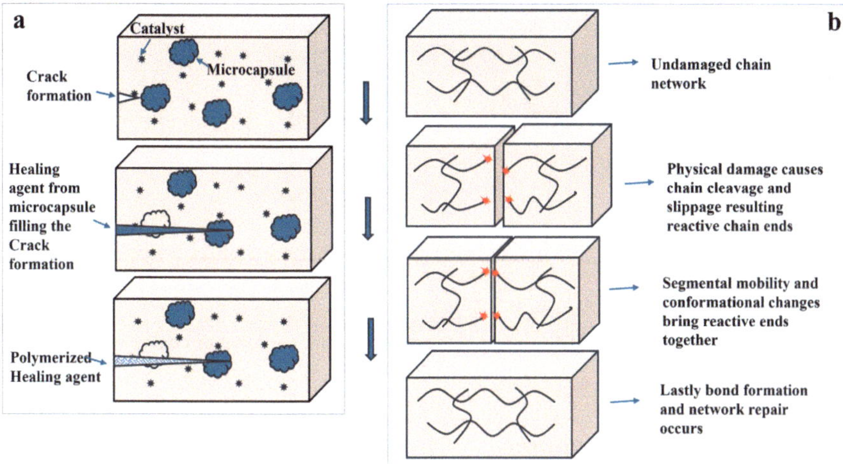

Fig. 4 Self-healing mechanisms: **a** extrinsic, using microcapsules, **b** intrinsic, through polymer re-bonding

capabilities often entail reversible bonding or molecular interplay, allowing the material to repair itself without external intervention [29, 30]. Self-healing polymer composites (PMCs) are lightweight, easily processed, and chemically consistent. These materials can be classified into thermosets and thermoplastics. Self-healing of microcracks is more documented for thermosetting polymers, whereas thermoplastics are more difficult to repair, however some study has been conducted in this domain [31]. Zhu et al. described a simple but effective technique for developing inherently self-healing and thermochromic fibrous membranes with a core–shell structure [32]. Wang et al. developed a self-healing solid-state polymer electrolyte for flexible lithium batteries that is highly stretchy, non-flammable, and not sensitive to temperature changes [33]. In flexible electronics, self-healing polymers enable a field-effect transistor (FET) to independently restore its mechanical and electrical characteristics upon injury, improving robustness and performance in applications such as skin-like sensors [34]. Sustainable anti-/de-icing technologies are made possible by an intrinsic self-healing poly-silicone-urea material that exhibits remarkable de-icing capabilities with a self-healing efficiency of about 99% and maintains low ice adhesion strength even after healing [35]. There are many excellent examples of this type of study. Polymer matrix composites (PMCs) are utilized in aircraft wings and fuselages, as well as load-bearing structural elements and intake propeller blades. Furthermore, self-healing polymers are being investigated for use in automobile parts and finishes. One example is Nissan's transparent self-healing car finish, "Scratch Guard Coat" [36]. Additionally, durable batteries, resealing tires, and resilient textiles use these polymers (Fig. 5).

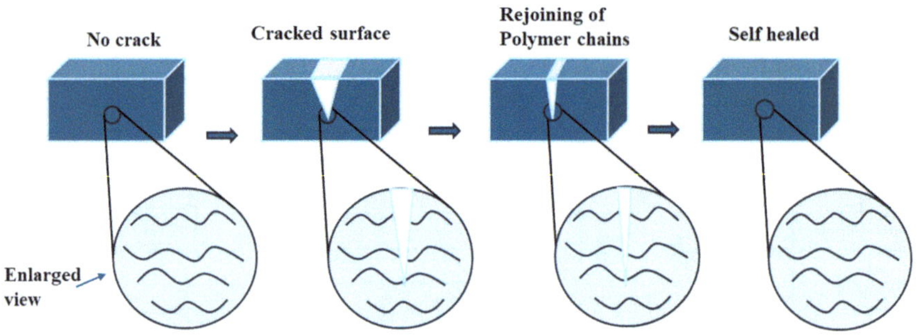

Fig. 5 Basic mechanism of self-healing polymers: a network of polymer chains with reversible bonds

2.4 Self-healing Ceramics

Despite their great strength and stiffness, ceramics are frequently fragile and prone to breaking. By including processes that can fix these cracks at the micro- or nanoscale, self-healing ceramics seek to overcome this constraint. Using hollow fibers or micro-capsules loaded with a therapeutic agent—like a liquid metal or a ceramic precursor—is one method. The capsules or fibers are ruptured when a crack forms, allowing the healing agent to enter the crack. The integrity of the material is subsequently restored when the healing agent fills the fracture by solidifying or reacting. Adding particles or whiskers that can bridge the crack and stop it from spreading is another tactic [37]. Another mechanism by which self-healing ceramics repair damage on their own is thermal activation. These materials work well in contexts where resilience and high temperatures are essential [38]. They could be utilized, for example, in the packaging of industrial equipment, critical electrical components, or even specific food items that need special preservation. Making these materials affordable and versatile for widespread packaging applications is still a hurdle, though (Fig. 6).

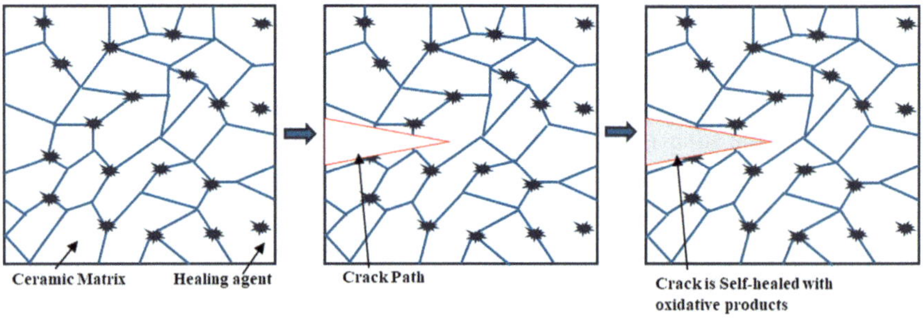

Fig. 6 Microscopic illustration shows healing agents repairing fractures in a ceramic matrix in self-healing ceramics

2.5 *Self-healing Metals*

Even though they are usually ductile, metals can eventually become fatigued and break. The goal of self-healing metals is to fix these fractures and increase the material's durability. Using microcapsules holding a liquid metal alloy that can enter cracks and solidify—thereby "welding" the crack closed—is one method. Shape memory alloys (SMAs), which can "remember" and return to their previous shape when heated, are another tactic. The SMA can be set to contract in response to the formation of a fracture, thereby sealing it [39, 40]. Self-healing metals may find use in specialty packaging in the domain of durability and impact resistance like reusable shipping containers and packaging for expensive items. But self-healing metal's high processing cost and added complexity prevent them from being widely used in traditional packaging. Researchers are still carrying out studies to develop more economical and adoptable methods for self-healing metals [41].

3 Focusing on Self-healing Materials Used in Packaging Composites

3.1 *Motivation of Emphasizing Composite*

Emphasizing composite materials in this domain of self-healing packaging is strategic and calculated. By combining the advantageous qualities of several component materials, composites provide a special chance to produce a synergistic impact that goes beyond the limitations of individual parts [42, 43]. Essentially, composites enable us to modify material characteristics to satisfy the unique requirements of packaging applications. There is a strong and complex case for concentrating on composites, especially in applications that call for a synthesis of qualities that are impossible for any single material to provide.

- Improved Mechanical Properties: Composites can be designed to outperform conventional materials in terms of strength-to-weight ratio. Researchers can produce composites with exceptional tensile strength, durability, and fatigue resistance by choosing the right fiber and matrix combinations. These properties are essential for high-performance applications in the automotive, aerospace, and construction sectors [44].
- Tailored Thermal Properties: Composites make it possible to control thermal characteristics like resistance and conductivity. This makes them perfect for uses where control over heat flow is crucial, such as thermal insulation in buildings or heat-resistant parts in electronics and automobiles [42].
- Better Corrosion Resistance: Composites have a far higher capacity to withstand corrosion-induced environmental deterioration. Composites, which combine materials like fiberglass with polymer matrices, are perfect for usage in outdoor,

chemical processing, and maritime applications because they can resist corrosive elements, dampness, and harsh chemicals [45, 46].

- Versatility in Design: In sectors like aerospace and automotive, composites provide unmatched design and fabrication flexibility. This may be unattainable using only metals or ceramics where customized designs can result in better aerodynamics, lighter, and fuel-efficient cars [47].
- Adjustable Electrical Properties: Depending on the requirements of the application, composites can be made to have electrical properties, such as conductive, semi-conductive, or insulating. This enable applications in the field of electrical engineering, such as conductive parts for electronic devices or insulators in electrical installations [48].
- Promoting Sustainability: Extending the service life of items and lessening the dependency on non-renewable resources, advancements in biodegradable and bio-based composites also promote environmental sustainability.

Focusing on composites materials, researchers can utilize the above-combined qualities for satisfying the specific need for their end-product applications. Thus, providing solutions that are highly efficient and better aligned to evolving environmental and sustainable goals. As technology advances, the ability to customize composites is expected to increase, making them more valuable across a wider range of industries.

3.2 Polymer Matrix Self-healing Composites

Self-healing polymer matrix composites (PMCs) are a breakthrough in materials science that provide workable ways to increase the lifespan of packaging materials by means of self-repairing mechanisms. In the packaging business, where material integrity is necessary for maintaining the properties of packed goods, these composites are especially pertinent. The kinds of polymers utilized in these composites, the typical healing agents they contain, and their uses in rigid and flexible packaging are examined below [8, 31, 49].

3.2.1 Types of Polymers Used as Composite Matrix

Polyolefins (PE, PP): Because of their affordability, adaptability, and strong barrier qualities, polyethylene (PE) and polypropylene (PP) are extensively utilized. They are frequently found in containers and flexible films. Certain grades of PE and PP that meet requirements are used for food contact [50, 51].

Polyethylene Terephthalate (PET): Polyethylene Terephthalate, is renowned for its strength, clarity, and recyclable nature. It is frequently found in containers and bottles. The usage of recycled PET (rPET) in packaging is also growing [52, 53].

Polylactic Acid (PLA): PLA is a bio-based and biodegradable polymer made from renewable resources. It is becoming more and more well-liked as a sustainable substitute for conventional plastics, especially in food packaging [54, 55].

Ethylene Vinyl Acetate (EVA): Copolymers of EVA are appropriate for flexible packaging applications because of their durability and flexibility [56].

Other Polymers: Depending on the needs of the packaging application, other polymers such as polyamides (PA), polyvinyl chloride (PVC), and different biopolymers may also be utilized.

3.2.2 Common Healing Agents

The type of polymer matrix and the intended self-healing process determine which healing agent is best. Here are a few typical examples.

Microencapsulated Healing Agents

- Dicyclopentadiene (DCPD): DCPD combines with a Grubbs catalyst to polymerize and close the fracture when microcapsules containing it burst. Compared to the basic specimen, the toughness of fracture regeneration may be enhanced by 75% [57, 58].
- Hydroxyl-Functionalized Polydimethylsiloxane (HOPDMS) and Polydiethylsiloxane (PDES): Hydroxyl-functionalized Polydimethylsiloxane (HOPDMS) and Polydiethylsiloxane (PDES) offer special characteristics when added to a synthetic ester base matrix. These monomers are present as droplets during a microphase separation process, which enables the matrix's properties—such as increased elasticity, durability, and resistance to environmental influences—to be improved. Better dispersion and integration within the matrix are made possible by the hydroxyl groups in PDES and HOPDMS, which improve their compatibility with the ester base. This combination not only improves the material's performance but also increases its potential uses in fields including coatings, sealants, and specialty synthetic rubbers that call for strong, flexible, and long-lasting materials [31, 59].
- Urea–Formaldehyde Microcapsules: These microcapsules, which include epoxy as a hardening molecule and binder material, can be released when fractures in epoxy matrix-based polymer composites begin and progress [60].

Vascular Networks

A vascular network can be formed by embedding hollow fibers or tubes in the polymer matrix. A healing agent, such as a liquid metal or a polymer solution, can be included into this network and released when injury occurs [61].

Reversible Chemical Bonds

These polymers may fix damage by rearranging themselves in response to outside stimuli like heat or light [62].

Shape Memory Polymers

In response to external stimuli like heat, SMPs could "remember" and regain to their original shape. This shape memory polymers have their application in seal punctures and cracks [63].

3.3 *Specific Applications in Packaging*

Flexible Packaging: Self-healing polymer composites could be used in the application of packaging in the form of flexible films, pouches, and wrappers. These films can be made to mend small rips or punctures on their own, increasing the packaged foods' shelf life and minimizing food waste [64].

Rigid Packaging: Bottles, jars, and tubs are examples of rigid containers that can employ self-healing polymers. The durability and reusability of these containers can be increased by designing them to self-heal cracks or impacts [65].

Coatings: To increase the scratch resistance and longevity of current packaging materials, self-healing polymer coatings can be used.

Seals and Closures: To guarantee their integrity and stop leaks, self-healing polymers can be added to seals and closures [66].

Overall Packaging: Epoxy resins might be among the best resin materials for self-healing composites when compared to other healing agents. When the epoxy resin attaches to the catalyst, the polymerization phase begins, enabling the curing agent to repair the internal break.

3.4 *Briefly Cover Other Matrix Materials (Ceramics, Metals)*

An informative summary of the unique characteristics, uses, and difficulties related to the use of ceramics and metals as matrix materials in packaging is given in this table.

Matrix material	Characteristics	Niche applications in packaging	Challenges	References
Ceramics	High thermal resistance, hardness	Heat-resistant containers for high-temperature goods	Cost, difficulty in processing complex shapes, brittleness even after healing. Limited widespread use in conventional packaging	[67, 68]
	Chemical stability	Protective coatings for barrier protection against corrosion		
		Advanced cosmetic and pharmaceutical packaging for enhanced protection		
Metals	High strength, durability	Rigid containers for industrial goods	High cost, complex processing, weight. Limited use in typical packaging. Research focuses on cost-effective, scalable solutions	[69]
	Conductivity	Specialty foils for electromagnetic shielding in electronic		
		Packaging premium beverage containers with enhanced durability		

4 Discussion of Applications and Case Studies: Potential of Self-healing Composites Materials

Even though the inception of self-healing composites is not much old, their absolute potential lies in their real-world applications. Several encouraging examples highlight its potential and progress including new commercial items that are starting to incorporate self-healing technology as well as research prototypes that show the concept's viability. These examples not only show how self-healing technology may be used practically, but they also raise the possibility of future mass use. Below, we discuss the applications of self-healing composites which explains their efficacy and practicality [70, 71].

4.1 Research Prototypes

Researchers from a top institution have created a food packaging film that contains healing agent-filled microcapsules. Upon tearing or puncturing the film, the microcapsules break down and release the agent that promptly seals the opening and preserves the barrier against oxygen and moisture. By reducing exposure to air and pollutants, the shelf life of perishable items might be extended [72, 73]. A fiber-reinforced polymer matrix composite capable of autonomously healing delamination (is a process where a material separates into layers) at room temperature utilizes woven graphite fiber and an epoxy matrix, incorporating 20 wt% microcapsules filled with a DCPD monomeric healing agent and 5 wt% of a ROMP catalyst (Grubbs' catalyst). Delamination areas were pre-filled with these microcapsules. Specimens were clamped and allowed to heal at room temperature for 48 h, achieving an average healing efficiency of 38%, with peaks up to 45%. The healing efficiency increases as the temperature rises to 80 °C, improved to an average of 66%, with a maximum of over 80%. Control experiments with manual injection of the healing agent showed a 100% recovery, demonstrating the potential for repairing significant delamination [74]. Researchers at the University of Bristol have shown the potential of biomimetic self-healing in sophisticated composite structures with hollow glass fibers in a groundbreaking study. The durability and endurance of composite materials used in a variety of applications, including packaging, are greatly increased by this novel technique, which enables the autonomous repair of delamination in fiber-reinforced polymers at ambient temperature [75]. Another recent study used in-situ polymerization to encapsulate two diene monomers, 5-ethylidene-2-norbornene (ENB) and dicyclopentadiene (DCPD), for use as self-healing agents in polymeric composites. Differential scanning calorimetry and dynamic mechanical research revealed that ENB in particular exhibits encouraging properties as a self-healing agent; it creates a resin with a higher glass transition temperature and cures more quickly and with less catalyst than DCPD. When detailed characterization and evaluation of their thermal and physical properties was carried out, these microcapsule's potential use in enhancing the tensile properties of composite materials was confirmed [76]. Using EPON 828 and DMP-30, researchers have created a self-healing epoxy composite that improves performance and heat resistance at temperatures as high as 200 °C by incorporating a microencapsulated epoxy/mercaptan healing agent. This novel composite is perfect for high-performance applications where thermal stability is essential since it exhibits notable autonomous restoration capabilities and performs well in fatigue tests and fracture mechanics [77]. A thermal-responsive polymer used in a beverage can prototype repairs dents or scratches in response to temperature changes. In order to stop leaks and preserve the product's quality while it is being transported and stored, this technology makes sure that the packaging is sturdy and long-lasting [78].

A novel class of bio-based, photo-crosslinkable monomers has been created by Sinha Roy et al. using syringaldehyde and vanillin, which are byproducts of lignin oxidation. These monomers contain p-hydroxycinnamic acids, which are essential for permitting [2 + 2] cycloaddition events that promote self-healing because of their photo-responsive reversible characteristics. Computer investigations were done to confirm the structure–activity correlations of these monomers which reveals their increasing potential for rapid use in self-healing technology [79].

4.2 Commercially Available Products

Self-healing polymer films incorporating microencapsulated healing agents was developed. Fabricated films showed the ability to repair minor damages, leading to extended shelf life and reduced waste of food products. In one study, the healing of punctures in flexible films and thus preserving the barrier properties highlights the potential for self-healing films to revolutionize food packaging. However, these prototypes while not yet widely commercialized but have huge future perspective [49, 72]. Self-healing coatings that can repair scratches and minor damage, improving the aesthetic appeal and extending the lifespan of the containers have been developed. Their real-world application could be rigid containers, such as bottles and jars [80]. To protect sensitive electronic equipment's during shipping and handling, self-healing coatings could be applied in outer surface of packaging. As an example, Zhu et al. reported a coating that can self-heal minor damages and reduce the chances of damage to the electronics [81]. Self-healing polymers in closures and seals for packaging were explored by Tavassoli et al. In pharmaceutical or food packaging, this application is particularly relevant as these maintain a hermetic seal [82].

However, there are still challenges in widespread adoption of self-healing composite packaging and as well as their commercialization. Two critical issues that must be overcome are the overwhelming cost of self-healing materials and the problem of scalability. Despite these challenges, self-healing composites appear to have a promise in future in the packaging industry, especially with the ongoing research and development initiatives and the growing demand for long-lasting and environmentally friendly packaging options.

4.3 Potential Applications of Self-healing Composites in Packaging

Emphasizing the methods, materials employed, and possible advantages, this table briefly summarizes the various uses of self-healing technologies throughout many kinds of packaging and industries.

Packaging type/industry	Self-healing approach	Material(s)	Potential benefits	Status/example	References
Food—flexible films	Microencapsulated healing agents (e.g., DCPD, epoxy)	PLA, biopolymers, polyolefins (PE, PP)	Improved barrier qualities, longer shelf life, less food waste, and self-sealing tears and punctures	Investigate the possibilities and prototypes for snacks, fresh fruit, etc.	[76, 77]

(continued)

(continued)

Packaging type/ industry	Self-healing approach	Material(s)	Potential benefits	Status/example	References
Films that are flexible (general)	Vascular networks with healing agents	Different polymers	Greater robustness, resistance to tearing, and content protection	Prototypes for research; possibilities for high-value products and industrial packaging	[83, 84]
Food and beverage—rigid containers	Self-healing coatings (microcapsules, reversible bonds)	PET, other plastics	Increased durability of containers, improved appearance, decreased waste, and improved scratch resistance	Research prototypes; the possibility of reusable jars and bottles	[49, 85]
Electronics—rigid containers	Self-healing coatings and integrated systems	Specialized plastics and composites	Protecting sensitive components and repairing minor damage during transportation	Research prototypes; potential for high-value electronics packaging	[66, 81]
Closures and seals	Self-healing elastomers, polymers	Integrated into packaging closures	Increased shelf life, improved seal integrity, and leak avoidance	Research prototypes; potential for food, pharmaceutical, packaging	[82, 86]
Coatings (various)	Self-healing polymers, silanes	Used with pre-existing packaging materials (plastics, paperboard)	Improved barrier qualities, increased scratch resistance, and longer packing life	Research prototypes; potential for various packaging types	[87]

5 Challenges of Self-healing Materials in Packaging

To take the full benefit of self-healing technology, it is important to understand and address the challenges and hurdles of self-healing materials before they can be widely adopted in the packaging sector.

Self-healing materials are frequently more expensive to produce than traditional materials, especially when they include complex chemical combinations or advanced manufacturing techniques. For instance, encapsulation, which essentially involves encasing healing agents in capsules and adding the material matrix with a layer of catalyst to activate the self-healing mechanism when a crack begins, is discovered to have significant drawbacks due to the high cost of the catalyst [88]. Mass adoption may be hampered in market areas where price sensitivity is high for the consumers. For wider use, cost reduction through more economical material sources or more effective production techniques will be essential.

There are many obstacles to overcome when producing self-healing materials on an industrial scale from lab settings. Significant advancements in process engineering and quality control are necessary to guarantee consistent quality and performance on a broad scale, such as those that use vascular networks or microencapsulated healing agents. The process of microencapsulation involves introducing microscopic capsules with a healing agent into the material matrix. Because of the necessity for exact control over capsule size, wall thickness, and dispersion throughout the composite, reaching uniform distribution and consistent capsule rupture behavior on an industrial scale is challenging. Comparably, complex network manufacturing is needed for composites with integrated vascular networks, which imitate biological systems to supply healing agents. This is challenging to accomplish on a wide scale without sacrificing structural integrity and functioning [39, 89]. Both of these technologies require high levels of precision and sophisticated manufacturing processes, which raise production costs and technical requirements considerably and make large-scale implementation difficult. Financially feasible scale-up as well as prevention of material's beneficial properties should be recognized. Those applications that involve food contact must take care of the material compatibility and to avoid contamination must adhere to strict health and safety regulations. This includes making certain that no hazardous materials are introduced into the food environment by the healing agents and procedures used [90].

Self-healing system's intricate chemistry can make recycling and end-of-life processing difficult, making these materials less recyclable. Among these materials which are easily segregable or compatible with ongoing recycling procedures is crucial for environmental sustainability. For example, the presence of non-biodegradable microcapsules in epoxy-based polymers containing microencapsulated therapeutic agents such as dicyclopentadiene (DCPD) makes mechanical recycling more difficult [74, 91]. In a comparable manner, thermoset composites that use reversible processes like Diels–Alder and dual-component systems with distinct hardeners and resins are challenging to dissociate and treat using conventional recycling techniques [42, 92]. Material deterioration results from the need for precise temperature controls for recycling shape memory materials, such as polyurethanes and nickel-titanium alloys, which are not common in present methods. To link the benefits of self-healing technologies with environmental sustainability goals, these issues underscore the necessity of creating self-healing materials that are either intrinsically biodegradable or more compatible with current recycling procedures.

When a new material dealing food and pharmaceuticals are launched into the market, then it involves not only navigates extensive regulatory landscapes but also consumes a lot of time and money for gaining regulatory approval. Several instances demonstrate the significance of combining innovation and regulatory strategy in the development of self-healing materials for industries such as food and medicines. For example, EU Regulation 10/2011 on plastic materials intended to encounter food must be followed by self-healing packaging materials in the European Union. To make sure that no dangerous materials are transported to food, this calls for thorough migration testing and risk assessment [93]. Under the Federal Food, Drug, and Cosmetic Act, which requires that materials be generally recognized as safe (GRAS) before they may be used in food contact applications, comparable products would have to satisfy FDA criteria in the US [94]. Getting around these rules not only makes sure safety and compliance, but it also makes entering the market easier. This shows how important it is to have a plan that includes both new technologies and proactive regulatory involvement.

Beyond the current challenges, the field of self-healing packaging is being accelerated by innovative new trends that hold the potential to expand its uses and improve its capabilities. These trends push the limits of what is feasible in terms of packaging design and usefulness by utilizing cutting-edge technologies and environmentally friendly methods.

The creative fusion of cutting-edge technologies and bio-inspired solutions is where self-healing packaging is headed. Researchers are laying the foundation for a new generation of packaging that is not only self-repairing but also intelligent and ecologically conscious by investigating cutting-edge trends like nanotechnology, smart materials, and bio-based systems.

This table offers a perceptive summary of new developments in self-healing packaging with an emphasis on bio-based systems, smart materials, and nanotechnology.

Trend	Description	Potential applications and benefits	Research and development focus	Challenges and considerations	References
Nanotechnology	Utilizing nanomaterials (nanoparticles, nanofibers, nanotubes) to enhance self-healing properties	Improved mechanical strength Enhanced healing efficiency Controlled release of healing agents Creation of nanoscale healing networks	Development of nano-encapsulation techniques, incorporation of nanomaterials into polymer matrices, creation of nanoscale vascular networks, and application of responsive nanomaterials	Ensure nanomaterial biocompatibility and safety, achieve uniform dispersion and integration, increase production capacity, and handle potential environmental concerns	[95, 96]
Smart materials	Using materials that respond to external stimuli (such as heat, light, pH, and stress) to initiate self-healing	On-demand healing, targeted repair, real-time damage detection, adaptive packaging solutions, and better shelf-life monitoring	Developing stimuli-responsive polymers, integrating sensors and actuators, creating smart coatings, and designing self-diagnostic packaging	Obtaining precise control over stimulus–response, assuring long-term stability and dependability, integrating sensors and actuators at a low cost, and resolving data security and privacy concerns	[97]
Bio-based self-healing systems	Utilizing healing agents and natural polymers made from sustainable resources	Reduced environmental impact, greater biodegradability and compostability, increased biocompatibility, and sustainable packaging options	Extraction and purification of natural healing agents, creation of bio-based microcapsules and vascular networks, design of bio-inspired self-healing processes, and application of bacterial cellulose and other natural polymers	Ensure consistent quality and performance of bio-based materials, solve potential constraints in mechanical characteristics, increase production from renewable resources, and ensure cost-effectiveness	[98, 99]

6 Conclusion and Future Perspective

From natural containers to advanced self-healing composite packaging materials have come a long way highlighting exponential progress that shows our constant drive for better sustainability and innovation. These cutting-edge materials provide a sustainable solution to increase packaging longevity, reduce waste, and guarantee safe delivery of goods. As mentioned in the earlier sections of this book chapter, they are inspired by how nature heals and repairs itself. Starting from examining the historical context, next highlighting innovative materials, and lastly projecting future developments, this book chapter has projected self-healing composites as a game-changer in the field of packaging rather than just a scientific curiosity.

This chapter underscores the versatility and trust of self-healing technology, spanning from the intricate mechanisms of polymer chemistry to the investigation of diverse applications. Based on their versatility and regenerative remark these materials are being developed into rigid and flexible packaging solutions extending their capacity to satisfy the ever changing needs of society. Once more, these materials present exceptional chances to prolong product lifecycles, improve economic efficiency, and lessen environmental effect, combining the molecular inventiveness of intrinsic healing with the engineering accuracy of extrinsic treatments.

Continued research of this domain into advanced fields like bio-based systems, smart materials, and nanotechnology further promises the future of packaging quite environment friendly and sustainable, providing ecologically friendly, sensitive, and intelligent packaging solutions rather than simply fixing damage. Future research and development in this area will surely open up new avenues, making self-healing packaging a vital part of a circular industrial economy.

However, there are many obstacles in the way of broad adoption, including financial viability, manufacturing scalability, material compatibility, recyclability, and strict regulatory environments. Researchers, industry stakeholders, and regulators must work together to overcome obstacles in order to promote technologically sophisticated, commercially feasible, and ecologically conscious solutions.

The exciting prototypes and new commercial uses covered here offer an idea of what might be achieved when consumer packaging and state-of-the-art science come together. These materials could revolutionize the packaging industry, whether it's keeping food fresh longer or making electronics and diverse material last.

Finally, self-healing composite packaging is not only a new technology but also have the capability to completely change the packaging industry. Groundbreaking change in our perspective to packaging is that from being one of consumption to disposal, it will become one of sustainability and renewable. The findings and advancements covered in this chapter pave the way for upcoming researches as we approach the future, in which packaging not only protects but also preserves.

Declaration of Competing Interest The authors declare that they have no known competing financial interests or personal relationships that could have appeared to influence the work reported in this paper.

References

1. Yam KL (2010) The Wiley encyclopedia of packaging technology. Wiley
2. Brody AL et al (2008) Innovative food packaging solutions. J Food Sci 73(8):107–116
3. Bevan A (2014) Mediterranean containerization. Curr Anthropol 55(4):387–418
4. Miller SW (2007) An environmental history of Latin America. Cambridge University Press
5. Rasmussen SC (2012) How glass changed the world: the history and chemistry of glass from antiquity to the 13th century. Springer Science & Business Media
6. Erdal G (2021) Metal and metal packaging. Sentez Yayıncılık
7. Twede D (2016) History of packaging. In: The Routledge companion to marketing history. Routledge, pp 115–129
8. Marsh K, Bugusu B (2007) Food packaging—roles, materials, and environmental issues. J Food Sci 72(3):R39–R55
9. Akbarzadeh J et al (2014) Timescales of self-healing in human bone tissue and polymeric ionic liquids
10. Enyedi B, Niethammer P (2015) Mechanisms of epithelial wound detection
11. Huffman MA (1997) Current evidence for self-medication in primates: a multidisciplinary perspective. Am J Phys Anthropol 40:171–200
12. Szpaderska AM, Zuckerman JD, DiPietro LA (2003) Differential injury responses in oral mucosal and cutaneous wounds
13. Emberts Z, Escalante I, Bateman PW (2019) The ecology and evolution of autotomy 3
14. Zhang G-y et al (2013) Chemical constituents of the starfish *Asterias rollestoni* Bell
15. Joven A, Elewa A, Simon A (2019) Model systems for regeneration: salamanders
16. Morgan TH (1902) Experimental studies of the internal factors of regeneration in the earthworm
17. Li Y et al. Sea cucumber genome provides insights into saponin biosynthesis and aestivation regulation
18. Benson H, Stark M (2009) Timeless healing. Scribner
19. Learn Genetics, Genetic Science Learning Center. What is blood? Available from: https://learn.genetics.utah.edu/
20. Toohey KS, Sottos NR, White ASR. Self-healing materials with microvascular networks
21. Hamilton BAR, Sottos NR, White SR. Self-healing of internal damage in synthetic vascular materials
22. Singh S, Young A, McNaught C-E (2017) The physiology of wound healing
23. Kahar NMN et al (2021) The versatility of polymeric materials as self-healing agents for various types of applications: a review
24. Speck O, Speck T (2019) An overview of bioinspired and biomimetic self-repairing materials
25. Diesendruck CE et al (2015) Biomimetic self-healing
26. Hager MD et al (2010) Self-healing materials
27. Li B et al (2022) Intrinsically self-healing polymers: from mechanistic insight to current challenges
28. Cremaldi JC, Bhushan B (2018) Bioinspired self-healing materials: lessons from nature
29. Yang Y, Urban MW (2013) Self-healing polymeric materials. Chem Soc Rev 42(17):7446–7467
30. Wang S, Urban MW (2020) Self-healing polymers. Nat Rev Mater 5(8):562–583
31. Paladugu SR et al (2022) A comprehensive review of self-healing polymer, metal, and ceramic matrix composites and their modeling aspects for aerospace applications
32. Zhu M, Li J, Yu J, Li Z, Ding B (2022) Superstable and intrinsically self-healing fibrous membrane with bionic confined protective structure for breathable electronic skin
33. Wang C, Li R, Chen P, Fu Y, Ma X, Shen T, Zhou B, Chen K, Fu J, Bao X, Yan W, Yang Y (2021) Highly stretchable, non-flammable and notch-insensitive intrinsic self-healing solid-state polymer electrolyte for stable and safe flexible lithium batteries
34. Khatib M, Huynh T-P, Deng Y, Horev YD, Saliba W, Wu W, Haick H (2019) A freestanding stretchable and multifunctional transistor with intrinsic self-healing properties of all device components

35. Chen J, Luo Z, An R, Marklund P, Björling M, Shi Y (2022) Novel intrinsic self-healing poly-silicone-urea with super-low ice adhesion strength
36. Chowdhury MIS et al (2022) Polymer nanocomposites for automotive applications, chap 11. In: Enamul HM, Kumar R, Ahmed S (eds) Advanced polymer nanocomposites. Woodhead Publishing, pp 267–317
37. Ilyaei S, Sourki R, Akbari YH (2021) Capsule-based healing systems in composite materials: a review. Crit Rev Solid State Mater Sci 46(6):491–531
38. Nakao W, Takahashi K, Ando K (2008) Self-healing of surface cracks in structural ceramics. In: Self-healing materials, pp 183–217
39. Ullah H, Azizli KAM, Man ZB, Ismail MBC, Khan MI (2016) The potential of microencapsulated self-healing materials for microcracks recovery in self-healing composite systems: a review. Polym Rev 56(3):429–485
40. Salonitis K et al (2010) Multifunctional materials: engineering applications and processing challenges. Int J Adv Manuf Technol 49(5):803–826
41. Gopalakrishnan K, Mishra P (2024) Self-healing polymer a dynamic solution in food industry: a comprehensive review. Food Biophys 19(1):1–17
42. Hia IL, Vahedi V, Pasbakhsh P (2016) Self-healing polymer composites: prospects, challenges, and applications. Polym Rev 56(2):225–261
43. Mauldin TC, Kessler MR (2010) Self-healing polymers and composites. Int Mater Rev 55(6):317–346
44. Rajak D et al (2019) Fiber-reinforced polymer composites: manufacturing, properties, and applications. Polymers 11(10):1667
45. Sahmaran M, Li VC, Andrade C (2008) Corrosion resistance performance of steel-reinforced engineered cementitious composite beams
46. Rubino F et al (2020) Marine application of fiber reinforced composites: a review. J Mar Sci Eng 8(1):26
47. Hemachandran E, Singh P, Singh VK (2024) Recent manufacturing approaches for composite materials. Composites. CRC Press, pp 33–52
48. Mylsamy G, Krishnasamy P (2023) A review on electrical properties of fiber-reinforced polymer material: fabrication, measurement, and performances. Trans Indian Inst Met 76(7):1691–1708
49. Lai W-F (2023) Design and application of self-healable polymeric films and coatings for smart food packaging. npj Sci Food 7(1):11
50. Trinh BM, Chang BP, Mekonnen TH (2023) The barrier properties of sustainable multiphase and multicomponent packaging materials: a review. Prog Mater Sci 133:101071
51. Ebnesajjad S (2012) Plastic films in food packaging: materials, technology and applications. William Andrew
52. Nisticò R (2020) Polyethylene terephthalate (PET) in the packaging industry. Polym Test 90:106707
53. Sarda P et al (2022) Sustainability performance of polyethylene terephthalate, clarifying challenges and opportunities. J Polym Sci 60(1):7–31
54. Taib N-AAB et al (2023) A review on poly lactic acid (PLA) as a biodegradable polymer. Polym Bull 80(2):1179–1213
55. Swetha TA et al (2023) A comprehensive review on polylactic acid (PLA)—synthesis, processing and application in food packaging. Int J Biol Macromol 234:123715
56. Mokwena KK, Tang J (2012) Ethylene vinyl alcohol: a review of barrier properties for packaging shelf stable foods. Crit Rev Food Sci Nutr 52(7):640–650
57. Zhu D, Rong M, Zhang M (2015) Microcapsule-based self-healing materials. In: Recent advances in smart self-healing polymers and composites, pp 101–127
58. Morrison CE, Laurent EM, Li ZH (2015) Damage detection in a microencapsulated dicyclopentadiene and Grubbs' catalyst self-healing system
59. Cho SH, Andersson HM, White SR, Sottos NR, Braun PV (2006) Polydimethylsiloxane-based self-healing materials

60. Yan X, Qian X, Chang Y (2019) Preparation and characterization of urea formaldehyde @ epoxy resin microcapsule on waterborne wood coatings. Coatings 9(8):475
61. An S et al (2018) A review on corrosion-protective extrinsic self-healing: comparison of microcapsule-based systems and those based on core-shell vascular networks. Chem Eng J 344:206–220
62. Amaral AJ, Pasparakis G (2017) Stimuli responsive self-healing polymers: gels, elastomers and membranes. Polym Chem 8(42):6464–6484
63. Caruso MM et al (2009) Mechanically-induced chemical changes in polymeric materials. Chem Rev 109(11):5755–5798
64. Kumar A ct al (2023) A comprchcnsivc rcvicw of various biopolymcr compositcs and thcir applications: from biocompatibility to self-healing. Mater Today Sustain 23:100431
65. Bera S, Haldar D (2016) A rechargeable self-healing safety fuel gel. J Mater Chem A 4(18):6933–6939
66. Ekeocha J et al (2021) Challenges and opportunities of self-healing polymers and devices for extreme and hostile environments. Adv Mater 33(33):2008052
67. Otitoju TA et al (2020) Advanced ceramic components: materials, fabrication, and applications. J Ind Eng Chem 85:34–65
68. Iqbal A, Moskal G (2023) Recent development in advance ceramic materials and understanding the mechanisms of thermal barrier coatings degradation. Arch Comput Methods Eng 30(8):4855–4896
69. Morris BA (2022) The science and technology of flexible packaging: multilayer films from resin and process to end use. William Andrew
70. Dallaev R (2024) Advances in materials with self-healing properties: a brief review. Materials 17(10):2464
71. Li G (2014) Self-healing composites: shape memory polymer based structures. Wiley
72. Baghi F et al (2022) Advancements in biodegradable active films for food packaging: effects of nano/microcapsule incorporation. Foods 11(5):760
73. Ullah H et al (2016) The potential of microencapsulated self-healing materials for microcracks recovery in self-healing composite systems: a review. Polym Rev 56(3):429–485
74. Kessler M, Sottos NR, White S (2003) Self-healing structural composite materials. Compos A Appl Sci Manuf 34(8):743–753
75. Trask R, Bond I (2006) Biomimetic self-healing of advanced composite structures using hollow glass fibres. Smart Mater Struct 15(3):704
76. Lee JK et al (2004) Characterization of dicyclopentadiene and 5-ethylidene-2-norbornene as self-healing agents for polymer composite and its microcapsules. Macromol Res 12:478–483
77. Yuan YC et al (2011) Self-healing epoxy composite with heat-resistant healant. ACS Appl Mater Interfaces 3(11):4487–4495
78. Zhong N (2020) Self-healing thermal interface materials
79. Roy PS et al (2021) Bio-based photo-reversible self-healing polymer designed from lignin. Green Chem 23(24):10050–10061
80. Chen Y et al (2023) "Smart" micro/nano container-based self-healing coatings on magnesium alloys: a review. J Magn Alloys
81. Zhou Y et al (2022) Self-healing polymers for electronics and energy devices. Chem Rev 123(2):558–612
82. Tavassoli M et al (2025) Self-healing packaging films/coatings for food applications; an emerging strategy. Adv Coll Interface Sci 103423
83. Lee MW et al (2018) Advances in self-healing materials based on vascular networks with mechanical self-repair characteristics. Adv Coll Interface Sci 252:21–37
84. Minnebo P et al (2017) A novel design of autonomously healed concrete: towards a vascular healing network. Materials 10(1):49
85. Zhu DY, Rong MZ, Zhang MQ (2015) Self-healing polymeric materials based on microencapsulated healing agents: from design to preparation. Prog Polym Sci 49:175–220
86. Wang D-P et al (2019) An ultrafast self-healing polydimethylsiloxane elastomer with persistent sealing performance. Mater Chem Front 3(7):1411–1421

87. Zhang F et al (2018) Self-healing mechanisms in smart protective coatings: a review. Corros Sci 144:74–88
88. Purohit R et al (2015) An investigation on manufacturing of self-healing materials. Mater Today Proc 2(4–5):3371–3377
89. Ghosh SK (2006) Functional coatings and microencapsulation: a general perspective. In: Functional coatings: by polymer microencapsulation, pp 1–28
90. Geueke B, Groh K, Muncke J (2018) Food packaging in the circular economy: overview of chemical safety aspects for commonly used materials. J Clean Prod 193:491–505
91. Naveen V, Raja S, Deshpande AP (2019) A two-step approach for synthesis, characterization and analysis of dicyclopentadiene–urea formaldehyde–siloxane-based double-walled microcapsules used in self-healing composites. Int J Plast Technol 23(2):157–169
92. Khan NI, Halder S, Wang J (2019) Diels–Alder based epoxy matrix and interfacial healing of bismaleimide grafted GNP infused hybrid nanocomposites. Polym Test 74:138–151
93. Plastic Recycling. Recycling of plastic intended for contact with food. Available from: https://food.ec.europa.eu/food-safety/chemical-safety/food-contact-materials/plastic-recycling_en#:~:text=Ensuring%20the%20safety%20of%20recycled,according%20to%20internationally%20accepted%20standards
94. Generally Recognized as Safe (GRAS) (2023)
95. Christian P (2009) Nanomaterials: properties, preparation and applications. In: Environmental and human health impacts of nanotechnology, pp 31–77
96. Yang Y et al (2019) Self-healing of electrical damage in polymers using superparamagnetic nanoparticles. Nat Nanotechnol 14(2):151–155
97. Cui X et al (2023) Sponge-like, semi-interpenetrating self-sensory hydrogel for smart photothermal-responsive soft actuator with biomimetic self-diagnostic intelligence. Chem Eng J 467:143515
98. Tziviloglou E et al (2016) Bio-based self-healing concrete: from research to field application. In: Self-healing materials, pp 345–385
99. Gupta S, Pang SD, Kua HW (2017) Autonomous healing in concrete by bio-based healing agents—a review. Constr Build Mater 146:419–428

Chapter 8
Polymer-Based Composites for Future Packaging Materials

Pragti Saini, Pankti Patel, Himadri Chatterjee, Subrata Biswas, Sampat Singh Bhati, and Dharm Dutt

1 Introduction

Packaging is a fundamental element of contemporary industries, and it serves a variety of purposes beyond enclosing and safeguarding a product. Packaging has an important role in maintaining product integrity, improving shelf life, being convenient and playing a role in brand identification. With the progression of industries and the growing demands of consumers, packaging has developed into a highly advanced science that involves integrating advanced materials, creative designs and intelligent functions. Among all the packaging materials, polymer-based composites have become a highly sought-after material because of their versatility, light weight, strength and capacity to achieve targeted performance [27]. With the ever-increasing environmental concerns regarding sustainability, high-performance, eco-friendliness is now an immediate need for packaging solutions. Polymer-based composites provide promising alternatives to traditional packaging materials, resolving issues such as waste generation, recyclability and improved barrier properties while maintaining product safety and industrial efficiency [49].

Packaging material selection has a direct influence on product quality, marketability and customer satisfaction. The packaging material must be tailored to the respective demands of various industries, such as food and beverage, pharmaceuticals, cosmetics, electronics and e-commerce [30]. One of the main functions of packaging is to shield products from outside influences like moisture, oxygen, light, microbial contamination and mechanical damage. For example, in the food sector, packaging materials with high-barrier properties avoid spoilage, keeping food fresh for a long time. Likewise, in the pharmaceutical industry, packaging

P. Saini · P. Patel · H. Chatterjee · S. Biswas (✉) · S. S. Bhati · D. Dutt
Department of Paper and packaging Technology, Indian Institute of Technology Roorkee, Roorkee, India
e-mail: subrata.biswas@pt.iitr.ac.in

A. Verma et al. (eds.), *Polymers and Composite Materials for Packaging: Smart Food Packaging and Solutions*, Materials Horizons: From Nature to Nanomaterials,
https://doi.org/10.1007/978-981-95-0677-4_8

protects medicines from environmental influences, keeping them effective and free from contamination. Aside from protection, packaging also increases functionality and convenience [35]. Resealability, easy opening, portability and light weight are some of the features that make packaging more consumer-friendly and usable. Intelligent designs like tamper-proof seals, child-resistant closures and QR codes for product tracking and authenticity checking are some of the innovations in packaging solutions. These innovations address the changing needs of industries and consumers, further emphasizing the need to select the appropriate packaging materials. Marketing and branding are also essential factors that are affected by packaging. Innovative and engaging packaging designs deliver a powerful visual presence, setting products apart in the face of competition. The capacity to imprint quality graphics, brand logos and product details on packaging materials contributes to an increased value of consumer interaction and buying decisions [50]. This is especially relevant in sectors like cosmetics and upscale products, where beauty plays a significant role in influencing consumer behaviour. The worldwide trend of e-commerce and online shopping has further highlighted the importance of packaging. E-commerce packaging needs to provide protection for the product while in transit, avoid damages and reduce returns. Cost-efficient and eco-friendly packaging solutions are essential to reduce shipping expenses, maximize storage space and lower environmental footprints. The demand for intelligent, adaptable and light packaging has promoted the use of polymer-based materials that are superior in performance to conventional materials.

As businesses develop, so too does the concern for packaging waste, environmental pollution and natural resource depletion. Traditional packaging products, including one-time use plastics, are primary contributors to landfill build-up and marine pollution. Governments, industry and consumers alike have elevated sustainable packaging alternatives as a necessity. High-performance packaging products must not only ensure outstanding functionality but also meet the principles of sustainability by minimizing waste, enhancing recyclability and reducing carbon prints. One of the major drawbacks of traditional plastic packaging is its extended degradation period and role in environmental pollution. Polyethylene (PE), polypropylene (PP) and polyethylene terephthalate (PET) plastics are among those that degrade in hundreds of years, resulting in serious ecological impacts. Industries are, therefore, turning towards biodegradable, compostable and recyclable packaging materials to counter this [37]. Biopolymers like polylactic acid (PLA), polyhydroxyalkanoates (PHA) and starch plastics provide environmentally friendly substitutes with similar performance to conventional plastics. These materials break down naturally, lowering the environmental footprint and enhancing circular economy values.

Besides biodegradability, advanced packaging solutions prioritize minimizing material usage and optimizing package design. Lighter-weight packaging materials reduce raw material consumption, decrease transportation costs and minimize total waste generation. Sophisticated manufacturing methods, like 3D printing and precision moulding, allow for the production of tailor-made packaging with minimal wastage of resources. Additionally, intelligent packaging technologies, such as active and intelligent packaging, increase product shelf life and minimize food wastage. Active packaging features antimicrobial and oxygen-scavenging functionalities, whereas intelligent packaging employs sensors and indicators to track product freshness and quality. Sustainability in packaging also entails the utilization of recycled and upcycled materials [9]. Post-consumer recycled (PCR) plastic and bio-based polymers help decouple the use of fossil fuels and lower greenhouse gas emissions. Post-consumer recycled content packaging, including PET bottles made from recycled plastic, assists in closing the loop in plastic waste management. The introduction of mono-material packaging structures, where a single material of plastic is utilized so that it can be easily recycled, is a major innovation in sustainable packaging. Governments and regulatory agencies across the globe are enforcing tight policies to support environment-friendly packaging habits. Single-use plastic bags, extended producer responsibility (EPR) initiatives and environmental incentives for eco-friendly packaging innovation are nudging industries towards greener options. Industries are making investments in research and development to investigate new materials, biodegradable coatings and novel packaging formats that are compatible with sustainability objectives while maintaining excellent performance [38].

Polymer-based composites are proving to be a game-changing answer to the needs of the packaging sector. Polymeric materials are mixed with fibres, nanoparticles, or bio-based fillers as reinforcing additives to produce composites with enhanced mechanical, thermal and barrier attributes. Polymer composites provide light, strong and versatile packages that surpass the performance of conventional materials in strength, flexibility and functionality. Among various benefits of polymer-based composites, one of the most important is that they can enhance barrier properties [10]. Several applications in packaging demand materials that can be able to block effectively the passage of moisture, oxygen, UV light and impurities. By adding nanomaterials like graphene, clay nanoparticles and silica, polymer composites are able to provide much better barrier performance. This is especially useful in food and pharma packaging, where product shelf life and safety are the main concerns. Along with greater barrier characteristics, polymer composites possess superior mechanical toughness and resistance. Brittleness, puncture ability and impact vulnerability of conventional plastic packaging are possible problems. Adding reinforcement of polymers using cellulose, glass, or carbon fibres enables composite materials to have more flexibility and toughness. It becomes possible to produce thinner and lighter-weight packages with reduced structure failure without a corresponding reduction in performance, thus lower material use and economic savings.

Yet another important benefit of polymer-based composites lies in their future possibilities for recyclability and biodegradability. By integrating bio-based polymers with sustainable reinforcements, composite materials can be designed to easily degrade or recycle effectively. Through the addition of plant-based fibres to biopolymer matrices, the composting properties of package materials are strengthened without compromising their functional performance. These developments lend momentum to increased market demand for sustainable packaging materials. The flexibility of polymer composites carries over to intelligent and functional packaging uses. Special composite materials may be engineered to have antimicrobial activity to suppress bacterial growth and are therefore best suited for medical and food packaging. Conductive polymer composites allow for embedding electronic devices like RFID tags and temperature sensors to track and monitor in real-time [15]. These innovations improve supply chain efficiency, product authenticity and consumer safety. In addition, polymer-based composites facilitate circular economy concepts through material recovery and reuse. In contrast to multi-layered packaging structures, which are difficult to recycle, composite materials can be designed for simple disassembly and separation. Closed-loop recycling systems, where polymer composites are recycled into new packaging materials, are in line with international sustainability efforts.

2 Types of Polymer-Based Composites for Packaging

The development of packaging materials has been influenced by the demand for enhanced performance, sustainability and cost savings. Among other advanced materials, polymer-based composites have become extremely popular because of their enhanced properties, including increased strength, flexibility, barrier properties and environmental advantages. These composites are generally created by mixing polymers with reinforcing agents or additives to improve their physical, mechanical and functional properties. With increasing concern regarding environmental sustainability and waste minimization, polymer composites have proven to be a suitable solution for contemporary packaging. They can be categorized into four broad types: bio-based polymer composites, recycled polymer composites, nanocomposites for packaging and hybrid polymer composites. All of these categories have specific functions in fulfilling different requirements related to packaging materials, from biodegradability to enhanced mechanical and barrier properties.

2.1 Bio-based Polymer Composites

Bio-based polymer composites have also received a lot of attention owing to their sustainable nature and their ability to replace traditional petroleum-based plastics [39]. They are mainly made up of biodegradable polymers reinforced with natural fillers, bio-based additives, or organic fibres. The most popular bio-based polymer composites used in packaging are polylactic acid (PLA), polyhydroxyalkanoates (PHA), starch-based composites and chitosan composites.

2.2 Polylactic Acid (PLA) Composites

PLA is one of the most widely applied bio-based polymers for green packaging. Made from renewable feedstocks like corn starch or sugarcane, PLA has good biodegradability, transparency and mechanical properties. When blended with natural fibres or fillers like cellulose, bamboo fibre, or lignin, PLA composites demonstrate improved thermal stability and mechanical properties. PLA composites are widely applied in food packaging, disposable cutlery and biodegradable films. Nonetheless, PLA's natural brittleness and diminished barrier properties when contrasted with conventional plastics have contributed to the creation of composite blends with other biodegradable polymers or reinforcements to enhance durability and flexibility.

2.3 Polyhydroxyalkanoates (PHA) Composites

PHA is another prospective bio-based polymer that is manufactured using bacterial fermentation of renewable resources. PHA-based composite materials have been noted for their superior biodegradability under marine and soil conditions, whereby they are a perfect match for eco-friendly packaging [33]. In order to improve their mechanical properties, PHA is frequently reinforced with organic fillers like chitosan, starch, or cellulose. Such composites find extensive applications in food packaging, agricultural films and biomedical applications. Their natural degradability without the release of toxic residues makes them very attractive in the drive towards a circular economy.

2.4 Starch-Based Composites

Starch-based composites are among the most extensively studied bio-composites because of their low cost, abundance and biodegradability. These composites usually contain starch mixed with other biodegradable polymers like polylactic acid (PLA) or polyvinyl alcohol (PVA) to improve their water resistance and mechanical strength. Starch-based packaging materials are widely employed in food wraps, biodegradable bags and disposable packages. While they are environmentally friendly, starch-based composites have limitations in their water sensitivity and mechanical weakness that are mitigated with cross-linking agents or plasticizers.

2.5 Chitosan Composites

Chitosan, a biopolymer obtained from chitin (present in shrimp and crab shells), is increasingly being used due to its antimicrobial and biodegradable nature. Chitosan composites are usually reinforced with nanoparticles, cellulose, or bio-fillers to enhance their mechanical properties and resistance to moisture. These composites find special use in food packaging, where they ensure longer shelf life by preventing microbial growth. Chitosan films and coatings offer a sustainable substitute to traditional synthetic films, diminishing food wastage and improving product storage.

2.6 Recycled Polymer Composites

Recycled polymer composites contribute significantly to plastic waste minimization and sustainability. The composites are produced by combining post-consumer recycled (PCR) plastics or post-industrial recycled polymers with the addition of reinforcing agents in order to improve properties. Examples of common ones include PCR plastics, rPET and rPE blends.

2.7 Post-consumer Recycled (PCR) Plastics

PCR plastics are made from post-consumer plastic packaging materials gathered from recycling programs. The recycled polymers are washed, processed and mixed with virgin materials or additives to regain their mechanical properties. PCR-based composites are extensively applied in rigid and flexible packaging, such as bottles, containers and films [21]. In spite of contamination and degradation issues during recycling, improvements in polymer processing technologies have enhanced the quality and uniformity of PCR-based packaging materials.

2.8 Recycled PET (rPET) and PE (rPE) Blends

Recycled PET (rPET) is a highly utilized recycled polymer in the packaging sector, especially for bottles of beverages and food packaging. rPET composites usually have reinforcing fibres or nano-additives to enhance mechanical properties and barrier characteristics. In the same vein, recycled polyethylene (rPE) blends are applied in flexible packaging formats like plastic bags, shrink films and agro-films. Recycled composites assist in mitigating virgin plastics dependency while facilitating circular economy applications.

2.9 Nanocomposites for Packaging

Nanocomposites are a new direction in packaging materials with better mechanical, thermal and barrier properties. The incorporation of nanomaterials like nano-cellulose, nano-clay, graphene and carbon nanotubes (CNTs) into polymer matrices leads to these composites having better functionality without jeopardizing biodegradability or recyclability.

2.10 Nano-cellulose Composites

Plant fibre-derived nano-cellulose is a potential green reinforcement material for sustainable packaging. Nano-cellulose composites have remarkable strength, lightweight nature and biodegradability. Nano-cellulose composites are applied in flexible films, coatings and bio-based laminates, providing enhanced moisture barrier and mechanical stability.

2.11 Nano-clay Composites

Nano-clay composites have broad applications in food packaging with outstanding barrier capability against oxygen, moisture and UV light. Adding nano-clay particles into polymer matrices maximizes the shelf life of consumable products. The nanocomposites are routinely applied in food wraps with superior performance, vacuum packaging and blister packs used in pharmaceutical products.

2.12 Graphene-Based Composites

Graphene composites have exceptional mechanical strength, electrical conductivity and antimicrobial properties. The nanocomposites are being investigated for smart packaging requirements, such as intelligent labels, sensors and active packaging that can sense food freshness or pick up contaminants.

2.13 Carbon Nanotube (CNT) Composites

Carbon nanotubes (CNTs) are another sophisticated nanomaterial that is utilized to strengthen polymer composites, providing enhanced mechanical properties and electrical conductivity. Composites of CNTs are applied in electronic packaging, antistatic coatings and smart packaging solutions for high-technology applications.

2.14 Hybrid Polymer Composites

Hybrid polymer composites involve the integration of two or more polymer types or reinforcement materials to yield multi-functionality. Such composites are expected to provide smart functions, improved performance and greater sustainability. Prominent examples include multi-layer polymer blends and smart composites.

2.15 Multi-layered Polymer Blends

Multi-layered polymer composites are made up of various polymer layers, each conferring unique functionalities like barrier protection, mechanical reinforcement, or thermal resistance. Such composites find widespread application in flexible packaging of food, drugs and industrial products. Multi-layered packaging is non-recyclable, and its recyclability poses a challenge, leading to efforts towards mono-material alternatives.

2.16 Smart and Functional Composites

Intelligent packaging solutions embed functional additives, e.g. temperature-sensitive compounds, moisture indicators and RFID tags, into polymer composites. These intelligent materials improve product safety, traceability and consumer interaction. Antimicrobial, self-healing, or shape-memory functional composites are also being developed for future-generation packaging applications.

3 Functional Requirements of Packaging Materials

See Fig. 1.

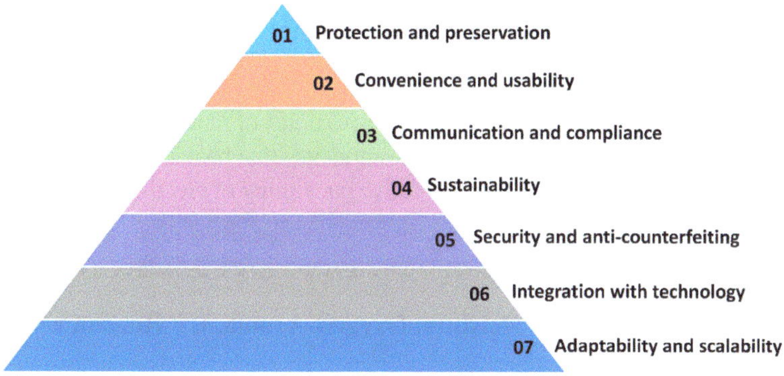

Fig. 1 Functional requirements of packaging materials

3.1 Important Characteristics of Polymer-Based Packages

Packaging is a vital component in the protection, preservation and presentation of products. Material selection for packaging mainly relies on their barrier, mechanical, thermal, biodegradability and aesthetic characteristics. Due to developments in polymer technology, polymer-based composites are now a preferred option for current packaging. These materials are designed to provide strength, flexibility, impact resistance, moisture and oxygen barrier, UV resistance, heat stability, biodegradable disposal and high printability. All of these characteristics contribute to the overall performance, durability and functionality of packaging materials.

3.1.1 Mechanical Properties (Strength, Flexibility, Impact Resistance)

The mechanical characteristics of a packaging material define its resistance to external forces, mechanical stress and environmental conditions without affecting the integrity of the packaged product. Mechanical performance is especially critical for rigid and flexible packaging uses, including bottles, containers, films and laminates. Tensile strength in a polymer composite determines its toughness against breaking or stretching when it is subjected to tension. Rigid packaging ensures that it remains undamaged under handling and shipping. Polymer composite materials with reinforcements of fibres, nanoparticles, or fillers possess greater tensile strength and hence are ideal for load-carrying applications. Carbon-fibre-reinforced polymer composites, for instance, offer maximum strength in lightweight packaging materials. Flexibility is a critical characteristic for pouches, films and flexible packaging containers. A packaging material should be flexible enough to bend, curve and conform to different shapes without cracking. Elastomeric polymers such as polyethylene (PE), polypropylene (PP) and polylactic acid (PLA) blends are excellent in terms of flexibility while also ensuring structural integrity as mentioned in Fig. 1 [42]. Flexible packaging finds extensive applications in food wraps, medical packaging and biodegradable shopping bags. Impact resistance is important for packaging that is frequently handled, stacked, or transported. Sensitive packaging materials can crack or deform under stress, resulting in product loss. Impact-modified toughened polymer composites, with rubber particles or nanofillers, improve toughness. An example is polycarbonate (PC) composites that have high impact resistance and find widespread application in protective packaging of electronic equipment and medical equipment.

3.1.2 Barrier Properties (Moisture, Oxygen, Aroma, UV Resistance)

Barrier properties define the ability of a packaging material to protect contents from exposure to the environment. Packaging needs to serve as a barrier to moisture, oxygen, UV radiation and aroma loss to keep product quality, shelf life and freshness intact.

Moisture Barrier

Moisture resistance is important for food, pharmaceutical and electronic packaging. Prolonged exposure to moisture can cause food spoilage, breakdown of pharmaceutical active ingredients and corrosion of electronic materials. Polymer composites filled with nano-clay or silica nanoparticles improve oxygen barrier properties. Polyvinyl alcohol (PVOH) composites, for instance, offer excellent moisture barrier and are employed in high-performance food packaging.

Oxygen Barrier

Oxygen exposure triggers oxidation, spoilage and deterioration in perishable items like dairy, meat and drinks. Ethylene vinyl alcohol (EVOH)-, nano-cellulose-, or graphene oxide-based oxygen scavenger composites provide higher gas barrier properties to avoid oxidation and improve shelf life. Multi-layer packaging films with high-barrier coatings, which are widely employed in vacuum-sealed food packaging, employ polyvinylidene chloride (PVDC) as an oxygen-blocking coating.

Aroma Barrier

Retaining the flavour and aroma of packaged items is crucial, particularly for coffee, spices and ready-to-eat food. Aroma barrier functionality holds volatiles within and keeps odours outside. Multi-layer polymer composites with aluminium foil or ethylene-vinyl alcohol (EVOH) are effective in maintaining aroma integrity.

UV Resistance

UV exposure leads to colour fading, nutrient loss and food and pharmaceutical product spoilage. UV-resistant packaging materials, including polyester (PET) with UV stabilizers, nano-TiO_2-reinforced polymer composites and UV-absorbing polycarbonate (PC), provide improved UV protection, inhibiting photodegradation and ensuring product quality.

3.1.3 Thermal Stability (for Heat-Sealable and Temperature-Sensitive Applications)

Thermal stability is an important concern for packaging for heat-sealing, microwave and high-temperature processing. The capacity of a polymer composite to resist heat without degrading or distorting guarantees product integrity and functional effectiveness.

Heat-Sealable Properties

Heat sealing is essential for food pouches, pharmaceutical blister packs and sterile medical packaging. Polymers such as polypropylene (PP), polyethylene (PE) and ethylene vinyl acetate (EVA) exhibit excellent heat-sealing properties, ensuring airtight closures. Composite laminates with low-melting-point layers enhance seal integrity and peel strength in flexible packaging.

Thermal Resistance

Some packaging uses need materials with high thermal resistance to be able to withstand high temperatures without degrading. Oven-ready meal packaging, boiling water-resistant food trays and industrial uses need to withstand high temperatures. Polyethylene terephthalate (PET), polyimide (PI) and high-temperature-resistant composites with ceramic fillers are commonly used in these applications.

Cold-Temperature Durability

For frozen food packaging and pharmaceutical storage, the material should not crack or become brittle at low temperatures. Cryogenic-resistant polymer composites like low-density polyethylene (LDPE) and thermoplastic polyurethane (TPU) blends retain flexibility and impact resistance even under sub-zero temperatures.

3.1.4 Biodegradability and Composability (Eco-Friendly Disposal)

Growing concerns for environmental sustainability have prompted the creation of biodegradable and compostable polymer composites that minimize plastic pollution and improve waste management.

Biodegradable Polymers

Biodegradable packaging material dissolves by itself under the influence of microorganisms, having lesser effects on the environment. The most favoured options for biodegradable packages include PLA, PHA and starch blends. All such materials under conditions of composting can dissolve into carbon dioxide, water and biomass.

Compostable Packaging

Compostable materials extend beyond that by degrading to harmless organic material in a predetermined period under industrial composting conditions. Chitosan composites, bio-polybutylene succinate (PBS) and PHA-starch blends exhibit very good composability with no compromise on mechanical properties.

Challenges and Innovations

Though beneficial, biodegradable and compostable products tend to possess reduced durability and barrier properties relative to traditional plastics. New advancements, including nano-enhanced bio-composites and bio-plasticized blends, have enhanced their water resistance and mechanical performance.

3.1.5 Printability and Aesthetic Appeal

Packaging is not just a utilitarian purpose but also an advertising tool. Quality printability and aesthetic attractiveness contribute to brand promotion, product differentiation and consumer interaction.

Printability

Packaging materials need to be compatible with high-resolution printing, embossing and labelling in order to convey product information and branding details efficiently. Polyester (PET), polyethylene (PE) and polyvinyl chloride (PVC) composites provide flat, printable surfaces for digital, flexographic and UV-curable printing.

Colour and Transparency

Appearance matters in consumer purchase decisions. UV-resistant coatings, coloured polymer laminates and transparent PET composites add appeal while preserving longevity. Metallic finishes, glossy coatings and matte finishes enhance aesthetic appeal.

Smart and Interactive Packaging

The latest developments have seen the introduction of intelligent packaging solutions, including QR codes, RFID tags and thermochromic inks for product authentication, real-time tracking and temperature monitoring. Conductive polymer composites provide printed electronics for interactive packaging to improve consumer experience.

4 Advanced Processing Techniques for Polymer-Based Composites

4.1 *Extrusion and Co-extrusion*

Polymer-based composites is manufactured by using processing techniques such as extrusion and co-extrusion, which offers high efficiency, scalability and versatility for designing the materials. Extrusion is widely used manufacturing process in which polymer material is mixed with fillers, and additives which process through the die to create products according to the requirements. This technique is commonly used to make products such as pipes, sheets and films from polymer raw material which can be tuned mechanical properties by controlling temperature, pressure, flow rate and design [1]. Moreover, co-extrusion is an advanced extrusion technique where multiple polymer layers can be combined in a single step through feeding disparate polymer materials in one die leads to multi-layer composites with remarkable functionalities. This method is specifically used for recycled raw materials to enhance the mechanical, thermal and rheological properties. For instance, co-extrusion is used to enhance barrier, flexibility, strength and durability in packaging, automotive and construction industries.

4.2 *Injection Moulding and Thermoforming*

Simply described, plastic injection moulding is a process of melting plastic pellets and injecting it into a mould or cavity to form a shape. Once the shape is formed, the material is cooled, ejected from the machine and a finished injection moulded product or part is ready. Most products in our world today items on desk, at home and work (vacuum cleaners, espresso machines, speakers, health aids, electric tools, smart home devices, electric transport, cashless payment system and virtual reality goggles) are precision plastic injection moulded, and almost 90% on everything produced in an automotive product is injection moulded [19]. The plastic material is heated above its melting point, resulting in the conversion of the solid polymer to a molten fluid with a reasonably low viscosity. It is then forced into a closed mould that defines the shape of the article to be produced. There the material is cooled until it reverts to a solid, then the mould is opened, and the finished part is extracted. Although the principle may be simple, the practice of injection moulding is anything but simple this is a consequence of the complex behaviour of plastics melts and the ability of the process to encompass complicated products. The essential mechanisms of injection moulding are heat transfer and pressure flow. The essential equipment is an injection moulding machine, sometimes known as a press, and a mould which may also be referred to as a tool or sometimes a die.

A mould is a tool for making a shaped article to compose of various parts different moulds are composed of different parts (see Fig. 2). It mainly realizes the processing of the shape of the article by changing the physical state of the formed material. There're seven systems of injection mould: (1) Gating system, (2) Guidance system, (3) Moulding system, (4) Core system, (5) Ejection system, (6) Cooling system and (7) Exhaust system.

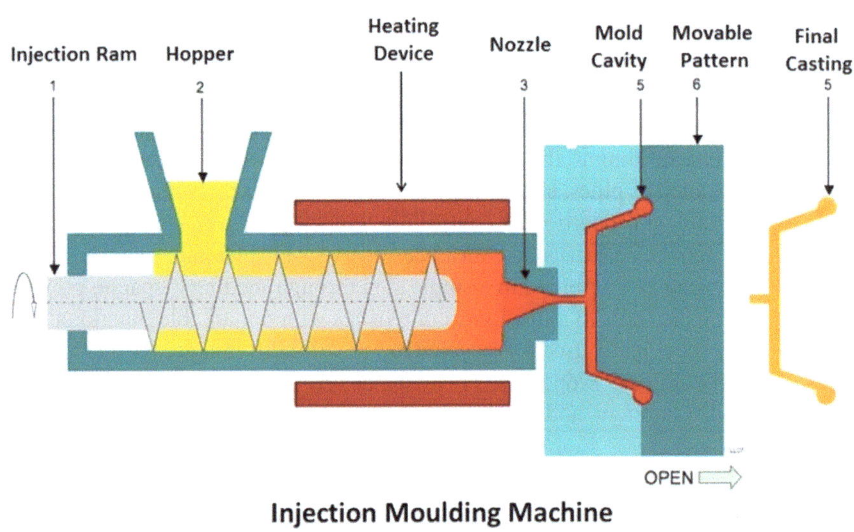

Fig. 2 Overview of injection moulding machine

4.2.1 Gating System

- Sprue: Where the molten plastic flows through from the injection machine to the mould.
- Runner: Where the molten plastic flows through before reaching different cavities.
- Gate: The last opening that allows the molten plastic to flow into the cavities.
- Cold Slug Well: When the molten plastic flows in the sprue and runner, it will solidify due to heat transfer with the mould. The cold slug well is where the solidified material is stored.

4.2.2 Guidance System

- In-mould opening/closing process, mould displacement will occur due to manufacturing tolerances and installation tolerances.
- When the mould is closing, it is required that the mould can return to its original position.
- This is where guidance and location (i.e. mould guidance system) fit in.

4.2.3 Moulding System

- Moulding System: A closed space that consists of the core, cavity and inserts, etc.

4.2.4 Core System

- Core is the side of the tool where the plastic part is injected from.

4.2.5 Ejection System

- The ejector of the injection machine applies force on the press plate of the mould, which drives the ejector pins, ejector blocks, lifters, etc. to push the product away from the core or the cavity and thus enable smooth mould release.

4.2.6 Cooling System

- The mould itself is a large cooling system.
- To achieve a better cooling effect, one or more groups of cooling channels will be designed for the mould.
- Usually, the distance between the cooling channel and the product is twice the diameter of the channel.
- The coolant is not necessarily cold water—it can be water at 100 °C depending on the mould temperature and sometimes it can also be hot oil.

4.2.7 Exhaust System

- The injection mould venting slot serves two purposes: (A) Expel air from the plastic injection mould cavity during the injection process of the molten plastic material; (B) Get rid of the various gases produced during injection moulding processing.
- The setup of venting slots is of great importance, especially for thin-walled products or the locations far away from the gate.
- In addition, close attention should also be paid to venting slots to produce small-size or precision parts, because they are able to protect the products from surface burns, insufficient filling, as well as other defects.

4.2.8 Thermoforming

Thermoforming is basically used for applying heat and pressure to shape polymer-based composite sheets. Thermoforming forms a mould using pressure, vacuum and mechanical force to make the plastic sheet after heating. After formation of sheets, it is trimmed and cut into final product. This technique is faster and more cost effective than injection moulding. The thermoforming process is used to make many different packaging articles, for example, pots, trays and blister packs, and can be broken down into seven basic steps:

1. Making the sheet—normally by the cast extrusion method.
2. Heating the sheet to soften it for further processing.
3. Forming the pack by stretching the sheet either into or over a mould.
4. Cooling the formed packaging.
5. Printing or decorating as required.
6. Cutting and trimming the multi-unit moulded sheet into individual units. This can be carried out at the same time as the forming process.
7. Stacking the individual units before packing and labelling for dispatch to the customer.

4.3 Film Blowing and Lamination

Film blowing and lamination are two basic processes in polymer processing that have enormous applications in the packaging, agricultural and industrial sectors. Film blowing is a common process used to create thin plastic films, while lamination enhances the performance of the films by combining different layers for enhanced functionality. Together, these processes play a major role in the manufacture of flexible packaging materials, protective coatings and barrier films. The film blowing process is extruding a molten polymer through a round die and inflating the extrudate with air to form a thin-walled bubble. The bubble is cooled and collapsed into a flat film, which is wound on rolls for further processing. The main advantage of film blowing is that it is capable of producing uniform and quality films of controlled thickness and enhanced mechanical properties. Polyethylene (PE), polypropylene (PP), polyvinyl chloride (PVC) and ethylene-vinyl acetate (EVA) are some of the usual polymers used in film blowing. The film properties can be tailored by adjusting process parameters like die gap, air pressure, cooling rate and draw-down ratio [31].

Film blowing is highly versatile and can be employed to produce monolayer and multi-layer films. Monolayer film blowing employs a single polymer to produce the film, whereas multi-layer films involve co-extrusion of several polymers in an effort to confer properties such as barrier resistance, mechanical strength and heat sealability. Multi-layer films find widespread use in food packaging, where they shield the food from moisture, oxygen and UV light. Film blowing technology has also made it possible to produce biodegradable and compostable films, which is part of the transition towards sustainability. Lamination is also a process used to seal two or more material layers to strengthen the final product, its barrier properties and durability. Lamination is also classified as dry lamination, wet lamination, extrusion lamination and thermal lamination based on the process of bonding. Dry lamination employs an adhesive on one substrate, dries and bonds another layer using heat and pressure. Wet lamination employs adhesive on one layer and directly pressing against another layer and then drying. Extrusion lamination involves using molten polymer as the bonding layer between two substrates, while thermal lamination involves heat-activated films to deliver adhesion.

Lamination is widely utilized in the packaging industry to make materials with a specific functionality such as moisture barrier, puncture resistance and printability. For example, laminated packaging films for the food industry could be composed of polyethylene terephthalate (PET) for good strength, aluminium foil for the barrier and low-density polyethylene (LDPE) as a seal. Laminated structures are utilized in medical packagings, pharmaceutical blister packaging and flexible electronics. For industrial applications, laminated films are used to provide insulation, corrosion protection and enhanced mechanical properties. Both lamination and film blowing involve some issues. In film blowing, making film thickness as uniform as possible and reducing the number of pinholes and gels is particularly important. It is important that the process conditions of cooling rate, bubble stability and the rheology of polymer are optimally controlled for making high-quality film. Adhesion failure, curling and interlayer compatibility are just some of the problems in lamination. Process condition optimization and appropriate adhesive or polymer selection must be done for forming strong and lasting lamination. The marriage of film blowing and lamination has given rise to high-performance packaging materials that are utilized in flexible packaging, agricultural film, industrial wraps and protective coatings. Research continues to enhance the process for greater sustainability with the use of biodegradable polymers, solvent-free adhesives and energy-efficient processing methods. With growing demand for lightweight, high-barrier and sustainable packaging materials, advances in film blowing and lamination will continue to push the limits of polymer processing technologies.

4.4 3D Printing for Customized Packaging

3D printing, or additive manufacturing, has transformed numerous sectors, such as packaging, through highly customizable, affordable and environmentally friendly solutions. Customization in the packaging sector is critical in increasing product appeal, functionality and material efficiency. Packaging traditionally employs mass production with standardized designs, which hinders flexibility. Yet, 3D printing makes it possible for manufacturers to develop customized packaging options that address exact product needs, enhance branding and reduce wastage. In its capacity for printing complex forms, light structure and signature designs, 3D printing is leading to the future of customized packaging. Perhaps the largest benefit of using 3D printing in packaging is that it can produce bespoke and personalized forms that cannot be done by mainstream manufacturing processes [5, 26]. This customization is especially worth its weight in gold in the luxury products, food, drug and cosmetic markets, where packaging is an essential component of brand identity. Businesses can now create complex packaging structures, embossed logos and interactive features without having to invest in costly moulds or tooling. For example, premium brands can employ 3D-printed packaging to produce special and limited-edition designs that add value to the consumer experience. Second, companies can easily prototype new packages and test them prior to large-scale production, saving time and money compared to conventional prototyping techniques.

Yet another important advantage of 3D printing packaging is that it is sustainable. Traditional packaging manufacturing creates a huge amount of waste material, where leftover materials from moulds and cuts are usually just thrown away. On the contrary, 3D printing is a process of additive manufacturing in the sense that it constructs products layer upon layer by using just the necessary quantity of material, hence minimizing wastage. Additionally, 3D printing provides room for using recyclable and biodegradable materials like polylactic acid (PLA) that comes from renewable sources of corn starch and sugarcane. With sustainability as a concern for consumers and businesses alike, 3D-printed packaging is an eco-friendly option compared to conventional plastic-based packaging solutions. The versatility of materials employed in 3D printing also broadens its usage in customized packaging. Various types of polymers, biopolymers, composites and even biodegradable products can be utilized to produce packaging that fulfils certain functional specifications. For instance, packaging food necessitates materials that are food-safe, water-resistant and biodegradable, while packaging electronics might need antistatic and shock-absorbing capabilities. With the proper selection of 3D printing materials, producers can create packaging that not only safeguards the content but also makes it easier to use. Furthermore, on-demand 3D printing of packaging eliminates the requirement of large storage facilities and overstocking, resulting in a streamlined supply chain.

Another fun innovation enabled by 3D printing is the integration of smart packaging capabilities. As the usage of the Internet of Things (IoT) expands, the package can now embed sensors, QR codes, NFC chips and RFID tags which give information related to the status of the product, whether original or counterfeit, and when the product should expire. As an example, packages for medication may have features like smart alert reminding users to consume their drug, or the detection of if anyone has touched it. Similarly, food packaging can incorporate freshness indicators that change colour when the food inside is no longer safe to consume. These smart packaging solutions, enabled by 3D printing, enhance user engagement, product safety and supply chain efficiency. In addition to functional improvements, 3D-printed packaging supports ergonomic and aesthetic enhancements. In contrast to traditional packaging, which tends to favour cost savings over consumer experience, 3D printing enables designers to develop packaging that is not only aesthetically pleasing but also comfortable to the touch and simple to open. This is especially useful for products marketed to older people or those with disabilities, where ease of use is an important consideration. Ergonomic packaging can include easy-to-grip surfaces, simple-to-open openings and resealable closures that improve convenience. This degree of customization makes packaging with the end consumer in mind, ultimately enhancing the overall consumer experience.

One of the most effective uses of 3D printing for packaging is in the food sector. Food packaging traditionally tends to be wasteful and non-biodegradable, and is a major cause of environmental pollution. With 3D printing, companies can make environment-friendly packaging that is customized to certain food products, minimizing the waste of excess materials and enhancing product conservation. Edible packaging, a new idea, is also being researched through 3D printing. For instance, food-safe materials like chocolate, sugar and seaweed can be utilized to produce biodegradable and even edible packaging, which completely eliminates waste. This technique is gaining momentum in the fast food and confectionery markets, where sustainability and innovation are prime movers of consumer interest [52]. Despite all the benefits, challenges are present to implement 3D printing for customized packaging at scale. The most significant constraint is the speed of production. Conventional packaging processes like injection moulding and thermoforming can create thousands of units per hour, while 3D printing is typically slower. Yet, new high-speed 3D printing technologies like continuous liquid interface production (CLIP) and multi-jet fusion (MJF) are overcoming these shortcomings and enabling additive manufacturing in high-volume production. Another issue is the expense of 3D printing materials, which may be more than traditional packaging materials. Yet, as the technology becomes more prevalent and material prices fall, the economic viability of 3D-printed packaging will increase.

The future of 3D printing for packaging is bright, with continued research and development fuelling its use. Companies are venturing into hybrid manufacturing strategies, with 3D printing combined with conventional methods for optimal output. Brands, for instance, may utilize 3D printing in crafting distinctive limited-series packaging design but apply standard manufacturing for volume production. Future advances in sustainable and bio-based materials for 3D printing will also enhance the environmental aspects of the technology. As demand for customized and sustainable products keeps growing, 3D-printed packaging will be instrumental in revolutionizing the business. In summary, 3D printing is revolutionizing tailored packaging by providing unmatched design freedom, sustainability and functionality. From designing intricate branding features to incorporating smart features and ergonomic designs, the technology enables companies to satisfy changing consumer aspirations without harming the environment. While issues like production time and cost of materials still exist, ongoing improvements in 3D printing technology are making it ever more feasible for bulk packaging solutions. As the industry becomes more open to innovation, 3D printing will be a game-changer in the future of personalized packaging, changing the way products are designed, packaged and consumed by the end-user.

5 Smart Packaging

Primary purpose of packaging is to protect the product; however, this is not enough to meet the demands in current scenario. Day-by-day cases of adulteration, tempering and counterfeiting are increasing. To mitigate these problems, smart packaging has brought a significant shift in the trends of packaging industry [38, 40]. Smart packaging interacts with the product in real-time and provides crucial data of product condition and security. There are various types of perishable products, such as fruit, vegetables, meats and fish and pharmaceutical drugs and medicines which requires specific conditions like temperature, humidity and pressure to remain fresh and active till they reach at their destination. In some cases, it takes several days to export or import these products. Hence, smart packaging play important role. Smart packaging can be divided into active packaging and intelligent packaging mentioned in Fig. 3.

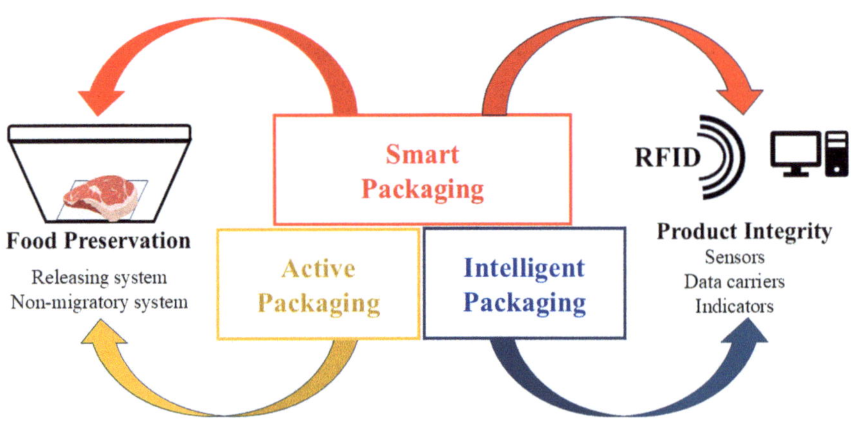

Fig. 3 Components of smart packaging

5.1 Active Packaging

Active packaging refers to incorporation of active ingredients or components into packaging container or films with the aim of maintaining and extending products shelf life. These active components interact with food or surrounding environment to inhibit or slow down the microbial growth, reduce respiration rate and monitor moisture, oxygen, carbon dioxide level in package. Thus, active packaging is advanced and innovative form of packaging which not only help to maintain food quality and freshness but goes beyond to ensure safety of product and extend its shelf life [44]. These active ingredients can be incorporated into package into many ways. The most employed approach is addition of active ingredient into sachets/pads. Secondly, incorporation of volatile and non-volatile components directly into package during converting process of package or applying coating or adsorbing onto package surface. Active ingredients can be immobilized onto the package by electrostatic or covalent linkage also [12].

Active packaging can be divided into two system based on their working principles: non-migratory active packaging and active releasing packaging. In non-migratory active packaging, the scavengers are used to eliminate undesirable elements from the environment inside the packaging. Scavengers do not directly migrate into package; instead, they enhance or changes the conditions within packaging by absorption of carbon dioxide, ethylene, oxygen, or moisture, depending on the application. Active releasing packaging, which is mainly related to emitters that provide controlled migration of desirable compounds into the package environment. Emitters work on the concept of releasing compounds into the package environment that are desirable and beneficial for product. These are used to restrict the growth of hazardous microorganisms and regulate humidity inside packaging [2, 47].

5.1.1 Non-migratory Active Packaging

Oxygen Scavenger

Oxygen scavengers are one of the most promising active packaging systems since the presence of oxygen is responsible for many foods and drug degradation. Presence of oxygen leads to oxidation, colour changes and microbial growth. In food products, oxygen oxidize the fats or lipid present in food leading to development of undesirable or rancid flavour. Oxygen scavengers can lengthen the shelf life significantly by lowering the oxygen content in the headspace. Limiting oxygen in packaging will reduce the growth of aerobic microbes, complying them to thrive in oxygen limiting environment. This limits their capacity to break down nutrients and slowing down spoiling by regulating oxidation [7]. There are several approaches of using oxygen scavengers. One is by addition of active oxygen absorbing components into sachets, pads, or adhesive labels kept into the container. This is the most easy and affordable approach. But there are chances of accidental consumption or leakage of sachets. As an alternative, oxygen scavengers can be included straight into the packing materials, either in mass through the use of injection or extrusion techniques, or as a coating on the packaging surface. The most effective scavenging methods to date are metal-based ones, whose reaction has to be activated by moisture and incorporating them through extrusion at the industrial level is a significant difficulty [20, 24]. In polyamide or polyolefin material, oxygen scavenging components are added that may be utilized in package walls. Other polymer-based oxygen scavenging agents include oxidation–reduction resins and polymer metallic complex [16].

Moisture and Ethylene Absorber

Moisture and ethylene absorber are commonly used in food packaging industry. Moisture is one the main factor which accelerates the growth of microbes after oxygen. By decreasing the amount of moisture within the container shelf life of food can be increased. To absorb extra moisture and reduce spoilage, moisture absorbing substances like silica gel, activated charcoal, polyvinyl alcohol (PVA), starch, cellulose and chitosan can be employed [48]. Fruit undergoes physiological and biochemical alterations during the ripening phase. The ripening process of climacteric and non-climacteric fruits is different. Ethylene, natural plant growth regulator controls the ripening of climacteric fruit, including the decrease of chlorophyll, the synthesis of carotenoid or anthocyanin, metabolism of sugar and production of volatile organic compounds (VOCs) [23]. High ethylene concentrations during storage causes respiration to speed up and eventually leads to senescence. Ethylene absorbers help increase shelf life, which lowers waste in the supply chain and benefits suppliers and retailers financially. For the storage and transportation of produce, these absorbers can be included into paper bags or corrugated fibreboard transit packaging. They are also available as sachets to retail produce packets to increase the shelf life of the food [11]. Ethylene regulation can be done through a chemical or physical adsorption mechanism. This can be done by reducing ethylene production by using modified atmospheric packaging (MAP) that alters the various gases in the headspace or using perforated packaging materials to permeate the gases to the inside and outside of the package or by removing ethylene by using ethylene scavenger. Some commonly used ethylene scavenger includes zeolite, potassium permanganate, palladium, tetrazines, activated charcoal, silver nitrate, kieselguhr, silica gel and aluminium oxide [17].

5.1.2 Active Releasing Packaging

Carbon Dioxide Emitter

Packages that combine carbon dioxide emitters and oxygen absorbers are most frequently used for products like peanuts and potato crisps that need a lot of packing space. Ferrous carbonate or a mixture of ascorbic acid and bicarbonate are the key components of these systems. Carbon dioxide adsorption prevents pressure building, swelling and rupture in packages containing respiring foods, which can reduce product shelf life. Carbon dioxide can hinder microbial activity. Therefore, a complementary approach to O_2 scavenging is the impregnating of a packing structure with a CO_2 producing device, or the insertion of the latter in the form of a sachet. The permeability of CO_2 in most plastic films is three to five times that of O_2, therefore it needs to be generated continually to keep the package's concentration at the appropriate level [6].

Antimicrobial Packaging

Antimicrobial packaging is a type of packaging in which antimicrobial agents can be added, using a variety of methods, to the packaging matrix of certain microbe species that cause food product contamination. Higher water activity foods are more prone to spoiling during manufacturing, storage and distribution when they come into contact with the outside environment. Antimicrobial packaging can use both natural and synthetic antimicrobial agents. However, because of cost-effectiveness and safety concerns, consumers prefer natural sources of antimicrobial agents [13].

5.2 Intelligent Packaging

Intelligent packaging materials can be defined as materials that monitor ongoing status or the environment surrounding the product without any interaction with product. These resources can help customers make informed decisions regarding the safety and quality of product. Intelligent packaging systems employ three major technologies: sensors, indicators and data carriers. Relative humidity sensors, gas leakage indicators and temporal temperature indicators are used to keep check on environmental factors. Biosensors and freshness sensors are used to directly monitor food quality. Data carriers like RFIDs and barcode are solely employed to store and convey data [29].

5.2.1 Sensors

Chemical sensors convert chemical information into a measurable output signal, which is processed by appropriate electronics or software device. The components of chemical sensors include a transducer, a receptor, signal processing electronics and a signal display unit [8]. Numerous sensors rely on polymers with distinct properties. The sensing technique may directly use polymers, or they may immobilize certain receptors. Polylactic-co-glycolic acid (PLGA), polyethylene terephthalate (PET), ethylene vinyl alcohol (EVOH), polyhydroxyalkanoates and polyaniline (PA) are a few examples [3]. Several polymer-based biosensors are used to detect allergen, antibiotics and heavy metals present in food [46]. Freshness indicators provide customers with real-time quality information. The information is presented by quick visual changes, such as varying colour intensities or dye diffusion or combination with different geometry. These types of indicators work by focusing on headspace, total volatile basic nitrogen (TVB), carbonyl chemicals (aldehydes and ketones), carbon dioxide and pH [32]. Colorimetric chemosensors are used to determine fruit freshness, ripeness and maturity based on their flavour profile [22].

5.2.2 Data Carries

Initially, RFID technology was used to track products during storage and delivery. Now, RFID sensors are extensive studied to track the quality of product within individual packages. There are three forms of RFID: chipped, chipless and near field communication (NFC). For food packaging chipless RFID are mostly used. All chipless RFID sensors rely on electromagnetic (EM) transduction. The electrical characteristics of the tag structure are altered by changes in physical factors, which in turn alters as electromagnetic responses. The most promising sensors for determining the state of food packaging in the food sector include temperature, pH level, gas, relative humidity and biosensors [14]. RFID tags and RFID readers are the two primary components of RFID systems. Included to a food package, the tag or label holds information as stored data. It transmits this information with the help of an antenna. The information is gathered from the tag by the reader or interrogator. Readers can be fixed at locations like transport vehicles or warehouses, or they can be portable. RFID tags are tiny and consist of two essential components: a memory-containing microchip and an antenna. While some tags just allow for reading, others let users edit or remove data [25].

Commercial RFID devices use silver inks and etched aluminium over polyethylene terephthalate (PET) substrates because metals are highly conductive and polymer materials are flexible. However, there are several drawbacks including expensive metals, material waste from chemical etching and significant pollution from the usage of non-biodegradable polymeric materials. As an alternative graphite-based materials are used for its flexibility and low weight. Antennas based on graphene offer an affordable, eco-friendly substitute that works with conventional production techniques. Without significantly altering their functionality, these tags can be fixed to non-rigid items like fabric and polymeric objects of various forms, such as cardboard boxes and plastic bottles. But due to low conductivity, graphene-based materials are not widely utilized in the commercial sector [36]. Carbon-fibre reinforced polymer (CFRP) components are also used in RFID in order to assess the readability of commercial UHF systems by changing their covers, which are partially made of CFRPS [4].

5.2.3 Indicators

It is challenging to continuously monitor perishable food items during storage and distribution. The significance of expiry date or best before dates are practical when the product is stored and transported at specific condition as specified during whole distribution supply. Hence, various time-temperature indicators (TTIs) are used which track changes in package on the basis of temperature, pH and light transmissivity [41]. Based on function, TTIs are of various types. These includes diffusional TTIs, enzymatic TTIs, polymer-based TTIs and bacteriological-based TTIs [51]. Another is thermochromic inks that can undergo colour changes when exposed to high temperatures. These inks help customers know whether a product is hot or cold before it is served, and they are mostly utilized in beverage packaging and microwave food goods. These inks have reversible or irreversible colour patterns depending on temperature. Packaging labels and foils applied to the package's surface can be printed with this ink [43].

6 Sustainability and Environmental Impact

6.1 Impact of Plastic on Environment

In last decade, production of plastic has increased and due to inappropriate facilities and waste management system excessive waste has been accumulated. This accumulated plastic has surged and become major environmental problem polluting air, water, land and ecosystem. For production of polymer-based packaging, huge amount of water and energy is used. Virgin plastic is manufactured by refining of petroproducts which involves release of greenhouse gases thus increasing carbon footprint. Secondly, unsegregated plastic is mostly incinerated or either dumped on land or in water bodies thus disturbing habitat of terrestrial and aquatic ecosystem [18]. Plastic waste is incinerated where due to incomplete combustion unburnt carbon, ash, toxic compounds and microplastic are released which worsens respiratory conditions. Aquatic ecosystem is disturbed due floating plastic waste dumped in nearby water body. They consume microplastic which get accumulated into their body causing internal obstruction and even death, which ultimately gets bioaccumulated in that trophic level and food chain. Contaminated water and seafood can introduce microplastics into the human food chain also. Plastic ended up in landfills persists in soil and remain unchanged for many years causing soil contamination which lowers fertility of soil.

6.2 Challenges in Plastic and Bio-based Plastic Waste Management

The fast growth in plastic manufacture and use has resulted in a worldwide trash issue, with plastic debris piling up in landfills, seas and ecosystems. Microplastic contamination, low recycling rates and improper handling of biodegradable plastics are among the issues that persist despite initiatives to enhance waste management and recycling. A study conducted by CIPET-CPCB on plastic waste in 60 major Indian cities found that 94% of the waste is recyclable (thermoplastics), while 6% is non-recyclable (thermosets). Of these, 67% are HDPE/LDP (used in multi-layer sacs for food packaging, tobacco and polybags), 10% are PP (polypropylene), 9% are PET (polyethylene terephthalate), 4% are each of PVC (polyvinyl chloride) and PS (polystyrene), and the remaining 6% are non-recyclable thermoset plastics, which include Sheet Moulded Composite (SMC), Fibre-Reinforced Plastic (FRP), Multi-layered Plastics and Expanded Polystyrene [34].

Plastics have played a major role in creating efficient packaging system that is economical and provide desired results and properties, energy-efficient, sanitary, sustainable and capable of maintaining a clean environment. But since last decade, plastic waste management has been a serious problem due to the challenges faced by a lack of plastic waste processing facilities [18, 28, 45]. Here are the few challenges faced while handling waste:

- It is difficult to deal with the rising diversity of waste materials as demands for better recycling and material circularity has emerged. Therefore, recycling trash into new raw materials is just as crucial as collecting it.
- The existence of biodegradable plastics may contaminate traditional plastic recycling streams, even if collection of packaging debris is a need for recycling.
- Proper segregation of plastic packaging according to recycling code is very important. Due to variation in mechanical properties of each material, the new material formed may does not get desired properties.
- Multi-layered plastic is made up of many layers of different polymeric material, each with distinctive properties and characteristics. These materials are difficult to recycle because of their complicated composition and difficult to separate each material. At the end these polymers usually end up in landfills or incinerated, raising environmental issues.
- Recycling efficiency decreases if collected waste is contaminated with leftovers food or adhesives, and colours.
- There is a misconception that all biodegradable plastics can degrade on its own. However, most require specific composting conditions like high temperature, humidity and microbial activity to break down organically.

6.3 Regulations and Policies Promoting Sustainable Packaging

India has been proactively enforcing laws and policies that emphasize in reduction plastic waste, encouraging recycling and the promotion of environmentally friendly packaging alternatives. Regulatory frameworks and inspections are conducted by the Food Safety and Standards Authority of India.

6.3.1 Plastic Waste Management (Amendment) Rules (2021)

Under this amendment, government has prohibited use of identified single-use plastic items which have low utility and high littering potential by 2022. The manufacture, import, stocking, distribution, sale and use of single-use plastic, including polystyrene and expanded polystyrene, commodities would be prohibited. To stop littering of light weight plastic bags, thickness of plastic carry bags was increased from 50 to 75 μm and to 120 μm. To strengthen the implementation of the law Plastic Waste Management Rules, 2016, waste management infrastructure was laid under Swachh Bharat Mission. Emphasis on implementation of EPR rules.

6.3.2 Draft Food Safety and Standards (Packaging) Amendment Regulation (2022)

This draft was formed under Plastic Waste Management (Amendment) Rules (2021), where FSSAI permits the use of recycled plastic as food contact material. This draft also provides guidelines regarding acceptance of recycles PET resins and recycling of post-consumer PET for food contact application.

6.3.3 Extended Producer Responsibility

For effective implementation of Extended Producer Responsibility, the Guidelines for Extended Producer Responsibility being brought out have been given legal force through Plastic Waste Management Amendment Rules, 2021. If the plastic packaging waste is identified as single-use plastic item, then it is the responsibility of producer to collect and manage it in sustainable way through the Extended Producer Responsibility of the Producer, importer and Brand owner (PIBO), as per Plastic Waste Management Rules, 2016.

7 Conclusion

The development of packaging materials is influenced by the demand for improved performance, sustainability and functionality. With industries seeking to reduce their environmental impact without compromising on product safety and performance, polymer-based composites present a revolutionary solution. The next-generation materials resolve some of the most important issues in packaging through enhanced barrier performance, mechanical properties, recyclability and biodegradability. Implementing sustainable packaging solutions is not just a matter of regulatory compliance but also a business strategy to satisfy consumer needs and corporate sustainability targets. With ongoing advances in research and innovation, polymer-based composites will be instrumental in determining the future of packaging, leading to environmentally friendly, high-performance and smart packaging solutions. Polymer-based composites have revolutionized the packaging industry by providing green, high-performance and innovative solutions. Bio-based polymer composites offer sustainable substitutes to traditional plastics, while recycled polymer composites support waste minimization and circular economy. Nanocomposites offer improved mechanical and barrier properties, and hybrid polymer composites bring in intelligent functionalities and multi-layer protection. While technology and research keep evolving, polymer composites are going to remain at the centre of solving global packaging issues with optimal performance, cost-effectiveness and ecocompatibility.

Declaration of Competing Interest The authors declare that they have no known competing financial interests or personal relationships that could have appeared to influence the work reported in this paper.

References

1. Abeykoon C, McMillan A, Nguyen BK (2021) Energy efficiency in extrusion-related polymer processing: a review of state of the art and potential efficiency improvements. Renew Sustain Energy Rev 147:111219. https://doi.org/10.1016/j.rser.2021.111219
2. Ahmed MW, Haque MA, Mohibbullah M, Khan MSI, Islam MA, Mondal MHT, Ahmmed R (2022) A review on active packaging for quality and safety of foods: current trends, applications, prospects and challenges. Food Packag Shelf Life 33:100913. https://doi.org/10.1016/j.fpsl.2022.100913
3. Alberti G, Zanoni C, Losi V, Magnaghi LR, Biesuz R (2021) Current trends in polymer based sensors. Chemosensors 9(5):108. https://doi.org/10.3390/chemosensors9050108
4. Ambrosini D, Pirondi A, Vescovi L, Arbucci F (2023) Performance evaluation of a UHF passive RFID tag embedded in a carbon fibre-reinforced polymer. Int J RF Technol Res Appl 13(1):37–51. https://doi.org/10.3233/RFT-221512
5. Arrigo R, Battegazzore D, Bernagozzi G, Cravero F, Ribero Pedraza DN, Frache A (2022) Recycled PP for 3D printing: material and processing optimization through design of experiment. Appl Sci 12(21):10840. https://doi.org/10.3390/app122110840
6. Awulachew MT (2022) A review of food packaging materials and active packaging system. Int J Health Policy Plan 1(1). https://doi.org/10.33140/ijhpp.01.01.03

7. Aydın A, Yüceer M, Ulugergerli EU, Caner C (2024) Improving food security as disaster relief using intermediate moisture foods and active packaging technologies. Appl Food Res 4(1):100378. https://doi.org/10.1016/j.afres.2023.100378
8. Azeredo HMC, Correa DS (2021) Smart choices: mechanisms of intelligent food packaging. Curr Res Food Sci 4:932–936. https://doi.org/10.1016/j.crfs.2021.11.016
9. Bhati SS, Saini P, Dutt D (2023) Applications of zeolites 13x to reduce the volatile organic compounds and odorous compounds in recycled polypropylene. https://doi.org/10.2139/ssrn.4530732
10. Bora S, Das K, Saini P, Katiyar V, Ghosh T (2025) Application of nanoemulsion for improving quality and safety of fruits and vegetables. In: Advancements in nanotechnology for food and packaging. Elsevier, pp 53–78. https://doi.org/10.1016/B978-0-443-21428-8.00003-5
11. Coles R (2013) Paper and paperboard innovations and developments for the packaging of food, beverages and other fast-moving consumer goods. In: Trends in packaging of food, beverages and other fast-moving consumer goods (FMCG). Elsevier, pp 187–220. https://doi.org/10.1533/9780857098979.187
12. Contreras CB, Charles G, Toselli R, Strumia MC (2017) Antimicrobial active packaging. In: Biopackaging. CRC Press, pp 36–58. https://doi.org/10.1201/9781315152349-3
13. Deshmukh RK, Hakim L, Gaikwad KK (2023) Active packaging materials. Curr Food Sci Technol Rep 1(2):123–132. https://doi.org/10.1007/s43555-023-00004-6
14. Fathi P, Karmakar NC, Bhattacharya M, Bhattacharya S (2020) Potential chipless RFID sensors for food packaging applications: a review. IEEE Sens J 20(17):9618–9636. https://doi.org/10.1109/JSEN.2020.2991751
15. Fernandez CM, Alves J, Gaspar PD, Lima TM, Silva PD (2023) Innovative processes in smart packaging. A systematic review. J Sci Food Agric 103(3):986–1003. https://doi.org/10.1002/jsfa.11863
16. Gaikwad KK, Singh S, Lee YS (2018) Oxygen scavenging films in food packaging. Environ Chem Lett 16(2):523–538. https://doi.org/10.1007/s10311-018-0705-z
17. Gaikwad KK, Singh S, Negi YS (2019) Ethylene scavengers for active packaging of fresh food produce. Environ Chem Lett 18(2):269–284. https://doi.org/10.1007/s10311-019-00938-1
18. Ganguly RK, Chakraborty SK (2024) Plastic waste management during and post COVID-19 pandemic: challenges and strategies towards circular economy. Heliyon 10(4):e25613. https://doi.org/10.1016/j.heliyon.2024.e25613
19. Itim B, Philip M (2015) Effect of multiple extrusions and influence of PP contamination on the thermal characteristics of bottle grade recycled PET. Polym Degrad Stab 117:84–89. https://doi.org/10.1016/j.polymdegradstab.2015.04.004
20. Juan-Polo A, Maestre Pérez SE, Monedero Prieto M, Tone AM, Sánchez Reig C, Beltrán Sanahuja A (2021) Double-function oxygen scavenger and aromatic food packaging films based on LDPE/polybutadiene and peanut aroma. Polymers 13(8):1310. https://doi.org/10.3390/polym13081310
21. Kaiser K, Schmid M, Schlummer M (2017) Recycling of polymer-based multilayer packaging: a review. Recycling 3(1):1. https://doi.org/10.3390/recycling3010001
22. Kim YH, Yang YJ, Kim JS, Choi DS, Park SH, Jin SY, Park JS (2018) Non-destructive monitoring of apple ripeness using an aldehyde sensitive colorimetric sensor. Food Chem 267:149–156. https://doi.org/10.1016/j.foodchem.2018.02.110
23. Kou X, Feng Y, Yuan S, Zhao X, Wu C, Wang C, Xue Z (2021) Different regulatory mechanisms of plant hormones in the ripening of climacteric and non-climacteric fruits: a review. Plant Mol Biol 107(6):477–497. https://doi.org/10.1007/s11103-021-01199-9
24. López-de-Dicastillo C, López-Carballo G, Vázquez P, Schwager F, Aragón-Gutiérrez A, Alonso JM, Hernández-Muñoz P, Gavara R (2023) Designing an oxygen scavenger multilayer system including volatile organic compound (VOC) adsorbents for potential use in food packaging. Polymers 15(19):3899. https://doi.org/10.3390/polym15193899
25. López-Gómez A, Cerdán-Cartagena F, Suardíaz-Muro J, Boluda-Aguilar M, Hernández-Hernández ME, López-Serrano MA, López-Coronado J (2015) Radiofrequency identification and surface acoustic wave technologies for developing the food intelligent packaging concept. Food Eng Rev 7(1):11–32. https://doi.org/10.1007/s12393-014-9102-y

26. Maldonado-García B, Pal AK, Misra M, Gregori S, Mohanty AK (2021) Sustainable 3D printed composites from recycled ocean plastics and pyrolyzed soy-hulls: optimization of printing parameters, performance studies and prototypes development. Compos Part C Open Access 6:100197. https://doi.org/10.1016/j.jcomc.2021.100197
27. Marsh K, Bugusu B (2007) Food packaging? Roles, materials, and environmental issues. J Food Sci 72(3):R39–R55. https://doi.org/10.1111/j.1750-3841.2007.00301.x
28. Mhaddolkar N, Astrup TF, Tischberger-Aldrian A, Pomberger R, Vollprecht D (2024) Challenges and opportunities in managing biodegradable plastic waste: a review. Waste Manage Res J Sustain Circ Econ. https://doi.org/10.1177/0734242X241279902
29. Müller P, Schmid M (2019) Intelligent packaging in the food sector: a brief overview. Foods 8(1):16. https://doi.org/10.3390/foods8010016
30. Oloyede OO, Lignou S (2021) Sustainable paper-based packaging: a consumer's perspective. Foods 10(5):1035. https://doi.org/10.3390/foods10051035
31. Pandya M, Dhadwal R, Valadi JK (2023) Support vector machines and random forest classification models for identification of stability in extrusion film casting process, pp 187–195. https://doi.org/10.1007/978-981-19-2600-6_13
32. Panjagari NR, Raman RK, Uma K, Suwalka R, Thomas E (2021) Freshness indicators for real-time quality evaluation of packaged animal origin foods: a mini-review. Indian J Anim Health 60(2-Spl). https://doi.org/10.36062/ijah.2021.spl.02421
33. Popa MS, Frone AN, Panaitescu DM (2022) Polyhydroxybutyrate blends: a solution for biodegradable packaging? Int J Biol Macromol 207:263–277. https://doi.org/10.1016/j.ijbiomac.2022.02.185
34. Rafey A, Siddiqui FZ (2023) A review of plastic waste management in India—challenges and opportunities. Int J Environ Anal Chem 103(16):3971–3987. https://doi.org/10.1080/03067319.2021.1917560
35. Reichert CL, Bugnicourt E, Coltelli M-B, Cinelli P, Lazzeri A, Canesi I, Braca F, Martínez BM, Alonso R, Agostinis L, Verstichel S, Six L, Mets SD, Gómez EC, Ißbrücker C, Geerinck R, Nettleton DF, Campos I, Sauter E et al (2020) Bio-based packaging: materials, modifications, industrial applications and sustainability. Polymers 12(7):1558. https://doi.org/10.3390/polym12071558
36. Roggero UFS, Nista SVG, Hernández-Figueroa HE, Mei LHI, Moshkalev SA (2022) Graphene-biopolymer-based RFID tags: a performance comparison. Mater Today Commun 31:103726. https://doi.org/10.1016/j.mtcomm.2022.103726
37. Saini P, Bhati SS, Dutt D (2023) Removal of volatile organic compounds and odorous compounds for multilayer packaging recyclates using heated air purging. J Phys Conf Ser 2652(1):012005. https://doi.org/10.1088/1742-6596/2652/1/012005
38. Saini P, Sharma K, Chauhan A, Bhati SS (2025) Sustainable smart packaging. In: Sustainable materials for food packaging and preservation. Elsevier, pp 285–300. https://doi.org/10.1016/B978-0-443-13567-5.00013-7
39. Shivakumar N, Raj S, Ahmed S, Rajeswari M (2023) Bio polymers and sensors used in food packaging—present and future prospects. In: Biobased materials. Springer Nature Singapore, pp 211–226. https://doi.org/10.1007/978-981-19-6024-6_10
40. Siddiqui J, Taheri M, Alam AU, Deen MJ (2022) Nanomaterials in smart packaging applications: a review. Small 18(1):2101171. https://doi.org/10.1002/smll.202101171
41. Simpson R, Almonacid S, Nuñez H, Pinto M, Abakarov A, Teixeira A (2012) Time-temperature indicator to monitor cold chain distribution of fresh salmon (*Salmo salar*). J Food Process Eng 35(5):742–750. https://doi.org/10.1111/j.1745-4530.2010.00623.x
42. Soares CTDM, Ek M, Östmark E, Gällstedt M, Karlsson S (2022) Recycling of multi-material multilayer plastic packaging: current trends and future scenarios. Resour Conserv Recycl 176:105905. https://doi.org/10.1016/j.resconrec.2021.105905
43. Thamrin ES, Warsiki E, Bindar Y, Kartika IA (2022) Thermochromic ink as a smart indicator on cold product packaging—review. IOP Conf Ser Earth Environ Sci 1063(1):012021. https://doi.org/10.1088/1755-1315/1063/1/012021

44. Versino F, Ortega F, Monroy Y, Rivero S, López OV, García MA (2023) Sustainable and bio-based food packaging: a review on past and current design innovations. Foods 12(5):1057. https://doi.org/10.3390/foods12051057
45. Vuppaladadiyam SSV, Vuppaladadiyam AK, Sahoo A, Urgunde A, Murugavelh S, Šrámek V, Pohořelý M, Trakal L, Bhattacharya S, Sarmah AK, Shah K, Pant KK (2024) Waste to energy: trending key challenges and current technologies in waste plastic management. Sci Total Environ 913:169436. https://doi.org/10.1016/j.scitotenv.2023.169436
46. Wang B, Huang D, Weng Z (2023) Recent advances in polymer-based biosensors for food safety detection. Polymers 15(15):3253. https://doi.org/10.3390/polym15153253
47. Wyrwa J, Barska A (2017) Innovations in the food packaging market: active packaging. Eur Food Res Technol 243(10):1681–1692. https://doi.org/10.1007/s00217-017-2878-2
48. Yadav S, Dutta PK (2024) Moisture-absorbent food packaging systems and the role of chitosan. In: Smart food packaging systems. Wiley, pp 169–193. https://doi.org/10.1002/9781394189595.ch7
49. Yam KL, Takhistov PT, Miltz J (2005) Intelligent packaging: concepts and applications. J Food Sci 70(1):R1–R10. https://doi.org/10.1111/j.1365-2621.2005.tb09052.x
50. Yu Z, Boyarkina V, Liao Z, Lin M, Zeng W, Lu X (2023) Boosting food system sustainability through intelligent packaging: application of biodegradable freshness indicators. ACS Food Sci Technol 3(1):199–212. https://doi.org/10.1021/acsfoodscitech.2c00372
51. Zabala S, Castán J, Martínez C (2015) Development of a time–temperature indicator (TTI) label by rotary printing technologies. Food Control 50:57–64. https://doi.org/10.1016/j.foodcont.2014.08.007
52. Zander NE, Gillan M, Burckhard Z, Gardea F (2019) Recycled polypropylene blends as novel 3D printing materials. Addit Manuf 25:122–130. https://doi.org/10.1016/j.addma.2018.11.009

Chapter 9
Smart Packaging, Smarter Inspection: A Deep Learning Approach

Deepali Verma, Hariome Sharan Gupta, Vineet Kumar Verma, Varun, and Akarsh Verma

1 Introduction

In today's interconnected world, polymer and polymer composites materials in packaging field, plays a pivotal role in food technology, far beyond merely containing food products [1, 2]. It serves as a vital point of interaction between food brands and consumers, providing crucial information, ensuring product freshness, safety, quality, and contributing to sustainability efforts. Simultaneously, precise and efficient inspection throughout the packaging lifecycle–from food production to distribution–is essential for maintaining product integrity, reducing food waste, and ensuring customer satisfaction. Traditional inspection methods often fall short when faced with the growing complexity of food packaging designs, materials, and production

D. Verma
Department of Computer Science and Engineering, Indian Institute of Technology (BHU) Varanasi, Varanasi, Uttar Pradesh, India

H. S. Gupta (✉)
Department of Polymer and Process Engineering, Indian Institute of Technology Roorkee (DPP IIT-R SRE Campus), Roorkee, Uttar Pradesh, India
e-mail: hgupta@pe.iitr.ac.in

V. K. Verma
Department of Electronics Engineering, National Institute of Technology, Patna, Bihar, India

Government Polytechnic Bighapur, Unnao, Uttar Pradesh, India

Varun
Department of Applied Science, Government Polytechnic Bighapur, Unnao, Uttar Pradesh, India

A. Verma
Department of Mechanical Engineering, UPES (University of Petroleum and Enery Studies), Dehradun, India

A. Verma et al. (eds.), *Polymers and Composite Materials for Packaging: Smart Food Packaging and Solutions*, Materials Horizons: From Nature to Nanomaterials,
https://doi.org/10.1007/978-981-95-0677-4_9

speeds, heavily relying on manual processes that can lead to errors and inconsistencies. This underlines the pressing need for smarter, more automated solutions in food packaging and inspection to meet the demands of the evolving food industry [3, 4].

This exploration delves into the exciting intersection of "smart packaging" and "smarter inspection," with a particular focus on the transformative potential of deep learning [5], as illustrated in Fig. 1. Deep learning plays a critical role in the multiple real-time applications, such as action recognition, captioning [6], surveilliance [7], defects in composite materials [5]. Smart packaging, which incorporates advanced technologies such as embedded sensors, QR codes, and NFC tags, adds layers of intelligence to traditional packaging, allowing for real-time tracking, condition monitoring, and enhanced consumer engagement with food products [8], as discussed in Fig. 2. However, unlocking the full potential of this data requires sophisticated analysis. This is where deep learning becomes an invaluable tool. By employing deep learning algorithms, we can automate and greatly enhance the food packaging inspection process, enabling precise defect detection, predictive maintenance, and optimized supply chain management [9]. This approach not only improves the efficiency and accuracy of food packaging inspection but also uncovers valuable insights from the enormous amounts of data generated by smart packaging, paving the way for a more intelligent, responsive, and efficient food packaging ecosystem. In the following sections, we will explore the different aspects of this synergy, investigating the specific applications of deep learning in food packaging inspection and discussing the benefits and challenges associated with its implementation.

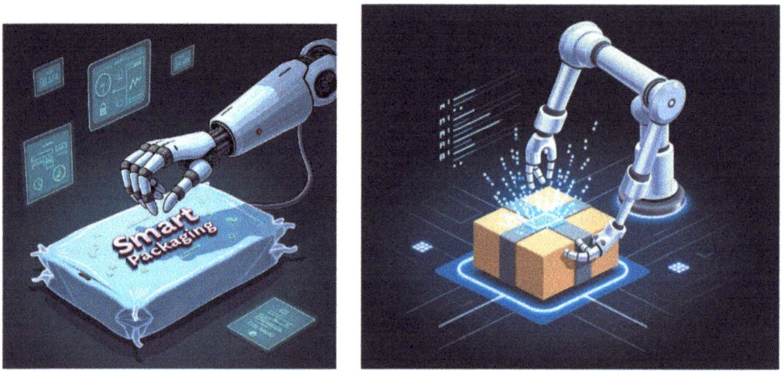

(a) smart packaging (b) smart inspection

Fig. 1 Overview of smart packaging, smarter inspection using deep learning approach

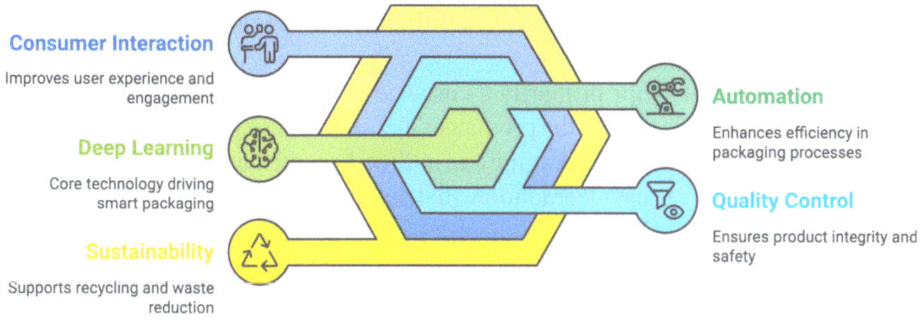

Fig. 2 Benefits of deep learning in smart packaging

2 Traditional Packaging and Inspection Methods

Packaging inspection has long been a critical aspect of quality control in manufacturing and supply chains. Traditional inspection methods rely heavily on human inspectors or classical computer vision techniques. Manual inspection, while effective in detecting complex or subjective quality issues, is inherently slow and prone to human error, especially in high-volume production environments [10–12]. Automate, introduced to overcome these limitations, typically involve rule-based algorithms that analyze images for predefined characteristics such as color, shape, and size [13]. However, these systems often struggle to adapt to new packaging designs and environmental variations, making them less robust in dynamic production settings.

Historically, food packaging was primarily designed for containment and protection. Materials such as glass, metal, and plastic were used to prevent contamination. Quality inspection was manual, relying on human vision and sensory analysis. Traditional methods included: Visual inspection for defects and contamination. Chemical and microbiological testing. Use of expiration dates and batch tracking. While these methods were effective, they lacked real-time monitoring and were labor-intensive, prone to human error.

3 Emergence of Smart Packaging

The advent of smart packaging has added a layer of complexity to inspection processes [14, 15]. Smart packaging integrates technologies such as Radio Frequency Identification (RFID), Near Field Communication (NFC), and printed sensors to enhance functionality, providing real-time tracking, freshness indicators, and anti-counterfeiting measures [16]. These advancements demand more sophisticated inspection techniques capable of verifying the functionality and integrity of embedded components. Traditional systems lack the flexibility to inspect such diverse and intricate packaging, necessitating more intelligent solutions. Smart packaging integrates sensors, indicators, and data transmission technologies to monitor food conditions. The two main types are:

Active Packaging: While primarily focused on extending shelf life and enhancing safety through interactions with the product or environment (e.g., oxygen scavengers, antimicrobial release), active packaging can indirectly contribute to quality assurance by preventing spoilage and degradation that might otherwise be detected during inspection [17].

Intelligent Packaging: This category directly addresses the need for enhanced information and monitoring capabilities [18]. The integration of sensors (e.g., temperature, humidity, gas sensors) [16], indicators (e.g., time-temperature indicators, freshness indicators) [19], and communication technologies (e.g., RFID, NFC, QR codes) [20] allows for real-time monitoring of product conditions and traceability throughout the supply chain. This data can be invaluable for identifying potential quality issues proactively and improving overall supply chain integrity.

4 Challenges in Smart Packaging Inspection

Real-life smart packaging inspection presents several challenges that hinder efficient and accurate quality control:

4.1 Complexity of Embedded Technologies

Smart packaging often includes electronic components such as: RFID/NFC tags, Printed electronics, Sensors (temperature, humidity, gas sensors), QR codes, and augmented reality (AR) markers. The main challenges involved:

Signal Interference: RFID and NFC tags may face interference from metallic or liquid-based packaging materials, leading to read errors.

Sensor Calibration: Embedded sensors require precise calibration to ensure accurate data collection, which complicates inspection.

Compatibility Issues: Different smart packaging technologies may not integrate seamlessly with existing inspection systems.

4.2 Complexity of Materials and Components

Smart packaging blends conventional materials such as polymers with sophisticated electronic components, sensors, and active agents to enhance functionality. Consequently, ensuring the reliability and performance of these disparate elements and understanding their intricate interplay presents a considerable challenge.

4.3 Data Overload and Interpretation

Smart packaging, particularly when equipped with sophisticated sensors capable of tracking environmental conditions like temperature and assessing qualities such as freshness, has the potential to generate substantial volumes of data. The crucial hurdle, however, lies in establishing effective mechanisms for the collection, processing, and insightful interpretation of this information to facilitate prompt and well-informed decision-making.

4.4 Environmental Impact and Safety

A thorough yet complex evaluation of smart packaging materials is essential to understand their environmental footprint and ensure consumer safety. This includes rigorously assessing the potential for components like sensors or active substances to migrate into food, a factor that demands careful scrutiny.

4.5 Cost and Scalability

Implementing robust inspection methods for smart packaging can present significant financial investments, particularly when incorporating cutting-edge technologies. Furthermore, ensuring the scalability of these inspection processes to accommodate high-throughput manufacturing environments poses a considerable challenge.

Smart packaging inspection faces challenges related to technology integration, material variability, data security, regulatory compliance, and cost. Addressing these issues requires advancements in sensor reliability, standardization, and AI-driven inspection systems.

5 Importance of ML and DL in Smart Packaging and Inspection

In an increasingly globalized world, ensuring food safety, quality, and freshness from farm to fork has become more complex and critical than ever. Consumers demand not only transparency and convenience but also longer shelf life and higher safety standards from the food products they purchase. Simultaneously, food manufacturers and supply chain stakeholders face mounting pressure to reduce waste, prevent contamination, and comply with stringent regulatory frameworks. Traditional methods of food inspection–primarily manual visual checks and chemical testing–have long been the industry standard. However, these methods often fall short in terms of efficiency, accuracy, and scalability. As food products become more diverse and supply chains more intricate, there is a growing need for smarter, faster, and more reliable inspection systems, as shown in Fig. 3. This need has paved the way for smart packaging, a new generation of packaging solutions equipped with sensors, indicators, and communication technologies that enable real-time monitoring of food quality and environmental conditions. Smart packaging not only extends shelf life but also acts as an interface between the product and the consumer or manufacturer, providing critical data throughout the product's lifecycle. To complement these advancements, deep learning–a subset of artificial intelligence (AI)–is revolutionizing the way food inspection is conducted. With its ability to analyze complex data patterns from images, videos, and sensor inputs, deep learning offers unprecedented accuracy and automation in identifying defects, contamination, spoilage, and packaging integrity issues. By learning from vast datasets, these models continuously improve and adapt, enabling proactive quality control and real-time decision-making. This chapter examines the dynamic convergence of smart packaging technologies and deep learning-driven inspection systems in the context of modern food technology. It provides an in-depth analysis of the foundational concepts, practical applications, existing challenges, and emerging opportunities associated with integrating artificial intelligence into food packaging and quality control. By the end of the chapter, readers will have a thorough understanding of how deep learning is revolutionizing food inspection–making it more intelligent, efficient, sustainable, and adaptable to future demands.

Machine learning and deep learning have brought major advancements to smart packaging and inspection, streamlining automation, improving quality control, and boosting efficiency across manufacturing and supply chains. Smart packaging now often includes sensors, IoT devices, and intelligent data analysis to track conditions

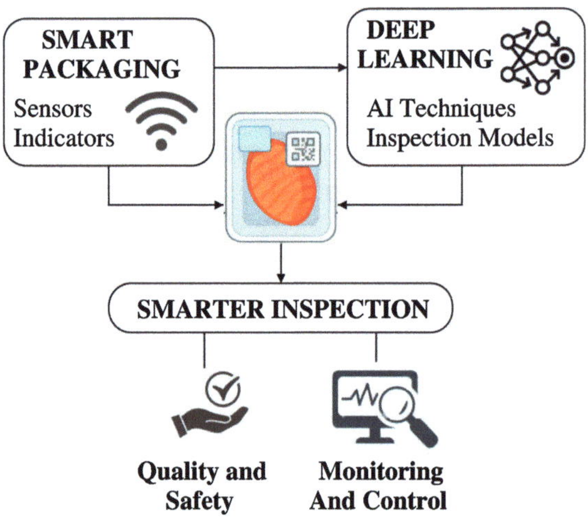

Fig. 3 Relation between smart packaging, deep learning and smart inspection

like temperature, humidity, and product freshness. At the same time, automate use these technologies to spot defects, verify labels, and ensure regulatory compliance. Together, they help reduce human error, cut down on waste, and enhance safety–making them essential tools in industries such as food and beverage, pharmaceuticals, and consumer goods.

A significant use of machine learning and deep learning in smart packaging lies in the real-time evaluation of product quality. For example, computer vision systems based on Convolutional Neural Networks (CNNs) are highly effective in identifying issues like leaks, misprints, or structural deformities with remarkable precision [21]. In the context of food packaging, deep learning techniques are also applied to hyperspectral imaging to detect early signs of spoilage or contamination, helping manufacturers maintain safety and regulatory compliance [22]. Additionally, machine learning models contribute to the design optimization of packaging by forecasting material performance and sustainability, ultimately supporting cost reduction and lowering environmental impact [23].

In inspection systems, DL-based object detection frameworks like You Only Look Once (YOLO) and Faster R-CNN enable high-speed defect classification in production lines [24]. These models outperform traditional rule-based systems by learning from vast datasets, adapting to new defects, and reducing false positives. Furthermore, reinforcement learning (RL) optimizes robotic packaging systems by improving pick-and-place accuracy and minimizing downtime [25].

The integration of ML/DL in smart packaging also enhances traceability and anti-counterfeiting measures. Blockchain-enabled AI systems verify product authenticity by analyzing packaging barcodes and RFID tags, combating fraud in pharmaceuticals and luxury goods [26]. As Industry 4.0 advances, ML/DL will continue to drive innovations in intelligent packaging, predictive maintenance, and automated quality assurance, making supply chains more resilient and consumer-friendly.

6 Deep Learning for Smart Packaging

Smart Food Packaging using deep learning is an advanced system that combines intelligent sensors, data analytics, and deep learning models to monitor and assess food quality in real time. Figure 4 shows a breakdown of how the system works.
Smart Food Package Components: The packaging itself includes embedded sensors that monitor: *Temperature:* Continuously monitors for deviations that could compromise food safety. *Gas levels:* Identifies the presence of spoilage indicators such as ethylene or carbon dioxide. *Humidity:* Tracks moisture levels that can impact food freshness and shelf life. *Freshness indicators:* Often employing visual cues like color-changing labels to provide an immediate indication of quality. These sensors operate continuously, capturing a stream of real-time data. These sensors collect data continuously.

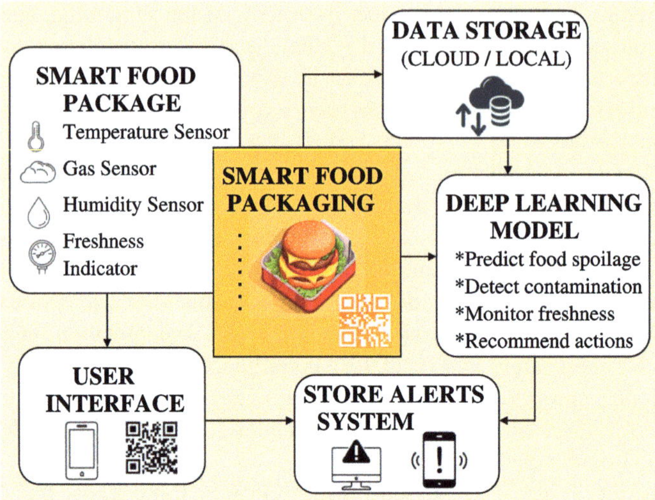

Fig. 4 Deep learning based smart packaging

Data Transmission and Storage: The wealth of data generated by these sensors is transmitted wirelessly via technologies like RFID/NFC or Bluetooth to secure cloud-based or local storage facilities. This infrastructure enables remote, round-the-clock monitoring and comprehensive data logging for in-depth analysis.
Deep Learning Model: The collected data then becomes the input for sophisticated deep learning models, such as Convolutional Neural Networks or Transformer architectures. These models are trained on extensive datasets correlating sensor readings with actual food quality outcomes, enabling them to: Predict spoilage or degradation based on sensor trends, Detect contamination from abnormal readings, Monitor freshness levels over time.

Recommend actions, such as whether to: Extend shelf life, Flag a product for recall, Trigger a discount or disposal. These models are trained on large datasets of sensor readings correlated with actual food quality conditions.
User Interface: The system outputs its decisions to: Retailers or inspectors via dashboards or store alert systems. Consumers, through QR codes or mobile apps, giving real-time information about the food's safety and freshness.

The benefits are Reduces food waste by detecting spoilage early. Improves food safety with automated alerts. Builds consumer trust through transparency. Enables predictive maintenance of cold chains.

7 Deep Learning Based Smarter Inspection

Deep learning, particularly Convolutional Neural Networks (CNNs), revolutionized the field by eliminating the need for manual feature extraction. CNNs excel at recognizing intricate patterns in visual data, making them well-suited for detecting defects in packaging materials, identifying misaligned labels, and ensuring the proper integration of smart components [27]. Recent studies have demonstrated CNN-based models achieving accuracy rates exceeding 95% in defect detection tasks, outperforming traditional approaches in both speed and reliability [28]. Furthermore, Recurrent Neural Networks (RNNs) and Transformer models have shown promise in analyzing sequential data from smart packaging sensors, providing insights into environmental conditions, tampering events, and supply chain anomalies [29]. Transfer learning and data augmentation techniques have further enhanced these models' robustness, enabling them to generalize effectively across various packaging types with limited training data. Deep learning, a subset of artificial intelligence (AI), has revolutionized food inspection by automating quality control and safety assessment. Figure 5 illustrates a smart food inspection method that leverages advanced technologies to automate and enhance quality control. Here's a breakdown of the process:

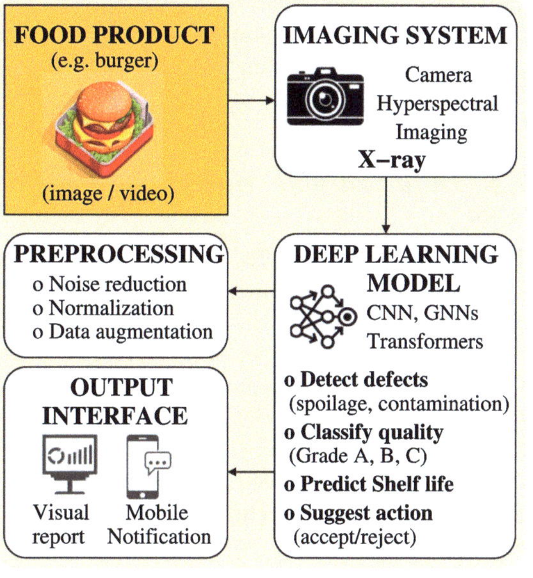

Fig. 5 Deep learning based smart inspection

7.1 Food Product Input

The process begins with a food product (illustrated by a burger) being presented for inspection. This input can be in the form of an image or video.

7.2 Imaging System

The food product is then analyzed by an imaging system. This system can employ various techniques, including a standard camera, hyperspectral imaging (which captures a wide spectrum of light to reveal detailed chemical information), or X-ray imaging (useful for detecting internal defects or foreign objects). The choice of imaging technique depends on the specific inspection requirements.

7.3 Preprocessing

The captured image or video undergoes preprocessing. This crucial step enhances the data for better analysis by the deep learning model. Common preprocessing techniques include noise reduction to improve image clarity, normalization to standardize the data range, and data augmentation to artificially increase the size and variability of the training data, making the model more robust.

7.4 Deep Learning Model

The preprocessed data is fed into a sophisticated deep learning model. The diagram specifically mentions Convolutional Neural Networks (CNNs), Graph Neural Networks (GNNs), and Transformers as potential model architectures. These advanced AI models are trained to perform several critical tasks (Fig. 6):
Detect defects Identify issues like spoilage or contamination.
Classify quality Categorize the food product based on predefined quality grades (e.g., Grade A, B, C).
Predict shelf life Estimate how long the product will remain fresh and safe.
Suggest action Recommend whether to accept or reject the food product based on the analysis.

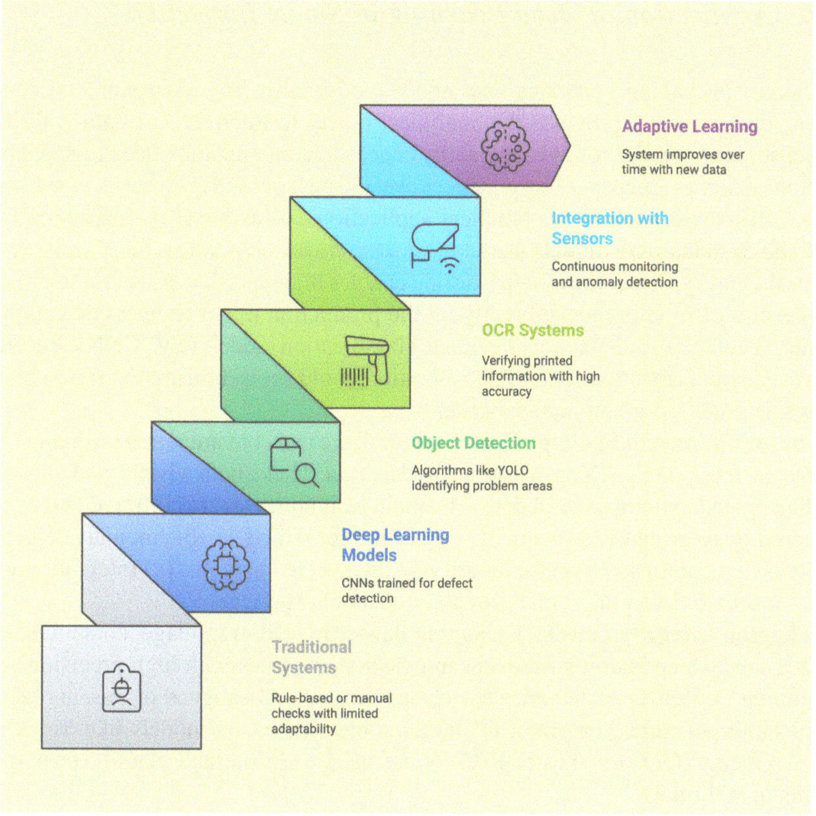

Fig. 6 Advance packaging inspection using deep learning

7.5 Output Interface

Finally, the results of the deep learning model are presented through an output interface. This can take various forms, such as a visual report displayed on a monitor or a mobile notification sent to personnel. This allows for timely and informed decision-making regarding the food product's quality and disposition.

In essence, this smart inspection method moves beyond traditional manual checks by using advanced imaging to capture detailed information, AI-powered deep learning to analyze this information objectively and efficiently, and a clear output interface to communicate the results effectively. This leads to more accurate, faster, and potentially non-destructive food quality assessment.

7.6 Application of Deep Learning in Smart Inspection

As smart packaging generates data about food quality and environmental conditions, deep learning provides the analytical power to interpret this data with high precision and speed. From visual quality checks to contamination detection and predictive analytics, deep learning enhances every aspect of food inspection, as shown in Fig. 7. This section explores practical applications, categorized by inspection type and the deep learning models used.

Visual Quality Inspection Deep learning enables high-speed, non-destructive visual inspection of food products directly on the production line. Freshness detection of fruits, vegetables, and meats using image classification models (e.g., CNNs, ResNet). Surface defect detection such as discoloration, mold spots, bruises, or cracks. Shape and size grading for automated sorting

Contamination and Spoilage Detection Spoilage and contamination are major concerns in food safety. Deep learning models can detect both visible and invisible indicators using multispectral data. The main techniques are: (a) Infrared and hyperspectral imaging analyzed using 3D CNNs or hybrid models, (b) Anomaly detection models (e.g., autoencoders) for identifying unexpected patterns, (c) Integration with gas sensors to detect microbial growth or chemical spoilage

Packaging Integrity Checks Packaging flaws can lead to spoilage, contamination, and recalls. Deep learning models can detect such issues with high precision. The Common Defects Detected are Poor or broken seals, Misaligned or missing labels, Leaks and punctures, Incorrect fill levels. Object detection models like You Only Look Once (YOLO) or Faster R-CNN are used to locate and classify packaging faults in real time.

Shelf-Life Prediction and Monitoring Combining deep learning with time-series sensor data enables predictive modeling of product shelf life. The Sensors Used:

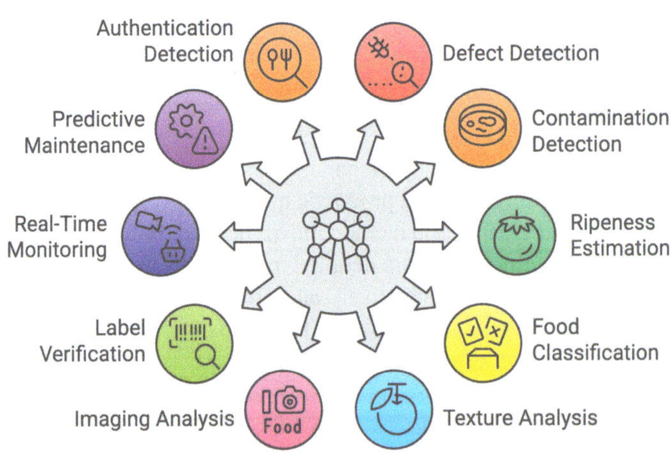

Fig. 7 Applications of deep learning in smart food inspection

Temperature, Humidity, Gas sensors (e.g., CO_2, ammonia), pH sensors The popular AI Model Types: Recurrent Neural Networks (RNNs) and LSTMs for sequential data, Regression models for time-to-expiry estimation, Fusion models combining images and sensor input

Integration with Smart Packaging Systems Deep learning models can be embedded into or connected with smart packaging systems for closed-loop quality control. Integration Benefits are: Real-time alerts on quality degradation, On-pack freshness scoring using mobile app interactions, Autonomous adjustment of storage conditions based on sensor input

8 Future Trends and Research Directions

As the global food industry becomes increasingly reliant on data-driven and sustainable solutions, smart packaging combined with deep learning is poised for transformative growth. Figure 6 shows the transformative growth from traditional systems to adaptive learning. This section outlines the emerging trends, research frontiers, and technological breakthroughs that are expected to shape the next decade of smart food inspection systems, as discussed in Fig. 8.

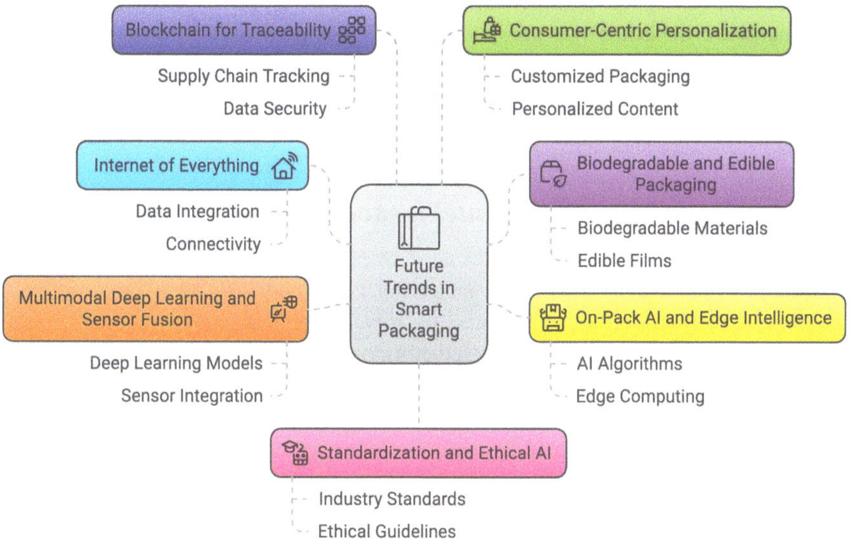

Fig. 8 Future trends and research directions

8.1 Integration with Internet of Everything (IoE)

The shift from IoT to Internet of Everything (IoE) will enable deeper integration of smart packaging systems with: (a) *People*: Real-time consumer feedback and behavior data. (b) *Processes*: Predictive maintenance and self-optimizing logistics. (c) *Data*: Automated analytics from packaging, supply chains, and retail shelves. (d) *Things*: Interconnected sensors, cameras, and devices across the food ecosystem.

Research Focus: Developing scalable, secure, and interoperable IoE frameworks tailored to perishable goods.

8.2 Biodegradable and Edible Smart Packaging

Driven by growing environmental consciousness, the field of smart packaging is witnessing exciting advancements in sustainability. Innovative designs now incorporate sensors and indicators that are: (a) fully biodegradable, offering a significant reduction in persistent plastic waste; and (b) composed of edible or food-grade materials, even allowing for direct consumption. Furthermore, these smart packaging solutions utilize (c) organic inks and sophisticated biosensors, often printed directly onto the material, which can visually signal food quality through color changes triggered by gas emissions or shifts in pH levels [30]. Future research is keenly focused on the synergistic integration of green chemistry principles with AI-compatible sensing materials, paving the way for truly intelligent and environmentally responsible packaging.

8.3 On-Pack AI and Edge Intelligence

Imagine a future where your food packaging is more than just a container. Envision intelligent packaging empowered by tiny, integrated microprocessors or on-pack edge AI chips. These sophisticated components will possess the ability to execute lightweight deep learning models directly on the package itself. This localized processing will enable real-time decision-making, such as instantly alerting you to potential spoilage. Furthermore, these smart packages will seamlessly interact with consumers through engaging interfaces like QR codes, dynamic LEDs, or informative e-ink displays. A key area driving this innovation forward is the development of energy-efficient AI hardware coupled with the implementation of federated learning techniques across a network of these distributed smart packages.

8.4 Synthetic Data and Simulation-Based Training

Accessing extensive and varied datasets for food inspection remains a significant obstacle. Looking ahead, the field could advance through several key strategies: (a) Employing generative artificial intelligence techniques, such as Generative Adversarial Networks (GANs), to create simulated instances of food spoilage. (b) Developing virtual testbeds that allow for the training and evaluation of models within artificially generated food environments. (c) Incorporating digital twins of food products to enable real-time predictions of shelf life and potential risks. A crucial area for future research lies in the development of dependable synthetic datasets that accurately reflect the variability observed in real-world conditions [31].

8.5 Blockchain for Data Integrity and Traceability

In an era where smart packaging and artificial intelligence generate crucial quality data, blockchain technology emerges as a powerful solution for ensuring data integrity and traceability. It provides: (a) Tamper-proof, auditable logs of inspection results, guaranteeing the reliability of quality assessments. (b) Unwavering transparency across the entire food supply chain, offering a clear lineage from farm to consumer. (c) Autonomous quality verification through smart contracts, enabling efficient and trustworthy validation processes. A significant research trend is exploring the synergistic integration of deep learning (DL) and blockchain to build next-generation food safety monitoring systems characterized by enhanced trust and security.

8.6 Multimodal Deep Learning and Sensor Fusion

Future AI models will revolutionize food analysis by integrating a variety of sensory inputs, such as visual, chemical, thermal, and olfactory data, into a unified decision-making process. This will unlock key capabilities, including enhanced accuracy through cross-sensor correlation, robust handling of incomplete or unclear information using multimodal transformers, and the ability to generalize across diverse food types and operational settings. A critical research focus is the development of resilient multimodal fusion architectures specifically designed to manage the dynamic and often noisy nature of food-related data.

8.7 Consumer-Centric Personalization

Smart packaging will evolve beyond basic monitoring to actively cater to individual consumer needs through embedded human-AI interaction models. This innovation will deliver: (a) Tailored freshness alerts, providing precise notifications based on personal consumption patterns and preferences. (b) Customized dietary guidance, proactively warning users about potential allergens or ingredients exceeding their specific sensitivities, such as sodium. (c) Dynamic shelf-life forecasting, accurately predicting spoilage based on both local environmental conditions and individual usage habits. This forward-thinking approach integrates AI into the very interface of our food, fostering a more intuitive and personalized experience.

8.8 Standardization and Ethical AI

As food technology continues to grow, it becomes increasingly important to develop industry-wide standards, certifications, and ethical frameworks. These measures are essential to promote fairness and reduce bias in automated quality assessments, clearly define who is responsible when AI systems make decisions—such as rejecting a product or initiating a recall—and ensure that smart packaging data can be used consistently across different platforms and regions. To move forward responsibly, more research is needed to guide the governance of AI in this field and to design inspection systems that are not only transparent but also ethically sound.

9 Conclusion

The convergence of deep learning and smart packaging technologies presents a paradigm shift in the food packaging and inspection landscape. As traditional inspection methods struggle to meet the increasing demands of modern food systems, intelligent, automated solutions are becoming not only beneficial but essential. This study underscores how deep learning, when integrated with data-rich smart packaging components such as sensors and digital tags, enables real-time monitoring, defect detection, and predictive analytics across the food supply chain.

By embedding deep learning models within inspection systems, the food industry can achieve significant improvements in accuracy, efficiency, and responsiveness. These technologies support the development of adaptive inspection frameworks that learn from data, identify patterns, and anticipate faults before they impact food quality or safety. Such innovations lead to reduced food waste, enhanced traceability, and increased consumer trust.

However, the implementation of deep learning-driven inspection systems is not without its challenges. Issues such as data standardization, model explainability, high initial costs, and the need for interdisciplinary expertise must be addressed to realize their full potential. Despite these hurdles, the benefits far outweigh the limitations.

This paper offers a forward-looking perspective on the future of food packaging–one that is intelligent, autonomous, and deeply integrated with emerging technologies. By embracing deep learning in inspection processes, stakeholders across the food industry can foster a more sustainable, transparent, and high-performing ecosystem that meets the evolving expectations of regulators and consumers alike.

References

1. Gupta HS, Gautam R, Raghuvanshi S, Sethi SK, Bembalge O (2025) Introduction to polymer composites. In: Advances in polymer composite research, pp 1–15
2. Gupta HS, Verma MS, Verma A (2024) Combination of different kinds of materials for hybrid composites: a brief review. In: Unraveling new frontiers and advances in bioinformatics, pp 331–363
3. Pålsson H (2018) Packaging logistics: understanding and managing the economic and environmental impacts of packaging in supply chains. Kogan Page Publishers
4. Malešević M, Stančić M (2021) Influence of packaging design parameters on customers' decision-making process. J Graphic Eng Design 12(4):33–38
5. Verma D, Verma A, Verma A, Gupta HS (2024) Applications of deep learning for composites materials. In: Hybrid composite materials: experimental and theoretical analysis. Springer, pp 195–213
6. Verma D, Haldar A, Dutta T (2023) Leveraging weighted cross-graph attention for visual and semantic enhanced video captioning network. Proceed AAAI Conferen Artific Intell 37(2):2465–2473
7. Verma D, Dutta T, Varshney M (2022) A low-cost driving risk prediction system for hilly roads via surveillance cameras and onsite webcams. In: Proceedings of the 9th ACM international conference on systems for energy-efficient buildings, cities, and transportation, pp 289–290
8. Yam KL, Takhistov PT, Miltz J (2005) Intelligent packaging: concepts and applications. J Food Sci 70(1):1–10
9. Islam MR, Zamil MZH, Rayed ME, Kabir MM, Mridha M, Nishimura S, Shin J (2024) Deep learning and computer vision techniques for enhanced quality control in manufacturing processes. IEEE Access
10. Natarajan S, Govindarajan M, Kumar B (2014) Fundamentals of packaging technology. PHI Learning Pvt. Ltd
11. Verma A, Gupta, HS, Sethi SK (2024) Hybrid composite materials: experimental and theoretical analysis. Springer
12. Sethi S, Gupta HS, Verma A (2024) Polymer composites: from computational to experimental aspects. Springer
13. Caon T, Martelli SM, Fakhouri FM (2017) New trends in the food industry: application of nanosensors in food packaging. In: Nanobiosensors. Elsevier, pp 773–804
14. Gupta HS, Sethi SK, Verma A (2023) Applications of coating materials: a critical overview. In: Coating materials: computational aspects, applications and challenges, pp 81–109
15. Gupta HS, Shankar U, Verma A, Gogoi R, Sethi SK (2023) Computational aspects: Self-clean coatings, plastics and polymers in coatings. In: Coating materials: computational aspects, applications and challenges. Springer, pp 167–184
16. Kuswandi B, Wicaksono Y, Jayus A, Abdullah A, Heng LY, Ahmad M (2011) Smart packaging: sensors for monitoring of food quality and safety. In: Sensing and instrumentation for food quality and safety, vol 5, pp 137–146
17. Hosseinnejad M (2014) Active packaging for food applications-a review

18. Raghuvanshi S, Khan H, Saroha V, Sharma H, Gupta HS, Kadam A, Dutt D (2023) Recent advances in biomacromolecule-based nanocomposite films for intelligent food packaging-a review. Int J Biologic Macromol 127420
19. Luo X, Zaitoon A, Lim L-T (2022) A review on colorimetric indicators for monitoring product freshness in intelligent food packaging: indicator dyes, preparation methods, and applications. Comprehen Rev Food Sci Food Saf 21(3):2489–2519
20. Ghaani M, Cozzolino CA, Castelli G, Farris S (2016) An overview of the intelligent packaging technologies in the food sector. Trends Food Sci Technol 51:1–11
21. Bochkovskiy A, Wang CY, Liao HYM (2020) Yolov4: optimal speed and accuracy of object detection. arXiv:2004.10934
22. Cheng J-H, Sun D-W (2015) Rapid and non-invasive detection of fish microbial spoilage by visible and near infrared hyperspectral imaging and multivariate analysis. LWT-food Sci Technol 62(2):1060–1068
23. Yakoubi S (2025) Sustainable revolution: ai-driven enhancements for composite polymer processing and optimization in intelligent food packaging. Food Bioprocess Technol 18(1):82–107
24. Redmon J, Divvala S, Girshick R, Farhadi A (2016) You only look once: Unified, real-time object detection. In: Proceedings of the IEEE conference on computer vision and pattern recognition, pp 779–788
25. Sutton RS, Barto AG et al (1998) Reinforcement learning: an introduction. MIT press Cambridge, vol 1
26. Yang Y, Du Y, Gupta VK, Ahmad F, Amiri H, Pan J, Aghbashlo M, Tabatabaei M, Rajaei A (2024) Exploring blockchain and artificial intelligence in intelligent packaging to combat food fraud: a comprehensive review. Food Packag Shelf Life 43:101287
27. LeCun Y, Bengio Y, Hinton G (2015) Deep learning. Nature 521(7553):436–444
28. Hu C, Wang Y (2020) An efficient convolutional neural network model based on object-level attention mechanism for casting defect detection on radiography images. IEEE Trans Industr Electron 67(12):10922–10930
29. Mustakim R, Mamat M, Wong F, Mohamad Dasuki SNAS, Johari A (2024) Artificial intelligence in time series prediction, classification, and sequence-to-sequence problems. In: Internet of things and artificial intelligence for smart environments. Springer, pp 103–118
30. Gautam R, Gupta HS, Chauhan A, Jaiswal AK, Verma A (2024) Composite materials for bioenergy. In: Hybrid composite materials: experimental and theoretical analysis. Springer, pp 91–104
31. Gupta HS, Kumar P, Sethi AK, Verma A, Sethi SK (2024) Software-based simulations of the creep recovery model of polymer composites. In: Dynamic mechanical and creep-recovery behavior of polymer-based composites. Elsevier, pp 271–289

Chapter 10
Green Packaging: Exploring Biodegradable and Eco-friendly Composite Materials

Arpan Tewary, Abhilasha Mishra, and Ishika Sharma

1 Introduction

In the past couple of decades, the global packaging industry has seen a revolution in conformity with growing environmental burden, increasing consumer awareness and governmental regulatory requirements [1]. Green packaging or eco-friendly packaging is becoming ever more desirable in comparison to conventional plastic packaging alternatives. The latest trend is dominated by a need to reduce carbon footprints, stop plastic pollution, and shift toward a circular economy [2]. The excessive use of traditional packaging materials, in particular single-use plastics, has brought about severe environmental effects including waste in landfills, pollution in oceans, and emissions of greenhouse gases (GHGs) [3]. Studies project that the packaging sector globally uses approximately 42% of plastic predominantly consisting of polyethylene (PE), polypropylene (PP), polystyrene (PS), polyvinylchloride (PVC), and polyethylene terephthalate (PET), making it the biggest contributor to environmental degradation [4].

According to history, the first synthetic plastic "Bakelite" was synthesized by a Belgian chemist Leo Hendrik Baekeland in 1907. Two years later he himself coined the term "plastic" to refer to any product made from macromolecules like resins, elastomers, and artificial fibers [5]. Erroneously referred to as "the contemporary symbol of modernity," plastics have fundamentally changed the modern world by taking over our everyday lives. Their use has become a huge problem and clearly displays the challenges being faced by the present and future society. Because plastics are the quintessential example of a substance that "does not go away," they have

A. Tewary
Indian Institute of Solar Energy, Mumbai, Maharashtra, India

A. Mishra (✉) · I. Sharma
Graphic Era University, Dehradun, Uttarakhand, India
e-mail: abhi1680@geu.ac.in

A. Verma et al. (eds.), *Polymers and Composite Materials for Packaging: Smart Food Packaging and Solutions*, Materials Horizons: From Nature to Nanomaterials,
https://doi.org/10.1007/978-981-95-0677-4_10

developed to be a major issue. They can be seen as an instinctual reflex since they are inexpensive, shared, and used so frequently. Identifying something that one uses or works with on a day-to-day basis that is not composed of or contains one or more types of plastic is difficult, if not impossible, in the contemporary environment [6].

Plastics are being utilized on a large scale for multiple purposes globally since long time. In 2019, the worldwide yearly manufacturing of plastic items hit 460 million tons with merely 9% being recycled and projections suggests that by 2060 it will rise to 1.2 billion tons [7]. High produce and reduced recovery lead to a significant amount of plastics entering the ecosystem and environment. Key commercial plastics available in the market comprise polyethylene, polyethylene terephthalate, polypropylene, polyvinylchloride, and polystyrene among others. Plastics generate tiny particles or fragments through crushing, splitting, and degradation for the duration of their usage. During 1970s, plastic particles were initially found in the Atlantic Ocean especially on the surface waters. Thompson and coworkers for the first time reportedly published introducing the idea of "microplastics." Microplastics are plastic particles and pieces that have a size of under 5 mm in diameter. They are referred to as "nanoplastics" when their diameter measures under 1 μm. In reality the size of microplastic particles varies from several microns to several millimeters. Unlike nanomaterials that are engineered, it is a blend of variously shaped heterogeneous plastic particles, which is becoming a rising concern for human health. Along with larger plastics, microplastics are specifically manufactured and incorporated into personal care items like cosmetics facial and cleansers. Industrial manufacturing, including the plastics sector, synthetic textile sector, and flocking sector, also generates a significant quantity of microplastics. It has been discovered that approximately 65 million particles of microplastic enter the water daily following treatment at the sewage treatment facility. Microplastics have been found ranging from the southern regions to Antarctica, northern areas up to the Arctic, summit of the Mount Everest, and even deeper in the Mariana Trench. Plastic constitutes the most significant portion of ocean wastes. In 2017, over 8 million tons of plastics found their way into the oceans, which is more than 33 times greater than the total plastic collected in the marine waters by 2015. Nevertheless, because of the constraints of collection techniques, the levels of microplastics in the environment could be undervalued.

The United Nations Environment Assembly (UNEA) has discussed plastic pollution and adopted a historic resolution captioned "End plastic pollution: Toward an international legally binding instrument." Microplastics can possibly reduce the accessibility of nutrients in microalgae or be directly consumed by the organisms, resulting in higher impact on the food web due to the bioaccumulation of the food chain. In the actual environment, microplastics have tremendously complex sources with more diverse occurrences and more concealed hazards, which makes it difficult to reveal their toxic effects and potential health hazards [8].

Governments, enterprises, and customers are thus striving for sustainable products that ensure efficiency in the use of resources as well as biodegradability. Advances in material science have enabled the production of new green packaging materials like biodegradable polymers, bio-based resins as well as natural fiber composites. Not only do these materials deliver the structural strength and protective role

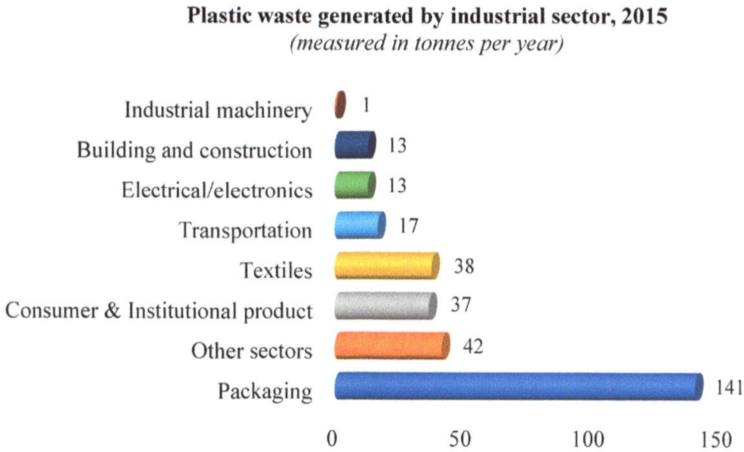

Fig. 1 Plastic waste generated by different industrial sectors. Reproduced with permission [10]

required, but they also minimize carbon footprint [9]. The push for green packaging is further supported by changing consumer attitudes, with opinion polls indicating that increasing number of consumers are now preferring to buy products packaged sustainably. With industries competing to harmonize sustainability and economic viability, the trend toward green packaging is no longer a choice but a necessity. The subsequent pages examine the role of sustainable packaging and the innovative variety of biodegradable and sustainable materials that are transforming the market. Figure 1 represents the statistical data of plastic waste generated by industrial sector in 2015.

1.1 Importance of Sustainable Packaging

Sustainable packaging is vital in addressing environmental issues while promoting economic and functional sustainability. Its significance cuts across various aspects, such as waste minimization, resource preservation, climate change mitigation, and consumer participation. Figure 2 gives an insight of the importance of sustainable packaging.

1.1.1 Minimizing Waste and Pollution

Conventional plastic packaging is infamous for its durability in the environment, lasting for multiple decades before it breaks down. The world plastic waste problem, compounded by poor recycling infrastructure, has caused serious pollution of marine and urban ecosystems. Eco-friendly packaging, especially biodegradable and

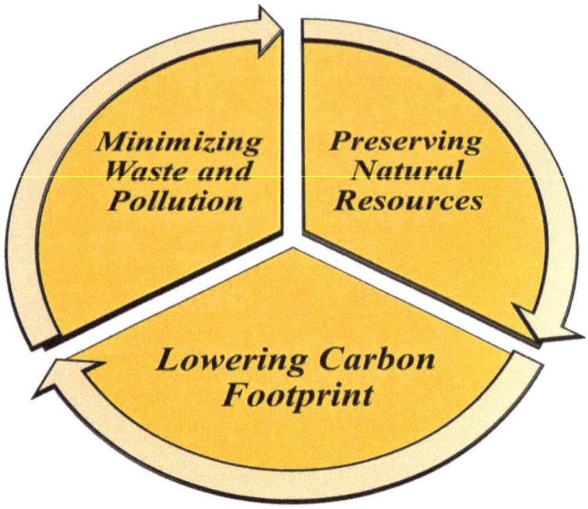

Fig. 2 Importance of sustainable packaging

compostable materials, solves this problem by decomposing naturally without any toxic residues [11].

1.1.2 Preserving Natural Resources

Sustainable packaging uses renewable resources like corn starch, sugarcane fibers, bamboo, and recycled paper to minimize dependence on limited fossil-based raw materials. Eco-friendly composites also make use of agricultural and industrial residues, reducing material extraction and facilitating circular material flows [12].

1.1.3 Lowering Carbon Footprint

Production, transportation, and waste disposal of packaging materials are among the largest contributors to carbon emissions worldwide. Sustainable packaging materials are engineered to have a lower carbon footprint as they use less energy to produce and dispose. They are made from low carbon, carbon neutral or carbon negative sources [13]. For instance, bio-based polymers like PLA produce around 70% less GHG's emissions compared to petroleum-based plastics.

1.1.4 Building Brand Image and Consumer Attraction

With growing environmental awareness, consumers are actively looking for brands that emphasize sustainability. Research shows that companies that adopt green packaging solutions have greater customer loyalty and brand confidence [14]. This phenomenon is most evident in sectors like food and beverage, cosmetics, and e-commerce, where packaging is a central factor in product perception.

Through the integration of sustainable packaging materials, business enterprises can seamlessly resolve environmental concerns in line with consumers' requirements and regulatory bodies. The subsequent part discusses the spectrum of available biodegradable, natural fiber, and eco-friendly composite-based packaging materials.

1.2 Biodegradable and Eco-friendly Materials

The shift to sustainable packaging is made easier by the developments in eco-friendly and biodegradable materials. These materials are designed to biodegrade naturally, either through microbial activity or under controlled composting conditions, with minimal long-term environmental impact. Figure 3 represents the classification of biodegradable and eco-friendly materials into biodegradable polymers, natural fiber-based materials and eco-friendly composite materials which are further classified into PHA, PLA, bamboo packaging, jute and coir packaging, hemp-based packaging, recycled fiber composite, bio-based resins, and mushroom packaging.

1.2.1 Biodegradable Polymers

PHA is a biopolymer that occurs naturally through bacterial fermentation. PHA packaging materials can degrade even under marine conditions, and hence PHA is emerging as a very viable option to curb ocean pollution [15]. PLA is a polymer of biodegradable nature sourced out of fermented sugarcane or corn starch. PLA is predominantly applied in compostable films, cutlery, and food containers because it is highly transparent and biodegradable [16]. Starch-based plastics is obtained from renewable starch resources (e.g., potatoes, maize), these plastics are combined with biodegradable polymers for increased mechanical strength and water resistance.

1.2.2 Natural Fiber-Based Materials

Bamboo Packaging: Biodegradable and very durable, bamboo packaging is used extensively for food containers, cutlery, and cosmetic packaging [17]. Jute and Coir Packaging: Used widely in retail and agriculture, these are biodegradable and supports moisture protection [18]. Hemp-Based Packaging: Hemp fibers possess high tensile strength and are utilized for environmentally friendly wrapping material and paperboard packaging [19].

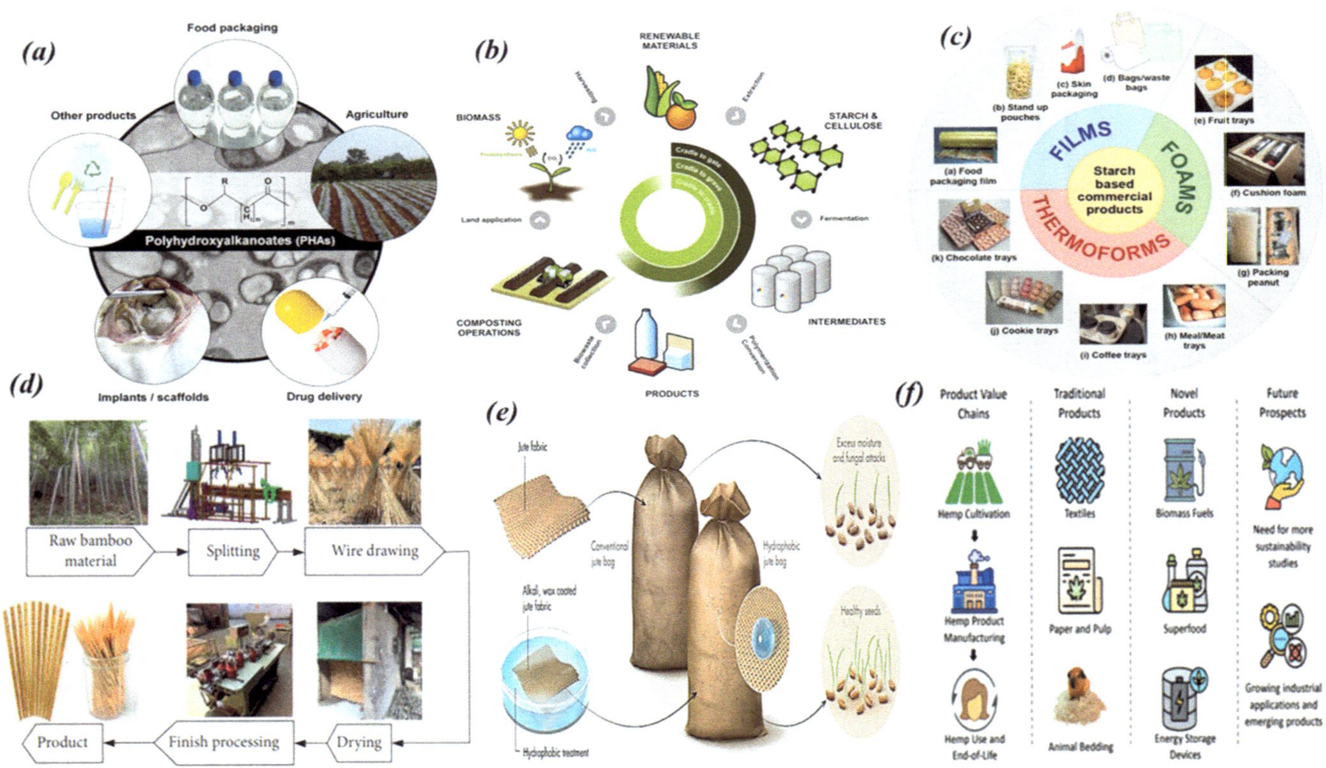

Fig. 3 Classification of biodegradable and eco-friendly materials with (**a–f**) representing PHA, PLA, starch-based plastics, bamboo, jute and coir, hemp-based packaging, respectively. Reproduced with permission [15–21]

1.2.3 Eco-friendly Composite Materials

Recycled Fiber Composites: Post-consumer paper waste is used to produce these composites, which minimize deforestation and offer long-lasting packaging solutions, Bio-Based Resins: Plant oil-derived, bio-based resins are employed as a substitute for petroleum-based adhesives in packaging [20], Mushroom Packaging: Mycelium (fungal roots) and agricultural waste are used to make mushroom packaging, which is completely biodegradable and employed as a substitute for Styrofoam in protective packaging [21].

1.3 Active and Smart Packaging Materials

They are intended to possess functional qualities that surpass the typical mechanical, optical, and barrier characteristics of traditional packaging materials. To increase the shelf life of foods, they might be made to, for example, limit enzyme activity, slow down unwanted chemical reactions, or stop microbial growth. Antimicrobials and/or antioxidants are usually added to packing materials to accomplish this, with a focus on using functional chemicals that are naturally produced from plants for clean labeling. Because the antioxidants and antimicrobials are found in the film rather than the food, active packaging materials may have the potential benefit of lowering the quantity of these chemicals needed to preserve the food during storage and lowering the amount consumed. Depending on their properties, the additives used for active packaging materials may extend the shelf life of packed foods in a variety of ways. Phytochemicals (like quercetin, curcumin, or anthocyanins), essential oils or other plant extracts (like thyme oil or tea extract), macronutrients (like carbohydrate molecules or certain protein), inorganic nanoparticles (like ZnO, TiO_2, clay, and Ag), and organic nanoparticles (like lipid-, protein-, or carbohydrate-based nanoparticles) are some of the most often used additives with antimicrobial or antioxidant qualities.

In the event of a systemic change (e.g., a change in the quality, safety, or maturity of a packaged food products), smart packaging materials for food products for instance are made to react in a certain way or to indicate (pH, gas, and temperature) changes. Consequently, the control of food safety and quality can be enhanced by these smart packaging materials [22].

1.4 Applications of Biodegradable and Eco-friendly Composite Materials

These materials have been incorporated into different packaging industries with success, such as: E-Commerce Packaging: Recyclable mailers, biodegradable bubble wrap, and paper-based cushioning [23], Food and Beverage Packaging:

Compostable trays, bioplastic films, and molded fiber containers [10], Personal Care and Cosmetics: Bamboo jars, paper tubes, and refillable containers [24].

The movement toward biodegradable and environmental-friendly packaging solutions is a fundamental shift in combating environmental threats arising from conventional packaging. With improvement in biopolymer technology, natural fiber composite materials, and the circular economy process, business firms are slowly integrating sustainable packing concepts that achieve functional needs coupled with environmental consideration. As compliance policies become tighter and consumer patterns shift, green packaging is well-set for serious development, bringing on board an environment-friendly and durable packaging market.

2 Environmental Impact of Traditional Packaging

The ecological footprint of conventional packaging, especially plastic materials, has become a serious global issue. Traditional packaging, mostly petroleum-based plastics, is a major source of pollution, and GHGs emissions. The widespread use of non-biodegradable resources has caused serious ecological degradation, posing a threat to marine life, air quality, and soil fertility. Traditional packaging has far-reaching negative effects on the environment, endangering global sustainability through plastic pollution, excessive carbon emissions, and ineffective waste management systems. Unlike conventional plastics, which remains in the environment for multiple decades, biodegradable materials are intended to degrade naturally over short span of time, reducing long-term pollution. This is indeed a significant advantage. It is essential to switch to eco-friendly and biodegradable materials as the need for packaging keeps growing [25] (Fig. 4).

2.1 Plastic Pollution

Plastics are still the material of choice, and many of them end up in the environment where they persist for centuries after serving their purpose. One of the most pressing issues of the current Anthropocene age is plastic pollution, which is acknowledged as a significant environmental and societal concern [26]. The amount of plastic trash of all sizes (nanoplastics and from macro- to micro) and the mix of polymers in the highly diverse environmental compartments (atmosphere, biosphere, lithosphere, and hydrosphere) have increased over the previous ten years, according to multiple sources. All environmental domains, especially distant ones like arctic regions and those with great biodiversity, are contaminated by plastic trash, which comes in a variety of colors, sizes, shapes, and primary and secondary origins. The prevalence, transfer, and distribution mechanisms of plastic particles among environmental matrices are the subject of several scientific investigations [27]. A more thorough understanding of their sources, transit routes, accumulation, and toxicity as well as

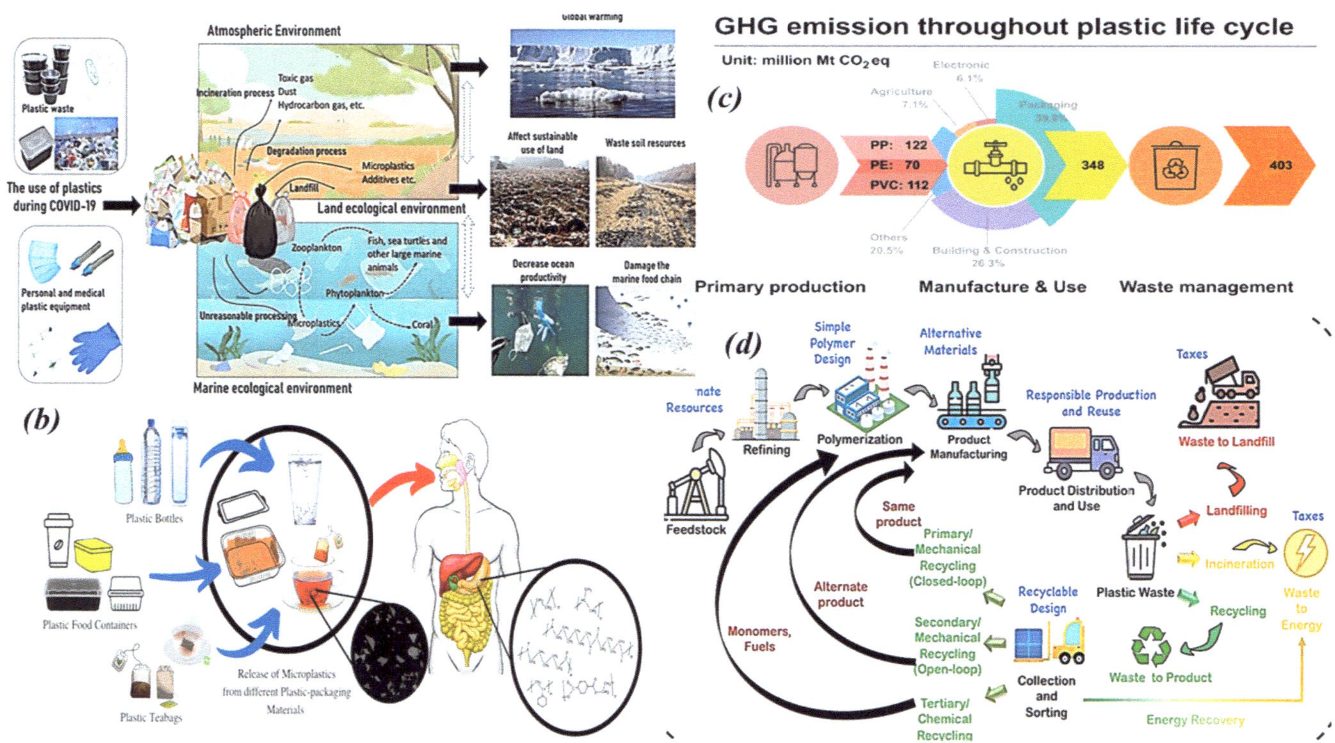

Fig. 4 Environmental impact of traditional packaging: **a** plastic pollution, **b** toxicity and health risks, **c** carbon footprint of conventional packaging, and **d** waste management challenges. Reproduced with permission [43–46]

their function as a vehicle for other chemical pollutants that are now present or linked to them, such as plastic additives is necessary in order to evaluate their effects on the health of the environment and mankind. Every year, a projected 8–12 million tons of plastic pass in to the oceans, posing a threat to marine life through entanglement and ingestion. Fish, drinking water, and even human blood have been shown to contain microplastics, which are made from broken-down plastic debris and may be harmful to one's health. Plastic trash degrades groundwater quality and soil fertility on land. Debris of non-biodegradable plastic hinders water infiltration, which has an impact on biodiversity and agricultural output. The packaging sector is the fastest growing and biggest market for synthetic plastics made using fossil fuels. Such plastics account for the bulk of plastic trash that are damaging the ecology. Moreover, the depletion of petroleum resources necessitates the production of substitute materials proficient to perform the same function as conventional plastics [28, 29]. Single-use plastics (such as plastic bottles, caps, grocery bags, cigarette butts, stirrers, lids, food wrappers, straws) are a sign of inadequate waste management systems, according to the United Nations Environment Program (UNEP).

2.2 *Toxicity and Health Risks*

Microplastic (<5 mm), mesoplastic (<2.5 cm), macroplastic (<1 m), and megaplastic (>1 m), are among the various sizes and kinds of plastic pollutants that are present in the ecosystem [30]. The sediment, biota, and water of marine and coastal environments are widely distributed with primary and secondary microplastics [31]. Majority of all commonly produced, commercial goods like personal care and cosmetics items contain microplastic particles. These microplastics are produced when the larger plastics are broken down by chemical, biological, and physical, processes in the environment [32]. Sea samples have lower incidence of microplastics in comparison to estuaries. Resultantly, plastic pollution has a variety of detrimental consequences in addition to ecological and socioeconomic ones [33]. Toxicological effects from plastic ingestion, entanglement, hunger, suffocation, dispersal, rafting of creatures, the introduction of exotic species, and the provision of new habitats are noteworthy environmental consequences that presents growing challenges to trophic connections and the biodiversity [34]. Changes in the condition of marine systems, known as degradation, are associated with a loss of ecological components and values. As a result, the socioeconomic aspects are impacted by plastic pollution through adverse effects on tourism, shipping, human health (carcinogenic, nervous, and respiratory), and fishing [35]. Practical strategies for tackling the problem of plastic pollution include preventing the accumulation of pollutants of plastics, implementing the 3Rs (Reduce-Reuse-Recycle), enhancing knowledge and capacity, and ensuring that manufacturers and consumers take responsibility for their actions. Reducing plastic waste in coastal and marine areas is greatly aided by current and enacted laws, policies, rules, and programs at the international, national, and regional levels.

The density of microplastics also affects how they disperse in the water column. Due to their low densities, polyethylene and polypropylene float in water, but polyvinyl chloride, polyethylene terephthalate, higher-density polystyrene, and polyamide, do not float in water but instead deposit through the water column at an inclination [36]. Therefore, microplastic pollution are extensively dispersed throughout all coastal and marine systems' layers and sub-zones, including benthic and pelagic. Salinity is one of the primary factors affecting the chemical breakdown of plastic. Therefore, marine and coastal environments, which have considerable salinity are particularly vulnerable to microplastic production. Hence, in order to determine the nature of the new problem, scientific indication on the durability and distribution of microplastic pollutants must concentrate on coastal ecosystems and ocean basins. These areas are seeing a large buildup of plastic pollutants, which has an adverse influence on the health of mankind, economic activity, and biodiversity. Various marine creatures consume microplastics causing fatalities. Evidence of microplastics in the watery environment raises concerns about plastic pollution's effects on the environment.

2.3 Carbon Footprint of Conventional Packaging

Conventional packaging's manufacture, shipping, and disposal greatly increase carbon emissions worldwide. The issues related to plastic waste and pollution are also tied to the significant emissions of GHGs linked to the generation and use of plastics. Global emissions of GHGs from the life cycles of plastics were cautiously estimated at 1.8 gigatons of carbon dioxide equivalent (Gt CO_2e) in 2015, and under typical future scenarios, they are projected to increase by three to five folds in the next few decades [37]. The extraction of raw materials is the first step. Deforestation for paper and petroleum extraction for plastics require massive amounts of energy. This process releases a lot of greenhouse gases, which greatly contributes to global warming. The extraction stage, which is both energy-intensive and environmentally damaging during the production stage, is one of the primary sources of carbon emissions throughout the package life cycle. At this point, manufacturing facilities mold, print, and put together package materials. These procedures use a lot of energy and frequently depend on fossil fuels like coal and natural gas. They thus release significant volumes of greenhouse gasses, including carbon dioxide, into the environment. Because fossil fuels provide the energy for these industries, this stage of manufacturing has a major impact on climate change [38, 39]. Utilizing lightweight packaging materials, careful material selection, efficient manufacturing techniques, and adopting optimized packaging design are ways to reduce carbon footprints. Also, adopting biodegradable packaging, recyclable packaging, reusable packaging and innovative materials are eco-friendly alternatives to traditional packaging.

2.4 Waste Management Challenges

Management of wastes involves the sorting, collection, handling, transportation, treatment, and ultimately disposing of the discarded materials and substances. Packaging wastes has accumulated in landfills, incinerators, and open spaces as a result of the ineffectiveness of worldwide waste management systems [40]. Degradation of the environment is made worse by inadequate recycling infrastructure and poor waste disposal methods. Waste packaging pertains to diverse packaging materials that are thrown away following industrial manufacturing, logistics transport, and the consumption of goods, such as foam, wood strips, cartons, wrapping tapes, plastics, and many more. Once utilized for packaging, the waste packaging is thrown away or rendered unusable for multiple reasons, turning into solid waste that adversely affects the surroundings. Typically, waste packaging materials are classified as cellulose or plastic types according to their chemical structure. Historically, various conventional methods like landfill, depolymerization and incineration, have been employed for disposing of plastic waste. Nonetheless, the real-world implementation of these strategies is frequently limited by legal frameworks and regulations because of the generation of microplastics and harmful emissions. The dangers to the health of mankind and the ecological surroundings that stem from the inadequate dumping of packaging waste must be taken into account. Lately, significant exertions have been concentrated on the waste plastics recycling through biorecycling and chemical methods [41]. Waste management practices helps in reducing burden on landfills, reduces recycling rates, and also reduces microplastic accumulation. Environmental sustainability has been integrated into the waste management industry and agenda through programs like the 3Rs "reduce, reuse, recycle" trash policy. But until now, recycling and other end-of-life solutions have dominated sustainability initiatives. Waste reuse and reduction have caught less attention and tangible policy support than recycling, which is the main focus of the 3Rs and has been endorsed through institutional and legislative channels throughout governments [42].

3 Biodegradable Materials in Packaging

An important step in resolving environmental concerns related to conventional plastic packaging is the transition to biodegradable materials. These materials spontaneously break down predominantly due to microbial activity, in to CO_2, CH_4, inorganic compounds, water, or biomass including some other environmentally acceptable constituents which lowers pollution and encourages sustainability [47]. In sectors like food, beverages, medicines, e-commerce, and consumer products, biodegradable packaging is becoming more and more popular as an environmentally acceptable substitute for plastics derived from petroleum. Materials like plastics are widely used, particularly in packaging. The extensive use of plastics made from fossil fuels has led to persistent issues with trash generation and environmental pollution [48].

Presently, almost 320 million tons of plastic are produced yearly worldwide, and growing requirement is a reflection of its many uses. Because these petroleum-based plastics are not biodegradable and were improperly disposed of, millions of tons of plastics keep moving into the ocean. Annually, greater than 30% of plastic garbage is dumped in landfills [49]. When packing has attractive graphics and component information tags, it can give customers a satisfying experience. The definition, traits, varieties, uses, benefits, and drawbacks of biodegradable materials in packaging are discussed in this section.

3.1 Types of Packaging [50]

3.1.1 Primary Packaging

It surrounds the product when it is taken home by the customer and ends up in the trash. It is the simplest form of packaging that has direct contact with the product. This kind of packaging serves as the initial layer that directly safeguards the product. For instance, a wrapper of five ballpoint pens, a chocolate wrapper, a metal can of soft drinks and label for fruit or vegetables, a jar of jam or sauces with the closure and label, to provide information about the product are all referred to as primary packaging.

3.1.2 Secondary Packaging

This kind of packaging is needed to combine or group primary packs. Secondary packaging is designed to provide protection for the primary packaging. Typically, they do not find its way into the household waste stream. A box made of cardboard, or plastic crates to hold dozens of wine bottles or soft drinks, hold a set of crockeries are examples of secondary packaging.

3.1.3 Tertiary Packing

This type of packaging keeps secondary packs together and enables movement along the supply chain. Pallets, roll cages, stretch or shrink wrap material that holds items firmly on a pallet, and the pallet label are examples of tertiary packaging. This kind of packaging is utilized for the bulk handling of goods during warehousing and transportation. The characteristics of tertiary packaging can influence efficiency of the logistics within supply chains and impose various requirements on the handling equipment's.

3.2 *Biodegradable, Degradable and Compostable [51]*

These terms biodegradable, degradable, and compostable have distinct meanings as elaborated below.

3.2.1 Biodegradable Materials

They can experience physical, chemical, thermal, or biological breakdown, resulting in the majority of the final compost eventually transforming into carbon dioxide, biomass, and water. They break down due to the activities of naturally existing microorganisms like algae, bacteria, and fungi. The breakdown of these plastics is influenced by several factors like moisture and temperature.

3.2.2 Degradable Materials

Over time, they will decompose into smaller particles when subjected to heat, oxygen, visible light or ultraviolet light. But it cannot be assured that it would completely turn out to be non-toxic.

3.2.3 Compostable Materials

They break down through biological processes, resulting in water, biomass, inorganic substances, and CO_2. Therefore, they are not detrimental to the environment as they do not emit any toxic elements into the environment and soil and do not damage plants after they decompose.

3.3 *Biodegradable Polymers*

3.3.1 Polylactic Acid (PLA)

It is a thermoplastic aliphatic polyester derived from the polymerization of lactic acid sourced from non-conventional resources, including corn starch, tapioca roots, chips or starch, and sugarcane. Environmental issues resulting from the extensive use of synthetic polymers derived from conventional resources will be lessened by the bio-based PLAs. In 1932, Carothers discovered polylactic acid. PLA is a popular biopolymer that has mechanical qualities that are competitive with those of petroleum-based polymers. PLA production offers several advantages over other biopolymer production methods, such as (a) producing lactide monomer from lactic acid, which is derived from fermenting renewable resources like corn, sugarcane, maize, and potatoes; (b) conserving energy; (c) lowering the amount of landfill space. (d) It offers a great deal of potential for producing compostable hybrid paper-plastic packaging and (e) it has a great deal of opportunity for material changes that optimize physical attributes [52]. The production of PLA was 0.2 million tons in 2015 and 0.3 million tons in 2019, making it one of the most widely used and commercial bioplastics in the world. PLA is a common polymer for packaging and is primarily used in the food sector to produce disposable tableware items such as cutlery, drinking cups, trays, food containers, food plates, and packing for sensitive food products. PLA is also utilized in a variety of other sectors, including packaging, textiles, and healthcare.

3.3.2 Polyhydroxyalkanoates (PHA)

PHAs, which are primarily utilized in biomedical applications at the moment, have garnered interest as possible food packaging materials due to their suitable qualities, which are comparable to those of synthetic plastics that have historically been used for this purpose. Polyethylene, polypropylene, nylon, and polyvinyl chloride are examples of petrochemical polymers that these biodegradable polymers are thought to be good alternatives to. They are used to come into contact with food products with high-moisture while retaining their structure because of their hydrophobic nature. They are a naturally occurring, renewable feedstock that can be processed using standard plastics equipment and biodegraded in a variety of environmental settings. The hydroxyl and carboxyl groups of two repeating units are merged by an ester link to form PHAs, which are linear polyesters of 3-hydroxy fatty acid monomers. They are categorized as (a) long chain length (15 or more carbon atoms), (b) medium chain length (6–14 carbon atoms), and (c) short chain length PHA (3–5 carbon atoms), according to the number of atoms of carbon in the side chains. Numerous PHAs with up to 150 diverse monomer units have been investigated. PHBV (poly(3-hydroxybutyrate-co-3-hydroxyvalerate) and PHB (Poly(3-hydroxybutyrate) are the two forms of PHA that are most studied. Because of the great chemical diversity of its radicals, PHAs can have physical qualities that are analogous to those of traditional synthetic polymers. PHA can be produced from soft elastomers, rubbers, and adhesives to stiff and brittle plastic products, depending on the monomer content. Because they are hydrophobic and have gas barrier qualities similar to those of PVC and PET, certain PHA packing materials provide good moisture resistance. PHA films have significantly superior moisture barrier qualities than polysaccharide-based materials, however their gas barrier capacity is lower [53].

3.3.3 Starch-Based Polymers

Starch serves as an excellent source of biodegradable material for food packaging, derived from rice, wheat, potatoes, and corn. It is broadly regarded as an eco-friendly alternative to plastics for food wrapping. Additionally, different foods like vegetables, fruits, dry items, and snacks, can be wrapped using starch as a compostable film. Starch can be utilized in producing biodegradable films in three ways: firstly, limited quantities of starch may be incorporated into the formulation of starch blends with other plastics. The core objective is to enhance the biodegradability of conventional, oil-derived starch substances. Secondly, the creation of starch composites where the starch accounts for over fifty percent of the weight. Thirdly, the application of extrusion, handling mixtures of granular starch in biodegradable production methods.

The rise in the utilization of starch-based biodegradable films for food packaging results from its many benefits. These consist of their role in decreasing fossil materials, absence of harmful substances, their plant-based origins, biodegradability and biocompatibility, the affordability and plentiful nature of starch, their safety for food packaging use, lower energy usage, their function as an environmentally friendly disposal option, and no overall rise in CO_2 within the global ecosystem. In spite of these advantages and benefits, drawbacks consist of inadequate mechanical properties, low water resistance, high-moisture susceptibility, the presence of an ineffective moisture barrier due to strong hydrophilic characteristics, their brittle nature at ambient temperature, and their elevated moisture levels. To address these drawbacks of using starch as a biodegradable substance, a thermoplastic starch matrix may be enhanced with some kind of nanofillers, thereby enhancing the characteristics.

Therefore, the application of MMT in food packaging is due to its lower expense, great durability, and high efficiency. Additional details regarding MMT nanoclay are provided in this review. Research has indicated that the characteristics of starch-based films contribute to their rigidity and decreased flexibility. Therefore, alternative polymers may serve as additive materials. Various origins of biopolymers, including proteins, fats, and, polysaccharides can function as biodegradable films. Starch primarily comes from plants, but it can also be derived from roots, tubers, grains, and legumes. Taking into account its natural biodegradability, plentiful supply, and yearly renewability, starch is an appealing natural polymer. Notably, earlier research has shown that starch comprises two types of microstructures: a) linear and b) branched. Therefore, it is considered a heterogeneous substance. The linear arrangement is referred to as amylose (the crystallizable type of starch consisting of lengthy unbranched polysaccharide chains), whereas the branched configuration is termed amylopectin (the non-crystallizable type of starch featuring a branched polysaccharide chain) [54].

4 Applications of Biodegradable Materials in Packaging

4.1 Food and Beverage Packaging

Biodegradable polymers provide a sustainable substitute for traditional plastic beverage packaging. It is possible to extract polysaccharides from plants, such as cellulose and starch, to create biodegradable films and bottles. For packaging applications, cellulose nanoparticles such as CNFs and CNCs that are derived from cotton, wood, and other plant sources offer beneficial qualities. Lactic acid is polymerized to create PLA, a commercially viable bioplastic. PLA is a useful material for beverage bottles and films because it is low toxicity, has good mechanical strength, and blocks gasses, water vapor, and odors [55].

4.2 Confections and Cereals

For the cereal and confectionery industry, biodegradable packaging materials derived from renewable resources provide a significant sustainable alternative. Cereals, such as rice, pasta, and flour, are extremely vulnerable to air exposure and moisture intrusion, which can cause microbial development, discoloration, and rancidity. Glass jars, cans, cartons, and plastic pouches are frequently utilized to offer the necessary oxygen and water vapor barrier [11]. However, the majority of traditional plastic packaging backs environmental pollution owing to its poor biodegradability and recyclability. While maintaining the necessary mechanical and barrier qualities, biodegradable materials can lessen dependency on exhaustible resources and plastic waste. Biodegradable films derived from proteins (whey, zein, glutten, etc.), lipids, and polysaccharides (starch, cellulose, chitosan, etc.) have demonstrated promise as long-term moisture and oxygen barriers for cereals. For cereal packaging, starch-based films in specific have been thoroughly studied. To improve their qualities, these films frequently use laminates with other biopolymers, crosslinking, or reinforcing using nanoclays [56].

4.3 Fresh Fruits and Vegetables

Fresh food should be packaged in biodegradable materials to support sustainability efforts and lessen plastic waste and pollution. Proteins, PLA, and polysaccharides are examples of biopolymers that can offer gas and moisture barrier qualities to preserve the nutritional value, texture, color, and flavor of fresh produce while also fully dissolving in the environment after disposal. For instances, PLA sheets are helpful in assisting to regulate the moisture loss fresh-cut pineapples and cucumbers' shelf life can also be increased by cellulose casings, which keep them firm and prevents weight loss [57].

4.4 Dairy, Seafood and Meat

The use of biodegradable packaging for meat and fish products derived from food waste and byproducts has showed immense promise. Grapefruit seed extract added to gelatin-based films shown significant antibacterial activity against typical meat spoilage germs including, Staphylococcus aureus, Salmonella Typhimurium, Listeria monocytogenes, and Escherichia coli [58]. Overall, there is good promise for improving the quality, shelf life of perishable meat and seafood items and safety using biodegradable packaging created from food waste. However, to go from laboratory-based research to massive commercial applications, more investigation and awareness is still required. Novel packaging materials of biodegradable nature generated from food waste to be successfully commercialized, requires more research into aspects like film forming and mechanical characteristics, the interaction between the packaging and the food matrix, viability and the economics of extensive production. Traditional dairy packaging can possibly be replaced with biodegradable packaging materials such as PLA and bio-based polyethylene including some kind of bio-nanocomposite materials [59].

5 Advantages and Limitations of Biodegradable Packaging

5.1 Advantages

5.1.1 Environmental Sustainability

Conventional plastic is notorious for its persistence in ecosystems, leading to pollution in oceans, waterways, and landfills. Biodegradable packaging, on the other hand, is engineered to degrade more quickly and safely, which helps reduce litter and plastic waste in the environment. Materials like PHA and PLA breakdown into non harmful components, reducing their harmful impact on marine life, wildlife, and ecosystems. Many biodegradable materials are derived from inexhaustible resources such as plant-based sugars (e.g., sugarcane and corn), which absorb CO_2 during their growth. This carbon sequestration can offset part of the emissions produced during their manufacturing process, resulting in a lower overall carbon footprint in comparison to petroleum-based plastics [60]. The raw materials used for biodegradable packaging are often derived from renewable plant-based sources, as opposed to conventional plastics, which rely on fossil fuels. This transition supports sustainability by reducing dependence on non-renewable resources and further reducing the carbon footprints.

Production of biodegradable packaging materials consume lesser energy compared to traditional plastics [61]. However the quantity of energy consumption may possibly vary depending on the production methods adopted. If renewable energy sources are used during production, the overall environmental impact can possibly be further reduced significantly.

5.1.2 Material Recyclability

Though biodegradable packaging is not always recyclable certain biodegradable materials can contribute to improved recycling systems when managed appropriately. For instance, recyclable bioplastics are distinct from purely biodegradable materials in that they can be processed within existing recycling streams. Some biodegradable packaging materials, particularly bio-based plastics like PLA, can be processed and recycled through specific recycling channels. This contributes to a circular economy, where materials are recycled, reused, and reintroduced into the production cycles, reducing the need for new raw materials [61]. PLA can be recycled into new products like compostable trays, bags, and even clothing fibers, provided the material is separated from other plastic types during waste collection [62]. Unlike conventional plastics that often contaminate the recycling stream, some biodegradable plastics are designed to be easily separated from other materials in the recycling process, reducing contamination risks and enhancing the effectiveness of recycling systems. In some cases, biodegradable materials are more suitable for industrial composting rather than recycling [63]. When composted, these materials break down into organic matter, providing nutrient-rich compost for soil, thus diverting waste from landfills.

5.1.3 Material Biodegradability

Biodegradability is one of the defining characteristics of biodegradable packaging, and it offers substantial environmental advantages. The ability to naturally degrade in the environment without leaving harmful residues is a core advantage of biodegradable materials over traditional plastics. Biodegradable materials are designed to break down relatively quickly when exposed to natural conditions, such as moisture, heat, and microbial activity. This lessens the demand for long-term waste management solutions and the persistence of garbage in the environment. In contrast to polyethylene or polypropylene, which might take very long time to disintegrate, bioplastics like PHA can do so in a matter of weeks to months in natural settings. Biodegradable packaging decomposes into safe byproducts including CO_2, water, and organic materials that don't damage wildlife or contaminate the environment. It is therefore a better choice than plastics that decompose into poisonous compounds or dangerous microplastics [64]. Biodegradable packaging materials are less dangerous to wildlife, especially in aquatic areas, because they are made to break down in an environmentally appropriate manner. Marine life can become entangled in traditional plastic, but biodegradable alternatives are less likely to do so. In industrial composting facilities, biodegradable materials can be safely composted. There, they decompose into organic matter that improves soil health. The materials may nevertheless break down more quickly in landfills than traditional plastics, preventing the buildup of waste over time. When given the proper conditions, products developed from starch-based bioplastics also break down into CO_2, water, and trace minerals, adding to the carbon cycle without harming the environment [65].

5.2 Limitations

5.2.1 Performance

In many cases, biodegradable materials may fall short of the strength, durability, heat resistance, gas resistance, moisture, and flexibility of conventional plastics. For certain applications, manufacturers can be forced to invest in more advanced and costly biodegradable products to achieve the required performance. This further boosts the overall price [66]. Biodegradable materials, especially those from plant origins, can have a shorter shelf life compared to conventional plastics. They can degrade or lose their efficiency over time, and this may lead to higher waste and disposal costs. Such factors can make biodegradable products less desirable for broad or long-term use [67].

5.2.2 Processability

Biodegradable packaging material suggests being alternative, more environmentally friendly substitutes to traditional plastics but they usually have different challenges of processing and production. Most biodegradable packaging materials consist of starch, cellulose, or proteins which make them highly vulnerable to water. Upon exposure to humidity, these substances can deteriorate, weaken, or be difficult to process due to swelling, variable quality, or clogging and blockage of extrusion or molding equipment due to moisture content [68]. The control of humidity and proper drying of raw materials prior to processing can help mitigate this issue. Some biodegradable materials, particularly those from renewable resources, show lower thermal stability than typical plastics such as polyethylene or polypropylene. This may possibly cause process troubles during injection molding, show extrusion complications, deformation, and sometimes even premature degradations. Thermal stability can be improved by the development of heat-resistant additives or blending biodegradable materials with other polymers [69].

5.2.3 Policies and Regulations

There is frequently a deficiency of clear regulatory policies and guidelines regarding biodegradable materials, complicating the task for manufacturers to identify which materials are appropriate to utilize. Certain biodegradable materials might fail to decompose as anticipated in specific environments (e.g., in landfills or oceanic waters), prompting doubts about their true environmental advantages. This uncertainty can deter companies from committing to biodegradable options. In several places, there are limited or nonexistent subsidies or tax benefits for businesses to adopt biodegradable materials. In the absence of such financial assistance, the elevated expenses associated with biodegradable materials pose an obstacle to their acceptance [70, 71].

5.2.4 Cost

Conventional plastics are still significantly cheaper than biodegradable products due to decades of development, high-volume production, and established supply chains. It will be hard for biodegradable products to gain widespread acceptance until the price gap between traditional plastics and biodegradable products narrows. Until then, biodegradable materials will not be able to gain widespread acceptance, especially in price-sensitive industries like packaging and consumer goods. Hence, the price of biodegradable material forms one of the main hurdles to their widespread acceptance [72]. While such materials are generally more environmentally friendly than traditional counterparts, their higher costs associated with their production, processing, and use can be a problem for consumers, industries, and governments. As already mentioned earlier that the biodegradable materials are often made from renewable resources like starch derived from plants, cellulose, or PLA from sugarcane or corn. Sometimes ready availability of such raw materials can be more expensive to acquire compared to petroleum-based materials like plastic. Growing and processing these materials require significant labor, land, and energy, which makes them costly. Sometimes the process of making biodegradable materials may be more complex than that of regular plastics. For example, preparation of PLA requires polymerization and fermentation processes [73] that can be more expensive and energy-intensive than traditional plastic production methods. Government assistance, innovation content, consumer consciousness, and enhanced recycling and waste management are measures that can be implemented to lower the price of such content.

Thus, we can confidently assert that biodegradable materials offer a viable route for environmentally friendly packaging solutions that suit consumer and regulatory requirements. However, for wider acceptance for mass deployment, they need to overcome their own set of cost, durability, and waste-handling challenges. Research in the areas of future material science, waste treatment plants, and biopolymer synthesis will be crucial in the journey to develop the biodegradable packaging technology.

6 Conclusions

The first critical step in addressing the environmental concerns brought up by conventional plastic packaging is investigating and developing sustainable, biodegradable, and eco-friendly composite materials for green packaging. The need to transition to sustainable packaging solutions has never been more urgent as the world battles resource depletion, plastic waste, and the growing demand to lower carbon footprints. The significance of biodegradable polymers like PLA, PHA, and starch-based materials, as well as eco-friendly composites made from natural fibers and bio-based resins, in reducing environmental damage has been underlined in this chapter. Numerous barriers prevent these materials from being widely used, despite their clear potential in terms of biodegradability, recyclability, and a reduced carbon footprint. A number of obstacles, including cost, material qualities, and processing capabilities, continue to hinder the scalability of green packaging. However, there is optimism because of advancements in material science and growing consumer and industry demand for sustainable products. Using these materials in a circular economy paradigm could lead to a more sustainable future. Furthermore, the government's push for stronger environmental rules and consumers' growing preference for sustainably packaged items have heightened the need for innovation in the packaging business. In the fight against plastic pollution and the greater ecological emergency, using eco-friendly and biodegradable materials is not only a trend; rather, it is an essential step. Future developments in green packaging are likely to depend on continued advancements in these materials and a greater commitment to sustainable supply chains. The packaging industry can make a substantial contribution to the development of a cleaner and more sustainable world with these initiatives.

Declaration of Competing Interest The authors declare that they have no known competing financial interests or personal relationships that could have appeared to influence the work reported in this paper.

References

1. Moshood TD, Nawanir G, Mahmud F, Mohamad F, Ahmad MH, AbdulGhani A, Kumar S (2022) Green product innovation: a means towards achieving global sustainable product within biodegradable plastic industry. J Clean Prod 363:132506. https://doi.org/10.1016/j.jclepro.2022.132506
2. Zhao X, Wang Y, Chen X, Yu X, Li W, Zhang S, Meng X, Zhao Z-M, Dong T, Anderson A, Aiyedun A, Li Y, Webb E, Wu Z, Kunc V, Ragauskas A, Ozcan S, Zhu H (2023) Sustainable bioplastics derived from renewable natural resources for food packaging. Matter 6(1):97–127. https://doi.org/10.1016/j.matt.2022.11.006
3. Chen Y, Awasthi AK, Wei F, Tan Q, Li J (2020) Single-use plastics: production, usage, disposal, and adverse impacts. Sci Total Environ 752:141772. https://doi.org/10.1016/j.scitotenv.2020.141772
4. Geyer R, Jambeck JR, Law KL (2017) Production, use, and fate of all plastics ever made. Sci Adv 3(7). https://doi.org/10.1126/sciadv.1700782

5. Crespy D, Bozonnet M, Meier M (2008) 100 years of bakelite, the material of a 1000 uses. Angew Chem Int Ed 47(18):3322–3328. https://doi.org/10.1002/anie.200704281
6. Williams AT, Rangel-Buitrago N (2022) The past, present, and future of plastic pollution. Mar Pollut Bull 176:113429. https://doi.org/10.1016/j.marpolbul.2022.113429
7. Humphreys G (2023) Towards a global plastics treaty. Bull World Health Organ 101(12):754–755. https://doi.org/10.2471/blt.23.021223
8. Li Y, Tao L, Wang Q, Wang F, Li G, Song M (2023) Potential Health impact of Microplastics: a review of environmental distribution, human exposure, and toxic effects. Environ Health 1(4):249–257. https://doi.org/10.1021/envhealth.3c00052
9. Shrivastava A, Dondapati S (2021) Biodegradable composites based on biopolymers and natural bast fibres: a review. Mater Today Proceed 46:1420–1428. https://doi.org/10.1016/j.matpr.2021.02.652
10. Pulikkalparambil H, Varghese SA, Chonhenchob V, Nampitch T, Jarupan L, Harnkarnsujarit N (2023) Recent advances in natural fibre-based materials for food packaging applications. Polymers 15(6):1393. https://doi.org/10.3390/polym15061393
11. Khoaele KK, Gbadeyan OJ, Chunilall V, Sithole B (2023) The devastation of waste plastic on the environment and remediation processes: a critical review. Sustainability 15(6):5233. https://doi.org/10.3390/su15065233
12. Hussain S, Akhter R, Maktedar SS (2024) Advancements in sustainable food packaging: from eco-friendly materials to innovative technologies. Sustain Food Technol 2(5):1297–1364. https://doi.org/10.1039/d4fb00084f
13. Amulya K, Katakojwala R, Ramakrishna S, Mohan SV (2021) Low carbon biodegradable polymer matrices for sustainable future. Compos Part C Open Access 4:100111. https://doi.org/10.1016/j.jcomc.2021.100111
14. Duarte P, Silva SC, Roza AS, Dias JC (2024) Enhancing consumer purchase intentions for sustainable packaging products: an in-depth analysis of key determinants and strategic insights. Sustain Fut 7:100193. https://doi.org/10.1016/j.sftr.2024.100193
15. Choi SY, Cho IJ, Lee Y, Kim YJ, Kim KJ, Lee SY (2020) Microbial polyhydroxyalkanoates and nonnatural polyesters. Adv Mater 32(35):1907138. https://doi.org/10.1002/adma.201907138
16. Castro-Aguirre E, Iniguez-Franco F, Samsudin H, Fang X, Auras R (2016) Poly (lactic acid)-Mass production, processing, industrial applications, and end of life. Adv Drug Deliv Rev 107:333–366.https://doi.org/10.1016/j.addr.2016.03.010
17. Surendren A, Mohanty AK, Liu Q, Misra M (2022) A review of biodegradable thermoplastic starches, their blends and composites: Recent developments and opportunities for single-use plastic packaging alternatives. Green Chemistr 24(22):8606–8636.https://doi.org/10.1039/D2GC02169B
18. Liu TH, Wen YL, Li GQ, Nie XN (2020) Optimization and experimental study of an intelligent bamboo-splitting machine charging manipulator. J Robot 2020(1):4675301.https://doi.org/10.1155/2020/4675301
19. Odokonyero K, Gallo Jr A, Mishra H (2021) Nature-inspired wax-coated jute bags for reducing post-harvest storage losses. Sci Rep 11(1):15354.https://doi.org/10.1038/s41598-021-93247-z
20. Enarevba DR, Haapala KR (2024) The emerging hemp industry: a review of industrial hemp materials and product manufacturing. AgriEngineering 6(3):2891–2925.https://doi.org/10.3390/agriengineering6030167
21. Raj H, Tripathi S, Bauri S, Choudhury AM, Mandal SS (2023)Green composites using naturally occurring fibers: a comprehensive review. Sustain Polym Energy 1(2):10010. https://doi.org/10.35534/spe.2023.10010
22. Zhou W, Bergsma S, Colpa DI, Euverink G-JW, Krooneman J (2023) Polyhydroxyalkanoates (PHAs) synthesis and degradation by microbes and applications towards a circular economy. J Environ Manage 341:118033. https://doi.org/10.1016/j.jenvman.2023.118033
23. De Luca S, Milanese D, Gallichi-Nottiani D, Cavazza A, Sciancalepore C (2023) Poly(lactic acid) and its blends for packaging application: a review. Clean Technol 5(4):1304–1343. https://doi.org/10.3390/cleantechnol5040066

24. Fazita MRN, Jayaraman K, Bhattacharyya D, Haafiz MKM, Saurabh CK, Hussin MH, SA KHP (2016) Green composites made of bamboo fabric and poly (Lactic) acid for packaging applications—a review. Materials 9(6):435. https://doi.org/10.3390/ma9060435
25. Yin Y, Woo MW (2024) Transitioning of petroleum-based plastic food packaging to sustainable bio-based alternatives. Sustain Food Technol 2(3):548–566. https://doi.org/10.1039/d4fb00028e
26. Promhuad K, Srisa A, San H, Laorenza Y, Wongphan P, Sodsai J, Tansin K, Phromphen P, Chartvivatpornchai N, Ngoenchai P, Harnkarnsujarit N (2022) Applications of hemp polymers and extracts in food, textile and packaging: a review. Polymers 14(20):4274. https://doi.org/10.3390/polym14204274
27. Chen H, Wu Q, Ren X, Zhu X, Fan D (2024) A fully bio-based adhesive with high bonding strength, low environmental impact, and competitive economic performance. Chem Eng J 494:153198. https://doi.org/10.1016/j.cej.2024.153198
28. Pohan JN, Kusumawati YA, Radhitanti A (2023) Mushroom mycelium-based biodegradable packaging material: a promising sustainable solution for food industry. E3S Web of Conferences 426:02128. https://doi.org/10.1051/e3sconf/202342602128
29. Sani MA, Azizi-Lalabadi M, Tavassoli M, Mohammadi K, McClements DJ (2021) Recent advances in the development of smart and active biodegradable packaging materials. Nanomaterials 11(5):1331. https://doi.org/10.3390/nano11051331
30. Xie G, Huang L, Apostolidis C, Huang Z, Cai W, Li G (2021) Assessing consumer preference for overpackaging solutions in E-Commerce. Int J Environ Res Public Health 18(15):7951. https://doi.org/10.3390/ijerph18157951
31. Semple KE, Zhou C, Rojas OJ, Nkeuwa WN, Dai C (2022) Moulded pulp fibers for disposable food packaging: a state-of-the-art review. Food Packag Shelf Life 33:100908. https://doi.org/10.1016/j.fpsl.2022.100908
32. Morel S, Mura G, Gallarate M, Sapino S (2024) Cosmetic packaging: European regulatory aspects and sustainability. Cosmetics 11(4):110. https://doi.org/10.3390/cosmetics11040110
33. Moshood TD, Nawanir G, Mahmud F, Mohamad F, Ahmad MH, AbdulGhani A (2022) Sustainability of biodegradable plastics: new problem or solution to solve the global plastic pollution? Curr Res Green Sustain Chem 5:100273. https://doi.org/10.1016/j.crgsc.2022.100273
34. Praet E (2023) Plastic pollution: archaeological perspective on an anthropocene climate emergency. World Archaeol 55(3):329–347. https://doi.org/10.1080/00438243.2024.2320123
35. Caruso G, Corsi I, Wu C, Bergami E, Corami F, Azevedo-Santos VM (2024) Plastic pollution in marine and freshwater biota. Water Biol Secur 100247. https://doi.org/10.1016/j.watbs.2024.100247
36. Jaikumar IM, Tomson M, Meyyazhagan A, Balamuralikrishnann B, Baskaran R, Pappuswamy M, Kamyab H, Khalili E, Farajnezhad M (2024) A comprehensive review of microplastic pollution in freshwater and marine environments. Green Analy Chem 100202. https://doi.org/10.1016/j.greeac.2024.100202
37. Thushari GGN, Senevirathna JDM (2020) Plastic pollution in the marine environment. Heliyon 6(8):e04709. https://doi.org/10.1016/j.heliyon.2020.e04709
38. Wang J, Zheng L, Li J (2018) A critical review on the sources and instruments of marine microplastics and prospects on the relevant management in China. Waste Manag Res J Sustain Circ Econ 36(10):898–911. https://doi.org/10.1177/0734242x18793504
39. Auta HS, Emenike CU, Fauziah SH (2017) Distribution and importance of microplastics in the marine environment: a review of the sources, fate, effects, and potential solutions. Environ Int 102:165–176. https://doi.org/10.1016/j.envint.2017.02.013
40. Lalrinfela P, Vanlalsangi R, Lalrinzuali K, Babu PJ (2024) Microplastics: their effects on the environment, human health, and plant ecosystems. Environ Poll Manag. https://doi.org/10.1016/j.epm.2024.11.004
41. Ricciardi M, Pironti C, Motta O, Miele Y, Proto A, Montano L (2021) Microplastics in the aquatic environment: occurrence, persistence, analysis, and human exposure. Water 13(7):973. https://doi.org/10.3390/w13070973

42. Gregory MR (2009) Environmental implications of plastic debris in marine settings—entanglement, ingestion, smothering, hangers-on, hitch-hiking and alien invasions. Philosophic Trans Royal Soc B Biolo Sci 364(1526):2013–2025. https://doi.org/10.1098/rstb.2008.0265
43. Wang Q, Zhang C, Li R (2023) Plastic pollution induced by the COVID-19: environmental challenges and outlook. Environ Sci Pollut Res 30(14):40405–40426. https://doi.org/10.1007/s11356-022-24901-w
44. Jadhav EB, Sankhla MS, Bhat RA, Bhagat DS (2021) Microplastics from food packaging: an overview of human consumption, health threats, and alternative solutions. Environ Nanotechnol Monit Manag 16:100608. https://doi.org/10.1016/j.enmm.2021.100608
45. An J, Wu F, Wang D, You J (2022) Estimated material metabolism and life cycle greenhouse gas emission of major plastics in China: a commercial sector-scale perspective. Resour Conserv Recycl 180:106161. https://doi.org/10.1016/j.resconrec.2022.106161
46. Singh N, Walker TR (2024) Plastic recycling: a panacea or environmental pollution problem. Npj Mater Sustain 2(1):17. https://doi.org/10.1038/s44296-024-00024-w
47. Kumar R, Verma A, Shome A, Sinha R, Sinha S, Jha PK, Kumar R, Kumar P, Shubham N, Das S, Sharma P, Prasad PVV (2021) Impacts of plastic pollution on ecosystem services, sustainable development goals, and need to focus on circular economy and policy interventions. Sustainability 13(17):9963. https://doi.org/10.3390/su13179963
48. Guo X, Wang J (2019) The chemical behaviors of microplastics in marine environment: a review. Mar Pollut Bull 142:1–14. https://doi.org/10.1016/j.marpolbul.2019.03.019
49. Bauer F, Nielsen TD, Nilsson LJ, Palm E, Ericsson K, Fråne A, Cullen J (2022) Plastics and climate change—breaking carbon lock-ins through three mitigation pathways. One Earth 5(4):361–376. https://doi.org/10.1016/j.oneear.2022.03.007
50. Del Rio DDF, Sovacool BK, Griffiths S, Bazilian M, Kim J, Foley AM, Rooney D (2022) Decarbonizing the pulp and paper industry: a critical and systematic review of sociotechnical developments and policy options. Renew Sustain Energy Rev 167:112706. https://doi.org/10.1016/j.rser.2022.112706
51. Shamoon A, Haleem A, Bahl S, Javaid M, Garg SB, Sharma RC, Garg J (2022) Environmental impact of energy production and extraction of materials—a review. Mater Today Proceed 57:936–941. https://doi.org/10.1016/j.matpr.2022.03.159
52. Ganguly RK, Chakraborty SK (2024) Plastic waste management during and post Covid19 pandemic: challenges and strategies towards circular economy. Heliyon 10(4):e25613. https://doi.org/10.1016/j.heliyon.2024.e25613
53. Huang S, Wang M, Dai Y, Deng C, Xue S, Qiu F, Zhang T (2024) Waste to treasure: upcycling waste express packing to sustainable packaging materials. J Environ Chem Eng 115016. https://doi.org/10.1016/j.jece.2024.115016
54. Sicotte DM, Seamon JL (2020) Solving the plastics problem: moving the U.S. from recycling to reduction. Soc Nat Resour 34(3):393–402. https://doi.org/10.1080/08941920.2020.1801922
55. Song JH, Murphy RJ, Narayan R, Davies GBH (2009) Biodegradable and compostable alternatives to conventional plastics. Philosoph Trans Royal Soc B Biol Sci 364(1526):2127–2139. https://doi.org/10.1098/rstb.2008.0289
56. Gao S, Shao B (2022) Why do consumers switch to biodegradable plastic consumption? the effect of push, pull and mooring on the plastic consumption intention of young consumers. Sustainability 14(23):15819. https://doi.org/10.3390/su142315819
57. Paletta A, Filho WL, Balogun A-L, Foschi E, Bonoli A (2019) Barriers and challenges to plastics valorisation in the context of a circular economy: case studies from Italy. J Clean Prod 241:118149. https://doi.org/10.1016/j.jclepro.2019.118149
58. Ibrahim ID, Hamam Y, Sadiku ER, Ndambuki JM, Kupolati WK, Jamiru T, Eze AA, Snyman J (2022) Need for sustainable packaging: an overview. Polymers 14(20):4430. https://doi.org/10.3390/polym14204430
59. Nizamuddin S, Chen C (2024) Biobased, biodegradable and compostable plastics: chemical nature, biodegradation pathways and environmental strategy. Environ Sci Pollut Res 31(6):8387–8399. https://doi.org/10.1007/s11356-023-31689-w

60. Swetha TA, Bora A, Mohanrasu K, Balaji P, Raja R, Ponnuchamy K, Muthusamy G, Arun A (2023) A comprehensive review on polylactic acid (PLA)–synthesis, processing and application in food packaging. Int J Biol Macromol 234:123715. https://doi.org/10.1016/j.ijbiomac.2023.123715
61. Atarés L, Chiralt A, González-Martínez C, Vargas M (2024) Production of polyhydroxyalkanoates for biodegradable food packaging applications using Haloferax Mediterranei and agrifood wastes. Foods 13(6):950. https://doi.org/10.3390/foods13060950
62. Xu Y, Wu Z, Li A, Chen N, Rao J, Zeng Q (2024) Nanocellulose composite films in food packaging materials: a review. Polymers 16(3):423. https://doi.org/10.3390/polym16030423
63. Bauer A-S, Leppik K, Galić K, Anestopoulos I, Panayiotidis MI, Agriopoulou S, Milousi M, Uysal-Unalan I, Varzakas T, Krauter V (2022) Cereal and confectionary packaging: background, application and shelf-life extension. Foods 11(5):697. https://doi.org/10.3390/foods11050697
64. Cheng J, Gao R, Zhu Y, Lin Q (2024) Applications of biodegradable materials in food packaging: a review. Alex Eng J 91:70–83. https://doi.org/10.1016/j.aej.2024.01.080
65. De Abreu DAP, Cruz JM, Losada PP (2011) Active and intelligent packaging for the food industry. Food Rev Int 28(2):146–187. https://doi.org/10.1080/87559129.2011.595022
66. Santhosh R, Nath D, Sarkar P (2021) Novel food packaging materials including plant-based byproducts: a review. Trends Food Sci Technol 118:471–489. https://doi.org/10.1016/j.tifs.2021.10.013
67. Ghobish SA, Motti CA, Bissember AC, Vamvounis G (2025) Microplastics in the marine environment: challenges and the shift towards sustainable plastics and plasticizers. J Hazard Mater 137945. https://doi.org/10.1016/j.jhazmat.2025.137945
68. Shaikh S, Yaqoob M, Aggarwal P (2021) An overview of biodegradable packaging in food industry. Curr Res Food Sci 4:503–520. https://doi.org/10.1016/j.crfs.2021.07.005
69. Rosenboom J-G, Langer R, Traverso G (2022) Bioplastics for a circular economy. Nat Rev Mater 7(2):117–137. https://doi.org/10.1038/s41578-021-00407-8
70. Trivedi AK, Gupta MK, Singh H (2023) PLA based biocomposites for sustainable products: a review. Adv Industr Eng Polym Res 6(4):382–395. https://doi.org/10.1016/j.aiepr.2023.02.002
71. Samir A, Ashour FH, Hakim AA, Bassyouni M (2022) Recent advances in biodegradable polymers for sustainable applications. Npj Mater Degrada 6(1). https://doi.org/10.1038/s41529-022-00277-7
72. Kumar R, Lalnundiki V, Shelare SD, Abhishek GJ, Sharma S, Sharma D, Kumar A, Abbas M (2023) An investigation of the environmental implications of bioplastics: recent advancements on the development of environmentally friendly bioplastics solutions. Environ Res 244:117707. https://doi.org/10.1016/j.envres.2023.117707
73. Fogašová M, Figalla S, Danišová L, Medlenová E, Hlaváčiková S, Vanovčanová Z, Omaníková L, Baco A, Horváth V, Mikolajová M, Feranc J, Bočkaj J, Plavec R, Alexy P, Repiská M, Přikryl R, Kontárová S, Báreková A, Sláviková M, Koutný M, Fayyazbakhsh A, Kadlečková M (2022) PLA/PHB-based materials fully biodegradable under both industrial and home-composting conditions. Polymers 14(19):4113. https://doi.org/10.3390/polym14194113

Chapter 11
Lignin-Based Biocomposites for Smart and Sustainable Food Packaging

Nnaemeka Ewurum, Sodiq Babatunde Yusuf, Harrison Appiah, and Jovale Vincent Tongco

1 Introduction

The global food packaging industry, which has, for decades, relied on conventional fossil-based plastics, has been facing increasing pressure to make a transition to more sustainable alternatives. This need is due to concerns about environmental degradation and the increasing awareness about sustainability among consumers [1, 2]. Traditional plastics rank high in terms of convenience and flexibility of application; however, they also make significant contributions to plastic pollution, greenhouse gas emissions, and the accumulation of persistent waste in both terrestrial and marine ecosystems [3]. As the demand for packaged food continues to grow as a result of urbanization, changing lifestyles, globalization, and population growth, the environmental footprint of packaging materials has gradually become an urgent concern. Single-use plastics are some of the biggest culprits. They have persisted in the environment for decades or even centuries, contributing to methane emissions in landfills, and the breakdown into microplastics causes contamination in the food chain, which affects marine life and human health [4]. It is quickly becoming evident to governments, industries, the scientific body, and consumers that sustainable packaging solutions are not merely a market trend but a necessary step for aligning industrial practices with broader goals of environmental stewardship, food security, and circular economy principles [5].

In the food sector, the transition to sustainable packaging is critical because the industry must maintain a delicate balance between extending shelf life, ensuring product safety, and minimizing environmental impact. It is therefore imperative that food packaging materials meet specific, strict performance requirements. Such properties include mechanical strength, barrier properties, chemical inertness, shelf-life

N. Ewurum · S. B. Yusuf · H. Appiah · J. V. Tongco (✉)
Department of Forest, Rangeland and Fire Sciences, University of Idaho, Moscow, ID, USA
e-mail: jtongco@uidaho.edu

A. Verma et al. (eds.), *Polymers and Composite Materials for Packaging: Smart Food Packaging and Solutions*, Materials Horizons: From Nature to Nanomaterials,
https://doi.org/10.1007/978-981-95-0677-4_11

extension, etc., all while also being sustainable and ecologically friendly. Certain biopolymers like polysaccharides, polylactic acid (PLA), polyamides, polyhydroxyalkanoates (PHA), polyhydroxybutyrates (PHB), and polybutylene succinate (PBS) have generally shown promise in this regard [2, 6]. However, their widespread adoption is currently hindered by limitations in cost, scalability, and functionality. The incorporation of lignin, which is an abundant, low-cost, and bioactive byproduct of the pulping and paper industry, into bioplastic matrices to create biocomposites, presents a compelling opportunity to improve the performance of biopolymers while also potentially reducing cost while also reducing the reliance on fossil fuels [7, 8]. Lignin possesses UV-blocking, antioxidant, and antimicrobial properties, which can enable smart packaging solutions that could actively preserve food quality and open new paths for sustainable, high-performance materials that are tailored to the food sector's changing needs [9, 10]. Figure 1 shows the different biopolymers and their specialized functions in cells.

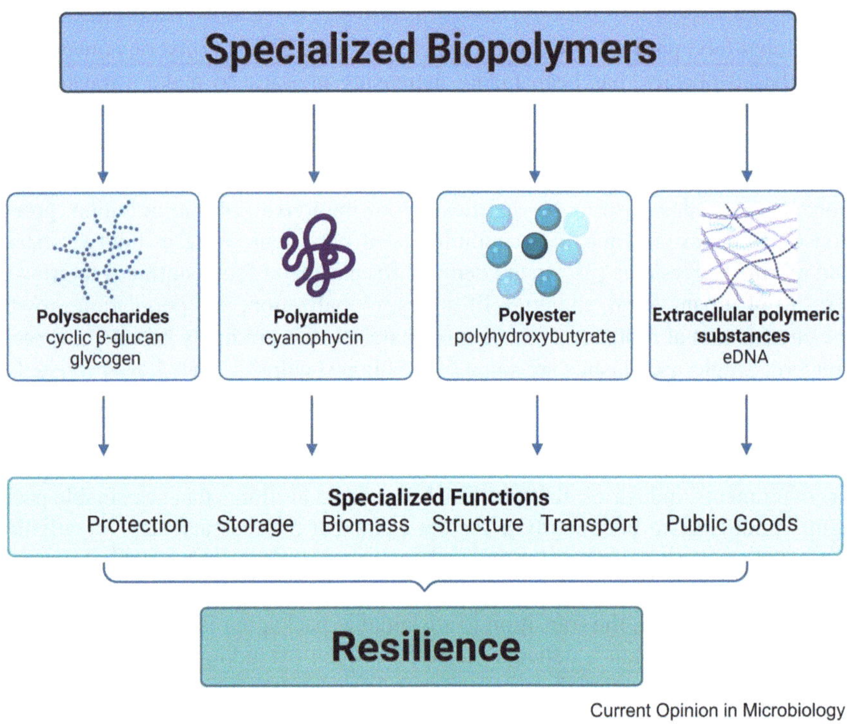

Fig. 1 Biopolymers and their specialized functions in cells (Reprinted with permission from reference [11])

2 Lignin: A Promising Filler for Food Packaging Applications

2.1 *Natural Role and Abundance of Lignin*

Lignin is one of the three vital structural components of the plant cell wall, along with cellulose and hemicellulose, and accounts for 15–30% of the dry weight of lignocellulosic biomass [7]. In plants, lignin serves as nature's primary reinforcement material as it provides mechanical strength, water resistance, and protection against microbial degradation, therefore enabling the upright growth and long-term survival of plants. This natural, complex three-dimensional phenolic network also allows plants to defend against environmental stressors such as pathogens, ultraviolet radiation, and oxidative damage [12]. Lignin is the second most abundant renewable polymer on Earth, after cellulose. It is present in all vascular plants and has a higher concentration in softwoods (25–35%) compared to hardwoods (18–25%) and agricultural residues (10–20%). The pulping and paper industry, as well as the biorefining sector, produce industrial lignin as a byproduct, globally producing over 70 million metric tons of lignin on an annual basis. Despite this abundance, lignin remains underutilized in industrial applications, with most of it being burned for low-value energy recovery.

The global supply of lignin is amplified by agricultural and forestry waste streams, such as wheat straw, sugarcane bagasse, hemp stalk, wood chips. These could be valorized by being used as low-cost feedstocks to produce packaging materials. However, lignin's complex, heterogeneous structure, which comprises three phenylpropane monomers (*p*-coumaryl alcohol, coniferyl alcohol, and sinapyl alcohol), varies significantly across plant sources as well as extraction processes [13, 14]. This influences its chemical reactivity and compatibility with polymers. An understanding of these natural and industrial variations in composition and chemistry is essential for tailoring lignin's functionality in biocomposites, where its inherent properties like UV absorption, oxidant capacity, etc., can be explored for smart food packaging solutions [15].

2.2 *Industrial Sources of Lignin*

Today, most of the lignin used in research and industrial applications comes from large-scale industrial processes like paper pulping and bioethanol production. These are processes where it is necessary to separate lignin from cellulose and hemicellulose. In pulp and paper mills, various chemical processes are used to break down the lignin in the feedstock biomass as a way to free the cellulose fibers. The extracted lignin is then separated as a waste stream. The most common chemical process in paper pulping is the kraft pulping method, which accounts for over 80% of pulp production globally [7]. Lignin recovered from the kraft pulping process is known as kraft lignin, which happens to be the biggest source of industrial lignin, and this is typically separated from what is known as "black liquor" and then recovered through precipitation or ultrafiltration [16]. Another major source of industrial lignin is the soda pulping process. This process is similar to kraft pulping but uses different chemicals and produces sulfur-free lignin, most commonly used for non-wood plants like wheat straw and bagasse [17].

Bioethanol production is also becoming a growing source of technical lignin. In this case, lignin is extracted from the biomass after the cellulose and hemicellulose have been converted into fermentable sugars through chemical or enzymatic processes. The residual solid that is left after this process usually contains a high concentration of lignin. One of the most common types of technical lignin derived from this process is called organosolv lignin, which is recovered using organic solvents like ethanol or acetic acid. Organosolv lignin is fast becoming popular because it is relatively pure, has a low sulfur content, and also retains a lot of the original aromatic structure that is characteristic of lignin, thus making it more suitable for material and chemical applications [18]. There is also another type of technical lignin called hydrolysis lignin, which is obtained from the acid hydrolysis processes, and that is often highly condensed and difficult to process, but still rich in carbon and functional groups [19]. It is important to understand that the type of industrial process used to extract the lignin from biomass has a big effect on the properties of the lignin itself. For example, kraft lignin contains sulfur and has a higher molecular weight, and this makes it more thermally stable but also more difficult to work with in some applications. On the other hand, soda and organosolv technical lignin are more chemically uniform and are easier to modify; however, they tend to have lower yield and their extraction processes can be more expensive [14]. Although there is an abundance of lignin coming from industrial processes, it can be said that not all technical lignin is created equal, as the type of process used determines not just how much lignin is available but also how suitable it is for different applications, whether in bioplastics, coatings, adhesives, or smart packaging materials.

2.3 *Types of Technical Lignin*

Different types of lignin are produced from different industrial processes. Each one of these lignin types comes with its own unique structure and chemical properties. The most widely produced and studied type happens to be kraft lignin, which comes from the kraft pulping process used in most paper mills. Kraft lignin has a characteristic dark color, and it is relatively thermally stable, although it does contain sulfur [7]. Because it is produced in large quantities, it is easily the most available form of technical lignin. However, the downside is that its sulfur content and high molecular weight make it a less attractive form of lignin for certain applications like bioplastic or high-purity chemicals. Instead, kraft lignin is often either chemically modified before being used in advanced materials, or it gets used in energy recovery systems. Despite these limitations, kraft lignin remains one of the most researched types because it is so abundant and accessible [17].

Another important form of technical lignin is soda lignin. This type is produced in a sulfur-free process by using sodium-based chemicals. It is more commonly obtained from non-wood biomass sources such as sugarcane bagasse, rice husk, wheat straw, hemp stalk. Soda lignin has a lower molecular weight than kraft lignin and is relatively easier to modify and process, especially in applications where sulfur content is a concern [20]. Because it is sulfur-free, soda lignin is also a better choice for such applications as adhesives, phenolic resins, and biodegradable films. However, it is not as widely available as kraft lignin, and this is because soda pulping is mostly used in small-scale or specialized operations. Still, soda lignin has been gaining more attention in bio-based material development because it has a cleaner composition and is more compatible with functionalization methods. There is also organosolv lignin. This is considered to be one of the highest quality technical lignins available. It is produced through the process of organosolv pulping, which uses organic solvents like ethanol, methanol, or acetic acid to break down the lignocellulosic structure of the feedstock biomass as fermentable sugars are isolated for biofuel production. Organosolv lignin has low ash content, contains no sulfur, and retains many of the functional groups that are needed for further chemical modification [19]. Its low molecular weight and high purity are what make it ideal for applications that require fine control over lignin reactivity, such as carbon fibers, smart coatings, functional packaging, etc. There are other types of lignin, like lignosulfonates, which are water-soluble lignins that are used mostly as dispersants, and hydrolysis lignin, which is often too condensed and therefore difficult to process, but which is still a source of valuable carbon [21]. Each type of technical lignin comes with tradeoffs, so the choice of which one to use depends heavily on the target application and the required processing steps.

2.4 Extraction and Isolation Method

The method that is used to extract lignin from the lignocellulosic biomass has a huge impact on its structure, purity, and how useful it will be in downstream applications. Traditionally, lignin extraction usually happens during paper pulping or bioethanol production, where the main goal is not to preserve the quality of the lignin byproduct but rather to prioritize the extraction of the cellulosic portion of the biomass. In most cases, the lignin portion of the biomass ends up being highly modified or partially degraded, which affects its reactivity and consistency. For example, the kraft and soda pulping processes use strong alkaline methods that break the lignin bonds aggressively, which leads to lignin with higher molecular weight, more condensed structures, and less uniformity [22]. The organosolv processes, on the other hand, are milder and so allow for the better preservation of the lignin's functional group, which is why organosolv lignin is considered more chemically useful than the kraft and soda lignin, even though the extraction process is more expensive [19].

Apart from these traditional methods, there are newer approaches that are being developed to isolate lignin under conditions that preserve its structure, making it easier to work with. Solvent-based methods, which employ organic acids or ionic liquids, are gaining attention because they offer more control over the extraction environment and therefore produce lignin that contains less ash and overall fewer impurities [23]. Enzymatic treatments are also being explored as lignin extraction methods because they work in milder conditions and reduce the chances of damaging the lignin backbone, although these methods are still relatively new and currently not yet scalable as large-scale industrial processes [24]. There are also steam and hydrothermal pretreatment processes, which use a more physicomechanical approach to separate biomass components without too many chemical alterations, but these often result in lignin that is still highly crosslinked and therefore difficult to solubilize [25]. The choice of extraction method matters a lot because it determines what kind of lignin you end up with, how pure it is, what functional groups are left intact, what color or odor it carries, and how compatible it will be with different polymers or different chemical pathways. For example, lignin that contains a lot of residual sulfur or ash may not be suitable for food-contact packaging or high-performance plastics. Also, the presence of certain lignin linkages like β-O-4, which are easy to break and modify, may be reduced or even completely lost, depending on how harsh the extraction process is [26]. When choosing a lignin source for food packaging applications, the focus should not be only on availability, but also on how the extraction process affects its performance and what tradeoffs you are willing to accept in terms of purity, cost, and reactivity.

2.5 *Impact on Packaging Applications*

The structural and chemical qualities of lignin play a big role in how it behaves when used in packaging materials. This means that not all lignin types perform the same way, and most of the differences come down to the type of lignin used and how it was extracted. For example, the lignin that still has functional groups like hydroxyl and methoxyl groups intact would potentially bond better with bio-based polymers like PLA and PBS, thereby improving compatibility, which is one of the biggest challenges in the production of biocomposites [8, 13]. But lignin that has been degraded or that contains too many impurities like residual sugars, sulfur, or ash can lead to poor dispersion in the polymer matrix, which would weaken the mechanical performance of the final product due to poor stress transfer between phases. The starting quality of the lignin matters a lot, especially when you are trying to develop materials that can compete with traditional plastics in terms of strength and quality [26].

Another important point to note is how lignin affects the thermal and processing behavior of bioplastics. In a lot of cases, lignin can increase the thermal stability of bio-based composites, which is useful when the end-product needs to withstand moderate heat, like during sealing or sterilization [22]. But that same thermal resistance could become a double-edged sword by making it harder to melt or thermally process the material, especially if the lignin has a high molecular weight or is highly crosslinked. Lignin exhibiting darker coloration tends to affect the final color of the composite material, as this may limit its use in applications where transparency or visual appeal is required. However, the same dark color and UV-absorption properties of lignin can become an advantage when designing packaging for light-sensitive products like oils, vitamins, or dairy, where UV protection helps to extend shelf life [19]. Lignin also brings with it functional benefits that go beyond the structure alone. Some forms of lignin have antioxidant and antimicrobial activity, which opens up possibilities for developing active packaging solutions that interact with food and slow down spoilage [9]. But then again, not all lignin types will perform well in this regard, and purification or chemical modification is often needed to activate these properties. When considering lignin for food packaging, it is not just about whether it can act as a filler, but it is also about how well its specific properties match the performance that is required of the packaging application. In other words, lignin has potential for food packaging applications, but only if the right type is picked, processed properly, and matched with the right polymer system.

2.6 Functional Properties Relevant to Packaging

What makes lignin especially interesting for food packaging applications is not just about being a bio-based filler, but it is because it has built-in properties that can improve the performance of the packaging itself. One of the most valuable properties is the ability to absorb UV light. Lignin's aromatic structure, which is made up of phenylpropane units, allows it to block UV radiation effectively, which is something that conventional plastics like PLA and PBS cannot do on their own [19]. This UV-blocking ability is important when it comes to packing light-sensitive products like cooking oils, dairy, beverages, or supplements because UV exposure can trigger oxidation and degrade both the product and the packaging material. Adding lignin into the polymer matrix, even in small amounts, can improve the UV resistance of the material without the need for additional chemical additives [9].

Another major benefit of lignin is its antioxidant activity. Because lignin contains phenolic hydroxyl groups, it can scavenge free radicals and, by doing so, slow down oxidation processes, which is a useful feature in active packaging systems [21]. In some cases, lignin can even slightly migrate into the food environment and act as a stabilizing agent, thereby helping to preserve the flavor, color, and nutritional value of packaged food. This feature is especially relevant in situations where food spoilage is a concern, such as the packaging of fresh produce, meats, or processed foods, where oxidative spoilage is a major concern. The antioxidant capacity of lignin depends on its purity and structure, which in turn is dependent on the extraction method, so not all lignin types will perform equally well in this regard. But with the right processing and surface modification, it is possible to tune this property and make the packaging more than just a passive barrier. It becomes an active, protective system.

Lignin can also influence the barrier properties of the packaging material, although this effect is a bit more complex. On its own, lignin is relatively hydrophobic and can therefore reduce the permeability of water vapor in some bioplastics, which is an important property in applications where moisture control is critical, like in dry food packaging or powdered products [10]. But the final effect depends a lot on how well the lignin is dispersed within the polymer and on whether it creates a uniform layer or causes agglomeration, which could disrupt the polymer matrix. Poor dispersion of lignin particles within the polymer matrix can create microchannels that allow moisture or gas through, thereby defeating the purpose. While lignin has the potential to improve both the UV-resistance and moisture barrier properties in biocomposites, the outcome depends on the processing, the type of lignin, and how it interacts with the base polymer [13].

3 Polymer Matrices Compatible with Lignin

Biopolymers are polymers (materials made from repeating units of monomers) that are of natural origin. They can be biodegradable (Polylactic acid, starch, etc.) or non-biodegradable (polyethylene). Biodegradable polymers degrade completely in soil, air, and water, mostly through microbial actions, over time. Biopolymers are eco-friendly because they do not persist and pollute the environment like most synthetic plastics. They consist of organic repeating units called monomers. The monomers are made mainly of carbon, hydrogen, oxygen, and nitrogen. The properties of biopolymers mainly depend on the monomer, degree of polymerization, and type of intramolecular bonds. Synthetic plastics have become a major threat to the environment and a source of great concern for scientific and industrial research over the years. Plastics are used in all aspects of our lives, from food packaging to everyday household materials. Biodegradable biopolymers have been considered a better alternative because they break down naturally through eco-friendly biological processes. Bio-based polymers can be classified broadly as primary and secondary. Primary biopolymers are derived directly from living organisms, while secondary biopolymers are derived from products primarily found in living organisms. Primary biopolymers can be classified based on their source as highlighted in Fig. 2.

3.1 Starch-Based Biopolymers

Starch or polysaccharide-based biopolymers have carbohydrate monomers joined together by glycosidic bonds. Polysaccharides are very abundant in nature because they form structural material (e.g., cellulose), energy storage material (e.g., starch

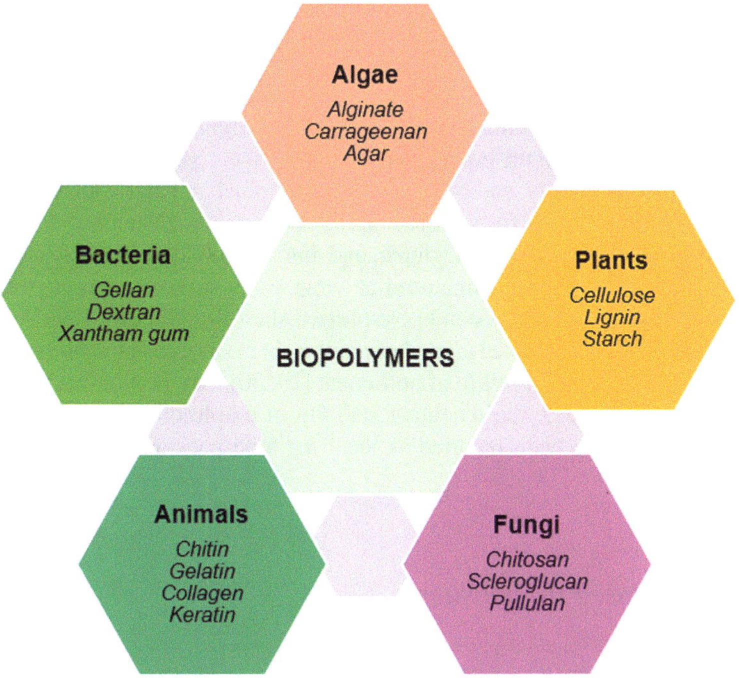

Fig. 2 Sources of primary biopolymers

in plants and glycogen in animals). Cellulose is the most abundant polymer worldwide. It is a homopolymer of β-D glucopyranose linked by β-1,4 linkage to form a semi-crystalline, high molecular weight, and insoluble structure. It has been widely used in various sustainable materials, including food packaging and preservation. Researchers have reported that low water absorption and water vapor permeability make it excellent for food packaging applications. Another major polysaccharide biopolymer is chitin, which is a heteropolymer consisting of 2-acetamide-2-deoxy-D-glucopyranose (N-acetyl-D-glucosamine) and 2-amino-2-deoxy-D-glycopyranose. It is primarily extracted from the exoskeleton of crustaceans, e.g., lobsters, shrimps, crabs, etc. Chitosan is a soluble derivative of chitin obtained by partial deacetylation. Both chitin and chitosan have antimicrobial and antioxidant properties, which make them suitable for several industrial applications, including food preservation and packaging. Chitosan has been approved by both the US Food and Drug Administration (FDA) and the European Union (EU) as sustainable and safe for food packaging.

3.2 Protein-Based Biopolymers

Proteins are heterogeneous polymers consisting of amino acid monomers joined by amide linkages. The properties of protein-based biopolymers depend on the number of amino acid residues in the polypeptide chain and their position in the chain. They have good foaming ability, and some have antimicrobial and antioxidant properties, which make them suitable for food packaging applications. Examples include plant proteins such as zein, soy protein, gluten, and animal proteins such as gelatin, whey, casein. Gluten-based edible films used to wrap fruits such as cherries, tomatoes, and mangoes were found to extend post-harvest shelf life [27, 28]. Zein is another protein-based biopolymer used in food packaging due to its good moisture resistance, reported to improve the shelf life of pork meat [29, 30]. Soy protein-based films are also reported to enhance the oxidative stability of edible oils [31]. Gelatin-based composites have also been reported as ideal for food packaging and preservation [32–34].

3.3 Lipid-Based Biopolymers

Lipids include phosphatides, mono-, di-, triglycerides, terpenes, fatty acids, and phospholipids. Their hydrophobic nature makes them an excellent candidate for food packaging applications [35]. Some lipids, such as eucalyptus oil, are also natural antioxidants and antimicrobials. The use of some lipids to improve the hydrophobicity of protein-based biopolymers has been reported, but they reduce the mechanical properties. Some researchers have also reported the use of beeswax (a lipid biopolymer), cinnamon oil, chitosan, gelatin, and carrageenan with corn starch for packaging and preservation in post-harvest fruit coatings [36–39].

3.4 Secondary Biopolymers

Primary biopolymers have several advantages, including their biodegradability, lower environmental footprint compared to synthetic plastic, and less energy demand. However, a major drawback is that each type has specific properties, merits, and demerits, and they have high production costs, making them more expensive than synthetic polymers. Secondary biopolymers are biodegradable polymers from renewable biomass. They are mostly polyesters, and the properties depend on the constituent monomers. Examples include polylactic acids (PLA), polybutylene succinate, polyglycoside, polyisoprene, and polybutylene adipate co-terephthalate. PLA, first discovered in 1932 by heating lactic acid under vacuum, has received much attention for food packaging applications due to its high molecular weight, moisture resistance, processability to films, containers, coatings, etc., the use of PLA with primary biopolymers such as chitosan, lipids, e.g., limonene have also been reported with enhanced properties.

3.5 Sustainable Composite Biopolymers

Recent research has focused on the use of composites that include both secondary and primary biopolymers as well as other additives for optimal properties [40]. This is because primary biopolymers are known for their abundance, while secondary biopolymers are known for their excellent properties, therefore, a blend of both can bring down cost while maintaining excellent properties for food packaging applications. These composites are classified mainly by the main components. Lignin, as the main natural protective component in biomass, has been extensively researched to produce sustainable bio-based polymers for food packaging applications [41]. The reason for this includes being the second most abundant plant-based polymer on earth, rich in polyphenols that provide excellent antioxidant, antibacterial, ultraviolet protection properties, and biocompatibility. 20% lignin improved water resistance by 41.2% and antioxidant properties by 288% in chitosan-based composite films for food packaging applications [42]. Lignin was also reported to improve oil resistance and mechanical properties when added to cellulose nanofibrils. Lignin-gelatin film shows improved UV-blocking and antioxidant properties with low cytotoxicity. Lignin has been used with other secondary biopolymers such as PLA, PBS, PBAT, and nanocellulose from primary biopolymers have been reported with applications in food packaging [41, 43–45].

3.6 Properties of Lignin that Make It Suitable for Food Packaging

The major properties of lignin that make it suitable for food packaging applications include the following:

- Antioxidant properties: Lignin has a complex chemical structure responsible for its antioxidant property [41]. This is due to the presence of methoxy and hydroxyl groups, which suppress oxidation through hydrogen donation and their ability to scavenge free radicals [46, 47]. However, it should be noted that the antioxidant properties of lignin can be influenced by molecular weight, polydispersity, and hydroxyl content [48]. This property has been exploited to use lignin in food packaging composites; however, high lignin content can cause cytotoxic effects [46].
- Antimicrobial ability: Lignin contains several polyphenolic compounds, which are mainly responsible for its antibacterial activity [49]. The presence of acidic functional groups, such as –COOH, also enhances this effect [50]. Lignin has

been found effective against gram-positive bacteria, with less effectiveness against gram-negative bacteria [41]. Antifungal and antiviral properties of lignin have also been reported [51–53].

- UV protective ability: Lignin can block ultraviolet (UV) rays due to the vinyl group at the para position of the phenolic OH, forming a UV chromophore that absorbs the rays [54]. Lignin can absorb UVA 9320–400 nm) and UVB (280–320 nm), which are harmful to humans [41]. This is one of the promising properties of lignin for sustainable food packaging films.
- Biocompatibility: Biocompatibility and safety are major issues to address before using a substance in food applications. Lignin-based materials were reported to be safe and compatible with human cells [55]. These findings confirm the biocompatibility of lignin; however, some lignin derivatives must be carefully examined for safety and biocompatibility for specific applications.

4 Processing of Lignin-Based Composites for Food Packaging

The main processes involved in food packaging to prevent food from physical, chemical, and microbial contamination include fabricating films or coatings.

4.1 Techniques for Film Fabrication

- Wet processing/solvent casting: Wet processing/solvent casting is the process of making films by preparing film film-forming solution, casting it in a thin layer, and lastly drying to evaporate the solvent. The properties of the film depend on the film-forming ability, solubility, compatibility with additives, film stability, and surface properties [56].
- Extrusion or Dry process: This process involves heating biopolymers above their glass transition temperature to form a uniform melt, shaped into its form on cooling [57]. Thermoset biopolymers undergo thermoplasticization and are obtained in pellet form after extrusion or pelletization. The pellets are processed either by injection molding, blow molding, fiber spinning, film extrusion, compression molding, fiber spinning, etc. Extrusion is used industrially in making petroleum-based synthetic plastics. It involves the use of shear and high temperature to effect some physicochemical changes in the material.
- Electrohydrodynamic technique: This method involves the use of electrospinning and electro-spraying for fiber production and encapsulation, respectively, without applying heat. The process applies high electrostatic forces to generate an electrically charged jet of polymer solution, which gives a particulate or ultrathin structure. High porosity, small pore size distribution, and high surface area make electrospun fibers very popular.

4.2 Methods of Coating

Another method of using biopolymers in food packaging and preservation is coating to increase shelf life. Some of the methods include:

- Brushing: This method involves the application of edible coatings using a brush directly on the food surface.
- Dipping: This method involves submerging food in a coating formulation to apply edible coating on it. It is the oldest and most widely practiced method of coating [58].
- Spraying: This method is widely used in industrial applications involving the use of hydraulic spray nozzles, ultrasonic atomizers, high-pressure spray guns, and air-spray atomizers [59]. It is efficient automation for homogeneous coating, effective thickness and temperature control, and contamination prevention of coating solution [37].
- Electro-spraying: This involves the use of electrostatic liquid for coating by passing the coating solution through the atomizer nozzle connected to a high electric power source. It has a special ability to produce a fine, uniform spray of droplets with a particle size of <100 nm, compared to normal spraying, which can only produce droplets of 20 μm [60].
- Layer-by-layer (LBL) electrostatic deposition: This approach is based on electrostatic interactions between biopolymers for the formation of coatings and films [61]. It involves alternate depositions of oppositely charged polyelectrolytes, polycations, and polyanions on a charged surface with subsequent rinsing after each deposition [62, 63]. This produces a multilayer coating and can be used for combining different types of biopolymers for composite use [64]. A study reported that the LBL method in mango stored at 25 °C for 14 days retained commercial value [65].
- Vacuum impregnation (VI): This method is based on diffusion, effective for applying thicker and uniform coatings, especially on fruits and vegetables [66, 67]. The amount of coating on a food product depends on the period of application and atmospheric conditions [66]. It was reported that VI has improved quality and firmness compared to dipping [68].

4.3 Effect of Fabrication Techniques on Properties

By adjusting the raw material's moisture/plasticizer concentration, the amylose to amylopectin ratio of the raw material, and the extruder's temperature and pressure, thermoplastic products with varying viscosities, water solubility, and water absorption can be produced [69]. The impact of processing parameters on film characteristics is demonstrated by the fact that starch film made by compression molding is less permeable to gas and water than solvent casting film [70]. This results from solvent evaporation and an open microstructure created over an extended drying period in solvent casting as opposed to a compact/denser structure in compression molding. Kurek et al. [71] used mix, bilayer, and monocomponent techniques to create a composite film based on whey protein (WP) and chitosan. Transparent films with uniform structures are produced by monocomponent and bilayer techniques. Compared to monocomponent and blend techniques, bilayer films exhibit reduced water vapor permeability and superior mechanical resistance. When processing starch/protein or starch composites, researchers found that wet casting produced better mechanical properties with a homogenous structure than dry casting [72–74]. However, other research indicates that films made by extrusion frequently have superior mechanical, barrier, and microstructural qualities to those made by solution casting [75, 76]. Krishna et al. [77] coupled extrusion and compression molding to create gelatin-based films, subsequently contrasted with solution-cast films. They found that extruded films had a greater WVP than solution-cast films, but that extruded films are more flexible, with a percentage elongation at break (%EAB), which is seven times higher. Using extrusion, blown extrusion, and casting, a gelatin-based film containing *Yucca schidigera* extract was produced [78]. With no discernible variation in solubility, the film produced by blown extrusion had a lower WVP than the other two techniques. For soybean protein concentrate, Ciannamea et al. [79] conducted a study akin to this. They observed that solution casting had a stronger oxygen barrier property, compression molding had a lower WVP, less solubility, and more stretchability and transparency.

5 Smart Packaging Potential and Functional Performance

5.1 Physical Properties

The physical properties include thickness and density, which affect the mechanical properties of packaging materials and how they will perform in actual applications. Another important property is water absorption, dimensional stability, and permeability. A major limitation of starch, protein, and cellulose bioplastics is their water absorption and hygroscopic nature, although this can be improved using plasticizers. Lignin-based composites for food packaging are mostly hydrophobic and stable over time [41, 43]. Optical properties, such as color changes on exposure to light, are also a very important indicator of material performance for applications in food packaging.

5.2 Mechanical Properties

The mechanical properties of a film will determine its final application. This is usually altered by using fillers and fibers to improve the mechanical strength, although factors such as chain orientation, compatibility of constituents, and external factors such as light, heat, moisture. [76]. Mechanical properties of interest include the tensile strength and modulus, percentage energy at break, compression test, etc. TS is the maximum tensile stress an object can take, while modulus is a measure of resistance to deformation or stiffness.

5.3 Thermal Properties

The glass transition temperature (T_g), melt temperature (T_m), melt enthalpy, specific heat, etc., are important thermal properties for packaging films. T_g is the transition temperature between the glassy and rubbery states. The T_g and T_m determine the crystallinity of a material. Techniques for determining this include dynamic mechanical analysis (DMA), differential scanning calorimetry (DSC), thermogravimetric analysis (TGA), and dynamic mechanical thermal analysis (DMTA).

5.4 Biodegradability and Environmental Impact

Up to 50% of plastics derived from fossil fuels are used in food packaging [80]. This huge amount of non-biodegradable plastics is dangerous for the environment, hence the need for sustainable materials for food packaging. Lignin, being the second most abundant biopolymer on earth, with great biodegradability and biocompatibility, has great prospects for sustainable food packaging materials [81]. Lignin has been widely used in food packaging applications, mostly along with other biopolymers.

5.5 Lignin's Role in Active or Intelligent Packaging

Lignin has been reported to give excellent properties in food packaging applications, both as an additive and as a base material to produce food packaging films. Argan nutshell-derived cellulose nanocrystal-reinforced lignin-based films were developed for food packaging applications, while Acacia lignin-gelatin blended films exhibited improved UV-blocking and antioxidant properties while exhibiting low cytotoxicity [82]. Lignin has also been used in the development of coatings for food packaging materials to improve their surface properties. Composite coatings made of lignin, tannin, and ZnO nanoparticles have been applied to paper substrates, enhancing their mechanical strength, thermal stability, and barrier properties [44]. These coatings have shown multifunctional properties, such as moisture resistance, high mechanical strength, antibacterial properties, and heat resistance. Lignin-based coatings have also been used as a co-mixing additive in PBAT-based coating formulations, providing a biodegradable alternative to conventional petroleum-based waxes and polymer film laminates. Lignin-based hydrogels have potential as sustainable food packaging materials [52].

6 Commercialization Challenges and Industrial Trends

6.1 Technical Barriers to Scale-Up

The scaling up of lignin-based packaging materials from lab experiments to real-world industrial production is not as simple as it sounds. In small-scale studies, researchers can fine-tune some parameters like lignin loading, particle size, polymer blending ratio, etc., to get good results. But in industrial settings, those same variables become a lot harder to control. For instance, lignin behaves very differently under high-shear extrusion or film blowing conditions, especially when it is not uniformly dispersed within the matrix. Poor dispersion causes uneven properties and internal stresses, also in some cases, it causes visible defects in the final product [26]. Also, because lignin is such a chemically diverse material, even slight differences in source or extraction method can lead to major inconsistencies in how it processes or bonds with the polymer [83].

Another issue is that of color and odor. These two are often overlooked in the lab during trials, but they become major deal-breakers during the commercial-scale product development. Most technical lignins are dark brown, and they tend to darken the appearance of the composite material even at low loadings [84]. This limits the range of products that they can be used for, especially in applications where clarity and color appeal matter. On top of that, some lignin types give off odors during storage or processing, especially if they have not been purified properly, which can be off-putting and raise concerns for food packaging, where even slight sensory changes can affect customer acceptance [85]. Unless these issues are addressed during the processing stages or through post-treatment, the final product may not meet the aesthetics or sensory requirements of the food industry.

Thermal processing is also a challenge. Lignin does not melt the way conventional thermoplastics do, and when heated beyond a certain point, it tends to degrade rather than flow. This creates a rather narrow processing window when blended into biopolymers like PLA or PBS, because if the temperature is too low, then the blend does not form properly, and if it is too high, the lignin can degrade and release volatile compounds [8, 86]. This is especially challenging in continuous processing like injection molding and extrusion, where temperature control is critical. Although some of these issues can be minimized by modifying the lignin or by using compatibilizers, this adds costs and complexity to the production process [87]. So, from a technical point of view, lignin's variability, thermal behavior, and process sensitivity are all barriers to commercial use.

6.2 Regulatory and Safety Considerations

Using lignin in food packaging naturally raises a whole new set of questions around safety, as well as compliance and regulatory approval. Most food packaging materials are required to meet strict standards for food contact safety, especially in situations where there is a risk of chemical migration into the food. Agencies like the U.S. FDA and the European Food Safety Authority (EFSA) have established guidelines on what substances are allowed, how much can migrate, and under what conditions [88, 89]. However, for lignin, especially technical lignin, there is not yet a well-defined regulatory status for direct food contact. This stands then as a gray area that makes manufacturers hesitant about using it in packaging that touches the food directly unless the lignin is encapsulated or buried in multilayer structures where contact is minimal or prevented entirely.

One of the biggest concerns is that the lignin derived from industrial processes would contain impurities or residual chemicals from the extraction process. For example, kraft lignin can contain sulfur, heavy metals, or trace pulping chemicals, depending on how it was separated and purified [90]. If these contaminants migrate into food, they could raise health concerns or trigger compliance issues. Even organosolv lignin, which is considered relatively cleaner, still needs to be tested for things like solvent residues and unknown byproducts. Most of the data on toxicology for lignin is related to environmental or medical use, there is very little on long-term dietary exposure through packaging [10]. Before lignin can be fully accepted into mainstream food packaging, more migration studies, toxicity evaluations, and standardized testing protocols will be needed.

Another complication is that new materials like lignin tend to have slow and fragmented regulatory pathways; in many cases, approval has to be sought on a case-by-case basis depending on the country, the food product, and the exact formulation of the packaging [91]. This adds another layer of uncertainty and cost, especially for startups or smaller companies that are trying to commercialize lignin-based packaging. There is also a lack of harmonization across jurisdictions, which makes it harder to scale a single product globally. So even though lignin has very clear sustainability advantages, these will not mean much unless the regulatory framework catches up with the science and gives manufacturers a clear, science-backed roadmap for safe and compliant use for food packaging applications.

6.3 Economic Viability

For all of lignin's technical and environmental appeal, there is still the major roadblock of cost when it comes to widespread adoption in food packaging [92]. At first glance, it may seem as though lignin should be cheap because it is produced in massive quantities as a waste byproduct of large-scale industrial processes, however, the problem is that most of the technical lignin is not directly usable in high-value materials [93]. So, to make it suitable for food packaging, it often needs to be extracted, purified, and modified in ways that increase the overall cost and complexity of the process. The more refined the lignin is, the higher the production cost. So even though the raw feedstock may be cheap, the refined, usable forms of industrial lignin are expensive and still hard to justify at a commercial scale [94].

Another economic factor is that lignin has to compete with other bio-based or mineral fillers that are already cheaper, more consistent, and easier to use. Materials like starch, talc, calcium carbonate, cellulose fibers, etc., already have mature and established supply chains and fewer processing challenges, and they are already being used in large volumes in packaging materials, as they do not carry the same technical baggage as lignin [95, 96]. So, from a manufacturer's point of view it makes more economic sense to stick with what already works, unless lignin can offer a clear functional advantage or cost saving that would justify the switch. Right now, lignin-based packaging materials are still mostly in the prototype or pilot stage, and the economies of scale just are not there yet to bring the prices down [97].

The final economic challenge is the consistency in the supply chain. Even if lignin happens to become technically viable, producers will need access to reliable and high-quality feedstock continuously. But the lignin that comes out of different mills or even different batches of the same mill can vary widely in terms of purity, ash content, molecular weight, color, etc., thereby making it a challenge to create standardized formulations or to predict performance across different production runs. Until a more reliable system of sourcing and certifying lignin for material use is established, it will be difficult to build a stable market or pricing structure. From an economic perspective, lignin still has a lot to prove before it can compete with more established materials in the packaging sector [92].

7 Summary and Conclusion

Lignin has come a long way from being treated as a waste byproduct to being recognized as a potentially valuable material for use in sustainable food packaging and other high-value applications because of its unique set of properties—aromatic structure, UV shielding, antioxidant activity, thermal stability, etc., which give it an edge over many other biofillers. And because it is produced in such large quantities as a waste stream from large-scale industrial processes, making it one of the most abundant and underutilized resources available for material innovation. When it is properly incorporated into biopolymers to make composites, lignin has been shown to improve mechanical, thermal, and barrier properties, as well as add functional benefits while supporting sustainability goals through circular resource use.

The road to commercialization is not straightforward. There are challenges like feedstock variability, processing complexity, color and odor control, regulatory uncertainties, etc., which make it difficult to go mainstream with lignin-based food packaging materials. The cost of refining technical lignin to meet the standards required for food-grade applications is also a major constraint, especially when compared to cheaper, more established, and less complex alternatives. At the same time, the regulatory landscape is still evolving, and more safety studies are needed to support a broader adoption, and until these issues are addressed in a coordinated way, from extraction to formulation, to processing, and policy, lignin's full potential as an additive for smart packaging material applications will remain largely untapped.

Still, the momentum is building. With continued research, cross-sector collaboration, and strategic investments, lignin could be positioned to become a key ingredient in the next generation of smart, sustainable packaging solutions. However, its success will not depend only on lab-level breakthroughs but on real-world integration where material science is aligned with supply chain readiness, industry standards, and consumer expectations. If done right, lignin could be a model for how we valorize industrial waste into high-value products that work for both the environment and the economy.

Conflicts of Interest There are no conflicts of interest to declare by the authors.

References

1. Zhao Q, Liu Y (2019) Is anaerobic digestion a reliable barrier for deactivation of pathogens in biosludge? Sci Total Environ 668:893–902. https://doi.org/10.1016/J.SCITOTENV.2019.03.063
2. Edo GI, Ndudi W, Ali ABM, Yousif E, Jikah AN, Isoje EF, Igbuku UA, Mafe AN, Opiti RA, Madueke CJ et al (2025) Biopolymers: an inclusive review. Hybrid Adv 9:100418. https://doi.org/10.1016/J.HYBADV.2025.100418
3. Benalaya I, Alves G, Lopes J, Silva LR (2024) A review of natural polysaccharides: sources, characteristics, properties, food, and pharmaceutical applications. Int J Mol Sci 25:1322. https://doi.org/10.3390/IJMS25021322
4. Singh S, Sharma PK, Malviya R, Gupta A (2022) Modified polysaccharides and their biomedical applications: advancement and strategies. Curr Mater Sci 16:316–351. https://doi.org/10.2174/2666145416666221208150926/CITE/REFWORKS
5. Righetti GIC, Faedi F, Famulari A (2024) Embracing sustainability: the world of bio-based polymers in a mini review. Polymers 16:950. https://doi.org/10.3390/POLYM16070950
6. Ramos M, Valdés A, Beltrán A, Garrigós MC (2016) Gelatin-based films and coatings for food packaging applications. Coatings 6:41. https://doi.org/10.3390/COATINGS6040041
7. Upton BM, Kasko AM (2016) Strategies for the conversion of lignin to high-value polymeric materials: review and perspective. Chem Rev 116:2275–2306
8. Ewurum N, McDonald AG (2025) Lignin reinforcement in polybutylene succinate copolymers. Polymers 17:194. https://doi.org/10.3390/POLYM17020194
9. Anushikha S, Gaikwad KK (2024) Lignin as a UV blocking, antioxidant, and antimicrobial agent for food packaging applications. Biomass Convers Biorefin 14:16755–16767. https://doi.org/10.1007/S13399-022-03707-3/METRICS
10. Boarino A, Schreier A, Leterrier Y, Klok HA (2022) Uniformly dispersed poly(lactic acid)-grafted lignin nanoparticles enhance antioxidant activity and UV-barrier properties of poly(lactic acid) packaging films. ACS Appl Polym Mater 4:4808–4817. https://doi.org/10.1021/ACSAPM.2C00420/ASSET/IMAGES/LARGE/AP2C00420_0008.JPEG

11. Schniete JK, Brüser T, Horn MA, Tschowri N (2024) Specialized biopolymers: versatile tools for microbial resilience. Curr Opin Microbiol 77:102405. https://doi.org/10.1016/J.MIB.2023.102405
12. Dixon RA, Puente-Urbina A, Beckham GT, Román-Leshkov Y (2024) Enabling lignin valorization through integrated advances in plant biology and biorefining. Annu Rev Plant Biol 75:239–263. https://doi.org/10.1146/ANNUREV-ARPLANT-062923-022602/CITE/REFWORKS
13. Shorey R, Salaghi A, Fatehi P, Mekonnen TH (2024) Valorization of lignin for advanced material applications: a review. RSC Sustain 2:804–831. https://doi.org/10.1039/D3SU00401E
14. Ullah M, Liu P, Xie S, Sun S (2022) Recent advancements and challenges in lignin valorization: green routes towards sustainable bioproducts. Molecules 27:6055. https://doi.org/10.3390/MOLECULES27186055
15. Brienza F, Cannella D, Montesdeoca D, Cybulska I, Debecker DP (2024) A guide to lignin valorization in biorefineries: traditional, recent, and forthcoming approaches to convert raw lignocellulose into valuable materials and chemicals. RSC Sustain 2:37–90. https://doi.org/10.1039/D3SU00140G
16. Lora JH, Glasser WG (2002) Recent industrial applications of lignin: a sustainable alternative to nonrenewable materials. J Polym Environ 10:39–48. https://doi.org/10.1023/A:1021070006895/METRICS
17. Vishtal A, Kraslawski A (2011) Challenges of lignins 6
18. Ragauskas AJ, Beckham GT, Biddy MJ, Chandra R, Chen F, Davis MF, Davison BH, Dixon RA, Gilna P, Keller M et al (1979) Lignin valorization: improving lignin processing in the biorefinery. Science 2014:344. https://doi.org/10.1126/SCIENCE.1246843/ASSET/C0F02F1A-D2FF-4261-946E-BD209F59D829/ASSETS/GRAPHIC/344_1246843_F5.JPEG
19. Sadeghifar H, Ragauskas A (2020) Lignin as a UV light blocker—a review. Polymers 12:1134. https://doi.org/10.3390/POLYM12051134
20. Bajwa DS, Pourhashem G, Ullah AH, Bajwa SG (2019) A concise review of current lignin production, applications, products and their environmental impact. Ind Crops Prod 139:111526. https://doi.org/10.1016/J.INDCROP.2019.111526
21. Ruwoldt J (2020) A critical review of the physicochemical properties of lignosulfonates: chemical structure and behavior in aqueous solution, at surfaces and interfaces. Surfaces 3:622–648. https://doi.org/10.3390/SURFACES3040042
22. Kai D, Tan MJ, Chee PL, Chua YK, Yap YL, Loh XJ (2016) Towards lignin-based functional materials in a sustainable world. Green Chem 18:1175–1200. https://doi.org/10.1039/C5GC02616D
23. Schutyser W, Renders T, Van Den Bosch S, Koelewijn SF, Beckham GT, Sels BF (2018) Chemicals from lignin: an interplay of lignocellulose fractionation, depolymerisation, and upgrading. Chem Soc Rev 47:852–908. https://doi.org/10.1039/C7CS00566K
24. Reshmy R, Athiyaman Balakumaran P, Divakar K, Philip E, Madhavan A, Pugazhendhi A, Sirohi R, Binod P, Kumar Awasthi M, Sindhu R (2022) Microbial valorization of lignin: prospects and challenges. Bioresour Technol 344:126240. https://doi.org/10.1016/J.BIORTECH.2021.126240
25. Sun Z, Fridrich B, De Santi A, Elangovan S, Barta K (2018) Bright side of lignin depolymerization: toward new platform chemicals. Chem Rev 118:614–678. https://doi.org/10.1021/ACS.CHEMREV.7B00588/ASSET/IMAGES/LARGE/CR-2017-00588D_0042.JPEG
26. Beckham GT, Johnson CW, Karp EM, Salvachúa D, Vardon DR (2016) Opportunities and challenges in biological lignin valorization. Curr Opin Biotechnol 42:40–53. https://doi.org/10.1016/J.COPBIO.2016.02.030
27. Wang J, Euring M, Ostendorf K, Zhang K (2022) Biobased materials for food packaging. J Bioresour Bioprod 7:1–13
28. Kumari M, Mahajan H, Joshi R, Gupta M (2017) Development and structural characterization of edible films for improving fruit quality. Food Packag Shelf Life 12:42–50. https://doi.org/10.1016/j.fpsl.2017.02.003
29. Kadam DM, Thunga M, Srinivasan G, Wang S, Kessler MR, Grewell D, Yu C, Lamsal B (2017) Effect of TiO_2 nanoparticles on thermo-mechanical properties of cast zein protein films. Food Packag Shelf Life 13:35–43. https://doi.org/10.1016/j.fpsl.2017.06.001

30. Cui H, Surendhiran D, Li C, Lin L (2020) Biodegradable zein active film containing chitosan nanoparticle encapsulated with pomegranate peel extract for food packaging. Food Packag Shelf Life 24:511. https://doi.org/10.1016/j.fpsl.2020.100511
31. de Souza KC, Correa LG, da Silva TBV, Moreira TFM, de Oliveira A, Sakanaka LS, Dias MI, Barros L, Ferreira ICFR, Valderrama P et al (2020) Soy protein isolate films incorporated with pinhão (*Araucaria angustifolia* (Bertol.) Kuntze) extract for potential use as edible oil active packaging. Food Bioproc Tech 13:998–1008. https://doi.org/10.1007/s11947-020-02454-5
32. Li JH, Miao J, Wu JL, Chen SF, Zhang QQ (2014) Preparation and characterization of active gelatin-based films incorporated with natural antioxidants. Food Hydrocoll 37:166–173. https://doi.org/10.1016/j.foodhyd.2013.10.015
33. Kumar S, Mudai A, Roy B, Basumatary IB, Mukherjee A, Dutta J (2020) Biodegradable hybrid nanocomposite of chitosan/gelatin and green synthesized zinc oxide nanoparticles for food packaging. Foods 9:1143. https://doi.org/10.3390/foods9091143
34. Kumar S, Shukla A, Baul PP, Mitra A, Halder D (2018) Biodegradable hybrid nanocomposites of chitosan/gelatin and silver nanoparticles for active food packaging applications. Food Packag Shelf Life 16:178–184. https://doi.org/10.1016/j.fpsl.2018.03.008
35. Martau GA, Mihai M, Vodnar DC (2019) The use of chitosan, alginate, and pectin in the biomedical and food sector-biocompatibility, bioadhesiveness, and biodegradability. Polymers 11:6542
36. Trinetta V, Cutter CN, Floros JD (2011) Effects of ingredient composition on optical and mechanical properties of pullulan film for food-packaging applications. LWT 44:2296–2301. https://doi.org/10.1016/j.lwt.2011.07.015
37. Mohsin A, Zaman WQ, Guo M, Ahmed W, Khan IM, Niazi S, Rehman A, Hang H, Zhuang Y (2020) Xanthan-Curdlan nexus for synthesizing edible food packaging films. Int J Biol Macromol 162:43–49. https://doi.org/10.1016/j.ijbiomac.2020.06.008
38. de Morais-Lima M, Carneiro LC, Bianchini D, Dias ARG, Zavareze ER, Prentice C, Moreira AS (2017) Structural, thermal, physical, mechanical, and barrier properties of chitosan films with the addition of Xanthan Gum. J Food Sci 82:698–705. https://doi.org/10.1111/1750-3841.13653
39. Feng Z, Wu G, Liu C, Li D, Jiang B, Zhang X (2018) Edible coating based on whey protein isolate nanofibrils for antioxidation and inhibition of product browning. Food Hydrocoll 79:179–188. https://doi.org/10.1016/j.foodhyd.2017.12.028
40. Agarwal A, Shaida B, Rastogi M, Singh NB (2023) Food packaging materials with special reference to biopolymers-properties and applications. Chem Afr 6:117–144
41. Priyadarshi R, Ghosh T, Purohit SD, Prasannavenkadesan V, Rhim JW (2024) Lignin as a sustainable and functional material for active food packaging applications: a review. J Clean Prod 469:5548
42. Ji M, Li J, Li F, Wang X, Man J, Li J, Zhang C, Peng S (2022) A biodegradable chitosan-based composite film reinforced by ramie fibre and lignin for food packaging. Carbohydr Polym 281:78. https://doi.org/10.1016/j.carbpol.2021.119078
43. Domínguez-Robles J, Larrañeta E, Fong ML, Martin NK, Irwin NJ, Mutjé P, Tarrés Q, Delgado-Aguilar M (2020) Lignin/poly(butylene succinate) composites with antioxidant and antibacterial properties for potential biomedical applications. Int J Biol Macromol 145:92–99. https://doi.org/10.1016/j.ijbiomac.2019.12.146
44. Tavares LB, Ito NM, Salvadori MC, dos Santos DJ, Rosa DS (2018) PBAT/kraft lignin blend in flexible laminated food packaging: peeling resistance and thermal degradability. Polym Test 67:169–176. https://doi.org/10.1016/j.polymertesting.2018.03.004
45. Chaubey A, Aadil KR, Jha H (2021) Synthesis and characterization of lignin-poly lactic acid film as active food packaging material. Mater Technol 36:585–593. https://doi.org/10.1080/10667857.2020.1782060
46. Roy S, Baishnab KL (2023) SPISE: a tiny, cost effective, speedy block cipher for low resourced devices. J Eng Res 11:106–113. https://doi.org/10.1016/j.jer.2023.100071
47. Azadfar M, Gao AH, Chen S (2015) Structural characterization of lignin: a potential source of antioxidants guaiacol and 4-vinylguaiacol. Int J Biol Macromol 75:58–66. https://doi.org/10.1016/j.ijbiomac.2014.12.049

48. Ponomarenko J, Dizhbite T, Lauberts M, Volperts A, Dobele G, Telysheva G (2015) Analytical pyrolysis: a tool for revealing of lignin structure-antioxidant activity relationship. J Anal Appl Pyrolysis 113:360–369. https://doi.org/10.1016/j.jaap.2015.02.027
49. Kai D, Tan MJ, Chee PL, Chua YK, Yap YL, Loh XJ (2016) Towards lignin-based functional materials in a sustainable world. Green Chem 18:1175–1200
50. Li K, Zhong W, Li P, Ren J, Jiang K, Wu W (2023) Antibacterial mechanism of lignin and lignin-based antimicrobial materials in different fields. Int J Biol Macromol 252:542
51. Chen W, Dong T, Bai F, Wang J, Li X (2022) Lignin–carbohydrate complexes, their fractionation, and application to healthcare materials: a review. Int J Biol Macromol 203:29–39
52. Gordts SC, Férir G, D'Huys T, Petrova MI, Lebeer S, Snoeck R, Andrei G, Schols D (2015) The low-cost compound lignosulfonic acid (LA) exhibits broad-spectrum anti-HIV and anti-HSV activity and has potential for microbicidal applications. PLoS ONE 10:1219. https://doi.org/10.1371/journal.pone.0131219
53. Fukuchi K, Koshikawa T, Asai D, Inomata M, Sakagami H, Takemura H, Kanamoto T, Aimi H, Kikkawa Y (2021) Lignosulfonate rapidly inactivates human immunodeficiency and herpes simplex viruses. Medicines 8:56. https://doi.org/10.3390/medicines8100056
54. Dean JC, Navotnaya P, Parobek AP, Clayton RM, Zwier TS (2013) Ultraviolet spectroscopy of fundamental lignin subunits: guaiacol, 4-methylguaiacol, syringol, and 4-methylsyringol. J Chem Phys 139:19. https://doi.org/10.1063/1.4824019
55. Gordobil O, Oberemko A, Saulis G, Baublys V, Labidi J (2019) In vitro cytotoxicity studies of industrial eucalyptus Kraft Lignins on mouse hepatoma, melanoma and Chinese hamster ovary cells. Int J Biol Macromol 135:353–361. https://doi.org/10.1016/j.ijbiomac.2019.05.111
56. Embuscado ME, Huber KC (2009) Edible films and coatings for food applications. Springer, New York
57. Rooney ML (2005) P A R T 2 introduction to active food packaging technologies
58. De Azeredo HMC, Rosa MF, De Sá M, Souza Filho M, Waldron K (2014) The use of biomass for packaging films and coatings. Advances in biorefineries: biomass and waste supply chain exploitation. Elsevier, Amsterdam, pp 819–874
59. Andrade RD, Skurtys O, Osorio FA (2012) Atomizing spray systems for application of edible coatings. Compr Rev Food Sci Food Saf 11:323–337
60. Bizymis AP, Tzia C (2022) Edible films and coatings: properties for the selection of the components, evolution through composites and nanomaterials, and safety issues. Crit Rev Food Sci Nutr 62:8777–8792
61. Arnon H, Granit R, Porat R, Poverenov E (2015) Development of polysaccharides-based edible coatings for citrus fruits: a layer-by-layer approach. Food Chem 166:465–472. https://doi.org/10.1016/j.foodchem.2014.06.061
62. Yan J, Luo Z, Ban Z, Lu H, Li D, Yang D, Aghdam MS, Li L (2019) The effect of the layer-by-layer (LBL) edible coating on strawberry quality and metabolites during storage. Postharvest Biol Technol 147:29–38. https://doi.org/10.1016/j.postharvbio.2018.09.002
63. Decher G, Hong J-D (1991) Buildup of ultrathin multilayer films by a self-assembly process, 1 consecutive adsorption of anionic and cationic bipolar amphiphiles on charged surfaces. Chem Macromol Symp 46:321–327
64. Md-Nor S, Ding P (2020) Trends and advances in edible biopolymer coating for tropical fruit: a review. Food Res Int 134:1154
65. Yin C, Huang C, Wang J, Liu Y, Lu P, Huang L (2019) Effect of chitosan- and alginate-based coatings enriched with cinnamon essential oil microcapsules to improve the postharvest quality of mangoes. Materials 12:39. https://doi.org/10.3390/ma12132039
66. Soares AS, Ramos AM, Vieira ÉNR, Vanzela ESL, de Oliveira PM, Paula DA (2018) Vacuum impregnation of chitosan-based edible coating in minimally processed pumpkin. Int J Food Sci Technol 53:2229–2238. https://doi.org/10.1111/ijfs.13811
67. Radziejewska-Kubzdela E, Biegańska-Marecik R, Kidoń M (2014) Applicability of vacuum impregnation to modify physico-chemical, sensory and nutritive characteristics of plant origin products—a review. Int J Mol Sci 15:16577–16610

68. Parreidt TS, Schott M, Schmid M, Müller K (2018) Effect of presence and concentration of plasticizers, vegetable oils, and surfactants on the properties of sodium-alginate-based edible coatings. Int J Mol Sci 19:742. https://doi.org/10.3390/ijms19030742
69. Mohanty AK, Misra M, Hinrichsen G (2000) Biofibres, biodegradable polymers and biocomposites: an overview
70. Fabra MJ, López-Rubio A, Cabedo L, Lagaron JM (2016) Tailoring barrier properties of thermoplastic corn starch-based films (TPCS) by means of a multilayer design. J Colloid Interface Sci 483:84–92. https://doi.org/10.1016/j.jcis.2016.08.021
71. Kurek M, Galus S, Debeaufort F (2014) Surface, mechanical and barrier properties of bio-based composite films based on chitosan and whey protein. Food Packag Shelf Life 1:56–67. https://doi.org/10.1016/j.fpsl.2014.01.001
72. Park JW, Scott Whiteside W, Cho SY (2008) Mechanical and water vapor barrier properties of extruded and heat-pressed gelatin films. LWT 41:692–700. https://doi.org/10.1016/j.lwt.2007.04.015
73. Fakhouri FM, Costa D, Yamashita F, Martelli SM, Jesus RC, Alganer K, Collares-Queiroz FP, Innocentini-Mei LH (2013) Comparative study of processing methods for starch/gelatin films. Carbohydr Polym 95:681–689. https://doi.org/10.1016/j.carbpol.2013.03.027
74. Galdeano MC, Grossmann MVE, Mali S, Bello-Perez LA, Garcia MA, Zamudio-Flores PB (2009) Effects of production process and plasticizers on stability of films and sheets of oat starch. Mater Sci Eng C 29:492–498. https://doi.org/10.1016/j.msec.2008.08.031
75. Dutta PK, Kumar Dutta P, Dutta J, Tripathi VS (2004) Chitin and chitosan: chemistry, properties and applications
76. Butler BL, Vergano PJ, Testin RF, Bunn JM, Wiles JL (1996) Mechanical and barrier properties of edible chitosan films as affected by composition and storage
77. Krishna M, Nindo CI, Min SC (2012) Development of fish gelatin edible films using extrusion and compression molding. J Food Eng 108:337–344. https://doi.org/10.1016/j.jfoodeng.2011.08.002
78. Khwaldia K, Basta AH, Aloui H, El-Saied H (2014) Chitosan-caseinate bilayer coatings for paper packaging materials. Carbohydr Polym 99:508–516. https://doi.org/10.1016/j.carbpol.2013.08.086
79. Ciannamea EM, Stefani PM, Ruseckaite RA (2014) Physical and mechanical properties of compression molded and solution casting soybean protein concentrate based films. Food Hydrocoll 38:193–204. https://doi.org/10.1016/j.foodhyd.2013.12.013
80. Jacob J, Linson N, Mavelil-Sam R, Maria HJ, Pothan LA, Thomas S, Kabdrakhmanova S, Laroze D (2024) Poly(lactic acid)/nanocellulose biocomposites for sustainable food packaging. Cellulose 31:5997–6042
81. Zevallos-Torres LA, Lorenci-Woiciechowski A, de Andrade-Tanobe VO, Karp SG, Guimarães-Lorenci LC, Faulds C, Soccol CR (2020) Lignin as a potential source of high-added value compounds: a review. J Clean Prod 263:1154
82. Priyadarshini SR, Krishnamoorthy S, Moses JA, Anandharamakrishnan C (2022) Starch-based edible films and coatings. In: Biopolymer-based food packaging: innovations and technology applications. Wiley, pp 147–177
83. Lievonen M, Valle-Delgado JJ, Mattinen ML, Hult EL, Lintinen K, Kostiainen MA, Paananen A, Szilvay GR, Setälä H, Österberg M (2016) A simple process for lignin nanoparticle preparation. Green Chem 18:1416–1422. https://doi.org/10.1039/C5GC01436K
84. Figueiredo P, Lintinen K, Hirvonen JT, Kostiainen MA, Santos HA (2018) Properties and chemical modifications of lignin: towards lignin-based nanomaterials for biomedical applications. Prog Mater Sci 93:233–269. https://doi.org/10.1016/J.PMATSCI.2017.12.001
85. Xu C, Ferdosian F (2017) Conversion of lignin into bio-based. Chem Mater 15:9. https://doi.org/10.1007/978-3-662-54959-9
86. Luo S, Cao J, McDonald AG (2017) Esterification of industrial lignin and its effect on the resulting poly(3-hydroxybutyrate-co-3-hydroxyvalerate) or polypropylene blends. Ind Crops Prod 97:281–291. https://doi.org/10.1016/J.INDCROP.2016.12.024

87. Thakur VK, Thakur MK, Raghavan P, Kessler MR (2014) Progress in green polymer composites from lignin for multifunctional applications: a review. ACS Sustain Chem Eng 2:1072–1092. https://doi.org/10.1021/SC500087Z/ASSET/IMAGES/MEDIUM/SC-2014-00087Z_0017.GIF
88. Manuel J, Baviera B, Bolognesi C, Chesson A, Cocconcelli PS, Crebelli R, Gott DM, Grob K, Lambré C, Lampi E et al (2020) Scientific panel on food contact materials, enzymes and processing aids (CEP) minutes of the 26th plenary meeting. In: EFSA: food ingredients and packaging (FIP) unit: 1. welcome and apologies for absence
89. Inventory of effective food contact substance (FCS) notifications
90. Guo H, Zhao Y, Chang JS, Lee DJ (2023) Lignin to value-added products: research updates and prospects. Bioresour Technol 384:294. https://doi.org/10.1016/j.biortech.2023.129294
91. Sethupathy S, Murillo Morales G, Gao L, Wang H, Yang B, Jiang J, Sun J, Zhu D (2022) Lignin valorization: status, challenges and opportunities. Bioresour Technol 347:126696. https://doi.org/10.1016/J.BIORTECH.2022.126696
92. Lignin market share industry forecast report. https://www.snsinsider.com/reports/lignin-market-1285. Accessed on 6 May 2025
93. Libretti C, Santos-Correa L, Meier MAR (2024) From waste to resource: advancements in sustainable lignin modification. Green Chem 26:4358–4386. https://doi.org/10.1039/D4GC00745J
94. Nadányi R, Ház A, Lisý A, Jablonský M, Šurina I, Majová V, Baco A (2022) Lignin modifications, applications, and possible market prices. Energies 15:6520. https://doi.org/10.3390/EN15186520
95. Ridho MR, Agustiany EA, Rahmi-Dn M, Madyaratri EW, Ghozali M, Restu WK, Falah F, Rahandi-Lubis MA, Syamani FA, Nurhamiyah Y et al (2022) Lignin as green filler in polymer composites: development methods, characteristics, and potential applications. Adv Mater Sci Eng 2022:1363481. https://doi.org/10.1155/2022/1363481
96. Fillers-fact-sheet
97. Basbasan AJ, Hararak B, Winotapun C, Wanmolee W, Leelaphiwat P, Boonruang K, Chinsirikul W, Chonhenchob V (2022) Emerging challenges on viability and commercialization of lignin in biobased polymers for food packaging: a review. Food Packag Shelf Life 34:969. https://doi.org/10.1016/J.FPSL.2022.100969

Chapter 12
Natural Fiber-Based Polymer Composites for Food Packaging Applications

Nikhil Chauhan, Mridul Pant, and Swati Sharma

1 Introduction

Nowadays, the concept of "carbon–neutral" has been widely accepted, also seen as the goal of reducing emission targets built by significant nations throughout the globe in an attempt to fight global warming. Packaging sector has become a major source of carbon emissions worldwide because 36% of packaging is made from petroleum-based plastics, which are both readily available and reasonably priced [1], and packaging is a combination of science, technology, and art that aims to guarantee a product's safety, integrity, cleanliness, and quality. Majority of plastic packaging is challenging to degrade, and the world is now struggling with how to handle the 15.09 billion tons of new plastic garbage that is produced each year. Main strategies for handling plastic waste include landfilling, incineration, and ocean disposal [2]. These inadequate disposal techniques not only seriously harm the surroundings, but microplastics that infiltrate the ecosystem also endanger the health of people and animals.

The food business faces challenges in complying with recently imposed strict food safety laws and is under increasing pressure to feed the world's population, which is rising at an exponential rate. Consumer desire for nutritious, sustainable, nominally managed, ready-to-eat food with "neat" labeling and "fresh" organic properties has been sparked by the abundance of conventional and digital media consumption. Global enforcement of food traceability mechanisms to assure genuineness and safety across the food supply network caused curiosity among food manufacturers,

N. Chauhan (✉)
Department of Design, Indian Institute of Technology Roorkee, Roorkee, Uttarakhand, India
e-mail: n_chauhan@design.iitr.ac.in

M. Pant · S. Sharma
Department of Polymer and Process Engineering, Indian Institute of Technology Roorkee, Roorkee, Uttarakhand, India

A. Verma et al. (eds.), *Polymers and Composite Materials for Packaging: Smart Food Packaging and Solutions*, Materials Horizons: From Nature to Nanomaterials,
https://doi.org/10.1007/978-981-95-0677-4_12

merchants, consumer rights representatives, and food supervisory bodies in creating accurate, fast, cost-effective, reliable, noninvasive, and nondestructive techniques or tools to measure the freshness of food products. Strong motivation has been generated by these recently developed issues to build innovative packaging solutions that may surpass the capabilities of the conventional passive barrier idea. Food packaging is dominated by traditional polymers like polypropylene (PP), polyethylene (PE), and polyethylene (PET), which are derived from petroleum sources because of their affordability and longevity. However, their energy-intensive manufacturing, non-renewable sources, and ecological persistence have had detrimental effects on the environment, including greenhouse gas emissions and microplastic contamination [3]. Therefore, the production of materials based on naturally occurring biodegradable resources to replace petroleum-derived plastic packaging has become essential [4]. Numerous investigations and discussions have been carried out for this aim by several researchers, particularly with regard to green-material-based functional packaging, such as intelligent and active packaging [5–7]. Figure 1 demonstrates a few relevant primary functional characteristics of the packaging materials that are required to satisfy various objectives [8].

Natural fiber-reinforced biopolymer composites may be an innovative solution to this problem and can be called green functional food packaging that combines recyclable or biodegradable polymer matrices with renewable natural fibers (e.g., bagasse, coir, flax, wood, hemp, sisal, pineapple, etc.) to produce materials that strike a balance between environmental responsibility and practical performance [4]. Researchers are reviving packaging paradigms to answer the demands of the circular economy, waste reduction, and carbon footprint reduction by incorporating lignocellulosic materials, i.e., plant bast fibers, and agricultural leftovers, into biocomposites. Figure 2 shows coconut (coir) fiber-reinforced natural latex composites are used by Enkev Manufacturer to produce biodegradable packaging [9], because coir fiber is tough, durable, recyclable, and disposable without facing any problem. This green functional packaging uses proactive measures and efficient management

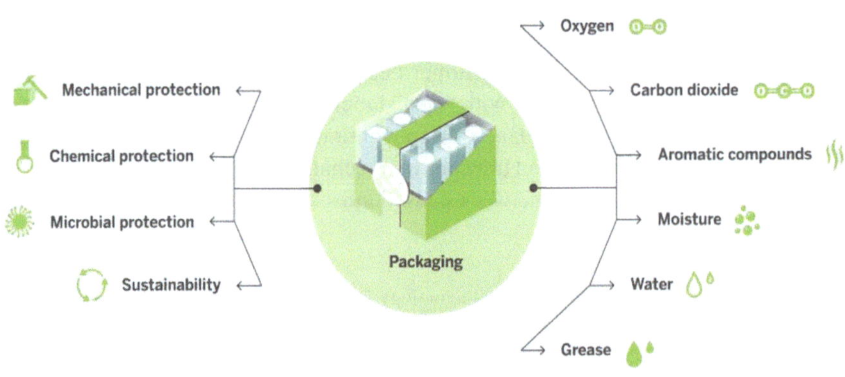

Fig. 1 Packing materials' desired qualities (left) in interaction with various components (right) and their ability to satisfy various objectives [8]

Fig. 2 Packaging manufactured by Enkev Manufacturer from coir fiber-reinforced natural latex composite [9]

to guarantee the safe distribution of food goods. Interestingly, it is important to note that a number of green functional food packaging have shown notable benefits over conventional food packaging when it comes to increasing shelf life, decreasing waste of food, and providing actual evaluation of food quality [10, 11].

2 Raw Materials and Components

2.1 Natural Fiber Types and Sources

The use of natural fibers obtained using renewable sources has intensified as reinforcements in polymeric composites used in the packaging of food. These fibers are divided according to their origins: plant-based, animal-based, and mineral-based [10]. The schematic categorization [11] of natural fibers is displayed in Fig. 3.

2.1.1 Plant-Based Fiber

The fundamental components of these fibers are plant cellulose, hemicellulose, plant lignin, pectin, and wax. Their relative amounts differ across species and even within a single species. Their mechanical properties are given in Table 1. Plant fibers are further subdivided according to their source genus:

Bast Fibers

They can be obtained from the stem or bark of plants, involving flax (*Linum usitatissimum*), hemp (*Cannabis sativa*), jute (*Corchorus species*), and kenaf (*Hibiscus*

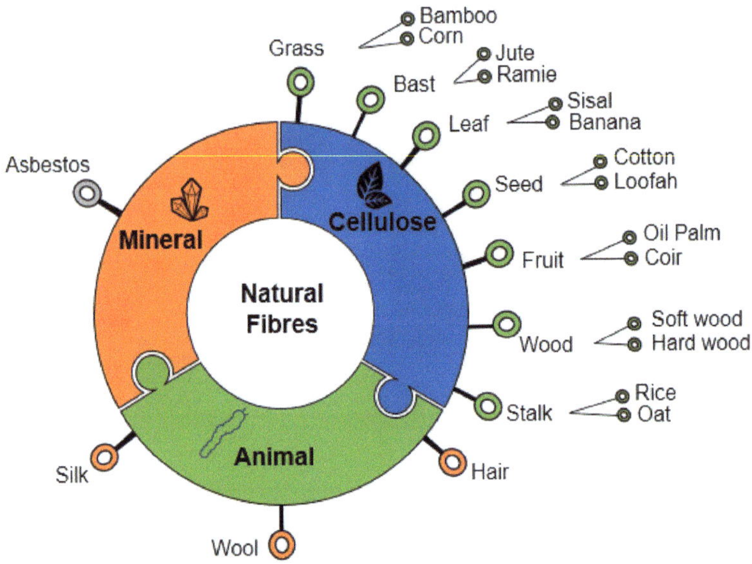

Fig. 3 Categorization of natural fibers [11]

Table 1 Mechanical characteristics of various plant fibers [16–26]

Type	Fiber	Density (kg/m^3)	Tensile strength (MPa)	Young's modulus (GPa)	Elongation at break (%)
Bast	Hemp	1.47–1.57	550–900	6.00–70.0	1.60–4.70
	Jute	1.30–1.50	320–800	10.0–30.0	1.00–1.80
	Flax	1.33–1.65	340–1500	18.0–80.0	1.20–3.20
	Kenaf	0.749–1.45	223–930	1.86–53.0	1.50–6.90
	Nettle	1.50	650–1590	38.0–87.0	1.00–6.00
	Ramie	1.30–1.80	400–1100	44.0–130	1.20–4.00
Leaf	Abaca	1.50	400–980	12.0–33.6	2.90–10.0
	Sisal	1.27–1.50	400–800	9.00–38.0	2.00–14.0
	Fique	0.970–1.33	411–509	8.60–8.73	4.90–9.40
	Henequen	1.20	500	13.2	4.80
	Pineapple	1.20–1.50	170–627	12.0–82.0	1.00–4.80
	Banana	0.750–1.50	355–914	7.70–32.0	1.50–5.30
Seed	Cotton	1.30–1.90	287–597	5.10–13.0	2.00–10.0
Fruit	Coir	1.15–1.50	106–593	1.27–6.00	15.0–47.0
Grass	Bagasse	1.20	20.0–290	19.7–27.1	1.10
	Bamboo	0.600–1.10	140–800	5.96–36.0	1.30–10.4
	Wheat straw	1.10–1.49	59.0–150	3.70–4.80	3.50–6.60
	Rice straw	0.900–1.50	100–200	0.300–12.5	5.40–10.6

cannabinus), etc. These fibers are well known for their being very strong, and, are widely used in the textile and composite industries.

Leaf Fibers

These are sourced from the leaves of monocotyledon plants, particularly sisal (*Agave sisalana*) and abaca (*Musa textilis*). Sisal is primarily raised in the arid to semi areas of China, Brazil, and East Africa, while abaca is chiefly grown in the Philippines and some parts of Central America [12, 13].

Seed Fibers

Cotton (*Gossypium species*), the most well-known example, is derived from seeds or seed cases and is utilized extensively in the textile industry. Around 26 million tons of textile fibers developed in 2018, cotton is the most extensively produced natural fiber worldwide [14].

Fruit Fibers

One such example is coir made from the husks of coconuts (*Cocos nucifera*), which is derived from plant fruit. As a byproduct of coconut harvesting, coir fiber is manufactured in tropical and subtropical nations, including Vietnam, Sri Lanka, and India. Coir fibers are often found in household goods like carpets, brushes, yarn, ropes, insulation and construction material [15].

2.1.2 Animal-Based Fibers

These fibers, which include alpaca hair, silk from silkworms, and sheep wool, are made of proteins like keratin and fibroin. The main protein component of animal fibers gives them qualities like toughness and flexibility. For example, dragline silk, which is derived from silkworms and certain spider species, is known for having exceptional tensile strength up to 1100 MPa, and the most recognized animal fiber is wool, which is made of keratin and has been extracted from sheep coats for ages to make textiles [14].

2.1.3 Mineral-Based Fibers

Non-metallic, inorganic fibers that are made from mineral-based materials or naturally occurring minerals are known as mineral fibers. They are useful in a variety of industrial applications due to their well-known superior fire resistance, thermal insulation, and soundproofing qualities. From some minerals exhibiting a crystalline needle-like shape (like asbestos) to synthetic threads produced using molten material (like basalt), the phrase "mineral fibers" encompasses a broad range of materials.

2.1.4 Modifications of the Natural Fibers

Weak fiber adhesion and an unequal spreading of fibers inside the matrix are caused by the inherent imbalance amid hydrophilic natural fibers and hydrophobic matrix polymers. However, a variety of treatments for natural fibers may effectively tackle this issue [27]. Modifications or treatments are necessary in the natural fibers in order to overcome problems of adhesion, weather dependence, moisture absorption, and boost their suitability with the matrix polymer and strengthen the overall effectiveness of composites containing natural fibers. Numerous methods have been developed, like chemical, physical and combined modifications [28] to evaluate the surface of natural fibers, mostly to increase their adherence to polymeric matrix and decrease their absorption of water. Physical treatment refines the interface zone without altering the chemical characteristics of the fiber, improving bonding at the interface between natural fibers and the polymeric matrix. Chemical treatments have a substantial effect on the mechanical characteristics of natural fibers because cellulose and lignin include hydroxyl groups. Cellulose is the main target for chemical modifications, as it is the main component of natural fiber. Chemically modifying the natural fibers can decrease moisture absorption, strengthening the interfacial connection and producing composites with better mechanical characteristics [29]. Figure 4a shows a schematic representation of various physical and (b) chemical treatments [29].

2.2 *Polymer Matrix and Additives*

The polymer matrix acts as the continuous phase in NFPCs, binding the natural fibers and promoting stress transmission, while additives are added to improve certain composite material characteristics. The matrix selection governs the stiffness and stability of the composite and plays a key role in figuring out mechanical property, thermal stability, and environment impact of the composite. Thermosetting resins like polyester and epoxy, etc., as well as thermoplastics like polyethylene and polypropylene, etc., are examples of common polymer matrices. Different additives are utilized to enhance fiber-matrix adhesion, and thereby the mechanical characteristics of the composite. These nanomaterials include coupling agents, plasticizers and flame retardants. Coupling agents like silanes and maleated anhydrides improve the interfacial adherence between hydrophilic natural fibers and hydrophobic polymeric matrices, resulting in enhanced mechanical properties. The addition of lubricants helps to improve the processability of the composites, and plasticizers increase their flexibility; the addition of flame retardants will also show their effect on improving the fire

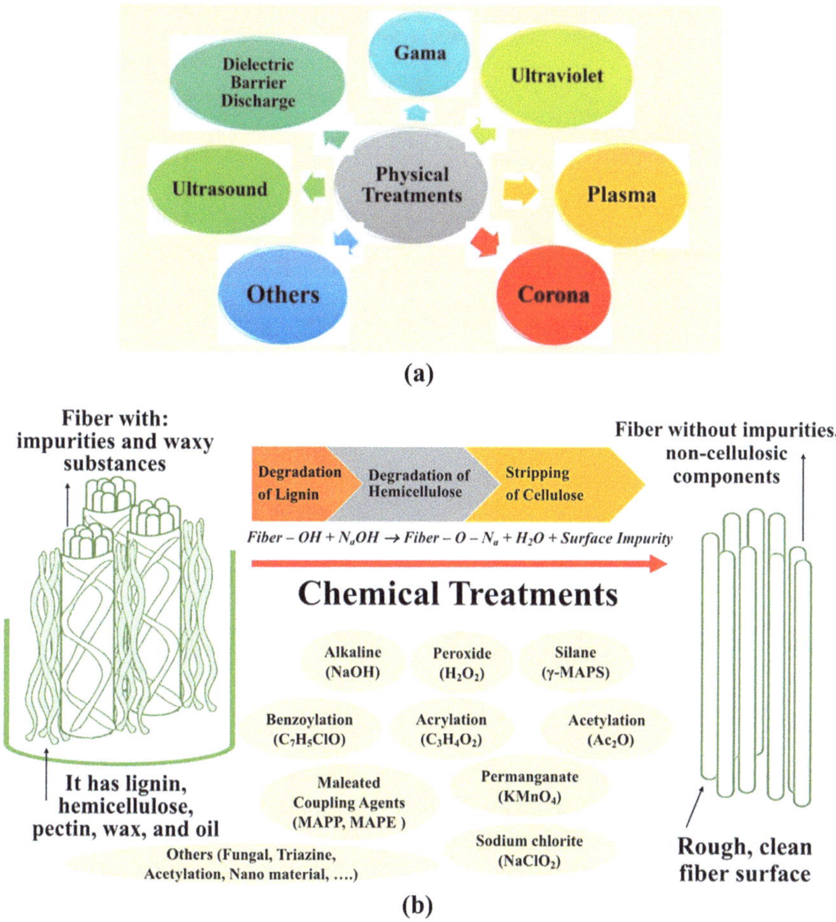

Fig. 4 **a** schematic representation of various physical and **b** chemical treatments

resistance performance of such composites. The specific application requirements for NFPCs are met by the synergistic selection and optimization of different polymer matrices and suitable additives. About 90% of plastic produced worldwide now is made from polymers generated from fossil fuels, as seen in Fig. 5, with just 1.5% coming from bio-based sources [30]. Therefore, research on biopolymer substitutes for commercial polymers has attracted a lot of attention recently in anticipation of future initiatives to lessen plastic pollution and the world's dependency on fossil fuels.

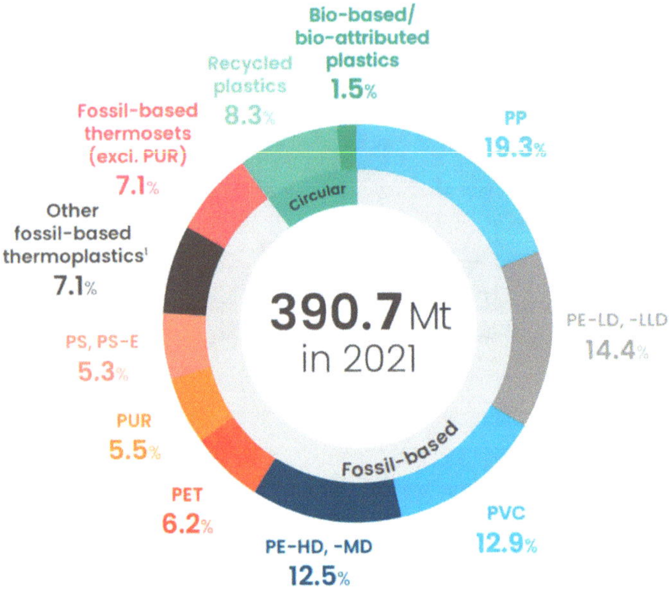

Fig. 5 Global plastic output by type and origin of polymers [30]

2.2.1 Biopolymers

A polymeric substance may only be categorized as a "biopolymer" if it satisfies one of two requirements: it must be biodegradable or generated from natural resources that are bio-based. Synthetic and biodegradable, natural and biodegradable, and natural and non-biodegradable are the three different categories into which a biopolymer may be classified [31]. Figure 6 summarizes the origin-based classification of various biodegradable polymers.

2.2.2 Compatibilizers and Functional Additives

Compatibilizers and functional additives are essential for improving interfacial adhesion and modifying material characteristics to satisfy particular application needs in NFPCs. Compatibilizers are those constituents that are added to polymer blends or in NFPCs to strengthen the interfacial interaction amid incompatible materials, such as hydrophobic matrix polymer and hydrophilic natural fibers. For example, Silane coupling agents, MAPP, and MAPE as compatibilizers are being demonstrated to improve the interface bonding between polymer matrices and natural fibers, and results in superior mechanical properties of composites [32]. Functional additives are a wide variety of materials added to polymeric composite materials to provide them specific properties. Adding carbon nanotubes as additives to the polymer composites may improve their mechanical strength and electrical conductivity therefore

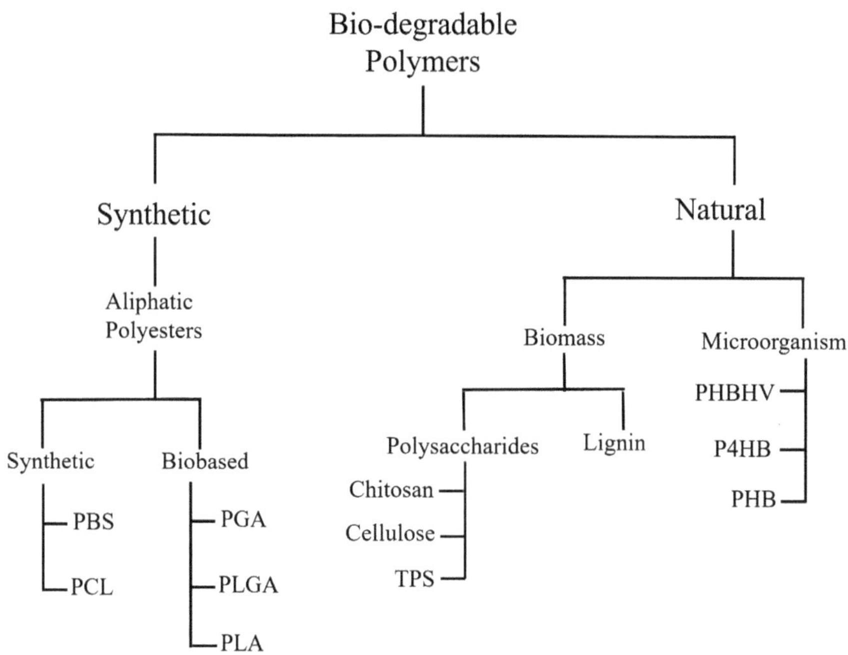

Fig. 6 Origin-based classification of various biodegradable polymers

increasing their potential uses in a variety of industries [33–35]. Additionally, it has been studied to include nanoclays into polymer composites to enhance their barrier qualities and flame retardancy, which prepares them for use in packaging and automobile applications [36, 37]. Optimizing the properties of NFPCs requires the thoughtful selection and integration of compatibilizers and functional additives, which makes it possible to produce materials with specific qualities for a range of industrial applications.

3 Processing and Manufacturing

The processing and manufacturing of NFRPCs are crucial for assessing their properties and functions in food packaging applications. This section examines conventional and advanced processing techniques together with their optimization methodologies for addressing scaling difficulties.

3.1 Conventional and Advanced Processing Methods

3.1.1 Extrusion

Extrusion is among the most widely used processes for the production of NFPCs. Natural fibers and polymer matrices are compounded in this method under high shear conditions and at high temperatures [38]. This process is usually employed for composites based on thermoplastic matrices. Twin screw extruders provide good mixing and proper fiber dispersion [39].

3.1.2 Compression Molding

NFPCs developed by compression molding involve placing the pre-mixed polymeric matrix and the fiber in a mold cavity and applying heat and pressure. This method is usually employed to develop thermoset matrix-based composites and allows for processing large, intricate parts and required fiber orientation [40]. Cellulose-reinforced PLA composites developed by compression molding showed better barrier properties than injection-molded cellulose-PLA composites owing to better fiber orientation and lesser void formation during processing [41].

3.1.3 Injection Molding

Injection molding processing involves heating the pre-mixed composite material around or just beyond its melting point, followed by pumping the liquefied mixture into a mold cavity under high pressure, where the molten material is allowed to cool and harden and conform to the mold cavity's shape and then ejected which allows for the production of intricate food packaging components with good dimensional stability. Nonetheless, fiber breakage during the injection molding process is quite challenging [42]. However, careful optimization of injection molding parameters can lead to a reduction in fiber breakage and provide good mechanical properties [43].

3.1.4 Hand Layup

Production of NFPCs by hand layup technique is a manual process where prepregs (layers of pre-impregnated fibers) are meticulously placed down and folded into a mold. Fibers are usually bundled into tows, which can be uni-directionally oriented or woven together as required. The hand layup technique is favored for small-scale prototyping and customized designs for the components in food packaging, such as trays and liners. This method is favored in specific applications because of its simplicity and cost-effectiveness, particularly when compared to more automated techniques. Nevertheless, it demonstrates reduced efficiency in large-scale product manufacturing.

3.1.5 Solution Casting

The solution-casting method for producing NFRPCs involves dissolving the polymer matrix in the solvent along with the reinforcing fiber, casting the solution onto a surface, and evaporating the solution to develop an NFRPC film or sheet. Solution casting is majorly used for the development of thin, flexible packaging applications and is particularly useful in active packaging systems, including oxygen scavengers [44].

3.1.6 Electrospinning

Composites developed by the electrospinning technique involve applying a high voltage to a solution with the dissolved polymer matrix and the reinforcement. This creates a charged jet that stretches and solidifies into ultra-thin films and membranes as they travel to a grounded collector. Composites developed by this technique have unique structural properties and higher surface areas. Chen et al. [59] demonstrated that electrospun PLA/cellulose nanocrystal membranes facilitate controlled release kinetics, thereby prolonging the shelf life of oxidation-sensitive foods and effectively serving as carriers for antioxidant compounds in active food packaging applications. Furthermore, the key parameters of the different processes have been summarized in Table 2.

3.2 Emerging Advanced Processing Methods

Recent advancements in manufacturing methods and technologies have presented various promising methods for the production of NFPCs which are discussed below.

3.2.1 3D Printing

3D printing or additive manufacturing enables the production of customized, intricate packaging designs characterized by exact geometrical specifications with reduction of voids and enhanced properties. PLA/cellulose nanocrystal composites have been successfully printed for food packaging applications with tailored mechanical and barrier properties [60].

Table 2 Key processes for the manufacturing of NFPCs for food packaging applications

Method	Advantages	Limitations	Common matrix materials	Key properties for food packaging	Examples in food packaging	References
Extrusion	Continuous process, high production rate, good uniformity, cost-effective, suitable for mass production	Limited to simpler shapes, potential fiber breakage during processing, temperature limitations with natural fibers, difficulty with high fiber loading	PP, PE, PLA, PBAT, PHB, starch blends	Barrier properties to O_2, CO_2, moisture, mechanical strength, thermal stability, biodegradability	Food containers, flexible films, sheets, trays, cups	[45, 46]
Compression molding	Good fiber orientation control, low fiber damage, higher fiber content possible, good mechanical properties, complex shapes possible	Batch process, longer cycle times, relatively lower production rate, higher energy consumption	PP, PE, epoxy, PLA, natural resins, phenolics	High strength, good barrier properties, impact resistance, thermal stability, stiffness	Rigid containers, trays, disposable plates, biodegradable cutlery, lids	[47–49]
Injection molding	High production rate, complex geometries, good dimensional accuracy, automation, consistent quality	Fiber orientation challenges, potential fiber breakage, higher initial investment, limited fiber length, fiber content limitations	PP,PE,PLA PBAT,PHA, starch blends	Dimensional stability, impact resistance, clarity and consistency, moisture resistance	Containers, caps and closures, cutlery, cups, thin-walled packaging	[50, 51]

(continued)

Table 2 (continued)

Method	Advantages	Limitations	Common matrix materials	Key properties for food packaging	Examples in food packaging	References
Hand layup	Versatile low tooling cost large parts possible, good for complex shapes, easy to implement	Labor intensive, lower production rate, inconsistent quality, batch process, worker exposure to resins	Epoxy, polyester, bio-based resins, PLA, starch-based resins	Customizable shapes, durability, lightweight, impact resistance, water resistance	Large containers, special food transport boxes, custom packaging solutions, catering trays	[52–54]
Solution casting	Minimal fiber damage, enhanced barrier properties, homogeneous distribution, bioactive compound integration	Solvent removal issues, slower process, limited to films/sheets, scale-up challenges, environmental concerns with some solvents	PLA, starch, cellulose derivatives, chitosan, alginate	Excellent barrier properties, transparency, biodegradability, controlled release, antimicrobial potential	Films, wraps, pouches, sachets, active packaging	[55, 56]
Electrospinning	Nanofiber production, high surface area, enhanced barrier properties, antimicrobial potential, controlled release functionality	Low production rate, limited thickness, typically needs combination with other processes, higher cost, technical complexity	PLA, PCL, gelatin, chitosan, cellulose acetate	Excellent barrier properties, antimicrobial activity, controlled release, functional properties, nanostructured surfaces	Active packaging films, sachets with controlled release, antimicrobial layers, intelligent packaging, moisture control packaging	[57, 58]

3.2.2 Layer-by-Layer (LbL) Assembly

Layer-by-layer (LbL) assembly produces ultra-thin films by the sequential deposition of complementary materials, generally facilitated by charged or molecular interactions. This method provides precise control over film thickness, structure, and properties. The simplicity, versatility, and scalability of this method can drive innovations in diverse food packaging applications [61].

3.2.3 Reactive Extrusion

Reactive extrusion is a technique for the chemical modification of polymers that occurs during with the extrusion process. This technique enhances fiber-matrix compatibility during extrusion by incorporating chemical reactions via in-situ grafting or functionalization. Reactive extrusion has been employed for starch esterification, a crucial method for producing biodegradable polymers utilized in food packaging applications [62].

3.3 Process Optimization and Scale-Up Considerations

Optimizing NFRPC's production processes and scaling up involves carefully considering material properties, optimizing various parameters for consistent quality and efficiency, and preserving fiber integrity for food packaging applications. Following is a discussion on these critical parameters and the challenges of industrial scaling up.

3.3.1 Critical Processing Parameters

Maintaining an adequate processing temperature is crucial to avert fiber deterioration while maintaining complete matrix melting and effective bonding with the fibers during the production of NFPCs [63]. Similarly, optimizing residence time mitigates the adverse effects of extended exposure to elevated temperatures and shear, hence maintaining fiber integrity [45]. Low shear rates promote an even fiber distribution while minimizing substantial structural damage [64]. Furthermore, regulating moisture content is essential; maintaining it at a low level avoids void formation and hydrolytic degradation, hence improving interfacial adhesion and total composite strength [65]. Optimizing these characteristics is critical for maximizing the efficiency of the reinforcing effect of natural fibers in NFPCs applications for food packaging.

3.3.2 Industrial Scale-Up Challenges

Scaling NFPCs from laboratory to industrial production poses considerable hurdles. Process consistency is a significant issue, as the heterogeneity of natural fibers impacts the quality and performance consistency of composites at higher production quantities [44, 66]. Modifications to equipment are crucial, as standard polymer processing gear must be adapted to accommodate the distinct properties of natural fibers, such as moisture sensitivity and the potential for thermal deterioration [67, 68]. Supply chain management poses a challenge, as securing regular, large-scale fiber supply is complicated by seasonal availability and quality fluctuations and quality variations [44].

The temperature stability of polymers restricts the widespread use of NFPCs, given that numerous natural fibers deteriorate at temperatures around standard polymer processing conditions [66, 69]. Cost-effectiveness poses a considerable challenge, as the inclusion of additional processing steps for fiber preparation raises production costs, even though raw material expenses are lower than those associated with synthetic reinforcements [44]. Moisture absorption challenges complicate manufacturing, necessitating specialized drying systems and specific processing parameters [69, 70]. Successful industrial implementation necessitates advanced processing technologies, specialized equipment configurations, and robust quality control systems [67, 68]. Ongoing research in chemical treatments and manufacturing innovations is enhancing the commercial viability of NFPCs in automotive, construction, and packaging applications, despite existing challenges [69, 70].

4 Performance Characteristics

4.1 Mechanical and Barrier Properties

NFPCs represent a significant advancement in food packaging materials due to their biodegradability, renewability, and favorable mechanical and barrier properties. Their effectiveness is heavily influenced by factors such as fiber treatment, matrix compatibility, and manufacturing methods. The addition of natural fibers, including cellulose and hemp, into biopolymer matrices such as polylactic acid (PLA) and chitosan has demonstrated notable enhancements in tensile strength, elongation at break, and water vapor resistance [71]. Previous literature suggests that natural fibers such as abaca, jute, and pineapple leaf fiber improve tensile strength (38.4 MPa at 40% fiber loading) and flexural properties when subjected to chemical modifications via alkali treatment or potassium permanganate immersion. These processes eliminate non-cellulosic components and enhance fiber-matrix adhesion [72]. However, the excessive fiber loading exceeding 40 wt.% may negatively impact mechanical performance due to aggregation and interfacial defects, as demonstrated in poly(3-hydroxybutyrate-co-3-hydroxyvalerate) composites reinforced with *Ceiba pentandra* fibers [73]. Furthermore, hybridization with nanomaterials such as ZnO and nanocellulose has shown enhanced barrier properties against oxygen and water vapor, which are essential for food preservation [74]. Nanocellulose reinforcement enhances the tensile strength of composite films by 38–200% and decreases water vapor permeability by up to 60% relative to unreinforced films. The strong hydrogen bonding between natural polymers and nanocellulose limits the movement of polymer chains, resulting in enhanced mechanical properties while reducing elongation at break [71].

4.2 Thermal, Surface, and Antimicrobial Properties

NFPCs represent a significant advancement in sustainable food packaging materials, providing a combination of thermal stability, adjustable surface characteristics, and embedded antimicrobial properties. The thermal properties of these composites are significantly affected by the natural fibers' restricted thermal degradation limit of 200–220 °C, requiring the careful selection of polymer matrices with suitable processing temperatures [36]. Surface functionalization techniques, including the coating of coconut fibers with essential oil-paraffinic oil solutions, improve fiber-matrix bonding and provide antimicrobial activity via controlled release mechanisms [75]. The antimicrobial potential of these composites arises due to bioactive agents like essential oils immobilized on fiber surfaces, which display dose-dependent inhibition against foodborne pathogens [73, 75]. Advanced composite architectures utilize the hygroscopic properties of natural fibers to develop moisture-activated antimicrobial release systems. However, achieving these demands precise control of fiber content, generally maintained between 10 and 30 wt.%, to optimize both barrier properties and active functionality [44, 73].

4.3 Biodegradability and Compostability

NFPCs provide a sustainable option for food packaging, attributed to their biodegradability and compostability. These composites incorporate renewable fibers, including cellulose and lignocellulosic materials, which improve microbial degradation under suitable conditions, as specified by ISO 14852:2018 and ISO 17088:2012 [76, 77]. Their integration into polymer matrices enables enzymatic breakdown while preserving mechanical attributes, as shown by Varghese et al. where prepared *C. pentandra* fiber-reinforced polyhydroxyalkanoate (PHA) composites for strawberry packaging [78]. Regulatory frameworks emphasize the ecological advantages of these materials, highlighting that their carbon footprints are 40–60% lower than those of petroleum-based plastics, especially when optimized for composting and anaerobic digestion [79]. These advancements define NFPCs as effective materials for food packaging solutions aligned with circular economy principles.

4.4 Testing Methods and Quality Control

The development and optimization of NFPCs for food packaging applications require comprehensive testing methods and multidimensional characterization to assess mechanical performance, barrier properties, thermal stability, and compliance with food safety standards. Mechanical properties are generally analyzed through standardized standards like ASTM D3039 for tensile, ASTM D790 for flexural, and ASTM D256 for impact tests, which measure critical parameters of composites like ultimate tensile strength, elastic modulus, and energy absorption capacity [80]. Chemical characterization techniques, such as Fourier-transform infrared spectroscopy (FTIR) and X-ray diffraction (XRD), provide compositional insights, including cellulose content (55.1–62.21% crystallinity index) and surface functional groups. Additionally, scanning electron microscopy (SEM) reveals morphological properties, including microfibril alignment and interfacial bonding quality [77]. The thermal degradation profile of composites can be studied by Thermal gravimetric analysis (TGA) and derivative thermogravimetry (DTG). Additionally, in the case of food packaging applications the barrier properties are very important for food preservation. These properties are tested in terms of including water vapor transmission rate (WVTR) by gravimetric methods and oxygen permeability by gas chromatography. The biodegradability of obtained composites can be measured according to ISO 14855 standards and leaching of additives by using migration tests ensures environmental and food contact safety, while antimicrobial studies against various microbes validate shelf-life extension potential [77, 81].

5 Functional Applications

The incorporation of natural fibers into polymer composites has facilitated innovative advancements in the development of functional food packaging systems. These materials tackle environmental issues and enhance packaging efficacy through active, intelligent, and food-specific designs. This section will explore the role of NFPCs in advanced packaging systems, concentrating on oxygen scavenging, moisture control, freshness indication, and tailored applications for various food types.

5.1 Active and Intelligent Packaging Systems

5.1.1 Oxygen Scavengers and Moisture Regulators

Active packaging systems can readily interact with food items to extend their shelf life by reducing oxidative degradation and moisture-induced spoilage. NFPCs are being developed to include oxygen scavengers that chemically bind residual oxygen in packaged environments. Oxygen scavengers, including iron-based compounds and bio-based alternatives such as ascorbic acid and polyphenols, can be incorporated into these composites to effectively eliminate oxygen, thus preventing oxidative spoilage and prolonging the shelf life of perishable foods [82]. In addition, the natural fibers exhibit moisture-regulating properties too owing to their hygroscopic characteristics, allowing them to absorb excess humidity and reduce condensation in packaged food products. The efficient functioning of these composites as oxygen scavengers and moisture regulators is influenced by several factors, including the type of fiber employed, the methods of fiber treatment, and the specific polymer matrix utilized [83, 84]. Research indicates that polymer composites containing fibers like cellulose nanocrystals and lignocellulosic fillers enhance moisture management, improve antimicrobial activity, and decrease lipid oxidation in food packaging applications [85]. In some cases, silica gel or cellulose-based desiccants can be combined with fiber matrices to improve moisture control, mainly for dried snacks and powdered foods [86].

5.1.2 Freshness Indicators and Smart Features

Intelligent packaging systems utilize NFPCs for real-time quality monitoring via visual or chemical indicators. Freshness indicators integrated into these composites identify volatile metabolites such as ammonia, hydrogen sulfide, or diacetyl, which are byproducts of microbial growth in perishable goods. Recent advancements involve palladium-catalyzed films that interact with hydrogen to remove oxygen, thereby prolonging the freshness of vacuum-sealed meats and cheeses [86]. Various electronic components like radio-frequency identification (RFID) tags or time–temperature indicators (TTIs) can be integrated into fiber-based packaging to impart smart features [73]. These systems monitor logistical conditions, such as temperature variations during transit, and transmit data via electrochromic displays. An outstanding case involves flax fiber-reinforced polylactic acid (PLA) films integrated with temperature–time indicators (TTIs), which assess cumulative heat exposure in frozen food products such as seafood and ready-to-eat meals.

5.2 *Food Category-Specific Applications*

5.2.1 Fresh Foods

Fresh foods including fruits, vegetables, and dairy products demand packaging that can effectively manage gas exchange and moisture retention to prevent spoilage. NFPCs obtained from banana stems or wheat straw demonstrate adjustable porosity, facilitating optimal diffusion of oxygen and carbon dioxide for respiring produce such as berries and leafy greens. In one of the studies, Varghese et al. reported the preparation of *C. pentandra* fiber-incorporated polyhydroxyalkanoate (PHA) films. The obtained composite films exhibited strong antibacterial activity against *S. aureus*, and owing to this, these films extended the shelf life of strawberries by 30% [73]. Similarly, in one of the works, chitosan-coated jute fiber-incorporated composite inhibited the mold growth in cheese and maintained the humidity levels required for its texture preservation.

5.2.2 Processed Foods

Processed foods such as baked goods, confectioneries, and cured meats demand packaging materials that exhibit robust barrier properties against grease, oxygen, and mechanical stress. The integration of natural fibers into polymer matrices has demonstrated significant improvements in these properties. Blending kenaf and hemp fibers with polypropylene (PP) enhances tensile strength and oxygen barrier performance, rendering them appropriate for oil-rich snacks such as potato chips [73]. Active packaging solutions for processed meats frequently incorporate sisal fiber-reinforced films alongside oxygen scavengers and antioxidants, including tocopherol, to mitigate lipid oxidation and color degradation [82]. Furthermore, ramie fiber composites functionalized with soy protein isolates have been utilized in vacuum-sealed sausages, exhibiting enhanced puncture resistance and elongation at break relative to synthetic alternatives [73].

5.2.3 Shelf-Stable Foods

Shelf-stable food products, including cereals, powdered soups, and canned goods, call for packaging materials that preserve integrity under ambient conditions. The European Union categorizes these composites according to their stability, indicating that non-shelf-stable variants necessitate controlled transportation conditions [73]. Packaging made from natural fibers, such as those reinforced with coir or rice husk fibers, demonstrates superior dimensional stability and reduced water activity, thereby effectively preventing clumping in hygroscopic products like instant coffee [87]. Recent developments in multilayer designs, including bamboo fiber/PLA laminates coated with aluminum oxide, improve moisture and light barrier properties, rendering them suitable for the preservation of nutrient-sensitive products such as vitamin-enriched cereals [74].

6 Conclusions and Future Outlook

NFPCs offer a promising path toward environmentally sustainable food packaging solutions. Their integration leverages the biodegradability and renewability of natural fibers with the functionality of polymer matrices, leading to enhanced mechanical strength, thermal stability, and barrier properties—all while significantly reducing the ecological footprint of conventional plastics [88]. Advances in fiber surface modification have notably improved compatibility with polymer matrices, resulting in better interfacial adhesion and overall composite performance [89]. However, ongoing challenges—such as variability in natural fiber quality, sensitivity to humidity, and cost-effective scalability—must still be addressed for broader industrial adoption [90–92].

Future research should focus on identifying innovative, low-impact fiber sources and enhancing eco-friendly extraction methods to improve functionality [77]. It is essential to devise advanced compatibilization technologies to enhance fiber-matrix adhesion while maintaining biodegradability [93]. The integration of nanotechnology may improve barrier characteristics while preserving mechanical strength and transparency [94]. Moreover, using active and intelligent packaging elements—such as antimicrobial agents and biosensors—can prolong shelf life and facilitate real-time quality assessment [95, 96]. These advancements will expedite the shift toward circular economy principles in the food packaging business, harmonizing sustainability objectives with performance requirements [97].

Acknowledgements Nikhil Chauhan, Mridul Pant, and Swati Sharma acknowledge the financial support provided by the MHRD, Government of India, in the form of a Fellowship of IIT Roorkee.

Credit Authorship Contribution Statement Nikhil Chauhan: Writing, conceptualization, and original draft; Mridul Pant: Review, conceptualization, and visualization; Swati Sharma: Review, conceptualization, and visualization;

Conflicts of Interest There are no conflicts of interest to declare by the authors.

References

1. De Vos B, Souza MF, Michels E, Meers E (2022) Industrial hemp (*C. sativa* L.) in a phytoattenuation strategy: remediation potential of a Cd, Pb and Zn contaminated soil and valorization potential of the fibers for textile production. Ind Crops Prod 178:4592. https://doi.org/10.1016/j.indcrop.2022.114592

2. Geyer R, Jambeck JR, Law KL (2017) Production, use, and fate of all plastics ever made. Sci Adv 3(7):25–29. https://doi.org/10.1126/sciadv.1700782
3. Srivastava V, Singh S, Das D (2023) Enhancing packaging sustainability with natural fiber reinforced biocomposites: an outlook into the future. E3S Web Conf 436:16. https://doi.org/10.1051/e3sconf/202343608016
4. Yue C, Wang M, Zhou Z, You Y, Wang G, Wu D (2024) Cellulose-based intelligent packaging films with antibacterial, UV-blocking, and biodegradable properties for shrimp freshness monitoring. Chem Eng J 488:150975. https://doi.org/10.1016/j.cej.2024.150975
5. Aghababaei F, McClements DJ, Martinez MM, Hadidi M (2024) Electrospun plant protein-based nanofibers in food packaging. Food Chem 432:137236. https://doi.org/10.1016/j.foodchem.2023.137236
6. Jafarzadeh S et al (2024) Green synthesis of nanomaterials for smart biopolymer packaging: challenges and outlooks. J Nanostruct Chem 14(2):113–136. https://doi.org/10.1007/s40097-023-00527-3
7. Wang Y et al (2023) Sustainable polysaccharide-based materials for intelligent packaging. Carbohydr Polym 313:120851. https://doi.org/10.1016/j.carbpol.2023.120851
8. Helanto K, Matikainen L, Talj R, Rojas OJ (2019) Bio-based polymers for sustainable packaging and biobarriers: a critical review. BioResources 14(2):4902–4951
9. Azman MA et al (2021) Natural fiber reinforced composite material for product design: a short review. Polymers 13:12. https://doi.org/10.3390/polym13121917
10. McKay I, Vargas J, Yang L, Felfel RM (2024) A review of natural fibres and biopolymer composites: progress, limitations, and enhancement strategies. Multidiscipl Dig Publ Instit 15:878. https://doi.org/10.3390/ma17194878
11. Neto JSS, de Queiroz HFM, Aguiar RAA, Banea MD (2021) A review on the thermal characterisation of natural and hybrid fiber composites. Polymers 13(24):4425. https://doi.org/10.3390/polym13244425
12. Hakeem KR, Jawaid M, Rashid U (2014) Abaca fiber: a renewable bio-resource for industrial uses and other applications. Springer, Cham
13. Broeren MLM, Dellaert SNC, Cok B, Patel MK, Worrell E, Shen L (2017) Life cycle assessment of sisal fibre: exploring how local practices can influence environmental performance. J Clean Prod 149(2017):818–827. https://doi.org/10.1016/j.jclepro.2017.02.073
14. Kozłowski RM, Mackiewicz-Talarczyk M (2020) Handbook of natural fibres: volume 2: processing and applications, 2nd edn. Woodhead Publishing
15. Wang B, Yan L, Kasal B (2022) A review of coir fibre and coir fibre reinforced cement-based composite materials (2000–2021). J Clean Prod 338:676. https://doi.org/10.1016/j.jclepro.2022.130676
16. Laverde-Sarmiento V, Benjumea JM, Rincón-Ortiz M (2025) Chemical and mechanical characterization of treated and untreated woven fique fiber textiles as an alternative external reinforcement for concrete. J Nat Fibers 22(1):975
17. Muñoz-Blandón O, Ramírez-Carmona M, Rendón-Castrillón L, Ocampo-López C (2023) Exploring the potential of fique fiber as a natural composite material: a comprehensive characterization study. Polymers 15(12):712. https://doi.org/10.3390/polym15122712
18. Aldroubi S, Kasal B, Yan L, Bachtiar EV (2023) Multi-scale investigation of morphological, physical and tensile properties of flax single fiber, yarn and unidirectional fabric. Compos B Eng 259:110732. https://doi.org/10.1016/j.compositesb.2023.110732
19. Dattatreya K, Sathees-Kumar S, Prasad VVSH, Ranjan-Pati P (2023) Mechanical properties of waste natural fibers/fillers reinforced epoxy hybrid composites for automotive applications. Mater Today Proc 15:1–7. https://doi.org/10.1016/j.matpr.2023.02.001
20. Ng LF et al (2024) Mechanical characterization, water absorption, and thickness swelling of lightweight pineapple leaf/ramie fabric-reinforced polypropylene hybrid composites. Polymers 16(13):1847. https://doi.org/10.3390/polym16131847
21. Pathan Y, Gb VK (2023) Potential of Agave angustifolia marginata for composite and textile applications: a new source of natural fibre. Ind Crops Prod 203:117213. https://doi.org/10.1016/j.indcrop.2023.117213

22. Shadhin M, Rahman M, Jayaraman R, Chen Y, Mann D, Zhong W (2023) Natural biomass and waste biomass fibers: structures, environmental footprints, sustainability, degumming methods, and surface modifications. Ind Crops Prod 204:117252. https://doi.org/10.1016/j.indcrop.2023.117252
23. Sriram M, Aswin-Sidhaarth KR (2022) Various properties of natural and artificial fibers with cementitious composites in hybrid form: a review. Mater Today Proc 60(2022):2018–2025
24. Stevens S, Raja-Dhas JE, Savio-Lewise KA, Mohammad A, Fahad M (2022) Investigations on chemical behaviours on mechanical properties of natural fiber composites: an evaluation. Mater Today Proc 64:410–415. https://doi.org/10.1016/j.matpr.2022.04.761
25. Pawar MD, Joshi R (2022) Vibrational behaviour and mechanical properties of hybrid polymer matrix composites reinforced with natural fibres: a review. Mater Today Proc 54:341–347. https://doi.org/10.1016/j.matpr.2021.09.298
26. Kumar S et al (2021) Physical and mechanical properties of natural leaf fiber-reinforced epoxy polyester composites. Polymers 13(9):369. https://doi.org/10.3390/polym13091369
27. Nlandu-Mayamba H et al (2024) Influence of surface treatments and addition of a reactive agent on the properties of PLA/flax and PLA/bamboo composites. J Thermoplast Compos Mater 15:1733. https://doi.org/10.1177/08927057241231733
28. Ramachandran AR, Mavinkere-Rangappa S, Kushvaha V, Khan A, Seingchin S, Dhakal HN (2022) Modification of fibers and matrices in natural fiber reinforced polymer composites: a comprehensive review. Macromol Rapid Commun 43(17):1–38. https://doi.org/10.1002/marc.202100862
29. Mohammadi M, Ishak MR, Sultan MTH (2024) Exploring chemical and physical advancements in surface modification techniques of natural fiber reinforced composite: a comprehensive review. J Nat Fibers 21(1):633. https://doi.org/10.1080/15440478.2024.2408633
30. Janssens V (2022) Plastics: the facts 2022. Plast Eur 10:81
31. Niaounakis M (2015) Introduction. Biopolym Process Prod 15:5. https://doi.org/10.1016/b978-0-323-26698-7.00001-5
32. Pokharel A, Falua KJ, Babaei-Ghazvini A, Acharya B (2022) Biobased polymer composites: a review. J Compos Sci 6(9):255. https://doi.org/10.3390/jcs6090255
33. Anjali S, Aishwaryalakshmi S, Ashwin V, Neeraja S, Rasana N, Jayanarayanan K (2018) Effect of compatibilizer and carbon nanotubes on blends of polypropylene and nylon 6. Mater Today Proc 5(11):25524–25533. https://doi.org/10.1016/j.matpr.2018.10.359
34. Nurazzi NM et al (2021) Mechanical performance and applications of CNTs reinforced polymer composites—a review. Nanomaterials 11(9):2186. https://doi.org/10.3390/nano11092186
35. Ali A, Rahimian-Koloor SS, Alshehri AH, Arockiarajan A (2023) Carbon nanotube characteristics and enhancement effects on the mechanical features of polymer-based materials and structures: a review. J Mater Res Technol 24:6495–6521. https://doi.org/10.1016/j.jmrt.2023.04.072
36. Mylsamy B, Shanmugam SKM, Aruchamy K, Palanisamy S, Nagarajan R, Ayrilmis N (2024) A review on natural fiber composites: polymer matrices, fiber surface treatments, fabrication methods, properties, and applications. Polym Eng Sci 64(6):2345–2373. https://doi.org/10.1002/pen.26713
37. Uddin MN et al (2024) Research and applications of nanoclays: a review. SPE Polym 5:507–535. https://doi.org/10.1002/pls2.10146
38. Khanam PN, Khalil HPSA, Jawaid M, Reddy GR, Narayana CS, Naidu SV (2010) Sisal/carbon fibre reinforced hybrid composites: tensile, flexural and chemical resistance properties. J Polym Environ 18(4):727–733. https://doi.org/10.1007/s10924-010-0210-3
39. Faruk O, Bledzki AK, Fink HP, Sain M (2012) Biocomposites reinforced with natural fibers: 2000–2010. Prog Polym Sci 37(11):1552–1596. https://doi.org/10.1016/J.PROGPOLYMSCI.2012.04.003
40. Väisänen T, Das O, Tomppo L (2017) A review on new bio-based constituents for natural fiber-polymer composites. J Clean Prod 149:582–596. https://doi.org/10.1016/J.JCLEPRO.2017.02.132

41. Xie F, Zhang T, Bryant P, Kurusingal V, Colwell JM, Laycock B (2019) Degradation and stabilization of polyurethane elastomers. Prog Polym Sci 90:211–268. https://doi.org/10.1016/J.PROGPOLYMSCI.2018.12.003
42. Pickering KL, Efendy MGA, Le TM (2016) A review of recent developments in natural fibre composites and their mechanical performance. Compos A Appl Sci Manuf 83:98–112. https://doi.org/10.1016/J.COMPOSITESA.2015.08.038
43. Höftberger T, Zitzenbacher G, Burgstaller C (2024) Influence of processing parameters in injection molding on the properties of short carbon and glass fiber reinforced polypropylene composites. Polymers 16(19):2745. https://doi.org/10.3390/POLYM16192745
44. Kamarudin SH et al (2022) A review on natural fiber reinforced polymer composites (NFRPC) for sustainable industrial applications. Polymers 14(17):3698
45. Gallos A, Paës G, Allais F, Beaugrand J (2017) Lignocellulosic fibers: a critical review of the extrusion process for enhancement of the properties of natural fiber composites. RSC Adv 7(55):34638–34654. https://doi.org/10.1039/C7RA05240E
46. Han-Lyn F, Ismail-Fitry MR, Noranizan MA, Tan TB, Nur-Hanani ZA (2024) Recent advances in extruded polylactic acid-based composites for food packaging: a review. Int J Biol Macromol 266:131340. https://doi.org/10.1016/J.IJBIOMAC.2024.131340
47. Jaafar J, Siregar JP, Tezara C, Hamdan MHM, Rihayat T (2019) A review of important considerations in the compression molding process of short natural fiber composites. Int J Adv Manuf Technol 105(7–8):3437–3450. https://doi.org/10.1007/S00170-019-04466-8/FIGURES/16
48. Uranga J, Etxabide A, Guerrero P, de la Caba K (2018) Development of active fish gelatin films with anthocyanins by compression molding. Food Hydrocoll 84:313–320. https://doi.org/10.1016/j.foodhyd.2018.06.018
49. Santos ACMQS, Monticeli FM, Ornaghi H, Santos LFP, Cioffi MOH (2021) Porosity characterization and respective influence on short-beam strength of advanced composite processed by resin transfer molding and compression molding. Polym Polym Compos 29(8):1353–1362. https://doi.org/10.1177/0967391120968452
50. Rabbi MS, Islam T, Islam GMS (2021) Injection-molded natural fiber-reinforced polymer composites: a review. Int J Mech Mater Eng 16(1):1–21. https://doi.org/10.1186/S40712-021-00139-1
51. Plasencia L, Arrieta MP, Lazaro-Hdez C, Gomez-Caturla J, Quiles-Carrillo L (2025) Innovative development and characterization of thermoplastic zein-based biopolymers for food packaging via injection molding. Food Hydrocoll 163:111134. https://doi.org/10.1016/J.FOODHYD.2025.111134
52. Phiri R, Mavinkere-Rangappa S, Siengchin S, Oladijo OP, Dhakal HN (2023) Development of sustainable biopolymer-based composites for lightweight applications from agricultural waste biomass: a review. Adv Ind Eng Polym Res 6(4):436–450. https://doi.org/10.1016/J.AIEPR.2023.04.004
53. Sasi Kumar M, Sathish S, Makeshkumar M, Gokulkumar S, Naveenkumar A (2023) Effect of manufacturing techniques on mechanical properties of natural fibers reinforced composites for lightweight products—a review. In: Springer proceedings in materials, vol 32. Springer, Singapore, pp 99–117. https://doi.org/10.1007/978-981-99-5567-1_8
54. Al-Mahmud MZ, Rabbi SMF, Islam MD, Hossain N (2025) Synthesis and applications of natural fiber-reinforced epoxy composites: a comprehensive review. SPE Polym 6(1):e10161. https://doi.org/10.1002/PLS2.10161
55. YousefniaPasha H, Mohtasebi SS, Tabatabaeekoloor R, Taherimehr M, Javadi A, Soltani-Firouz M (2021) Preparation and characterization of the plasticized polylactic acid films produced by the solvent-casting method for food packaging applications. J Food Process Preserv 45(12):e16089
56. Othman SH, Kahar NS, Nordin N, Ahmad-Shapii R (2023) Properties and food packaging applications of solvent casting-made starch-based films incorporated with essential oils: a review. Starch Stärke 75(5–6):2200152
57. Aman-Mohammadi M, Hosseini SM, Yousefi M (2020) Application of electrospinning technique in development of intelligent food packaging: a short review of recent trends. Food Sci Nutr 8(9):4656–4665

58. Zhang Y, Min T, Zhao Y, Cheng C, Yin H, Yue J (2024) The developments and trends of electrospinning active food packaging: a review and bibliometrics analysis. Food Control 160:110291. https://doi.org/10.1016/J.FOODCONT.2024.110291
59. Chen P, Cao Y, He S, Tian X, Cao Z (2023) Improvement of mechanical and thermal properties of polylactic acid electrospun films by incorporating L-Lactide functionalized cellulose nanocrystals. Fuller Nanotub Carbon Nanostruct 31(5):395–403. https://doi.org/10.1080/1536383X.2023.2168267
60. González-Henríquez CM, Sarabia-Vallejos MA, Rodriguez-Hernandez J (2019) Polymers for additive manufacturing and 4D-printing: materials, methodologies, and biomedical applications. Prog Polym Sci 94:57–116. https://doi.org/10.1016/J.PROGPOLYMSCI.2019.03.001
61. Richardson JJ, Cui J, Björnmalm M, Braunger JA, Ejima H, Caruso F (2016) Innovation in layer-by-layer assembly. Chem Rev 116(23):14828–14867. https://doi.org/10.1021/ACS.CHEMREV.6B00627/ASSET/IMAGES/LARGE/CR-2016-00627G_0047.JPEG
62. Miladinov VD, Hanna MA (2000) Starch esterification by reactive extrusion. Ind Crops Prod 11(1):51–57. https://doi.org/10.1016/S0926-6690(99)00033-3
63. Nurazzi NM et al (2021) Thermogravimetric analysis properties of cellulosic natural fiber polymer composites: a review on influence of chemical treatments. Polymers 13:16. https://doi.org/10.3390/polym13162710
64. Kidane A, Gowtham HL, Naik NK (2017) Strain rate effects in polymer matrix composites under shear loading: a critical review. J Dyn Behav Mater 3(1):110–132. https://doi.org/10.1007/S40870-017-0098-2/TABLES/10
65. Zuhudi NZM, Zulkifli AF, Zulkifli M, Yahaya ANA, Nur NM, Aris KDM (2021) Void and moisture content of fiber reinforced composites. J Adv Res Fluid Mech Therm Sci 87(3):78–93
66. Kumar R, Ul-Haq MI, Raina A, Anand A (2019) Industrial applications of natural fibre-reinforced polymer composites–challenges and opportunities. Int J Sustain Eng 12(3):212–220
67. Mohammed L, Ansari MNM, Pua G, Jawaid M, Islam MS (2015) A review on natural fiber reinforced polymer composite and its applications. Int J Polym Sci 2015(1):243947. https://doi.org/10.1155/2015/243947
68. Khan T, Sultan MTH, Ariffin AH (2018) The challenges of natural fiber in manufacturing, material selection, and technology application: a review. J Reinf Plast Compos 37(11):770–779
69. Islam T et al (2024) Advancements and challenges in natural fiber-reinforced hybrid composites: a comprehensive review. SPE Polym 5(4):481–506. https://doi.org/10.1002/PLS2.10145
70. Rajendran-Royan NR, Leong JS, Chan WN, Tan JR, Shamsuddin ZSB (2021) Current state and challenges of natural fibre-reinforced polymer composites as feeder in FDM-based 3D printing. Polymers 13(14):2289. https://doi.org/10.3390/POLYM13142289
71. Ponnusamy PG, Mani S (2022) Material and environmental properties of natural polymers and their composites for packaging applications: a review. Polymers 14(19):4033. https://doi.org/10.3390/POLYM14194033
72. Arun-Kumar K, Sabarish P, Shabrish-Bala G, Sudarsen A (2017) Investigation of mechanical properties of natural fibers: a review. Int J Innov Sci Res Technol 2(10):147–156
73. Pulikkalparambil H, Varghese SA, Chonhenchob V, Nampitch T, Jarupan L, Harnkarnsujarit N (2023) Recent advances in natural fibre-based materials for food packaging applications. Polymers 15(6):1393. https://doi.org/10.3390/POLYM15061393
74. Dejene BK, Gudayu AD (2024) Eco-friendly packaging innovations: integrating natural fibers and ZnO nanofillers in polylactic acid (PLA) based green composites: a review. Polym Technol Mater 63(12):1645–1681. https://doi.org/10.1080/25740881.2024.2350686
75. Torres-Giner S (2017) Development of PHA/fiber-based composites with antimicrobial performance for active food packaging applications. Eur Coop Sci Technol 15:1134
76. Arman-Alim AA, Mohammad-Shirajuddin SS, Anuar FH (2022) A review of nonbiodegradable and biodegradable composites for food packaging application. J Chem 2022:819. https://doi.org/10.1155/2022/7670819
77. Thapliyal D et al (2023) Natural fibers composites: origin, importance, consumption pattern, and challenges. J Compos Sci 7(12):506. https://doi.org/10.3390/JCS7120506

78. Varghese SA, Pulikkalparambil H, Rangappa SM, Siengchin S, Parameswaranpillai J (2020) Novel biodegradable polymer films based on poly(3-hydroxybutyrate-co-3-hydroxyvalerate) and *C. pentandra* natural fibers for packaging applications. Food Packag Shelf Life 25:100538. https://doi.org/10.1016/J.FPSL.2020.100538
79. Cinelli P et al (2019) Biocomposites based on polyhydroxyalkanoates and natural fibres from renewable byproducts. Appl Food Biotechnol 6(1):35–43
80. Krishna G, Shivaji S (2018) Fabrication and testing of natural fibre reinforced composite (sun hemp and E-glass). www.ijcrt.orgwww.ijcrt.org. Accessed on 15 March 2025
81. Salwa HN, Sapuan SM, Mastura MT (2019) Green bio composites for food packaging. Int J Recent Technol Eng 8:2S4
82. Dey A, Neogi S (2019) Oxygen scavengers for food packaging applications: a review. Trends Food Sci Technol 90:26–34. https://doi.org/10.1016/J.TIFS.2019.05.013
83. Nirmal-Kumar K et al (2022) Mechanical and thermal properties of bamboo fiber-reinforced PLA polymer composites: a critical study. Genet Res 2022:2157. https://doi.org/10.1155/2022/1332157
84. Birniwa AH et al (2023) Recent trends in treatment and fabrication of plant-based fiber-reinforced epoxy composite: a review. J Compos Sci 7(3):120. https://doi.org/10.3390/JCS7030120
85. Mu H, Gao H, Chen H, Tao F, Fang X, Ge L (2013) A nanosised oxygen scavenger: preparation and antioxidant application to roasted sunflower seeds and walnuts. Food Chem 136(1):245–250. https://doi.org/10.1016/J.FOODCHEM.2012.07.121
86. Gupta P (2023) Role of oxygen absorbers in food as packaging material, their characterization and applications. J Food Sci Technol 61(2):242–252. https://doi.org/10.1007/S13197-023-05681-8/TABLES/1
87. Bernstad-Saraiva A et al (2016) Comparative lifecycle assessment of mango packaging made from a polyethylene/natural fiber-composite and from cardboard material. J Clean Prod 139:1168–1180. https://doi.org/10.1016/J.JCLEPRO.2016.08.135
88. Weligama-Thuppahige VT, Karim MA (2022) A comprehensive review on the properties and functionalities of biodegradable and semibiodegradable food packaging materials. Compr Rev Food Sci Food Saf 21(1):689–718. https://doi.org/10.1111/1541-4337.12873
89. Mohanty AK, Misra M, Drzal LT (2001) Surface modifications of natural fibers and performance of the resulting biocomposites: an overview. Compos Interf 8(5):313–343. https://doi.org/10.1163/156855401753255422
90. Kamarudin SH et al (2022) A review on antimicrobial packaging from biodegradable polymer composites. Polymers 14(1):174. https://doi.org/10.3390/POLYM14010174
91. Peças P, Carvalho H, Salman H, Leite M (2018) Natural fibre composites and their applications: a review. J Compos Sci 2(4):66. https://doi.org/10.3390/JCS2040066
92. Wu F, Misra M, Mohanty AK (2021) Challenges and new opportunities on barrier performance of biodegradable polymers for sustainable packaging. Prog Polym Sci 117:101395. https://doi.org/10.1016/J.PROGPOLYMSCI.2021.101395
93. Shahdan D, Rosli NA, Chen RS, Ahmad S, Gan S (2023) Strategies for strengthening toughened poly(lactic acid) blend via natural reinforcement with enhanced biodegradability: a review. Int J Biol Macromol 251:126214. https://doi.org/10.1016/J.IJBIOMAC.2023.126214
94. Arrieta MP, Peponi L, López D, López J, Kenny JM (2017) An overview of nanoparticles role in the improvement of barrier properties of bioplastics for food packaging applications. Food Packag 64:391–424. https://doi.org/10.1016/B978-0-12-804302-8.00012-1
95. Dodero A, Escher A, Bertucci S, Castellano M, Lova P (2021) Intelligent packaging for real-time monitoring of food-quality: current and future developments. Appl Sci 11(8):3532. https://doi.org/10.3390/APP11083532
96. Azman NH, Khairul WM, Sarbon NM (2022) A comprehensive review on biocompatible film sensor containing natural extract: Active/intelligent food packaging. Food Control 141:109189. https://doi.org/10.1016/J.FOODCONT.2022.109189
97. Karimah A et al (2021) A comprehensive review on natural fibers: technological and socio-economical aspects. Polymers 13(24):4280. https://doi.org/10.3390/POLYM13244280

Chapter 13
Lignocellulosic Biomass as Feedstock for Paper and Packaging Applications

Jovale Vincent Tongco

1 Introduction

The utilization of lignocellulosic biomass originating from forestry and agricultural waste as a potential alternative feedstock for paper manufacturing and packaging production represents an opportunity to address the environmental challenges that the world has been facing in recent years. By repurposing biomass residue from routine forestry operations and agricultural produce, technological innovators and manufacturers can reduce the dependence of making pulp for paper and packaging products on virgin wood resources while diverting the biomass waste from being burned or decomposed, processes that are detrimental to the environment. The concept of circular economy was introduced as an alternative economic model in terms of product manufacturing that encourages sustainable growth and development [1]. This model attracts enormous political, academic, social, and commercial interest. As such, it is imperative to develop technologies that are deemed necessary to address the growing issues of raw material use and proper waste management and how they are connected. The main goal of this model is the accomplishment of synergistic relationships between growing societies and economies and their effect on the environment [2]. With an increasing material footprint and industrial production globally, manufacturers and producers are struggling to adapt to dynamic and ever-changing policies that directly affect their operations and reduce their dependence on the use of cheaper and readily available but non-renewable materials. The use of lignocellulosic biomass, defined as plant dry matter, consisting of lignin, cellulose, and hemicellulose, is the most abundant natural raw material [3], which can be used in different applications, predominantly in sustainable energy and materials

J. V. Tongco (✉)
Department of Forest, Rangeland and Fire Sciences, University of Idaho, Moscow, ID, USA
e-mail: jtongco@uidaho.edu

A. Verma et al. (eds.), *Polymers and Composite Materials for Packaging: Smart Food Packaging and Solutions*, Materials Horizons: From Nature to Nanomaterials,
https://doi.org/10.1007/978-981-95-0677-4_13

production, such as biofuels, bio-oil, biochar, biocomposites, and bioplastics [4–7]. Biodegradability of the resulting products for paper and smart packaging use is another advantage of utilizing lignocellulosic biomass [8–10]. According to recent statistical data, the annual production of lignocellulosic biomass in nature is approximately 181 billion tons, with only around 4.5% currently utilized and around 3% originating from forests, agriculture, and grasses, which is around seven billion tons and the remaining 1.2 billion tons coming from agricultural residues [11, 12]. These values are substantial in terms of waste management, mainly if the biomass residues are just incinerated or subjected to open burning, producing fine particulates that affect human health and the environment [13, 14]. This chapter explores the sources of lignocellulosic biomass and the processes, advantages, and challenges involved in converting the biomass residual waste streams into value-added products, such as paper and packaging materials.

2 Valorization of Lignocellulosic Biomass

Lignocellulosic biomass represents the most abundant source of carbon-based material in nature. Unlike conventional plastics sourced from fossil fuels, lignocellulosic biomass materials are sustainable, renewable, and widely available from various plant sources. The transition toward bio-based packaging materials aligns with key principles of bioeconomy, advocating for the use of biodegradable materials. Bio-based packaging materials are mostly from natural fibers obtained from lignocellulosic biomass, including wood from forests and the timber industry, agricultural crops focusing on fiber production (flax, hemp, sisal), and agricultural by-products of crops that have other primary uses (rice, corn, sugar, wheat). The properties and characteristics of the lignocellulosic biomass are different depending on the processing technologies employed, which affect the morphology and physicochemical attributes of the material. Therefore, after processing and sometimes surface modifications, the applications in packaging are also different. Besides the aforementioned cellulose, hemicellulose, and lignin, lignocellulosic biomass from plant sources is also made up of a variety of minor constituents, including pectins, suberins, proteins, fatty acids, lipids, waxes, and extractives (resins, terpenes, terpenoids, tannins, steroids, flavonoids, and other phenolics), especially if the biomass material includes bark, roots, stems, or dried leaves [4, 15–23]. Lignocellulosic biomass is a renewable and carbon–neutral resource that can be used to produce value-added products, such as construction materials, packaging materials, biofuels (bioethanol, bio-oil, biogas, briquettes), platform and specialty chemicals, bioplastics, biocomposites, and nanomaterials [4, 15]. Figure 1 shows the conversion of lignocellulosic biomass from agricultural waste into value-added products with emphasis on circular economy and global warming mitigation. The two main components of lignocellulose (cellulose and lignin) make up the majority of the valorized finish products due to their chemical and mechanical properties. These two important components change their composition according to the source of biomass, geographical origin, and even climate and

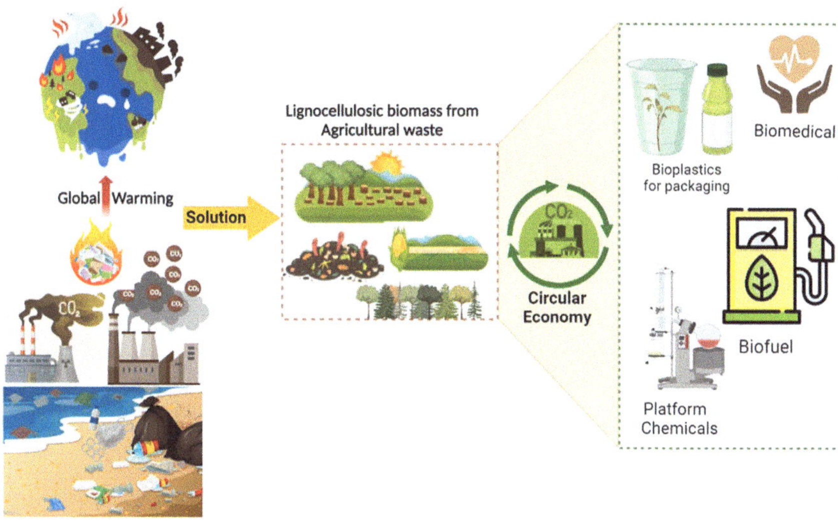

Fig. 1 Conversion of lignocellulosic biomass from agricultural waste into value-added products with emphasis on circular economy and global warming mitigation (reprinted with permission from reference [4])

weather conditions [24]. In terms of biofuels production, the carbohydrate moiety (cellulose and hemicelluloses) is the most important constituent because of its ability to be a feedstock in bioconversion processes to produce biogas and bioethanol through anaerobic digestion and fermentation, respectively [25–29]. In some lignocellulosic biomass, consisting of considerable amounts of proteins and lipids, such as agricultural waste and primary sludge from the effluent of pulp and paper industries, enzymatic treatment methods are employed to hydrolyze recalcitrant components and isolate the energy-rich carbohydrates prior to bioconversion [30–32].

Traditionally viewed as a low-value by-product of the pulp and paper industry, lignin is gaining a lot of attention recently as a source of valuable chemicals and as feedstock for a variety of applications. The complex structure of lignin, composed of phenylpropane units, makes its utilization challenging. The use of lignin for different applications depends on whether it is used in its polymeric form, its degraded and fragmented form, or broken down into its constituent monomers. In its polymeric form, lignin imparts mechanical properties to various materials and could be found as adhesives, binders, fillers, reinforcements, dispersants, surfactants, and asphalt additives [9, 33–36]. Lignin can be depolymerized or degraded into fragments or smaller aromatic molecules through chemical (acid/base catalysis, oxidative depolymerization, hydrogenation, deep eutectic solvents, ionic liquids), thermal (pyrolysis, gasification, liquefaction, gasification), and biological (enzymatic) methods [37–41]. Figure 2 shows the summary of processes for the conversion of lignin. Breaking down lignin into smaller aromatic molecules produces a wide range of specialty and industrial chemicals, such as aromatic chemicals (vanillin, syringaldehyde, catechols,

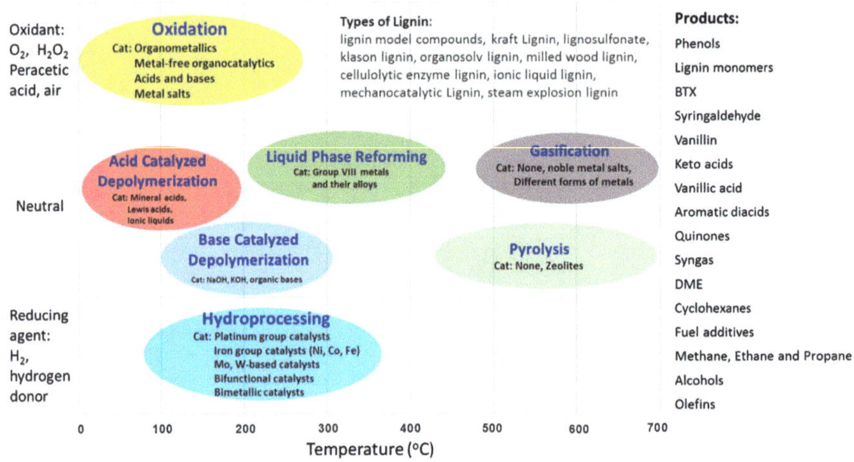

Fig. 2 Processes for lignin conversion and their corresponding temperature range requirements (reprinted with permission from reference [39])

phenols) and biofuels (ethanol, cycloalkanes, methane) [42–44]. These chemicals, especially aromatic chemicals, are important in manufacturing paper and packaging materials. They can be used as sizing agents (improving printability, making paper more water resistant) and as dyes and colorants [45–47].

3 Forestry and Agricultural Waste as Alternative Feedstock

Due to their lignocellulosic nature, forestry and agricultural operations generate substantial amounts of residues and waste materials suitable for paper production. Lignocellulosic waste can be further categorized into forestry waste (tree remains, shrubs, logging residue, sawdust), forest woods (spruce, oak, poplar), energy crops (jatropha, sorghum), agricultural waste (rice husk, wheat straw, sugarcane bagasse, coconut husks and shells), and algae [48]. Forest lignocellulosic biomass consists of primary biomass (forest) residues, secondary biomass (wood processing industries) residues, fresh biomass, and post-consumer wood products. The forest products industry, particularly the ones involved with processing softwood conifers, is the highest producer of lignocellulosic biomass [49]. Energy grasses can be cultivated for their lignocellulosic biomass. Good examples are switch grass, sorghum, and coppice [50]. Agricultural waste and by-products of related industries are also reliable and sustainable sources of lignocellulosic biomass. Previous work on the use of agricultural lignocellulosic biomass, such as rice straw, sugarcane bagasse, wheat bran, and coconut fibers and shells, has been utilized in producing enzymes (cellulase, protease, amylase, lipase, and pectinase) that are important in different

Table 1 Composition of various forestry and agricultural lignocellulosic biomass

Biomass	Cellulose, %	Hemicellulose, %	Lignin, %	References
Corn cob	42–45	35–39	14–15	[54]
Corn stover	37	31	13–26	[55]
Coconut shell	54–65	4–5	30–42	[56]
Coconut husk	38.86	30	34.96	[57]
Rice husk	37.1	29.4	24.1	[58]
Rice straw	28–47	19–28	5–24	[54]
Wheat straw	29–45	20–32	20–45	[54]
Sugarcane bagasse	42–45	35–39	14–15	[59]
Sorghum bagasse	45	27	21	[60]
Switchgrass	31	24	18	[61]
Bermuda grass	30	29	23	[61]
Bamboo	35–50	15–30	20–30	[62]
Poplar	44.5	22.5	19.5	[63]
Pine	42	21	30	[64]

industries, and are available in literature [51–53]. Table 1 shows the lignocellulosic compositions of important forest and agricultural plant sources.

4 Processing Forestry Lignocellulosic Waste

Converting forestry waste into paper and packaging materials represents a significant role in sustainable production by using underutilized biomass to reduce reliance on felling trees for wood. This process involves multiple stages, starting from residue and byproduct collection, pulping, drying, and specific usage. Forestry waste is made up of materials generated during logging, sawmill operations, and forest management activities. Logging residues include branches, stems, and tree tops left after timber harvest. By-products include sawdust and wood chips. Bark is a good example of another forestry waste, but is seldom used for paper and packaging applications due to its high extractive content [65]. Lastly, there are considerable amounts of residual materials removed during thinning operations or wildfire mitigation that could potentially be used in producing packaging materials [66]. In Idaho, over 90% of paper production utilizes these waste materials, which would otherwise be incinerated or decomposed [67]. Modern forestry operations employ integrated biomass recovery systems for streamlining residue collection. Balers or chippers aggregate forestry residues into compact and easily transportable biomass units, reducing handling costs. Piling the residues during harvesting operations has increased collection productivity by 20–30% [66]. Researchers from North Carolina State University worked on the design and engineering of the next generation of sustainable multilayer packaging from wood materials. Their approach involved the creation of a database of structures and properties of renewable resources from published literature, devised nanoscale modeling framework for the prediction of material performance for sustainable packaging development, testing and validation of the prediction models through experimental research, and refining the models through an iterative process to develop them further for use in various packaging applications [68].

5 Processing Agricultural Lignocellulosic Waste

Agricultural biomass can also be called a non-wood raw material and is an essential source of fiber and feedstock in regions where forest sources are limited. The current use of non-wood plant biomass such as rice straw, corn stalks, cotton stalks, and sugarcane bagasse plays a key role in the increase in feedstock supply and quality in making paper and packaging materials. One significant advantage of using agricultural waste is its lower lignin content compared to wood, making it easier to convert into pulp and subsequent bleaching [69]. The agricultural biomass waste is first collected, cleaned, and pre-treated prior to pulping and bleaching. This process involves the removal of contaminants, cutting or sizing the material to produce uniformly sized units, and pretreatment using chemical, mechanical (physical), or biological methods to facilitate fiber separation [70, 71]. The purpose of treating agricultural biomass is to produce a wide variety of products with specialized applications. It allows the isolation of lignocellulosic components through chemical means, including the disruption of chemical bonds, such as hydrolysis and dissolution (solvents). Physical treatment involves defibrillation and surface area improvements, such as grinding, sieving, and steam explosion. Thermal treatment transforms the biomass into other forms of materials with unique properties, such as pyrolysis, liquefaction, torrefaction, and carbonization. Lastly, biological treatments involve the use of microbes and enzymes, and are mostly applied in bioenergy industries, such as enzymatic treatment, digestion, and fermentation. Figure 3 shows the processes involved in isolating the different components of lignocellulose, i.e., cellulose, hemicellulose, and lignin.

6 Paper and Packaging Materials from Lignocellulose

Cellulose, the major component of paper and packaging materials, can be extracted from lignocellulosic fibers. The process starts when lignocellulosic biomass is converted into smaller units, which are then fed into pulp digesters where they are broken down by steam and pulping/bleaching chemicals (i.e., sodium hydroxide, chlorine dioxide, sodium bisulfite, sulfuric acid, and hydrogen peroxide) into a mixture of cellulosic fibers and other lignocellulosic components. These chemicals, extractives, resins, and lignin are removed through various processes prior to the

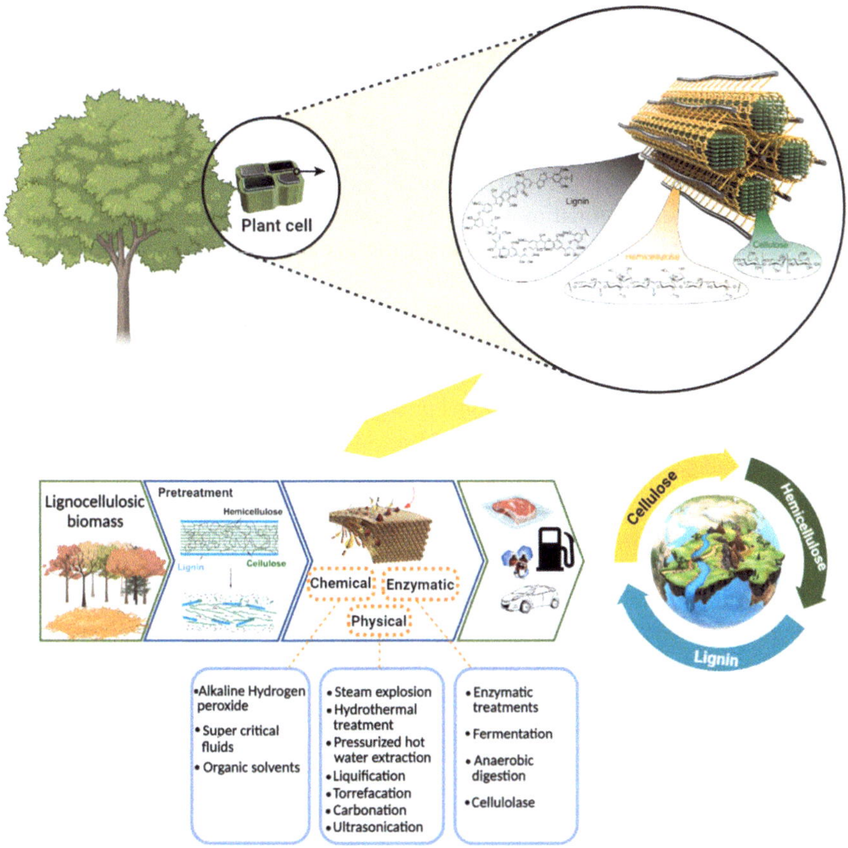

Fig. 3 Processes involved in isolating the different components of lignocellulose (reprinted with permission from reference [4])

isolation of cellulose. Pulping is the most crucial step in the production of paper and packaging materials, and the most common pulping method is called the kraft pulping process. This process is also known as the sulfate process and involves the use of hot alkaline solution (white liquor) to separate the cellulose fibers from wood, dissolving lignin and hemicellulose along the way. Approximately 75% of pulp produced is through the kraft pulping process [8]. Wood chips are placed in large vessels that are pressurized, called digesters. Then, the digester is filled with the hot alkaline solution consisting of sodium hydroxide (NaOH) and sodium sulfide (Na_2S). The temperature inside the digester is then increased to around 150–180 °C, and the pressure to around 800 kPa [72]. During this step of the process, the hydroxide and hydrosulfide anions react with lignin, causing fragmentation and cellulose fiber isolation. However, paper products produced using this process can sometimes be mechanically weak and prone to breakage, in addition to the amount of pulp produced, which is minimal due to pulping conditions being harsh. Recent studies have found that utilizing cellulosic nanofibers (CNF) and cross-linking agents and incorporating them into the pulp make much stronger paper products, suitable for packaging applications [73]. The mechanical strength of the paper and packaging materials made from cellulose depends on the fiber arrangements in the pulp (fiber length and orientation), density, covalent, and intramolecular bonds between the cellulose fibers, the nature of the fibers based on the plant source, and if the fibers themselves are tightly packed and contain minimal moisture. The paper pulp is then processed in papermaking (Fourdrinier) machines, where it is sprayed as a thin layer onto a moving wire mesh or screen that can travel at speeds of up to 60 mph (96.5 kph). As the pulp moves along the screen, it is dehydrated, allowing the cellulosic fibers to bond to each other, forming paper. The produced paper is subsequently fed through heated rollers that apply pressure and heat to compress and dry it before spooling it into rolls for further processing. The conventional process for paper production has significant environmental implications, such as increased energy and water consumption. Also, with paper and paperboard products making up approximately 13–15% of total global consumption and future demands forecasted to double by 2050, there are increasing concerns regarding the contribution of the pulp and paper industry to deforestation and ecosystem degradation [67, 74]. Figure 4 shows the summarized diagram for cellulose isolation and subsequent conversion into packaging materials. Cellulose can be further processed to produce nanocellulosic fibers with improved properties, such as chemical stability, biocompatibility, biodegradability, non-toxicity, and low cost, suitable for packaging applications [75].

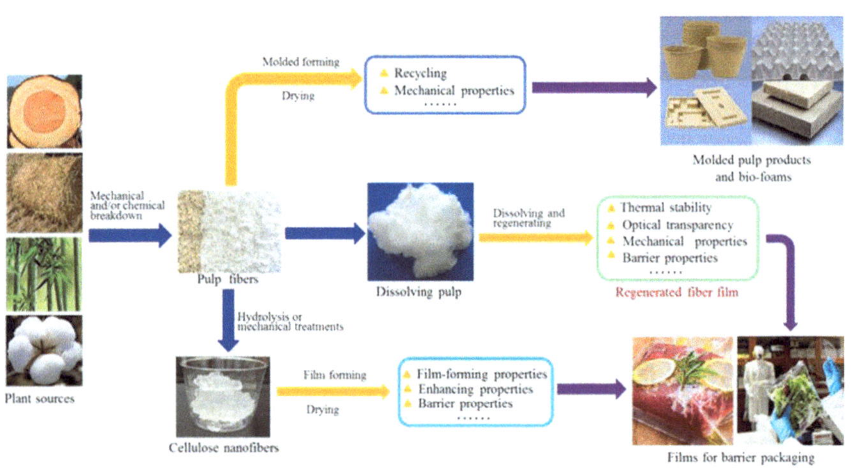

Fig. 4 Conversion of lignocellulosic biomass from plant sources to cellulose, a precursor to packaging materials (reprinted from reference [76])

7 Summary and Conclusion

The versatility of lignocellulosic biomass as feedstock for the production and development of diverse types of paper products for packaging has been the focus of research for the past few years, contributing to the growing list of renewable and sustainable alternatives to petroleum-based materials. The use of lignocellulosic biomass, being renewable and biodegradable, helps in minimizing the environmental effects of plastic manufacturing, contributing to a more green and circular economy. Cellulose is the most vital component of lignocellulose in terms of pulp, paper, and packaging production. It is abundant and renewable, as it is readily available after timber production or agricultural harvest. It is also eco-friendly since it reduces the reliance on wood-based pulp. The use of cellulose is also economically viable since it can open new value chains for the timber industry and crop farmers. However, extraction of lignin and other components also leads to other value-added products, such as fillers for composites, platform and specialty chemicals, and substrate for energy generation. While challenges remain in processing, moisture resistance, and economic issues, ongoing research and innovations continue to address the limitations imposed by utilizing lignocellulosic biomass, ranging from more efficient biomass collection, pretreatment with bio-based and environmentally friendly solvents and chemicals, and surface modifications. The future of the pulp and paper industry, including packaging materials production and development, appears to be increasingly reliant on lignocellulosic biomass due to increased consumer demand and collective consciousness toward the use of sustainable materials. The successful implementation of using lignocellulosic biomass as feedstock for paper and packaging applications significantly reduces the environmental impacts associated with the use of non-renewable and non-recyclable materials while supporting the transition toward a more sustainable and bio-based economy.

Conflicts of Interest There are no conflicts of interest to declare by the authors.

References

1. Gregorio VF, Pié L, Terceño A (2018) A systematic literature review of bio, green and circular economy trends in publications in the field of economics and business management. Sustainability 10:4232. https://doi.org/10.3390/SU10114232
2. Crespo N, Acerbi F, Sassanelli C, Taisch M (2022) Bioeconomy and circular economy approaches need to enhance the focus on biodiversity to achieve sustainability. Sustainability 14:10643. https://doi.org/10.3390/SU141710643
3. Tongco JV (2024) Physicochemical properties of delignified hemp fiber cellulose acetate (HFCA). Arab J Chem Environ Res 11:71–83
4. Mujtaba M, Fernandes-Fraceto L, Fazeli M, Mukherjee S, Savassa SM, Medeiros GA, Pereira AES, Mancini SD, Lipponen J, Vilaplana F (2023) Lignocellulosic biomass from agricultural waste to the circular economy: a review with focus on biofuels, biocomposites and bioplastics. J Clean Prod 402:136815. https://doi.org/10.1016/J.JCLEPRO.2023.136815

5. Wang F, Ouyang D, Zhou Z, Page SJ, Liu D, Zhao X (2021) Lignocellulosic biomass as sustainable feedstock and materials for power generation and energy storage. J Energy Chem 57:247–280. https://doi.org/10.1016/J.JECHEM.2020.08.060
6. Lawoko M, Berglund L, Johansson M (2021) Lignin as a renewable substrate for polymers: from molecular understanding and isolation to targeted applications. ACS Sustain Chem Eng 9:5481–5485. https://doi.org/10.1021/ACSSUSCHEMENG.1C01741
7. Sharma A, Kaur P, Singh G, Arya SK (2021) Economical concerns of lignin in the energy sector. Clean Eng Technol 4:100258. https://doi.org/10.1016/J.CLET.2021.100258
8. Tao S, Zhang C, Chen Y, Qin S, Qi H (2022) High strength holocellulose paper from bamboo as biodegradable packaging tape. Carbohydr Polym 283:119151. https://doi.org/10.1016/J.CARBPOL.2022.119151
9. Mili M, Hashmi SAR, Ather M, Hada V, Markandeya N, Kamble S, Mohapatra M, Rathore SKS, Srivastava AK, Verma S (2022) Novel lignin as natural-biodegradable binder for various sectors: a review. J Appl Polym Sci 139:51951. https://doi.org/10.1002/APP.51951
10. Nath PC, Sharma R, Debnath S, Sharma M, Inbaraj BS, Dikkala PK, Nayak PK, Sridhar K (2023) Recent trends in polysaccharide-based biodegradable polymers for smart food packaging industry. Int J Biol Macromol 253:127524. https://doi.org/10.1016/J.IJBIOMAC.2023.127524
11. Singh B, Korstad J, Guldhe A, Kothari R (2022) Editorial: emerging feedstocks and clean technologies for lignocellulosic biofuel. Front Energy Res 10:917081. https://doi.org/10.3389/FENRG.2022.917081/BIBTEX
12. Ashokkumar V, Venkatkarthick R, Jayashree S, Chuetor S, Dharmaraj S, Kumar G, Chen WH, Ngamcharussrivichai C (2022) Recent advances in lignocellulosic biomass for biofuels and value-added bioproducts: a critical review. Bioresour Technol 344:126195. https://doi.org/10.1016/J.BIORTECH.2021.126195
13. Suriyawong P, Chuetor S, Samae H, Piriyakarnsakul S, Amin M, Furuuchi M, Hata M, Inerb M, Phairuang W (2023) Airborne particulate matter from biomass burning in Thailand: recent issues, challenges, and options. Heliyon 9:e14261. https://doi.org/10.1016/J.HELIYON.2023.E14261
14. Weichenthal S, Kulka R, Lavigne E, Van Rijswijk D, Brauer M, Villeneuve PJ, Stieb D, Joseph L, Burnettf RT (2017) Biomass burning as a source of ambient fine particulate air pollution and acute myocardial infarction. Epidemiology 28:329. https://doi.org/10.1097/EDE.0000000000000636
15. Haldar D, Purkait MK (2021) A review on the environment-friendly emerging techniques for pretreatment of lignocellulosic biomass: mechanistic insight and advancements. Chemosphere 264:128523. https://doi.org/10.1016/J.CHEMOSPHERE.2020.128523
16. Tongco JV, Ewurum N, Rubiyatno S (2025) Suberin: chemistry and applications. J Mater Environ Sci 16:1–14
17. Tongco JV, Añis A, Tamayo J (2015) Nutritional analysis, phytochemical screening, and total phenolic content of *Basella alba* leaves from the Philippines. Int J Pharm Phytochem Res 7:1031–1033
18. Tongco JV, Aguda R, Razal R (2014) Proximate analysis, phytochemical screening, and total phenolic and flavonoid content of Philippine bamboo Schizostachyum lumampao. J Chem Pharm Res 6:709–713
19. Ragasa C, Torres O, Tongco JV, Razal R (2014) Chemical constituents of *Petersianthus quadrialatus* (Merr.). Res J Pharm Biol Chem Sci 5:181–186
20. Tongco JV, Rodriguez E, Abasolo W, Mun SP, Razal R (2016) Mineral, nutritional, and phytochemical profile, total phenolic content, and radical scavenging activity of philippine bamboo "bolo" *Gigantochloa levis* (Blanco) Merr. leaves. Nat Product Sci 22:60–63
21. Ragasa C, Torres O, Tongco JV, Razal R, Shen C-C (2013) Resorcinols from *Myristica philippensis* Lam. J Chem Pharm Res 5:614–616
22. Panti A, Aguda R, Razal R, Belina-aldemita MD, Tongco JV (2014) Proximate analysis, phytochemical screening and total phenolic and flavonoid content of the ethanolic extract of molave *Vitex parviflora* Juss. leaves. J Chem Pharm Res 6:1538–1542

23. Tongco JV, Villaber RA, Aguda R, Razal R (2014) Nutritional and phytochemical screening, and total phenolic and flavonoid content of *Diplazium esculentum* (Retz.) Sw. from Philippines. J Chem Pharm Res 6:238–242
24. Wu D, Wei Z, Mohamed TA, Zheng G, Qu F, Wang F, Zhao Y, Song C (2022) Lignocellulose biomass bioconversion during composting: mechanism of action of lignocellulase, pretreatment methods and future perspectives. Chemosphere 286:131635. https://doi.org/10.1016/J.CHEMOSPHERE.2021.131635
25. Lee J, Kim E, Han G, Tongco JV, Shin SG, Hwang S (2018) Microbial communities underpinning mesophilic anaerobic digesters treating food wastewater or sewage sludge: a full-scale study. Bioresour Technol 259:388–397. https://doi.org/10.1016/j.biortech.2018.03.052
26. Kim E, Shin SG, Jannat MAH, Tongco JV, Hwang S (2017) Use of food waste-recycling wastewater as an alternative carbon source for denitrification process: a full-scale study. Bioresour Technol 245:1016–1021. https://doi.org/10.1016/j.biortech.2017.08.168
27. Tamilselvan R, Immanuel Selwynraj A (2024) Enhancing biogas generation from lignocellulosic biomass through biological pretreatment: exploring the role of ruminant microbes and anaerobic fungi. Anaerobe 85:102815. https://doi.org/10.1016/J.ANAEROBE.2023.102815
28. Vasić K, Knez Ž, Leitgeb M (2021) Bioethanol production by enzymatic hydrolysis from different lignocellulosic sources. Molecules 26:753. https://doi.org/10.3390/MOLECULES26030753
29. Editors A, Kumar Patel A, Magalhães Pires JC, Beluhan S, Mihajlovski K, Šantek B, Ivanči M (2023) The production of bioethanol from lignocellulosic biomass: pretreatment methods, fermentation, and downstream processing. Energies 16:7003. https://doi.org/10.3390/EN16197003
30. Srivastava N, Shrivastav A, Singh R, Abohashrh M, Srivastava KR, Irfan S, Srivastava M, Mishra PK, Gupta VK, Thakur VK (2021) Advances in the structural composition of biomass: fundamental and bioenergy applications. J Renew Mater 9:615–636
31. Eraky M, Elsayed M, Qyyum MA, Ai P, Tawfik A (2022) A new cutting-edge review on the bioremediation of anaerobic digestate for environmental applications and cleaner bioenergy. Environ Res 213:113708. https://doi.org/10.1016/J.ENVRES.2022.113708
32. Tongco JV, Kim S, Oh BR, Heo SY, Lee J, Hwang S (2020) Enhancement of hydrolysis and biogas production of primary sludge by use of mixtures of protease and lipase. Biotechnol Bioprocess Eng 25:132–140. https://doi.org/10.1007/s12257-019-0302-4
33. Wu Z, Lin X, Teng J, Li M, Song J, Huang C, Wang R, Ying H, Zhang L, Zhu C (2023) Recent advances of lignin functionalization for high-performance and advanced functional rubber composites. Biomacromol 24:4553–4567. https://doi.org/10.1021/ACS.BIOMAC.3C00606/ASSET/IMAGES/LARGE/BM3C00606_0011.JPEG
34. Shorey R, Salaghi A, Fatehi P, Mekonnen TH (2024) Valorization of lignin for advanced material applications: a review. RSC Sustain 2:804–831. https://doi.org/10.1039/D3SU00401E
35. Lora JH (2016) Lignin: a platform for renewable aromatic polymeric materials. In: Lau PCK (ed) Quality living through chemurgy and green chemistry. Springer, Berlin, pp 221–261
36. Yao H, Wang Y, Liu J, Xu M, Ma P, Ji J, You Z (2022) Review on applications of lignin in pavement engineering: a recent survey. Front Mater 8:803524. https://doi.org/10.3389/FMATS.2021.803524/BIBTEX
37. Zhang C, Wang F (2020) Catalytic lignin depolymerization to aromatic chemicals. Acc Chem Res 53:470–484. https://doi.org/10.1021/ACS.ACCOUNTS.9B00573/ASSET/IMAGES/LARGE/AR9B00573_0018.JPEG
38. Li P, Ren J, Jiang Z, Huang L, Wu C, Wu W (2022) Review on the preparation of fuels and chemicals based on lignin. RSC Adv 12:10289–10305. https://doi.org/10.1039/D2RA01341J
39. Li C, Zhao X, Wang A, Huber GW, Zhang T (2015) Catalytic transformation of lignin for the production of chemicals and fuels. Chem Rev 115:11559–11624. https://doi.org/10.1021/ACS.CHEMREV.5B00155/ASSET/IMAGES/MEDIUM/CR-2015-00155G_0039.GIF
40. Ji Q, Yu X, Chen L, Mustapha AT, Okonkwo CE, Zhou C, Liu X (2023) Comprehensive depolymerization of lignin from lignocellulosic biomass: a review. Crit Rev Environ Sci Technol 53:1866–1887. https://doi.org/10.1080/10643389.2023.2190314

41. Lu X, Guo H, Wang D, Xiu P, Qin Y, Chen J, Xu C, Gu X (2021) A review on catalytic conversion of lignin into high-value chemicals over Ni-based catalysts. Biomass Convers Biorefin 13:11339–11381. https://doi.org/10.1007/S13399-021-01903-1
42. Kocaturk E, Salan T, Ozcelik O, Alma MH, Candan Z (2023) Recent advances in lignin-based biofuel production. Energies 16:3382. https://doi.org/10.3390/EN16083382
43. Patel R, Dhar P, Babaei-Ghazvini A, Nikkhah Dafchahi M, Acharya B (2023) Transforming lignin into renewable fuels, chemicals, and materials: a review. Bioresour Technol Rep 22:101463. https://doi.org/10.1016/J.BITEB.2023.101463
44. Lu X, Gu X (2022) A review on lignin pyrolysis: pyrolytic behavior, mechanism, and relevant upgrading for improving process efficiency. Biotechnol Biofuels Bioprod 15:1–43. https://doi.org/10.1186/S13068-022-02203-0
45. Wicochea-Rodríguez JD, Chalier P, Ruiz T, Gastaldi E (2019) Active food packaging based on biopolymers and aroma compounds: how to design and control the release. Front Chem 7:452563. https://doi.org/10.3389/FCHEM.2019.00398/BIBTEX
46. Deshwal GK, Panjagari NR, Alam T (2019) An overview of paper and paper based food packaging materials: health safety and environmental concerns. J Food Sci Technol 56:4391. https://doi.org/10.1007/S13197-019-03950-Z
47. Sharma S, Barkauskaite S, Jaiswal AK, Jaiswal S (2021) Essential oils as additives in active food packaging. Food Chem 343:128403. https://doi.org/10.1016/J.FOODCHEM.2020.128403
48. Velvizhi G, Goswami C, Shetti NP, Ahmad E, Kishore Pant K, Aminabhavi TM (2022) Valorisation of lignocellulosic biomass to value-added products: paving the pathway towards low-carbon footprint. Fuel 313:122678. https://doi.org/10.1016/J.FUEL.2021.122678
49. Lindner M, Dees MG, Anttila P, Verkerk PJ, Fitzgerald J, Datta P, Glavonjic B, Prinz R, Zudin S (2017) Assessing lignocellulosic biomass potentials from forests and industry. In: Panoutsou C (ed) Modeling and optimization of biomass supply chains: top-down and bottom-up assessment for agricultural, forest and waste feedstock. Academic Press, New York, pp 127–167
50. Marriott PE, Gómez LD, Mcqueen-Mason SJ (2016) Unlocking the potential of lignocellulosic biomass through plant science. New Phytol 209:1366–1381. https://doi.org/10.1111/NPH.13684
51. Sooch BS, Lugani Y, Singh RS (2019) Agro-industrial lignocellulosic residues for the production of industrial enzymes. In: Yadav M, Kumar V, Sehrawat N (eds) Industrial biotechnology: plant systems, resources and products. De Gruyter, pp 31–50
52. Tripathi M, Srivastava N, Tripathi SC, Singh R, Ahmad I, Rai AK, Abdel-razik NE, Mishra PK, Kumar Gupta V (2024) Co-fermentation of acid treated coconut wastes using mixed Bacillus cultures for enhanced production of extracellular enzymes: application in bioconversion of raw coconut fibers. Food Bioprod Process 146:177–184. https://doi.org/10.1016/J.FBP.2024.05.011
53. Wang Q, Sarkar J (2018) Pyrolysis behaviors of waste coconut shell and husk biomasses. Int J Energy Prod Manag 3:34–43. https://doi.org/10.2495/EQ-V3-N1-34-43
54. Shahzadi T, Mehmood S, Irshad M, Anwar Z, Afroz A, Zeeshan N, Rashid U, Sughra K, Shahzadi T, Mehmood S, Irshad M, Anwar Z, Afroz A, Zeeshan N, Rashid U, Sughra K (2014) Advances in lignocellulosic biotechnology: a brief review on lignocellulosic biomass and cellulases. Adv Biosci Biotechnol 5:246–251. https://doi.org/10.4236/ABB.2014.53031
55. Pordesimo LO, Hames BR, Sokhansanj S, Edens WC (2005) Variation in corn stover composition and energy content with crop maturity. Biomass Bioenergy 28:366–374. https://doi.org/10.1016/J.BIOMBIOE.2004.09.003
56. Wang Q, Sarkar J (2018) Pyrolysis behaviors of waste coconut shell and husk biomasses. Int J Energy Product Manag 3:34–43. https://doi.org/10.2495/EQ-V3-N1-34-43
57. Fatmawati A, Nurtono T, Widjaja A (2023) Thermogravimetric kinetic-based computation of raw and pretreated coconut husk powder lignocellulosic composition. Bioresour Technol Rep 22:101500. https://doi.org/10.1016/J.BITEB.2023.101500
58. Kalita E, Nath BK, Deb P, Agan F, Islam MR, Saikia K (2015) High quality fluorescent cellulose nanofibers from endemic rice husk: isolation and characterization. Carbohydr Polym 122:308–313. https://doi.org/10.1016/J.CARBPOL.2014.12.075

59. Rajendran K, Drielak E, Sudarshan Varma V, Muthusamy S, Kumar G (2018) Updates on the pretreatment of lignocellulosic feedstocks for bioenergy production: a review. Biomass Convers Biorefin 8:471–483. https://doi.org/10.1007/S13399-017-0269-3/TABLES/5
60. Kim M, Day DF (2011) Composition of sugar cane, energy cane, and sweet sorghum suitable for ethanol production at Louisiana sugar mills. J Ind Microbiol Biotechnol 38:803–807. https://doi.org/10.1007/S10295-010-0812-8
61. Lee JM, Shi J, Venditti RA, Jameel H (2009) Autohydrolysis pretreatment of Coastal Bermuda grass for increased enzyme hydrolysis. Bioresour Technol 100:6434–6441. https://doi.org/10.1016/J.BIORTECH.2008.12.068
62. Rusch F, Wastowski AD, de Lira TS, Moreira KCCSR, de Moraes LD (2023) Description of the component properties of species of bamboo: a review. Biomass Convers Biorefin 13:2487–2495. https://doi.org/10.1007/S13399-021-01359-3/TABLES/3
63. Guerra A, Filpponen I, Lucia LA, Argyropoulos DS (2006) Comparative evaluation of three lignin isolation protocols for various wood species. J Agric Food Chem 54:9696–9705. https://doi.org/10.1021/JF062433C/ASSET/IMAGES/LARGE/JF062433CF00005.JPEG
64. Sannigrahi P, Miller SJ, Ragauskas AJ (2010) Effects of organosolv pretreatment and enzymatic hydrolysis on cellulose structure and crystallinity in Loblolly pine. Carbohydr Res 345:965–970. https://doi.org/10.1016/J.CARRES.2010.02.010
65. Fradinho DM, Neto CP, Evtuguin D, Jorge FC, Irle MA, Gil MH, Pedrosa de Jesus J (2002) Chemical characterisation of bark and of alkaline bark extracts from maritime pine grown in Portugal. Ind Crops Prod 16:23–32. https://doi.org/10.1016/S0926-6690(02)00004-3
66. Ghaffariyan MR (2023) Short review of collecting technologies and methods in forest harvesting residues recovery. Silva Balcanica 24:55–68. https://doi.org/10.3897/silvabalcanica.22.e97620
67. Idaho Forests Products Commission (2025) How paper is made. https://www.idahoforests.org/content-item/how-paper-is-made-2/. Accessed 26 March 2025
68. Lavoine N, Pasquinelli M (2025) Design and engineering of the next generation of sustainable multilayer packaging from wood: an integrated experimental-computational research project. https://cnr.ncsu.edu/research/mcintire-stennis-program/packaging-wood/. Accessed 1 April 2025
69. Abd El-Sayed ES, El-Sakhawy M, El-Sakhawy MAM (2020) Non-wood fibers as raw material for pulp and paper industry. Nord Pulp Paper Res J 35:215–230. https://doi.org/10.1515/NPPRJ-2019-0064/MACHINEREADABLECITATION/RIS
70. Abolore RS, Jaiswal S, Jaiswal AK (2024) Green and sustainable pretreatment methods for cellulose extraction from lignocellulosic biomass and its applications: a review. Carbohydr Polym Technol Appl 7:100396. https://doi.org/10.1016/J.CARPTA.2023.100396
71. Arce C, Kratky L (2022) Mechanical pretreatment of lignocellulosic biomass toward enzymatic/fermentative valorization. iScience 25:104610. https://doi.org/10.1016/J.ISCI.2022.104610
72. Fernández-Rodríguez J, Erdocia X, Hernández-Ramos F, Alriols MG, Labid J (2019) Lignin separation and fractionation by ultrafiltration. Separation of functional molecules in food by membrane technology. Academic Press, New York, pp 229–265
73. LakshmiBalasubramaniam SP, Howell C, Tajvidi M, Skonberg D (2022) Characterization of novel cellulose nanofibril and phenolic acid-based active and hydrophobic packaging films. Food Chem 374:131773. https://doi.org/10.1016/J.FOODCHEM.2021.131773
74. World Wildlife Fund (2025) Pulp and paper. https://wwf.panda.org/discover/our_focus/forests_practice/forestry/pulp_and_paper/. Accessed 26 March 2025
75. Abdul Khalil HPS, Davoudpour Y, Saurabh CK, Hossain MS, Adnan AS, Dungani R, Paridah MT, Mohamed ZIS, Fazita MRN, Syakir MI, Haafiz MKM (2016) A review on nanocellulosic fibres as new material for sustainable packaging: process and applications. Renew Sustain Energy Rev 64:823–836. https://doi.org/10.1016/J.RSER.2016.06.072
76. Su Y, Yang B, Liu J, Sun B, Cao C, Zou X, Lutes R, He Z (2018) Prospects for replacement of some plastics in packaging with lignocellulose materials: a brief review. BioResources 13:4550–4576

Chapter 14
Advances in Sustainable 3D Printing Filaments: Innovations, Challenges, and Industrial Applications

Km Akanksha, Abhash Mohan Singh, Nitin Kumar, Parth Kaushik, Naman Jain, Vineet Kumar Sinaha, and Akarsh Verma

1 Introduction

The increasing demand for additive manufacturing worldwide suggests that a revolution is taking place in a number of industries. Layer-by-layer material deposition is the hallmark of additive manufacturing, a relatively new and extensively researched production technique [1, 2]. It contains terms like solid freeform fabrication (SFF), direct digital manufacturing (DDM), rapid manufacturing (RM), rapid prototyping (RP), and 3D printing (3DP). 3D printing, also known as additive manufacturing (AM), is a rapidly developing technology that finds use in a wide range of industries, including the automotive, food and fashion, medical and healthcare, construction, and aerospace sectors [1]. Because AM technology can produce lighter structures, which lead to lighter airplanes and spacecraft, the aerospace industry has begun using it. Similar to this, the automotive sector has realized that additive manufacturing has the ability to recreate hard-to-find parts and create intricate, high-quality pieces that were previously impossible. AM enables the healthcare industry to carry out intricate transplant procedures and create accurate anatomical models for surgical planning. The feasibility and limitations of the approach have also been highlighted by studies assessing the strength of objects made using additive techniques, including those made of metallic and non-metallic materials. Large-scale industrial applications, everyday household use, and small-scale manufacturing businesses may now access

K. Akanksha · A. M. Singh · N. Kumar · P. Kaushik · N. Jain (✉) · V. K. Sinaha
Department of Mechanical Engineering, ABES Engineering College, Ghaziabad, India
e-mail: namanjainyati@gmail.com

A. Verma
Department of Mechanical Engineering, UPES (University of Petroleum and Energy Studies), Dehradun, India

A. Verma et al. (eds.), *Polymers and Composite Materials for Packaging: Smart Food Packaging and Solutions*, Materials Horizons: From Nature to Nanomaterials,
https://doi.org/10.1007/978-981-95-0677-4_14

3D printing thanks to its decentralization, which is a major shift in the manufacturing sector.

The aerospace sector has started embracing AM technology because it can create lighter structures, which result in lighter spacecraft and airplanes. In a similar vein, the car industry has come to recognize that additive manufacturing can replicate difficult-to-find parts and produce sophisticated, superior items that were previously unattainable. AM makes it possible for the medical field to do complex transplant operations and provide precise anatomical models for surgical planning. Studies evaluating the robustness of things created using additive techniques, particularly those constructed of metallic and non-metallic materials, have also brought attention to the approach's viability and limitations. The decentralization of 3D printing has made technology accessible to small-scale manufacturing enterprises, everyday household use, and large-scale industrial applications—a significant change in the manufacturing industry. Rapid prototyping is a noteworthy application of 3D printing, which is becoming an essential tool in many different industries. Engineers, designers, and researchers may test and improve their ideas quickly thanks to an iterative design process made possible by the speedy conversion of digital designs into physical prototypes. This gives businesses like the consumer electronics, automotive, and aerospace sectors a competitive edge by speeding up the product development cycle and drastically cutting lead times. Rapid prototyping made possible by 3D printing's accuracy and efficiency boosts creativity and reduces costs, underscoring its significance in influencing future product design and development [1].

The main objective of additive manufacturing (AM) is to cut down on the amount of time and procedures needed in the manufacturing process. Rapid prototyping technologies, which use 3D modeling tools, like computer-aided design (CAD), to speed up product design, are used to accomplish this goal [3–5]. AM uses information from design software to create objects by applying layers of material one after the other [6–9]. Single step manufacture, which uses material fusion [9] to achieve the basic geometry, and multistep manufacturing, which uses an adhesion principle and is carried out through a sequence of sequential procedures [10], are the two main categories into which AM lies.

Figure 1 shows the layered manufacturing process and a 3D-printed item. Laminated object manufacturing (LOM), fused deposition modeling (FDM), stereo lithography (SLA), selective laser sintering (SLS), and other AM processes show how technology is developing to attain product shape and optimize manufacturing [11]. AM is well known for printing metals, alloys, polymers, and biomedical products with the least amount of material [12]. Researchers used AM's transdisciplinary capacity to combine materials for consolidated mechanical, optical, and physical properties [13–15]. It has streamlined supply chains and reduced lead times for essential replacement parts [16].

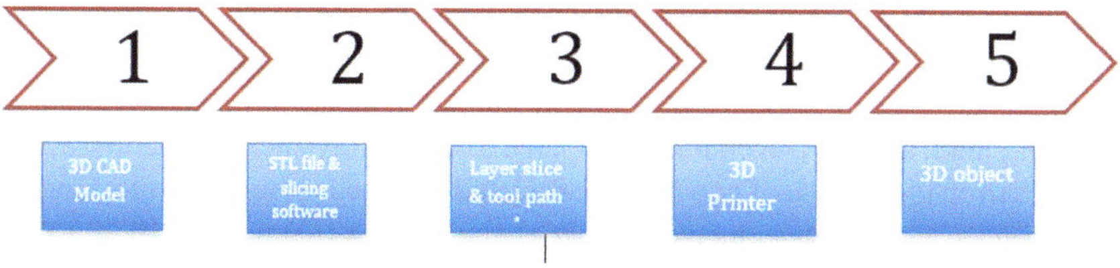

Fig. 1 Layered manufacturing of a 3D-printed component

2 Evolution of Additive Manufacturing for Decentralized Production

Charles "Chuck" Hull created stereolithography (SLA) in the 1980s, which is related to the development of additive manufacturing. SLA builds patterns layer by layer to produce a three-dimensional object by focusing ultraviolet (UV) light on a UV photo-curable liquid polymer solution. This innovation made it possible for 3D systems to be established in 1986, which in turn facilitated the development and manufacturing of 3D printers. Rapid prototyping systems were then developed, and AM was propelled to the forefront of manufacturing technology with the issuance of the United States Patent and Trademark Office (USPTO) patent award in 1986 [16].

Additive manufacturing, which combines materials utilizing a range of techniques like fusion, binding, and solidification, is a significant technical achievement despite its early difficulties. Layer by layer, the components are fitted together with the use of 3D CAD modeling, which uses Standard Tessellation Language (STL) files or 3D computer data to hold complex item shapes. The three main phases of AM are design, processing, and testing. Direct metal laser sintering (DMLS), stereolithography (SLA), fused deposition modeling (FDM), laminated object manufacturing (LOM), and selective laser sintering (SLS) are just a few of the many techniques employed in today's industries [17, 18]. Despite its many benefits, additive manufacturing has drawbacks, such as lengthier build times and less precision than CNC machines. The way different elements are oriented greatly affects accuracy, build times, the quantity of supports needed and, ultimately, the cost of manufacturing.

The ability to tailor production at various scales and locations—whether at the point of consumption, the point of sale, or within production facilities using local resources—defines decentralized manufacturing, also known as distributed manufacturing. User participation in product design, manufacturing, and supply is increased by this method. This decentralization is frequently fueled by digitalization and new production technologies like additive manufacturing. This strategy has a lot of potential, especially when it comes to allowing for tailored demands and manufacturing near locations with the highest demand. Decentralized manufacturing, especially in the healthcare sector, allows for greater accessibility to new markets and customers and allows for large-scale customization using inventory-light production methods [19].

Although additive manufacturing has received a lot of attention, many people are not aware of its capacity to decentralize production processes. It is anticipated that the ensuing localization movement will supplant globalization and significantly curtail air, land, and sea travel. It is anticipated to drastically alter international trade, causing a drop in the trade of goods and associated macroeconomic shifts in countries [20]. As fewer companies engage, supply chains will shift significantly as a result of streamlined logistics. Production will be closer to the consumer and driven by consumer demand, which will facilitate distribution and reduce issues with global supply chains. It is also expected that the new economy's occupational needs and skill requirements will shift, with jobs in installation, retail, packaging, transportation, delivery, and construction being replaced by more rational, integrative, creative, and independent roles like designers, consultants, engineers, and product developers. This change requires careful consideration and should be handled through education and training in order to handle sociocultural challenges [21]. AM can advance technology, giving developing countries a different route to economic success. The revolutionary potential of AM, which enables small companies to launch new goods without depending on international supply chains, is recognized by both big and small businesses [21]. Additive manufacturing, which enables customized mass production that requires less material to save weight, can be used to manufacture extremely complicated or even hybrid products. Thus, AM technologies have greatly aided in the production of specialty materials. Fewer production steps and resources are required to construct a desired structure since printed materials can now be accurately deposited in three dimensions. Thermoplastics are the main materials used in fused deposition modeling (FDM) 3D printing, which increases the use of plastic in general. The main issue with plastic consumption is that it has a negative effect on both people and the environment. Growing sustainable production is therefore becoming more and more important, especially in light of the modern world's dense population and the rising living standards brought about by global industrialization [22].

3 Current Recycling Methodologies

The 3D printing sector is one that is growing quickly. A variety of thermo the 3D printing sector is one that is growing quickly. Printable filaments can be made from a variety of thermoplastic compounds, including recyclable ones [23]. Recycling such polymeric materials typically entails a number of processes, such as material separation, remelting, grinding, extrusion, and purification and disinfection. The most common recycling strategy in AM is shown in Fig. 1. This section describes a number of 3D printer filament recycling procedures. Printable filaments can be made from plastic compounds, including recyclable ones [23]. Recycling such polymeric materials often entails a number of processes, such as material separation, decontamination and purification, grinding, remelting, and extrusion.

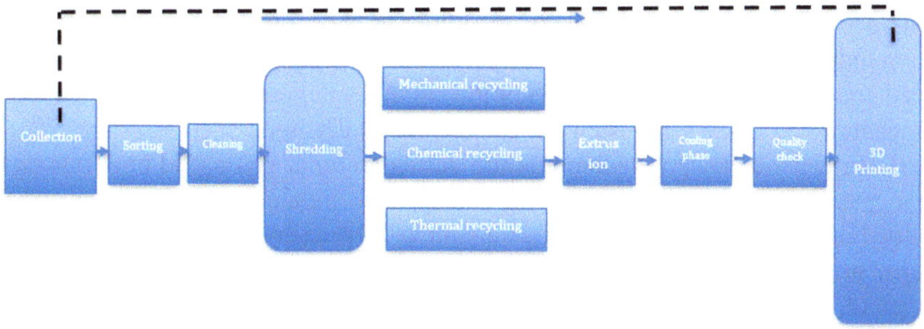

Fig. 2 General recycling methodology for additive manufacturing

Figure 2 illustrates the most common recycling technique utilized in AM. This section describes a number of 3D printer filament recycling procedures.

Depending on the classification, appropriate plastic materials for recycling are gathered using various methods in the first step, "collection." Municipal solid waste (MSW), industrial waste, agricultural waste, municipal sludge, and other wastes are among the various categories of solid waste that can be categorized based on their sources [24]. The efficiency of waste.

In educational institutions where trash generation is continuous, management becomes essential. Compared to other approaches, collecting waste at the source and recycling it locally is more efficient and much simpler to handle. However, in centralized regions such as municipalities and industry, a particular type of garbage collecting method is needed. This approach makes advantage of current waste management practices, such as curbside collection, vehicle collection, drop-off waste disposal, buy-back facilities, and deposit/refund programs, which are employed by municipalities worldwide [25].

The second phase is called "sorting," and it makes up around 25% of the recycling process. All of the collected plastic waste is divided into various categories according to its properties during this procedure. Although the majority of plastics are categorized as HDPE, PP, or PET, other polymers that cannot be recycled are isolated and put to different uses. Colors and types are then used to further categorize the sorted plastic [26].

The recycling procedure varies greatly based on the kind of polymer and is not consistent. It's interesting to note that polymers cannot melt together even if they have comparable chemical structures, such the CEC bond in LDPE and PP. These polymers present a significant separation issue once combined [27]. Because commercial products usually contain several different types of polymers in one product, this problem is frequently made worse. Because of this, sorting polymers is a difficult task. The effectiveness of mechanical recycling depends on the recovered polymer's uniform quality and ease of handling, which is ensured by automating the sorting process [28]. Usually, near-infrared spectroscopy (NIR) is used to do this. For bigger volumes, however, X-ray fluorescence (XRF) is the recommended technique. The shredded polymer bits' subsequent sizes vary according to their intended purpose. The recycling industry's increasing need for high-efficiency sorting methods emphasizes the need for creative ways to boost the efficacy and efficiency of polymer recycling procedures expenses for wastewater treatment and drying.

Since caustic cleaning techniques are unsuccessful at getting rid of components that cause odors, effectively removing contaminants related to food is a major difficulty. For example, PET goes through a number of processes to get rid of things like paper or metal, and it is cleaned with hot water to get rid of dirt or label residue. However, in order to completely eliminate all impurities, detergent bottles frequently need to be mechanically cleaned. In order to facilitate handling and subsequent processing, "shredding" entails reducing used or damaged prints and other plastic debris into smaller pieces. The surface area is greatly increased by shredding. This aids in melting, promoting efficient heat transport and guaranteeing a thorough and reliable melting process.

It helps to increase material melting homogeneity and processing efficiency. For example, the 3D-printed specimens were shredded using a FriendTM plastic mill. Since the goal of the study was to simulate the closed-loop recycling of 3D-printed products, washing and sorting were not included in the experiment; instead, the PLA's source and end use were analyzed [29]. Figure 2 gives the process stages for the preparation of the collected material, showing (A) a schematic representation outlining the process of material preparation, (B) the materials utilized in the process, and (C) the resultant material [30].

4 Recyclable Plastics for Extrusion-Based 3DP

To prolong their life cycle and make FDM plastics sustainable and environmentally beneficial, they must be recycled. AM Thermoplastics are recyclable due to their linear molecular chain structure, which causes them to soften when heated and solidify when cooled [31]. Plastics that are thermoset cure irreversibly. This essential distinction is necessary for reusability. Common thermoplastics for 3D printing, such as PLA and ABS, are listed in Table 1. Young's modulus, which gauges tensile elasticity, and tensile strength are essential. PLA is better for prosthetics and healthcare, while ABS is best for high-stress tooling parts [31–33].

Thermoplastics can be recycled chemically or mechanically. Shredded plastic is melted mechanically to create feedstock filament for 3D printers. Jagadeesh et al. found that each recycling cycle reduces the material's molecular weight by 46% and its viscosity by 80% due to chain-scission reactions caused by impurities, even though it is economically advantageous. Additionally, the recycled part's tensile strength is found to be lower than that of its virgin counterpart [34, 35], as shown in Table 1 with ABS. Other characteristics, such natural frequencies, can also be affected by material attributes [35]. Chemical recycling, on the other hand, reproduces plastic by depolymerizing it through a chemical reaction [36]. Compared to conventional filament manufacturing, the open-source Recyclebot reduces embodied energy and

Table 1 Material properties of extruded and recycled plastics (ABS, PLA, Nylon-6)

Material	Yield tensile strength (MPa)	Young modulus (GP)	Melting temperature (C)	Source
ABS, extruded	13.0–65.0	1.00–2.65	177–320	[40–43]
ABS recycled	32	2.125	177–320	[44]
PLA, extruded	30	2.3	205	[44, 45]
Nylon-06 extruded	35.0–186	0.450–3.50	205	[65.66]
Nylon-06 recycled	55.79–86.91	1.64	205	[44, 45]

environmental impact by recycling plastic trash into 3D printing filament [36, 37]. Material properties of extruded and recycled plastics as shown in Table 1. Physical characteristics are deteriorated by the melt-extrude cycle. The material qualities of FDM filaments are better preserved when nylon-6 waste is recycled and purified [38, 39].

5 Applications and Industrial Relevance

Figure 3 illustrates how 3D printing changes manufacturing, increasing sustainability and altering the environmental impact of the product life cycle. Additive manufacturing consumes less resources and generates less waste since it creates products layer by layer without cutting or reshaping. In most 3D printing techniques, support structures are typically taken out after production and repurposed, resulting in minimal material loss [39]. With 3D printing, the manufacturing process is quicker and more straightforward, which lowers CO_2 emissions and energy use [39]. Carbon emissions associated with shipping may be decreased by technology that enables on-site manufacture.

Although it needs to be used in mass production, manufacturing speed must be increased, and printable materials made more widely available, 3D printing has the potential to lower industrial net CO_2 emissions and energy consumption. It's also critical to take into account a "rebound effect," in which efficiency boosts activity [39, 46]. Compared to FDM, which consumes less energy but emits emissions, some 3D printing techniques, like laser metal deposition, are better for reusing materials [47–50].

One of the most significant environmental problems facing humanity today is the management of plastic garbage. The pervasiveness of non-biodegradable plastics in the waste stream raises serious concerns about the ongoing environmental harm they pose. The buildup of these waste plastics has caused serious problems, such as littering in the environment, clogged drains, and serious health problems. Many nations engage in negligent dumping without effective waste management and without one of the most significant environmental problems facing humanity today is the management of plastic garbage. The pervasiveness of non-biodegradable plastics

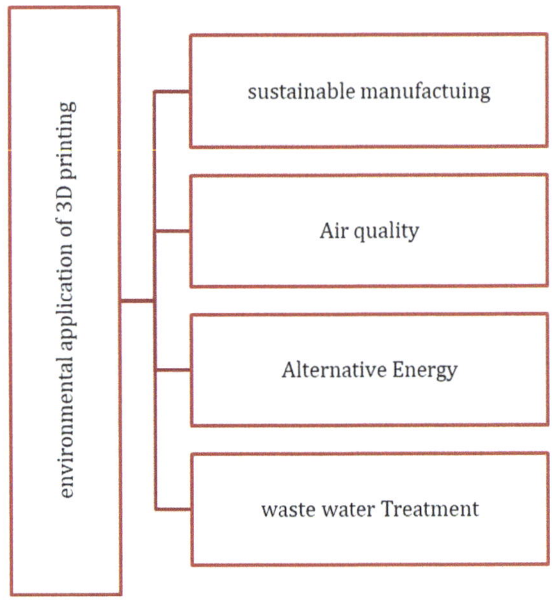

Fig. 3 3D printing applications for sustainable environment

in the waste stream raises serious concerns about the ongoing environmental harm they pose.

The buildup of these waste plastics has caused serious problems, such as littering in the environment, clogged drains, and serious health problems. Many nations engage in reckless dumping without effective waste management or adherence to scientific trash disposal techniques. As a result, discarded plastic pyramids accumulate in landfills, taking up a lot of room and exacerbating existing environmental issues. Low profitability and the considerable scientific difficulty of disassembling plastics into their component chemicals present additional obstacles to the recycling process. Handling plastics recovered from MSW is getting harder because of the rising demand for plastic-related products, their non-biodegradable nature, and the social risks they pose utilizing scientific techniques for disposing of garbage. As a result, discarded plastic pyramids accumulate in landfills, taking up a lot of room and exacerbating existing environmental issues. Low profitability and the considerable scientific difficulty of disassembling plastics into their component chemicals present additional obstacles to the recycling process. Handling plastics recovered from MSW is getting harder because of the rising demand for plastic-related products, their non-biodegradable nature, and the social risks they pose.

Environmentally friendly plastic recycling techniques are being researched as a result of these environmental issues. Notwithstanding its efficiency, mechanical recycling has disadvantages, including the potential for molecular weight degradation and the small quantity of recyclable polymers. One technique for turning plastic trash into useful feedstock materials for the production of fuel and monomers is chemical recycling. Despite its potential, chemical recycling is more expensive, labor- and energy-intensive, and may result in the production of extra pollutants than mechanical recycling. These environmental concerns are prompting research into environmentally friendly plastic recycling methods. Despite its effectiveness, mechanical recycling has drawbacks, such as the limited amount of recyclable polymers and the possibility of molecular weight degradation. Chemical recycling is one method of converting plastic waste into valuable feedstock materials for the synthesis of fuel and monomers. Chemical recycling has potential, but it is more costly, requires more energy and effort, and may produce more pollutants than mechanical recycling.

According to a study on the manufacturing of 3D-printed personal protective equipment (PPE) during the COVID-19 epidemic, 34% of the materials utilized in 3D printing ended up as waste, mostly from unsuccessful prints and support structures. However, the technique supported the circular economy and significantly reduced the overall environmental imprint by recycling this trash into new filaments and producing useful products like personal protective equipment [46]. Although additive manufacturing technology is being used more and more, its environmental effects have not been well examined. Steps like casting, molding, bending, and welding are commonly combined in conventional manufacturing processes, and each has a distinct environmental impact. On the other hand, AM usually does away with the requirement for such a combination when producing parts. One notable feature of additive manufacturing is its capacity to create incredibly accurate, elaborately formed objects while drastically cutting down on material waste layer by layer. AM technologies provide a number of advantageous environmental benefits when compared to conventional manufacturing techniques. The capacity to persist is further improved by combining innovative, clever materials and drastically minimizing wastage from raw materials. By reducing material waste, energy use, and machine emissions, additive manufacturing places a strong emphasis on component production efficiency [47]. Three environmental factors can be taken into account while assessing AM procedures: air pollution, waste disposal, and energy consumption. Each process's own set of parameters directly impacts how much energy is used in AM procedures. For example, higher printing resolution results in longer manufacturing times and higher energy use. Heating components in 3D printers is one example of an energy-intensive process that uses a lot of electricity. The overall energy load is increased by certain AM procedures that call for temperatures as high as 200 °C for particular plastics. In the long run, AM is a more sustainable choice since it may counteract these energy demands by localizing output and reducing transportation distances [48]. Reuse and recycling techniques can be used in additive manufacturing to handle waste materials effectively. Reusing waste polymer materials to create filaments is an effective technique to lessen the environmental impact. Minimizing waste materials can enhance the energy efficiency of AM operations. Additive manufacturing is pivotal in metal recycling due to its use in metallic additive processes. Producing aluminum from recycled materials consumes 95% less energy than using raw materials, while recycling copper saves between 75 and 90% in terms of energy.

By encouraging the use of closed-loop materials and lowering dependency on virgin resources, AM has the potential to create a more sustainable industrial environment, as demonstrated by these significant energy savings [49]. AM technologies affect air pollution directly as well as indirectly. Biopolymer filaments and non-toxic materials can be used to directly reduce air pollution. Air quality is indirectly improved by the decentralization brought about by additive manufacturing, which lowers shipping costs and brings items closer to consumers. Useful elements, including the ability to create replacement components using 3D printing, help prolong the life of devices without support from their original producers.

6 Sustainable Environmental Technologies for Air, Water, and Energy

Air quality monitors are made via 3D printing. nEMoS, a nano environmental monitoring system that gauges indoor air quality, was 3D printed by Salamone et al. [51] CO_2 concentration and other environmental data are reported by nEMoS, which is inexpensive and trustworthy. Cases for various air quality monitors, such as iAir for indoor air quality and HOPE for outdoor air quality, have been made possible by 3D printing's customization capabilities [52, 53]. Xu et al. [54] advanced miniaturized sensors by using 3D printing to develop a compact, portable wearable particulate matter monitor.

Scrubbers and pollutant filters are 3D printed. NO is eliminated from the air by Kumar et al. [55] using a flexible air filter and a photocatalyst. Furthermore, the circular channel of the Vortecone scrubber is one example of a scrubber component with unique shape made possible by 3D printing [56]. New techniques for treating water and wastewater have been made possible by advanced 3D printing technology. The ability to customize 3D printing may result in less expensive membranes, which would be an economical and effective substitute for traditional techniques [57, 58]. For ceramic membrane-based treatment materials, 3D printing works well [40], but it has trouble manufacturing structures with submicron precision and material compatibility [41, 42]. Research has been done on oil–water separation meshes and 3D-printed ceramic water filters [43, 44]. To enhance the removal of pollutants, air filters and super hydrophilic membranes can be 3D printed [45]. Renewable energy technologies such as photovoltaic (PV) cells, wind turbine blades, and 3D-printed microbial fuel cells are being tested. 3D printed anodes with improved microbial adhesion and area are advantageous for microbial fuel cells, which produce electricity and oxidize organic contaminants in wastewater [59, 60]. 3D printing is used to create flexible solar cells on translucent polymers and metal foils. Additionally, this approach produces flexible front electrodes for ultra-thin microcell arrays that function similarly to solar cells [61].

Compared to flat, stationary panels, 3D-printed photovoltaic cells have better energy densities because their geometries may be tuned [62, 63]. Researchers have developed self-heating mesh for blade de-icing and turbine blades that resemble plant leaves using 3D printing [64, 65]. Small, affordable residential wind turbines can be built using 3D printing, providing a sustainable power source [66, 67].

7 Research Gap and Future Trends

In a number of areas, including research, construction, architecture, automotive, aerospace, medicine, healthcare, and the food and fashion sectors, additive manufacturing, often known as 3D printing, shows promise [1]. It enables a variety of sectors to produce intricate designs that were previously deemed unfeasible and reduce product weight. The structural integrity of items made using additive processes, involving both metallic and non-metal materials, has been the subject of numerous studies in recent years in an effort to clarify the viability and constraints of additive manufacturing. However, a more thorough.

To fully utilize this technology, especially in terms of improving material qualities, research is necessary. In order to meet the Sustainable Development Goals of the UN, the research community has recently shown a special interest in additive manufacturing, with an emphasis on sustainable methods including recycling material waste. By comparing the number of publications annually, one may observe this pattern, which shows a notable increase over the past ten years. The increase in interest in this field has been mostly attributed to research communities in the US, China, India, and Italy. The need to improve the effectiveness of polymer recycling processes is one of the particular issues facing the additive manufacturing sector that necessitates specialized methods and solutions. Future research is needed to better understand how the number of reuse cycles affects the material's characteristics to guarantee appropriate quality requirements while recycling.

Furthermore, certain optimization conditions—like the impact of changing the refresh rate for ABS—are a crucial matter that requires further research since they could significantly impact the overall efficacy of 3D printing techniques [64, 66]. There is a notable dearth of study on how PLA and ABS materials lose their mechanical properties after numerous reuses in 3D-printed constructions. It highlights the need for more research and the gap in the body of knowledge regarding this particular subject. Investigating substitute materials is necessary to create a circular economy in the context of 3D printing in order to lower the demand for virgin feedstock and reused resources. To reduce the impact on the environment when it degrades, the ideal material should be biodegradable. To find techniques that shorten manufacturing times and prevent thermal damage, more research 3D printing technology is essential [67]. This evaluation highlights how crucial it is to carry out in-depth research in order to confirm the potential of emerging 3D printing trends. It also emphasizes how important it is to conduct thorough research in order to lay a strong basis for achieving the potential of these emerging trends. We may steer the advancement of additive manufacturing technology in the direction of sustainability by looking into these factors more thoroughly. Investigating and creating novel materials and composite formulations that can improve the mechanical characteristics of 3D-printed parts while resolving drawbacks like brittleness is imperative.

This can involve adding reinforcements, mixing polymers, or creating brand-new bio-based materials with enhanced durability, strength, and flexibility. These developments will increase additive manufacturing's use in a variety of industries, such as biomedical engineering, aerospace, and automotive. Achieving consistent quality is made more difficult by the variety of biodegradable materials used in additive manufacturing. Standardizing printing settings, including temperature.

For the printing process to be efficient, repeatable, and dimensionally accurate, layer height, speed, and cooling profiles will be crucial. Additionally, a consistent framework will promote wider corporate usage and improve comparisons between research investigations. Internal voids and porosity, which are key causes of structural problems, can be considerably decreased by research into sophisticated bonding processes between layers and optimum infill patterns. Enhancing the overall structural integrity and functionality of 3D-printed items will require research into new infill geometries, gradient infill densities, and thermal or chemical post-processing techniques. A multidisciplinary approach is necessary to address the intricate issues of sustainability and performance in 3D printing. Technological breakthroughs can be accelerated by fortifying partnerships among university researchers, industrial firms, and regulatory or policy-making organizations. In addition to promoting innovation in materials and techniques, these collaborations will aid in the creation of regulatory guidelines and lifecycle evaluation frameworks that will guarantee the safe, effective, and ecologically conscious application of additive manufacturing technologies.

8 Conclusions

A number of material and technical issues must be resolved to guarantee consistent structural quality and functional performance, even though the incorporation of biodegradable materials into Fused Deposition Modeling (FDM) 3D printing offers a promising avenue to improve environmental sustainability. Because of their potential for compostability and reduced carbon footprint, these materials—like PLA (polylactic acid) and other bio-based polymers—offer ecological benefits. The integrity and dependability of printed parts may be jeopardized by their use in additive manufacturing, which is frequently accompanied by problems with brittleness, poor heat stability, uneven layer adhesion, and degradation during processing.

Overcoming these constraints effectively necessitates a multidimensional and cooperative strategy. First and foremost, it is imperative to create standardized printing settings that are especially suited for biodegradable materials. These parameters should include ideal temperatures, layer heights, print speeds, and cooling rates. This kind of standardization can enhance part quality, allow for reproducibility across many machines and environments, and produce more predictable results. Second, to enhance the mechanical and thermal characteristics of biodegradable filaments, material science breakthroughs are crucial. This could entail creating new polymer blends, composite materials with nano- or natural-fillers, and chemically altered biopolymers that provide improved strength, flexibility, and wear resistance without sacrificing biodegradability.

Research into these materials' aging characteristics, recyclability, and long-term performance is also essential for their use in practical applications outside of prototyping, like consumer goods, packaging, and biomedical devices. Third, the inherent drawbacks of biodegradable materials can be lessened with ongoing advancements in additive manufacturing technology, such as enhanced printer hardware, slicing software algorithms, real-time monitoring, and post-processing methods. The performance of bio-based prints can be greatly improved by increased accuracy, improved control over extrusion parameters, and flexible printing environments. Furthermore, in order to create recommendations, exchange data, and advance best practices, academic researchers, industry stakeholders, and regulatory agencies must work closely together to address these issues. By working together, it is feasible to create a strong ecosystem that supports the technological viability of biodegradable 3D printing as well as its economic and environmental viability.

The additive manufacturing community can open the door for the wider adoption of sustainable practices across a variety of application domains by recognizing the existing constraints and aggressively seeking focused solutions. In the end, such advancements will make a substantial contribution to the worldwide endeavor to lessen environmental impact while preserving the creativity and adaptability that characterize the 3D printing sector. With special focus to the environmental issues related to waste generation during 3D printing processes, this analysis thoroughly examines the crucial and expanding problem of recycling waste materials in the field of additive manufacturing (AM). The amount of polymer waste and unsuccessful prints has grown along with the breadth and size of additive manufacturing's expansion across a number of industries, including consumer goods, healthcare, automotive, and aerospace. In addition to posing serious environmental dangers, this runs counter to the sustainability objectives that many contemporary manufacturing processes seek to uphold. The review examines a variety of recycling strategies, with a particular emphasis on mechanical and chemical recycling processes, in order to evaluate how the industry is addressing this issue. Usually, mechanical recycling entails breaking up old or unsuccessful prints into tiny granules, which are subsequently remelted and extruded into fresh filament. Although this process is economical, the material's mechanical and thermal qualities, like its strength, toughness, and elasticity, may deteriorate with repeated thermal cycles.

Chemical recycling, on the other hand, breaks down polymers at the molecular level to produce base monomers or oligomers that can subsequently be re-polymerized into superior materials. Although this approach delivers greater material realism, it necessitates a more sophisticated infrastructure and more energy. Evaluating the characteristics and possible uses of recycled polymers frequently used in 3D printing, particularly polylactic acid (PLA), acrylonitrile butadiene styrene (ABS), and polyethylene terephthalate glycol (PETG), takes up a sizable amount of the paper. Since these materials are widely used in both consumer and industrial settings, their recyclability is a highly relevant concern. The evaluation emphasizes the need for more recycling protocol optimization to maintain functionality and safety standards by highlighting the effects of frequent recycling on their tensile strength, surface quality, and printability.

This review explores the social and economic ramifications of employing recycled materials in additive manufacturing in addition to its technical features. On the societal front, more people support green technologies as a result of growing awareness of and adoption of sustainable practices. Opportunities in education, small-scale manufacturing, and isolated areas with limited access to virgin materials are also made possible by the democratization of AM with recycled materials. Utilizing recycled filaments can be cost-effective lessen reliance on international supplier chains and production expenses. But issues including inconsistent recycled filament quality, a lack of uniformity, and restricted scalability continue to broad adoption.

The review also highlights how 3D printing facilitates decentralized manufacturing, which advances environmental sustainability. Decentralized solutions reduce logistical waste and transportation emissions by facilitating localized output. Such technologies offer a paradigm for closed-loop production, where waste is continuously repurposed, greatly lowering the environmental impact, when combined with efficient recycling techniques. To sum up, this review provides insightful information about the recyclability of waste materials in additive manufacturing, laying the groundwork for future research as well as existing knowledge. In order to promote the circular economy within the AM ecosystem, it emphasizes how vital it is to incorporate sustainable design concepts, material reuse techniques, and cross-sector collaboration. To overcome present obstacles and guarantee that additive manufacturing develops in a way that is not only technologically sophisticated but also environmentally and financially sustainable, ongoing innovation and thorough research are crucial.

Data Availability Not applicable.

Declaration of Competing Interest The authors have no conflicts to disclose.

References

1. Wong KV, Hernandez A (2012) A review of additive manufacturing. ISRN Mech Eng 2012:208760
2. Amrita A, Manoj A, Panda RC (2022) Biodegradable filament for three-dimensional printing process: a review. Eng Sci 18:11–19
3. Ul-Haq MI, Khuroo S, Raina A et al (2020) 3D printing for development of medical equipment amidst coronavirus (COVID-19) pandemic—review and advancements. Res Biomed Eng 38:305–315. https://doi.org/10.1007/s42600-020-00098-0
4. Aziz R, Ul Haq MI, Raina A (2020) Effect of surface texturing on friction behaviour of 3D printed polylactic (PLA). Polymers 16:6434. https://doi.org/10.1016/j.polymertesting.2020.106434

5. Chadha A, Ul Haq MI, Raina A et al (2019) Effect of fused deposition modelling process parameters on mechanical properties of 3D printed parts. World J Eng 6:550–559. https://doi.org/10.1108/WJE-09-2018-0329
6. Naveed N (2020) Investigate the effects of process parameters on material properties and microstructural changes of 3D-printed specimens using fused deposition modelling (FDM). Mater Technol 36:317–330. https://doi.org/10.1080/10667857.2020.1758475
7. Naveed N (2021) Investigating the material properties and microstructural changes of fused filament PLA and tough-PLA parts. Polymers 13:1487. https://doi.org/10.3390/polym13091487
8. Ashrafi N, Duarte JP, Nazarian S et al (2018) Evaluating the relationship between deposition and layer quality in large-scale additive manufacturing of concrete. Virt Phys Prototyp 14:135–140. https://doi.org/10.1080/17452759.2018.1532800
9. Kumar MB, Sathiya P (2021) Methods and materials for additive manufacturing: a critical review on and challenges. Thin Walled Struct 23:1009
10. Ul Haq MI, Raina A, Ghazali MJ et al (2021) Potential of 3D printing technologies in developing applications of polymeric nanocomposites, In: Jena H, Katiyar JK, Patnaik A (eds) Tribology of polymer and polymer composites for industry 4.0, pp 193–210
11. Clarissa WHY, Chia CH, Zakaria S et al (2022) Recent advancement in 3-D printing: nanocomposites with added functionality. Prog Addit Manuf 7:325–350. https://doi.org/10.1007/s40964-021-00232-z
12. Birosz MT, Andó M, Jeganmohan S (2021) Finite element method modeling of additive manufactured compressor wheel. J Inst Eng Ser D 102:79–85. https://doi.org/10.1007/s40033-021-00251-8
13. Andó M, Birosz M, Jeganmohan S (2021) Surface bonding of additive manufactured parts from multi-colored PLA materials. Measurement 169:108583. https://doi.org/10.1016/j.measurement.2020.108583
14. Saini JS, Dowling L, Kennedy J et al (2020) Investigations of the mechanical properties on different print orientations in SLA 3D printed resin. Proc Inst Mech Eng Part C 234:2279–2293. https://doi.org/10.1177/0954406220904106
15. Węgrzyn N (2022) The use of additive manufacturing for production of commercial airplane power plants components: a review. Saf Def 8:2
16. Bose S, Sarkar N, Vahabzadeh S, Ke D, Bandyopadhyay A (2019) Additive manufacturing of ceramics. In: Additive manufacturing, 2nd edn. Taylor & Francis Group, London
17. Singh S, Ramakrishna S, Singh R (2017) Material issues in additive manufacturing: a review. J Manuf Process 25:185–200
18. Ngo TD, Kashani A, Imbalzano G, Nguyen KTQ, Hui D (2018) Additive manufacturing (3D printing): a review of materials, methods, applications and challenges. Compos B Eng 143:172–196
19. Srai JS, Kumar M, Graham G, Phillips W, Tooze J, Ford S, Beecher P, Raj B, Gregory M, Tiwari MK et al (2016) Distributed manufacturing: Scope, challenges and opportunities. Int J Prod Res 54:6917–6935
20. Zuboff S (2024) Creating value in the age of distributed capitalism artwork by Celia Johnson. https://www.mckinsey.com/capabilities/strategy-and-corporate-finance/our-insights/creating-value-in-the-age-of-distributed-capitalism. Accessed on 4 Aug 2024
21. Ben-Ner A, Siemsen E (2017) Decentralization and localization of production: the organizational and economic consequences of additive manufacturing (3D printing). Calif Manag Rev 59:5–23
22. Schneevogt H, Stelzner K, Yilmaz B, Abali BE, Klunker A, Völlmecke C (2021) Sustainability in additive manufacturing: Exploring the mechanical potential of recycled PET filaments. Compos Adv Mater 30:26349833211 10000
23. Murugan R, Mitilesh RN, Singamneni S (2019) Influence of process parameters on the mechanical behaviour and processing time of 3Dprinting. Int J Mod Manuf Technol 1:21–27
24. Balwada J, Samaiya S, Mishra RP (2021) Packaging plastic waste management for a circular economy and identifying a better waste collection system using analytical hierarchy process (AHP). Proced CIRP 98:270–275

25. AlRashid A, Koç M (2023) Additive manufacturing for sustainability and circular economy: needs, challenges, and opportunities for 3Dprinting of recycled polymeric waste. Mater Today Sustain 24:100529
26. Hlaváciková S, Omaníková L, Horváth V, Alexy P, Jancovicová V, Baco A, Mikolajová M, Fogašová M, Tomanová K, Feranc J et al (2023) The possibility of using the regranulate of a biodegradable polymer blend based on polylactic acid and polyhydroxybutyrate in FDM 3D printing technology. Results Mater 21:100511
27. Sanchez FAC, Boudaoud H, Camargo M, Pearce JM (2020) Plastic recycling in additive manufacturing: a systematic literature review and opportunities for the circular economy. J Clean Prod 264:121602
28. Gonzalez CS, Sanchez FAC, Boudaoud H, Nouvel C, Pearce JM (2024) Multi-material distributed recycling via material extrusion: recycled high density polyethylene and poly (ethylene terephthalate) mixture. Polym Eng Sci 64:1555–1570
29. Camacho DD, Clayton P, O'Brien WJ et al (2018) Applications of additive manufacturing in the construction industry—a forward-looking review. Autom Constr 89:110–119
30. Sepasgozar SME, Shi A, Yang LM et al (2020) Additive manufacturing applications for industry 40: a systematic critical review. Buildings 10:231
31. Paolini A, Kollmannsberger S, Rank E (2019) Additive manufacturing in construction: a review on processes, applications, and digital planning methods. Addit Manuf 30:100894. https://doi.org/10.1016/j.addma.2019.100894
32. Jagadeesh P, Rangappa SM, Siengchin S et al (2022) Sustainable recycling technologies for thermoplastic polymers and their composites: a review of the state of the art. Polym Compos 43:5831–5862. https://doi.org/10.1002/pc.27000
33. Sethi B (2016) Methods of recycling. Recycl Polym Methods Charact Appl 65:55–114. https://doi.org/10.1002/9783527689002.ch3
34. Dogu O, Pelucchi M, Vijver RV et al (2021) The chemistry of chemical recycling of solid plastic waste via pyrolysis and gasification: state-of-the-art, challenges, and future directions. Prog Energy Combust 84:100901. https://doi.org/10.1016/j.pecs.2020.100901
35. Jubinville D, Esmizadeh E, Saikrishnan S et al (2020) A comprehensive review of global production and recycling methods of polyolefin (PO) based products and their post-recycling applications. Sustain Mater J Technol 25:e00188
36. Zhang F, Zhao YT, Wang DD et al (2021) Current technologies for plastic waste treatment: a review. Cleaner Prod 282:124523
37. Markandeya N, Joshi AN, Chavan NN et al (2023) Plastic recycling: challenges, opportunities, and aspects. Adv Mater Recycled Waste 68:317–356
38. Kumar M, Bolan S, Padhye LP et al (2023) Retrieving back plastic wastes for conversion to value added petrochemicals: Opportunities, challenges and outlooks. Appl Energy 345:121307. https://doi.org/10.1016/j.apenergy.2023.121307
39. Maffia S, Chiappini F, Maggiani G et al (2023) Enhancing productivity and efficiency in conventional laser metal deposition process for Inconel 718—part II: advancing the process performance. Int J Adv Manuf Technol 129:279–298. https://doi.org/10.1007/s00170-023-12197-0
40. Kotz F, Helmer D, Rapp BE (2020) Emerging technologies and materials for high-resolution 3D printing of microfluidic chips. https://doi.org/10.1007/10_2020_141
41. Jonhson W, Xu X, Bian K et al (2022) 3D-printed hierarchical ceramic architectures for ultrafast emulsion treatment and simultaneous oil-water filtration. ACS Mater Lett 4:740–750. https://doi.org/10.1021/acsmaterialslett.2c00147
42. Jin Z, Mei H, Liu H et al (2022) High-strength, superhydrophilic/underwater superoleophobic multifunctional ceramics for high efficiency oil-water separation and water purification. Mater Today Nano 18:100199
43. Sreedhar N, Kumar M, Al Jitan S et al (2022) 3D printed photocatalytic feed spacers functionalized with β-FeOOH nanorods inducing pollutant degradation and membrane cleaning capabilities in water treatment. Appl Catal B 300:120318. https://doi.org/10.1016/j.apcatb.2021.120318

44. Sreelekshmy BR, Rajappan AJ, Basheer R et al (2020) Tuning of surface characteristics of anodes for efficient and sustained power generation in microbial fuel cells. ACS Appl Bio Mater 3:6224–6236. https://doi.org/10.1021/acsabm.0c00753
45. Cai T, Meng LJ, Chen G et al (2020) Application of advanced anodes in microbial fuel cells for power a review. Chemosphere 248:125985
46. Thakkar H, Eastman S, Hajari A et al (2016) 3D-printed zeolite monoliths for CO_2 removal from enclosed environments. ACS Appl Mater Interface 8:27753–27761
47. Ligon SC, Liska R, Stampfl J et al (2017) Polymers for 3D printing and customized additive manufacturing. Chem Rev 117:10212–10290. https://doi.org/10.1021/acs.chemrev.7b00074
48. Soliman A, AlAmoodi N, Karanikolos GN et al (2020) A review on new 3-D printed materials' geometries for catalysis and adsorption: paradigms from reforming reactions and CO_2 capture. Nanomaterials 10:2198. https://doi.org/10.3390/nano10112198
49. Sola A (2022) Materials requirements in fused filament fabrication: a framework for the design of next-generation 3D printable thermoplastics and composites. Macromol Mater Eng 307:2200197. https://doi.org/10.1002/mame.202200197
50. Nazir MH, Al-Marzouqi AH, Ahmed W et al (2023) The potential of adopting natural fibers reinforcements for fused deposition modeling: characterization and implications. Heliyon 9:e15023
51. Salamone F, Danza L, Meroni I et al (2017) A low-cost environmental monitoring system: how to prevent systematic errors in the design phase through the combined use of additive manufacturing and thermographic. https://doi.org/10.3390/s17040828
52. Aizlewood C, Dimitroulopoulou C (2006) The HOPE project: the UK experience. Indoor Built Environ 15:393–409. https://doi.org/10.1177/1420326X06069578
53. Wang Y, Mackenzie FV, Ingenhut B et al (2018) AP4. 1-miniaturized 3D printed particulate matter sensor for personal monitoring. In: Proceedings of the 17th international meeting on chemical sensors. https://doi.org/10.5162/IMCS2018/AP4.1
54. Xu X, Xiao SN, Willy HJ et al (2020) 3D-printed grids with polymeric photocatalytic system as flexible air filter. Appl Catal B 262:118307. https://doi.org/10.1016/j.apcatb.2019.118307
55. Kumar AR, Arya S, Levy A et al (2020) Scale and numerical modeling to determine operating points of a non-clogging vortecone filter in mining operation. Prog Scale Model Int J 1:7
56. Aghaei A, Firouzjaei MD, Karami P et al (2022) The implications of 3D-printed membranes for water and wastewater treatment and resource recovery. Can J Chem Eng 100:2309–2321. https://doi.org/10.1002/cjce.24488
57. Tijing LD, Dizon JRC, Ibrahim I et al (2020) 3D printing for membrane separation, desalination and treatment. Appl Mater Today 18:100486
58. Ye YY, Du Y, Hu TY et al (2021) 3D printing of integrated ceramic membranes by the DLP method. Ind Eng Chem Res 60:9368–9377. https://doi.org/10.1021/acs.iecr.1c02224
59. Liu CH, Xiao CY, Xie CC et al (2021) Flexible organic solar cells: materials, large-area fabrication techniques and potential applications. Nano Energy 89:106399. https://doi.org/10.1016/j.nanoen.2021.106399
60. Tian YX, Wang XQ, Li J et al (2022) Rapid manufacturing of turbine blades based on reverse engineering and 3D printing technology. In: Proceedings of 2022 Chinese intelligent systems conference, pp 540–553. https://doi.org/10.1007/978-981-19-6203-5_53
61. Rahimizadeh A, Kalman J, Fayazbakhsh K et al (2021) Mechanical and thermal study of 3D printing composite filaments from wind turbine waste. Polym Compos 42:2305–2316. https://doi.org/10.1002/pc.25978127
62. Dzogbewu TC, Beer DJ (2023) Additive manufacturing of selected ecofriendly energy devices. Virtual Phys Prototyp 18:e2150230. https://doi.org/10.1080/17452759.2023.2276245
63. Browne MP, Redondo E, Pumera M (2020) 3D printing for electrochemical energy applications. Chem Rev 120:2783–2810. https://doi.org/10.1021/acs.chemrev.9b00783
64. Javaid M, Haleem A, Singh RP, Suman R, Rab S (2021) Role of additive manufacturing applications towards environmental sustainability. Adv Ind Eng Polym Res 4:312–322
65. Wang H, Xiong BD, Zhang ZT et al (2023) Small wind turbines and their potential for internet of things applications. iScience 26:107674

66. Kravchenko M, Pigosso DC, McAloone TC (2020) Circular economy enabled by additive manufacturing: potential opportunities and key sustainability aspects. In: Proceedings of the NordDesign 2020, DS 101, Lyngby, Denmark
67. Giurco D, Littleboy A, Boyle T, Fyfe J, White S (2014) Circular economy: questions for responsible minerals, additive manufacturing and recycling of metals. Resources 3:432–453

Chapter 15
Circular Bioeconomy in the Packaging Industry Utilizing Coconut Processing Waste and By-products

Remil Aguda, Reniel Pamplona, Ben Jeuck, and Jovale Vincent Tongco

1 Introduction

Plastic plays a significant role in different industries around the world. Plastic was originally coined from the Greek word "Plasticos" (πλαστικός), which means capable of being formed or molded. Plastic product materials are made of different polymers like polyethylene terephthalate (PET), polyethylene (PE): high-density (HDPE) and low-density (LDPE), polypropylene (PP), polystyrene (PS), and polyvinyl chloride (PVC) [1–3]. This was first created in 1860 and mass-produced for commercialization in 1907, until an estimated 400.3 million tons of plastic were produced by 2022 [4]. Nowadays, plastic products are being used in different fields of science, including, but not limited to, medicine, agriculture, biotechnology, food science, and engineering [2, 5–9]. Throughout the field of studies, plastic products are commonly used as packaging materials due to their versatility, long-lasting performance, lightweight, and lower costs. Despite its beneficial utility, poor management, negligence in proper disposal, and lack of plastic waste recycling practices have created plastic disposal into a more serious global warming concern. This problem arises from the negligence of plastic waste disposal of different packaging in manufacturing and industrial setups, which now continuously poses a threat to public health and environmental safety [10–14].

R. Aguda · R. Pamplona · B. Jeuck · J. V. Tongco (✉)
BYCO Philippines, San Lorenzo, Makati City, Philippines
e-mail: jtongco@uidaho.edu

R. Pamplona
Crop Biotechnology Division, Institute of Crop Science, College of Agriculture and Food Science, University of the Philippines Los Baños, College, Laguna, Philippines

J. V. Tongco
Department of Forest, Rangeland and Fire Sciences, University of Idaho, Moscow, ID, USA

A. Verma et al. (eds.), *Polymers and Composite Materials for Packaging: Smart Food Packaging and Solutions*, Materials Horizons: From Nature to Nanomaterials,
https://doi.org/10.1007/978-981-95-0677-4_15

To resolve the growing risk, different strategies and policies were crafted, including experiments on mitigating plastic pollution. Iroegbu and coworkers reported that ~459,000 journal publications have been published since plastic's creation [15]. The publication comprises information about technology and the issues resulting from its mismanagement. Moreover, it also emphasizes the mitigation process and policies on how this pollution will be controlled. One of the interesting solutions to plastic pollution is the implementation of "green alternatives." Plastics are being replaced by materials that are eco-friendly, recyclable, and not harmful to humans. This idea led to the transition to utilize sustainable, biodegradable, and plant-based materials like agricultural waste (AW) for the packaging industry. This includes AW that are abundant in starch, cellulose, lignocellulose, and chitosan [16]. The application of chemical treatment, microorganisms, and enzymes on AW was also studied to produce biodegradable polymers, making it easier to dispose of while reducing the harm to the environment, and contributing to the circular economy [17–19].

Coconut (*Cocos nucifera* L.) is referred to as a "Tree of Life" because of its numerous known benefits. This crop can generate various products and by-products from various parts: coconut leaves for woven handicrafts and brooms, coconut fibers from the husks for soil amendment, coconut shell for activated carbon, coconut meat for oil and baking ingredients, coconut trunk for lumber for construction, and coconut roots for medicine. In 2024, the global production of coconut reached 62.4 million metric tons, while Indonesia tops as the largest producer of coconut, followed by the Philippines, India, Brazil, and Sri Lanka, respectively. Coconut meat and water are usually consumed as food, leaving the other parts of the fruit as waste or raw material for other processes. These materials, in the form of coir fibers, coconut husk, shell, pith, and copra meal, contain high lignocellulose content. The coconut husk is composed of 15–17% hemicellulose, 26–28% cellulose, and 32–41% lignin, which are suitable for making packaging, cardboard, or insulation boards [20].

In this chapter, the potential use of agricultural waste is explored, especially coconut processing wastes and by-products, as another potential supply chain and value chain for packaging materials, other than plastics. Coconut wastes are materials without any economic value and are thrown away in landfills or dumped in farms for biodegradation, with their fate in the soil or bodies of water. Coconut by-products are materials that can be sold to an open market at a low value and considered to have a by-product economic credit to the raw material cost with regard to the overall cost of production. The review will also give an overview of the impact of these agricultural wastes and by-products on the circular economy globally.

2 Traditional Uses of Coconut Waste Products

Coconut is known for its wide range of applications, primarily in food and cooking but also in medicine, cosmetics, paper and packaging, soil conditioner and fertilizer, building materials, and as an energy source. Figure 1 shows the diverse uses of

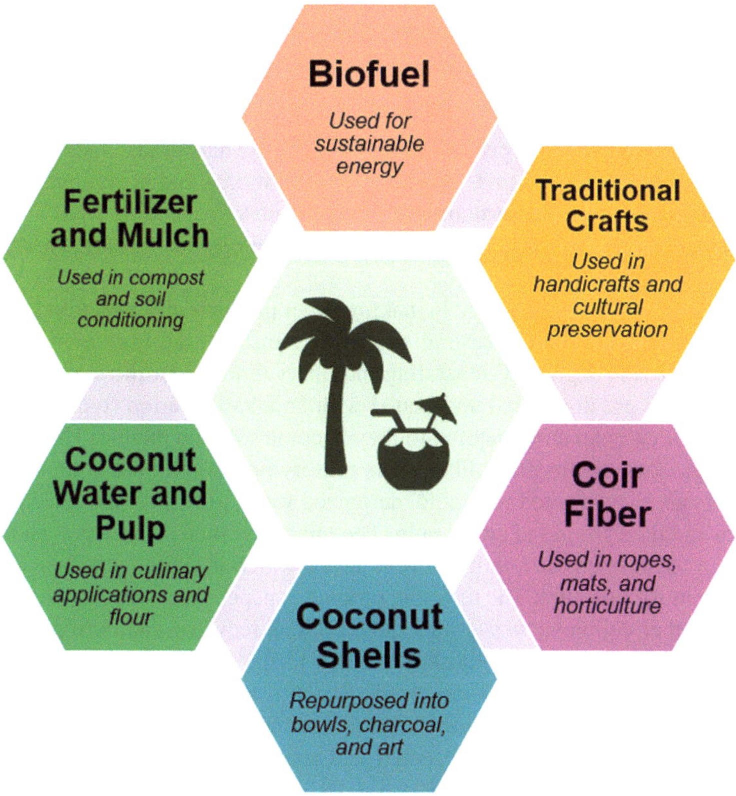

Fig. 1 Diverse uses of coconut

coconut after taking out coconut water and pulp for food applications, including the use of coconut waste such as biofuel, fertilizer, mulch, and coir fiber. The primary use of coconut is for its culinary purposes, such as the use of coconut milk as an ingredient in dishes, coconut water as a hydrating drink, and coconut pulp and meat for cooking and baking [21]. The coconut residue (coconut waste), after utilizing the aforementioned edible parts, is composed mostly of lignocellulosic biomass. Coconut waste utilization, mostly coming from the residual coconut husks and shells (particularly coconut coir), represents one of the most important examples of reusing waste biomass, with traditional processes now being improved by applying modern technologies to create viable alternatives to wood-based products, from paper and packaging materials exhibiting durability and flexibility, to fiberboards exhibiting higher density, stiffness, and water resistance. Similar to wood fiber, coir fiber is a renewable resource with similar physical characteristics and chemical properties. Only a portion of the coconut husk and shell waste is being utilized to produce fiber and other value-added products. Coir fiber is made up of lignocellulose, which consists of lignin, cellulose, and hemicellulose. The lignin component can be used as biomass for energy generation and production of platform chemicals, while the cellulose component can be used as feedstock for bioenergy (fermentation and anaerobic digestion) and a strong and renewable source of pulp, which produces paper products and packaging materials [22]. Extractives, with the likes of resins, terpenes, terpenoids, tannins, steroids, flavonoids, and other phenolics, pectins, suberins, proteins, fatty acids, lipids, waxes, and are among the minor constituents that make up waste biomass from plant sources, aside from lignocellulose. This is particularly true if the biomass material consists of bark, roots, stems, or dried leaves [23–32].

Traditional processes involved in making paper products and packaging materials from coconut waste involve several steps that transform the fibrous material into pulp. The process begins by separating the fibers (coir) from the husk and shells. The coir fibers are macerated and boiled with an alkali solution (typically sodium hydroxide or calcium carbonate) until they become soft and pliable. This treatment helps break down lignin in the fiber while preserving the cellulose component. The soft fibers are then washed with cold, deionized water to remove the residual alkali solution and then ground to create pulp. The mixture is then poured over and passed through a mesh to create a uniform fiber layer, which is subsequently pressed and dried to remove moisture. The resulting product is paper sheets from coconut waste. [33–35]. Other cellulosic waste materials, such as recycled and corrugated paper, are also used to reinforce the finished product [36]. A study showed that 60 gm^{-2} papers were produced from coconut husk. Starch was added as a binder to improve the properties of the resulting paper samples, and they were evaluated using physical and mechanical tests. The unreinforced and starch-reinforced paper samples had respective mean thicknesses of 0.267 ± 0.012 and 0.282 ± 0.016 mm. With the highest tensile and burst indexes of 19.25 Nm/g and 2.266 $kPam^2/g$, respectively, the reinforced 100% coconut husk paper specimens required the most 125 spins to be abraded. The absorbency rate of 60% unreinforced coconut fibers was the lowest, suggesting that it is 35 times more absorbent than regular paper with 60 g/m^2. The paper specimens with the highest mean recovery angle (77.6°) that were appropriate for writing materials were 40% wastepaper and 60% reinforced coconut [37].

Another application of using coconut waste is the production and manufacture of fiberboards. A recent study involved the use of coconut fiber reinforced with wood particles [38]. The fibers of waste coconut shells were separated, followed by air-drying for 2 months until they reached ~10% moisture content. The wood particles (sourced from a hybrid *Eucalyptus grandis* and *Eucalyptus urophylla* plantation) were mixed with coconut fibers in varying amounts by weight at 0, 10, 20, 30, and 40%. Urea–formaldehyde resin was added at 9% of the dry mass of the mixed lignocellulosic material. After mixing, the combined particles were placed in molds. Afterward, the panel was pressed at 130 °C and 0.98 MPa for 12 min. The fiberboard panels were conditioned in an environmental chamber at 20 °C and 65% relative humidity for 7 days. Figure 2 shows the flow diagram of fiberboard production using eucalyptus wood and coconut waste.

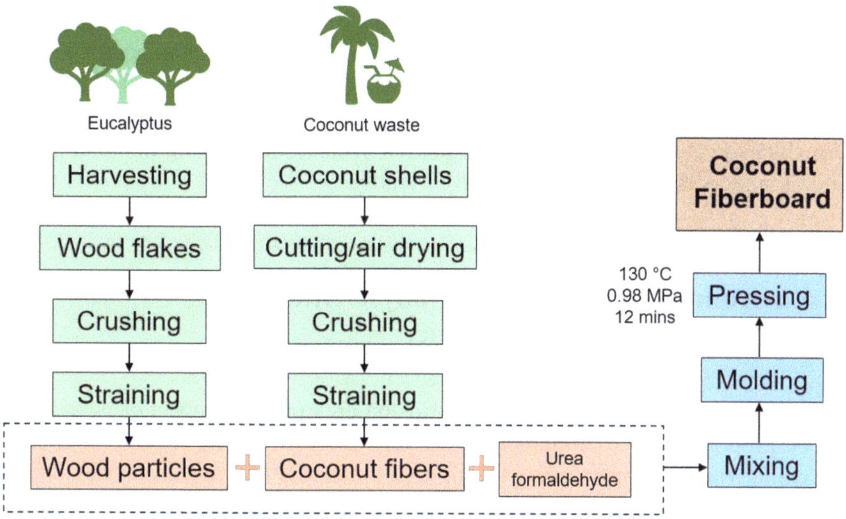

Fig. 2 Flow diagram of coconut fiberboard production

The presence of lignin in coconut husk and shell fibers acts as a natural binder in fiberboards under high pressure and temperature [39]. The use of traditional chemical binders and resin sourced from non-renewable and petroleum-based sources is potentially carcinogenic, environmentally hazardous, and not cost-effective [40]. Urea–formaldehyde is one of the most used resins in the fiberboard industry, and with a low-cost binder, the addition of around 10% of it in the fiberboard corresponds to more than 50% of the final cost [41, 42]. Exposing the fibers to elevated temperatures and pressures promotes the breaking of chemical bonds in the polymer matrix, resulting in depolymerization and fragmentation. Higher temperatures result in degradation of the cellulosic structure along with the aromatic rings in lignin, leading to a decline in mechanical properties of the fiberboard [43]. In recent years, research regarding the use of lignocellulosic materials such as sugarcane bagasse, poplar, banana, oil palm, kenaf, and even coconut has been explored [41, 43–48]. Binderless fiberboards are composed of processed particles having various shapes and sizes. The internal binding between the particles occurs through the activation of chemical components, predominantly lignin aromatic compounds, by hot pressing using temperature and pressure [49]. Lignin is a natural and renewable compound that is already present or can be supplemented for fiberboard production [50]. In the case of coconut biomass, a research study showed that it was feasible to produce fiberboards from unripe husks without the addition of synthetic binders [48]. White coir fiber and pith were pressed at 220 °C for 4 min at a pressure of 160 kgf/cm^2 to produce fiberboards with desirable properties. Fiberboards with adequate cure, high water uptake, and swelling were produced at temperatures lower than 220 °C. Prominent levels of degradation were seen at higher temperatures, which led to surface damage and stress points and, ultimately, poor mechanical characteristics for the fiberboards. The study also revealed that pressing temperature is crucial in the production of fiberboards from white coir materials. In the absence of external (natural or synthetic) binders, the lignin found in lignocellulose materials is cured when the right temperature is used. As the application of harmful binders like urea–formaldehyde can be avoided, the production of fiberboards using lignin and other natural binders is more environmentally friendly.

3 Environmental Life Cycle Analysis Studies as Frameworks for Circularity

This section is about the different scenarios on environment impact categories that can be proposed for environmental life cycle analysis (LCA) studies on coconut processing waste and by-product technologies. Circular bioeconomy systems focus on strategies to expand the use of biomass from agricultural crops like coconut. The future direct applications of these LCA studies on coconut waste are on product development and improvement (whether environmentally beneficial or not), strategic planning for communities of practice or stakeholder satisfaction (whether profitable or not), public policy making for areas vulnerable to climate change, drought or pollution (for corporate social responsibility programs and encourage social scientists to pursue further research to recommend policy), and material sourcing specifications for procurement managers who prefer to buy products with lower carbon footprint, water footprint, or less air pollution (as managers make decisions not only on price, quality, and supply chain reliability but also long-term sustainability).

Examples of LCA studies are given in Table 1. The LCA study on the palm empty fruit bunch offers a framework using open LCA for the greenhouse gas emissions from harvesting the coconut husk and coconut shell from the emptied coconut fruit when the coconut meat is harvested for copra processing. A proposed LCA study on the allocations for different scenarios on how the coconut husk and coconut shell are used for chemical production, fuel, planting material, or construction boards would allow strategic planning for farming communities vulnerable to climate change. Similarly, the LCA study conducted on lignin extraction from a Kraft pulp mill can also be applied to coconut-derived lignin products that can displace petroleum-derived phenolic resins. Pallets made from coconut shells can also be the functional unit, the basis for an LCA study, using the framework of the LCA study on the wood pallet sector in the USA. This study on coconut shell-based pallets necessitates the collection of data from regional pallet production facilities to analyze sensitivities in the adopted LCA wood pallet framework. Lastly, coco-paper and coco-cardboard can also be studied for product development and the improvement of recycling cardboard in island communities where there is an abundance of coconut processing waste and an increasing demand for cardboard self-sufficiency. An LCA study on corrugated cardboard box reuse suggested that the reuse is environmentally feasible and economically viable to pursue—the same concept applies to incorporating coconut process

Table 1 Environmental life cycle analysis (LCA) frameworks for coconut-based packaging

Goal and scope	Life cycle inventory database	Environmental impact category	Interpretation	References
Goal: To assess the environmental impact of commercial-scale production of sustainable green chemicals from empty fruit bunch biomass Scope: cradle-to-gate	Open LCA	Global warming potential (CO_2 eq)	Environmental burdens were sensitive to the allocation scenario, covered area, and land use change Delivery cost was also sensitive to transport distance, coverage, and yield	[51]
Goal: To assess the environmental impact of the commercial-scale lignin extraction process from the Kraft pulp mill Scope: cradle-to-gate	TRACI in WinGEMs	Global warming potential (CO_2 eq)	Environmental impact reduction was primarily due to the displacement of petroleum-derived phenolic resins by the lignin co-product. Sodium chlorate and natural gas were identified as process streams contributing significantly to environmental burdens at the mill	[52]
Goal: To assess the environmental implications of the wood pallet supply chain and identify optimization strategies that can be implemented Scope: cradle-to-grave	Data collected from U.S. pallet manufacturers were used to develop the first industry-average life cycle inventory	Global warming potential (CO_2 eq)	Total environmental impact was significantly affected by two main parameters: reference service life and load-bearing capacity Pallet repair was also found to be a vital component of the wood pallet supply chain, which has a lower environmental footprint compared with the overall impact of a pallet and enables mitigation of the overall impact by extending the reference life	[53]
Goal: To evaluate the carbon emissions in each phase of the lifecycle of a typical 1 kg corrugated cardboard box in the United States. Carbon emission for the proposed "reuse" phase is also calculated, and the results are compared Scope: cradle-to-gate and cradle-to-grave	Data from the United States environmental protection agency from 2018 and from surveys in the United States	Global warming potential (CO_2 eq)	The results indicate that the "reuse" method for CCBs is economically and environmentally feasible. It also supports the approach of using analytics, economics, and LCA to create a model that can be used for other products and processes as an evaluative process to determine if businesses can benefit from the reduction (or removal) of material extraction costs from the supply chain	[54]

waste into this LCA model. Overall, the application of these LCA models and frameworks from similar wood-derived or agricultural by-product-derived materials can be useful for strategic planning for mitigating climate change and also supporting the livelihood of coconut-growing communities vulnerable to climate change.

When the production systems utilizing coconut processing waste are expanded into other industries, the listed environmental impact categories are also the criteria for comparison of different scenarios using LCA frameworks. For example, carbon-capturing plants have higher greenhouse gas (carbon dioxide) utilization than the greenhouse gas emissions from their chemical, biological, or thermal processes. The redesign of coconut waste processes, including chemical, biological, or thermal processes to reduce the environmental impacts, can be measured in terms of global warming potential, human toxicity, water use, land use, and particulate matter. Table 2 shows the environmental impact categories in environmental LCA studies for the system expansion of coconut-based processing.

Scenarios for life cycle assessment for redesigning smart packaging systems using coconut waste processing are worthy of investigation:

- *Using coconut waste as an alternative to polystyrene foam for storing seafood by fishermen* [55]

Table 2 Environmental impact categories in environmental LCA studies for system expansion of coconut-based processing (based on environmental management systems reports and corporate sustainability reports)

Global warming potential: Greenhouse gases that contribute to climate change such as carbon dioxide and methane	Ozone depletion potential: Assesses the impact on the ozone layer, which is crucial protective shield against harmful UV irradiation	Acidification potential: Evaluates the potential for emissions like sulfur dioxide to form acid rain, which harm soil, water and ecosystems	Eutrophication potential: Measures the potential for nutrient pollution in aquatic ecosystems, often caused by excessive nitrogen or phosphorus emissions
Human toxicity: Assesses the potential for toxic substances released into the environment to harm human health	Ecotoxicity: Evaluates the potential for chemical emissions to harm non-human organisms in various ecosystems	Resource depletion: Considers that use of non-renewable resources like fossil fuels and minerals which can lead to sustainability risks	Land use: Evaluates the environmental impacts of converting land for different purposes, which affect biodiversity and ecosystem services
Water use: Assesses the amount of water in a product's life cycle, which can have implications for water availability and quality	Photochemical ozone creation potential: Measure the potential of emissions to contribute to the formation of ground level ozone, which can cause respiratory problems	Ionizing radiation: Evaluates the potential for radiation from radioactive materials to harm human health and ecosystems	Particulate matter: Assesses the impact of particulate matter emissions on human health and the environment

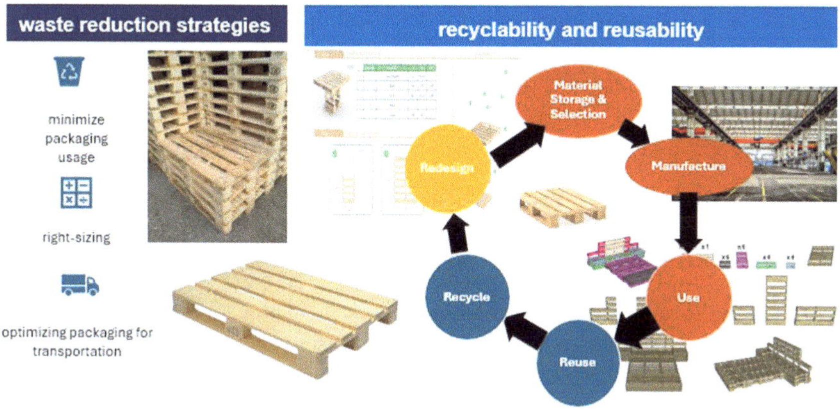

Fig. 3 Circular approach to material storage, selection, use, reuse, recycle, and redesign before the pallets are disposed of in landfills or burned as fuel

The application is to create a strategic plan for prioritizing recycled, renewable, and compostable materials like cardboard, paper, starch-based plastic, and cellulose for sustainable packaging redesign. Results of the LCA study are another case of transparency and communication for providing clear information about the packaging's sustainability and carbon footprint to consumers.

- *Using coconut waste pallets for agricultural commodities in warehouses* [56]

This case of reducing waste by implementing strategies to minimize packaging usage and material waste through right-sizing and optimizing packaging for efficiency of handling and transportation. This is also a compelling case of recyclability and reusability for promoting a circular economy for packaging materials, as shown in Fig. 3.

- *Using coconut waste for coco-board packaging*

This is a comparison case of utilizing Kraft paper and corrugated cardboard as sustainable smart packaging solutions with compostable coco-board packaging: end-of-life considerations when there are already packaging cardboard recycling instructions from sourcing in materials recovery systems to final disposal and addressing the environmental impact on each stage. Results of this study also open other opportunities for embracing innovative coconut-based composite materials other than cardboard, such as bio-based plastics, plant protein-based biodegradable packaging, and automated processes for efficiency in packaging. In addition to the analysis of the environmental impact on coconut waste processing into packaging, scenarios on cost model metrics for packaging products from coconut waste are presented. The next section lists the needed metrics for pursuing redesign and system expansion for a circular bioeconomy. System expansion happens when the coconut processing plant has either added more process units for its by-products or has included the utilization of its waste or by-products as part of its supply chain and value streams to other businesses.

4 Cost for Redesign and System Expansion for a Circular Bioeconomy

Research and development efforts across the circular bioeconomy space require a wide breadth of facilities, equipment, and expertise, requiring the convergence of chemistry, agriculture, biotechnology, forestry, and manufacturing. The costs can be summarized as follows:

Start-up Costs for Research and Development Initiatives

- Fundamental research on elucidating complex interactions between constitutive elements in man-made or natural polymeric composites (e.g., lignocellulosic biomass)
- Filling gaps in facilities that require interface with internal research programs to support existing institutional mission and goals
- Expand activities in renewable carbon materials development through the recently established university-wide institute for teaching, research, and extension
- Enhancing current teaching modules in engineering knowledge, complying with global quality standards of education, while involving students to engage in design projects for reliable preprocessing and recycling systems for biomass and bio-based materials and composites

Intermediate Costs for Further Research and Development

- Developing new tools to experimentally measure the mechanical properties of these materials during preprocessing and recycling in relevant industrial systems
- Building partnerships and collaborations through the university-wide institute leadership team, university administrators, and global engagement
- Establishing a project management life cycle for chemical and biological preprocessing and recycling the biopolymers back into their monomer building blocks.

Long-Term Costs for Sustaining Research Institutions

- Maintenance and operation of equipment with leases from technical vendors, with the mutual benefit of workforce development
- Establishing a group of professionals in academia, industry, and government agencies on content creation on social media for continuing education, outreach, and extension
- Participation in professional organizations for the dissemination of research data while maintaining a repository of engineering knowledge for public use

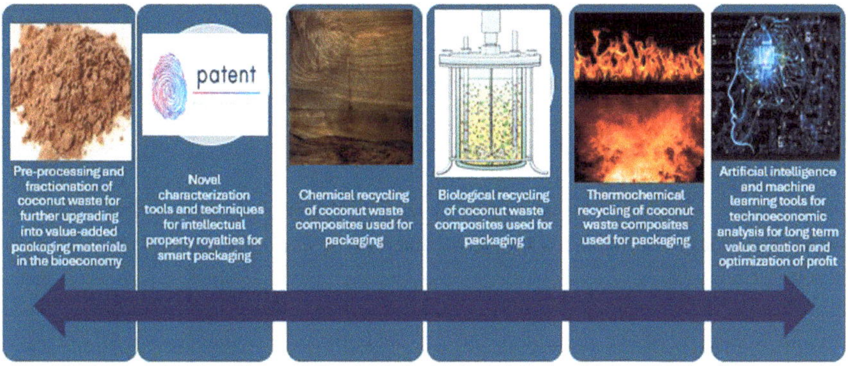

Fig. 4 Packaging research and development projects for circularity goals

- Establishing a staff of analytical chemists, materials scientists, service engineers, professors of practice, and business mentors for attracting the best global STEM talent and keeping them for workforce development, research productivity, and industry consulting.

Research and development project management decision-making tools with sustainability metrics for the circular bioeconomy are illustrated by compilations on systems utilizing lignocellulosic biomass by Aguda e al. [57]. Applying lean tools for process development prevents wasted time, resources, technician idle time, and saves material costs [58]. There are promising routes for gathering data while meeting sustainability criteria in the project management phases shown in Fig. 4. Metrics for sustainable investment for long-term value creation and maximizing profit include payback period, net present value, and future worth. When these prototypes, tools, techniques, and processes have been developed, the next step is for early adoption by companies, who are in the manufacture, distribution, marketing, and consumer education of packaging materials. The next section cites packaging companies that have reported their efforts in using sustainable materials globally while aligning themselves to the global sustainability goals set by the United Nations.

5 Sustainable Development Goals

Several packaging companies have shared their efforts in their sustainability reports since they have publicly traded stocks, for example, on the S&P 500. Sustainability reports are a way to attract more investors and also to provide transparent communication with stakeholders and stockholders. Currently, stock slices allow the public to invest in a portion of the company's shares. SDG reports of packaging companies have reported metrics from their circular bioeconomy systems in their global operations. These companies have reported metrics on aligning with the 17 Sustainable

Development Goals set by the United Nations, known as UN SDGs [59]. Having this baseline guide for the private sector allows them to track their progress and also provides a blueprint for those research and development efforts in universities, industries, and academia.

Avery Dennison reported advancing the circular economy setting their goal for 2025 in their operations with 95% landfill-free, with 75% of their waste reused, repurposed, or recycled. Through product lifecycle management programs, they plan to increase recyclability with products that enable the recyclability of plastics in their product portfolio. Labels show that these plastic components in their packaging liners contain recycled content. Furthermore, another measurable metric in the circular economy is the revenue from products that are reusable, recyclable, and/or compostable. This is possible through products designed for use-phase resource efficiency. An example is that at least 40 and 36% of their lifetime gross margin, and retail business intelligence solutions, respectively, were from products that enable recyclability, contain recycled content, or use less material while maintaining or exceeding material performance and specifications. This effort by Avert Dennison supports the goal of responsible consumption and production as a UN SDG [60].

Amcor has addressed end-of-life packaging and designed products to be recyclable and more efficient. This company also encouraged consumer participation and infrastructure development for circularity in their operations. Sustained efforts to advocate for clear, consistent regulations, while education of both customers and consumers is essential for their company to contribute positively to a more circular society. One example is turning plastic waste into plastic wood in Colombia, where more than 1500 metric tons of post-consumer plastic waste are generated per day, of which only 1% is incorporated back into the circular economy. These figures prompted the technical team in the plant in Cali, Colombia, to partner with a foundation and the members of the local community to develop a solution for sending less flexible packaging waste to landfills, while also benefiting the community. The team exceeded its initial goal for collecting 230 metric tons of material, slowing to create extra plastic wood to both build the classroom and furnish it with desks made from the material. Volunteers wrapped up the project by lending a hand in cleaning up and improving the school's playground, delivering a win-win situation for all involved. This project contributed to the activity metrics reported by the company in terms of the amount of production, by substrate, and percentage of production as paper and wood, and the number of employees involved in the project [61].

International Paper reported their plans to analyze and consolidate sustainability data and roadmaps to develop a new framework that reflects their current and future identity. This work has its 4 pillars of circularity: (1) source sustainable raw materials, (2) use of circular manufacturing processes, (3) create innovative products, and (4) collaborate on solutions. International Paper reported that it is a major global user of recovered fiber; the company uses over 5 million tons to make new fiber-based products in 2024. They also have a commitment to circularity yielded tangible results in 2024, with 49% of our manufacturing waste diverted from disposal for beneficial uses. Innovation involves design with the end of the product in mind, so that the material from the product is easily recovered, recycled, reused, or composted. Partnerships with the customers, suppliers, and thought leaders would foster innovation and gather a broad range of perspectives. This effort supports the goal of responsible consumption and production as a UN SDG [62].

Packaging Corp of America has a sustainability committee that formalized sustainable business principles to align their strategies, set goals for the organization, and report their progress through the year 2050. One of these principles is to manufacture high-quality, high-performance products from responsibly sourced renewable materials that are recyclable or reusable. PCA is the third-largest producer of container board and corrugated products in North America. PCA manufactures many grades of Kraft linerboard and corrugating medium at our container board mills and produces a wide variety of corrugated containers and displays at our converting facilities. Boise Paper is a leading producer of uncoated freesheet in North America. Their team provides high-quality products, outstanding customer service, and industry-leading supply chain performance, with a product portfolio that includes office papers and printing and converting papers. Their packaging segment reported 7 container board mills, 86 converting operations, 4.5 million tons of container board, and 60.5 billion square feet of corrugated products. Paper segment reported 1 white paper mill with 472 thousand tons of uncoated freesheet. The recycled content of our finished products is calculated based on the proportion of recovered fiber to overall production after considering production yield for each stream. In 2023, the average recycled content of our container board was 23% (8% pre-consumer/post-industrial, 15% post-consumer), with the substantial majority in our corrugating medium. Their white paper division sells products under a brand that specifies a minimum percentage of post-consumer recycled content (30, 50, and 100%). Overall, these data consolidation in the future also positions the company to meet the UN SDG goal of responsible consumption [63].

Smurfit West Rock revealed in their first sustainability report for the year 2024, after bringing together two highly complementary businesses and sets capabilities for circularity, that they manufactured 204.4 billion square feet of corrugated packaging and have key supply positions in solid boards, folding carton (consumer), and business-in-business markets. The company consumes ~14 million tons of recovered fiber every year globally, bringing valuable fiber back into the packaging production loop. Several case studies have been presented in their 2024 sustainability report. One scenario was on creating a circular packaging solution for a sustainable recipe box brand. They designed a fully recyclable box tailored to keep ingredients safe and cool during transit. Another solution was ingenious corrugated cardboard insulation packs in a smug two-piece telescopic design, reducing airflow, ideal for maintaining storage temperatures. The box size was also optimized, allowing more boxes to fit in a delivery truck. This investment enabled the production of 1200 boxes per hour, reducing manual labor, cutting costs, and providing faster delivery to customers. Another case study is on paper pallet wrap production. The craft paper was designed to be fully recyclable. Its capabilities also include high-energy adsorption and storage performance during humid conditions. The paper pallet wrap is manufactured in their paper mill in Spain. The company partnered with a leading glass supplier, which is the first company in the United Kingdom to use the new paper plastic wrap. An interesting case study was also presented in helping cherry distributors use paper baskets, made with a thinner e-flute corrugated board from renewable and recyclable materials. Overall, these product redesigns enable the company to meet the UN SDG goal of responsible consumption [64].

Global sustainable development goals from industry report on social, governance, and policy supporting coconut waste processing, product distribution, and procurement for packaging technologies incorporating coconut waste. Partnerships with these companies reporting their efforts aligning with sustainable development goals can be proposed on smart packaging using coconut waste to meet UN SDGs (Fig. 5). An example would be clean water, and sanitation can be achieved through reduced waste and by-products converted to activated carbon for water clean-up. Poverty, gender equality, and decent work for all can be achieved through policies that support programs to address these goals, as well as corporate social responsibility programs that report fair labor practices. Furthermore, ensuring access to reliable, sustainable energy for all by using end-of-life packaging as feedstock for energy from chemical, biological, and thermochemical conversion processes. Circularity is demonstrated when their by-products are used for soil amendment in coconut farming communities, educating the public on green chemistry practices from coconut chemical production, or reducing pollution in communities vulnerable to gaseous emissions in industrial areas located near coconut plantations.

Fig. 5 United Nations sustainable development goals (SDGs) can be met by packaging systems from coconut waste processing

6 Summary and Conclusion

Overall, this chapter has delved into the innovative use of coconut agricultural waste in packaging technologies, emphasizing the critical role in promoting sustainable, circular bioeconomics. By leveraging the expertise of specialists such as agriculturists, packaging engineers, wood chemists, chemical engineers, and biotechnology entrepreneurs, pathways for the utilization of coconut processing waste and by-products have been proposed. These proposals aim at developing and commercializing coconut waste-based packaging technologies, aligning with circular bioeconomy principles and the United Nations Sustainable Development Goals (UN SDGs). The integration of coconut waste into global supply chains, research, and lifecycle management of packaging materials enhances sustainability and reduces the overall environmental impact. As previously mentioned, studies have shown that the utilization of coconut husks can effectively reduce pollutants in water sources, thus contributing to cleaner and safer water. Additionally, the use of coconut by-products in packaging reduces reliance on non-renewable and primary material sources, thereby lowering the carbon footprint associated with both production and disposal. Findings, such as the study conducted by Halim et al., have indicated that incorporating coconut husk material into pallet production enhances the mechanical properties of the pallets, contributing to both an increase in physical properties and a reduction in environmental degradation [56]. Studies by Aguda et al. emphasize the application of lean tools in chemical processing development, demonstrating how coconut waste can be converted into value-added materials and promoting circular bioeconomy principles [57]. Another significant aspect highlighted is the transformation of coconut waste into activated carbon for water clean-up, contributing to the goal of cleaner water and sanitation. Collectively, these and other research findings help to underscore the importance of coconut waste in creating sustainable packaging and material solutions that align with global environmental goals. Policies supporting fair labor practices can be achieved through social responsibility programs of companies involved in coconut waste processing. These programs ensure decent work conditions and help alleviate poverty and reduce socioeconomic divide in the coconut farming communities.

Further, utilizing end-of-life packaging as feedstock for energy production through chemical, biological, and thermochemical conversion processes ensures access to reliable, sustainable energy for all. Finally, circularity in coconut waste processing is demonstrated by using by-products for soil amendment, educating the public on green chemistry practices, and reducing pollution in industrial areas near coconut plantations. This chapter also provides a comprehensive overview of how packaging technologies incorporating coconut waste can meet various UN SDGs, thus aligning efforts in research, development, and industry practices toward a more sustainable future. Ultimately, embracing coconut waste in packaging not only supports responsible consumption and production but also fosters innovation and community well-being, positioning these technologies as pivotal in the quest for global sustainability.

Conflicts of Interest There are no conflicts of interest to declare by the authors.

References

1. Torres-Agullo A, Karanasiou A, Moreno T, Lacorte S (2021) Overview on the occurrence of microplastics in air and implications from the use of face masks during the COVID-19 pandemic. Sci Total Environ 800:149555
2. Bai F, Chen G, Hu Y, Liu Y, Yang R, Liu J, Hou R, Li H, Wan X, Cai H (2024) Understanding the effect of plastic food packaging materials on food flavor: a critical review. Trends Food Sci Technol 43:104502
3. Tongco JV (2025) Effects of low concentration coupling agent addition on the physicomechanical behavior of wood fiber/HDPE composite. Med J Chem 15:187–195
4. Pilapitiya PGCNT, Ratnayake AS (2024) The world of plastic waste: a review. Clean Mater 11:100220
5. Rizan C, Mortimer F, Stancliffe R, Bhutta MF (2020) Plastics in healthcare: time for a re-evaluation. J R Soc Med 113:49–53
6. Hofmann T, Ghoshal S, Tufenkji N, Adamowski JF, Bayen S, Chen Q, Demokritou P, Flury M, Hüffer T, Ivleva NP (2023) Plastics can be used more sustainably in agriculture. Commun Earth Environ 4:332
7. Gilani IE, Sayadi S, Zouari N, Al-Ghouti MA (2023) Plastic waste impact and biotechnology: exploring polymer degradation, microbial role, and sustainable development implications. Bioresour Technol Rep 24:101606
8. Yates J, Deeney M, Muncke J, Carney-Almroth B, Dignac M-F, Castillo AC, Courtene-Jones W, Kadiyala S, Kumar E, Stoett P (2025) Plastics matter in the food system. Commun Earth Environ 6:176
9. Minde P, Kulkarni M, Patil J, Shelake A (2024) Comprehensive review on the use of plastic waste in sustainable concrete construction. Discov Mater 4:58
10. Kibria MG, Masuk NI, Safayet R, Nguyen HQ, Mourshed M (2023) Plastic waste: challenges and opportunities to mitigate pollution and effective management. Int J Environ Res 17:20
11. Gopinath PM, Parvathi VD, Yoghalakshmi N, Kumar SM, Athulya PA, Mukherjee A, Chandrasekaran N (2022) Plastic particles in medicine: a systematic review of exposure and effects to human health. Chemosphere 303:135227
12. Li Y, Tao L, Wang Q, Wang F, Li G, Song M (2023) Potential health impact of microplastics: a review of environmental distribution, human exposure, and toxic effects. Environ Health 1:249–257
13. Kumar R, Verma A, Shome A, Sinha R, Sinha S, Jha PK, Kumar R, Kumar P, Shubham DS (2021) Impacts of plastic pollution on ecosystem services, sustainable development goals, and need to focus on circular economy and policy interventions. Sustainability 13:9963
14. Saadu I, Farsang A (2023) Plastic contamination in agricultural soils: a review. Environ Sci Eur 35:13
15. Iroegbu AOC, Ray SS, Mbarane V, Bordado JC, Sardinha JP (2021) Plastic pollution: a perspective on matters arising: challenges and opportunities. ACS Omega 6:19343–19355
16. Baneshi M, Aryee ANA, English M, Mkandawire M (2024) Designing plant-based smart food packaging solutions for prolonging the consumable life of perishable foods. Food Chem Adv 53:100769
17. Samir A, Ashour FH, Hakim AAA, Bassyouni M (2022) Recent advances in biodegradable polymers for sustainable applications. NPJ Mater Degrad 6:68
18. Arruda EH, Melatto RAPB, Levy W, de Melo CD (2021) Circular economy: a brief literature review (2015–2020). Sustain Oper Comput 2:79–86
19. Duguma HT, Khule P, McArdle A, Fennell K, Almenar E (2023) Turning agricultural waste into packages for food: a literature review from origin to end-of-life. Food Packag Shelf Life 40:101166

20. Wijaya C, Sangadji NL, Muharja M, Widjaja T, Riadi L, Widjaja A (2025) An integrated green fractionation of coconut husk: hydrothermal and deep eutectic solvent pretreatment for enhanced sugar and lignin production. Bioresour Technol Rep 29:102078
21. Ahuja SC, Ahuja S, Ahuja U (2014) Coconut-history, uses, and folklore. Asian Agrihist 18:221
22. Poornachandhra C, Jayabalakrishnan RM, Balasubramanian G, Lakshmanan A, Selvakumar S, Maheswari M, John JE (2023) Coconut husk fiber: a low-cost bioresource for the synthesis of high-value nanocellulose. Biointerface Res Appl Chem 13:504
23. Haldar D, Purkait MK (2021) A review on the environment-friendly emerging techniques for pretreatment of lignocellulosic biomass: mechanistic insight and advancements. Chemosphere 264:128523. https://doi.org/10.1016/J.CHEMOSPHERE.2020.128523
24. Mujtaba M, Fernandes-Fraceto L, Fazeli M, Mukherjee S, Savassa SM, Medeiros GA, Pereira AES, Mancini SD, Lipponen J, Vilaplana F (2023) Lignocellulosic biomass from agricultural waste to the circular economy: a review with focus on biofuels, biocomposites and bioplastics. J Clean Prod 402:136815. https://doi.org/10.1016/J.JCLEPRO.2023.136815
25. Tongco JV, Ewurum N, Rubiyatno S (2025) Suberin: chemistry and applications. J Mater Environ Sci 16:1–14
26. Tongco JV, Añis A, Tamayo J (2015) Nutritional analysis, phytochemical screening, and total phenolic content of Basella alba leaves from the Philippines. Int J Pharm Phytochem Res 7:1031–1033
27. Tongco JV, Aguda R, Razal R (2014) Proximate analysis, phytochemical screening, and total phenolic and flavonoid content of Philippine bamboo *Schizostachyum lumampao*. J Chem Pharm Res 6:709–713
28. Ragasa C, Torres O, Tongco JV, Razal R (2014) Chemical constituents of *Petersianthus quadrialatus* (Merr.). Res J Pharm Biol Chem Sci 5:181–186
29. Tongco JV, Rodriguez E, Abasolo W, Mun SP, Razal R (2016) Mineral, nutritional, and phytochemical profile, total phenolic content, and radical scavenging activity of philippine bamboo "bolo" *Gigantochloa levis* (Blanco) Merr. leaves. Nat Prod Sci 22:60–63
30. Ragasa CY, Torres OB, Tongco JVV, Razal RA, Shen C-C (2013) Resorcinols from *Myristica philippensis* Lam. J Chem Pharm Res 5:614–616
31. Panti A, Aguda R, Razal R, Belina-aldemita MD, Tongco JV (2014) Proximate analysis, phytochemical screening and total phenolic and flavonoid content of the ethanolic extract of molave Vitex parviflora Juss. Leaves. J Chem Pharm Res 6:1538–1542
32. Tongco JV, Villaber RA, Aguda R, Razal R (2014) Nutritional and phytochemical screening, and total phenolic and flavonoid content of *Diplazium esculentum* (Retz.) Sw. from Philippines. J Chem Pharm Res 6:238–242
33. Phadnis P, Dongarjal A, Patil R, Bhave S, Runwal S, Ketkale C, Godbole R, Nagnath V (2018) Paper from coconut husk. Int J Adv Res Sci Eng 7:567–569
34. Arsyad M (2019) Sodium hydroxide and potassium permanganate treatment on mechanical properties of coconut fibers. IOP Conf Ser Mater Sci Eng 619:012011. https://doi.org/10.1088/1757-899X/619/1/012011
35. Arsyad M, Kondo Y, Arman S, Anzarih AM, Wahyuni N (2019) Effect of sodium hydroxide concentration on the tensile strength of coconut fiber. J Phys Conf Ser 1341:052001. https://doi.org/10.1088/1742-6596/1341/5/052001
36. Kitirojapn P, Teeravarunyou S, Kamolwit K, Pattanachan P (2025) Development of sustainable packaging cushions from coconut waste using 3D printing techniques. Environ Nat Resour J 23:132–141
37. Jeetah P, Jaffur N (2022) Coconut husk, a lignocellulosic biomass, as a promising engineering material for non-wood paper production. J Nat Fibers 19:5622–5636. https://doi.org/10.1080/15440478.2021.1889428
38. de Souza MJC, de Melo RR, Guimarães JB, Mascarenhas ARP, de Oliveira Paula EA, Pedrosa TD, Maskell D, Mensah P, Rodolfo F (2023) Eco-friendly particleboard production from coconut waste valorization. Environ Sci Pollut Res 30:15241–15252. https://doi.org/10.1007/S11356-022-23273-5/FIGURES/7

39. Van Dam JEG, Van Den Oever MJA, Teunissen W, Keijsers ERP, Peralta AG (2004) Process for production of high density/high performance binderless boards from whole coconut husk: part 1—lignin as intrinsic thermosetting binder resin. Ind Crops Prod 19:207–216. https://doi.org/10.1016/J.INDCROP.2003.10.003
40. Fan B, Zhang L, Gao Z, Zhang Y, Shi J, Li J (2016) Formulation of a novel soybean protein-based wood adhesive with desired water resistance and technological applicability. J Appl Polym Sci 133:43586. https://doi.org/10.1002/APP.43586
41. Hashim R, Nadhari WNAW, Sulaiman O, Sato M, Hiziroglu S, Kawamura F, Sugimoto T, Seng TG, Tanaka R (2012) Properties of binderless particleboard panels manufactured from oil palm biomass. BioResources 7:1352–1365
42. Curling SF, Loxton C, Ormondroyd GA (2012) A rapid method for investigating the absorption of formaldehyde from air by wool. J Mater Sci 47:3248–3251. https://doi.org/10.1007/S10853-011-6163-7/FIGURES/4
43. Quintana G, Velásquez J, Betancourt S, Gañán P (2009) Binderless fiberboard from steam exploded banana bunch. Ind Crops Prod 29:60–66. https://doi.org/10.1016/J.INDCROP.2008.04.007
44. Ali I, Jayaraman K, Bhattacharyya D (2015) Dimensional stability improvement of kenaf panels by post-manufacturing hygrothermal treatments using response surface methodology. Ind Crops Prod 67:422–431. https://doi.org/10.1016/J.INDCROP.2015.01.071
45. Mohareb ASO, Hassanin AH, Badr AA, Hassan KTS, Farag R (2015) Novel composite sandwich structure from green materials: mechanical, physical, and biological evaluation. J Appl Polym Sci 132:42253. https://doi.org/10.1002/APP.42253
46. Jayamani E, Hamdan S, Kok Heng S, Rahman MR, Bin Bakri MK (2016) Acoustical, thermal, and morphological properties of zein reinforced oil palm empty fruit bunch fiber bio-composites. J Appl Polym Sci 133:4164. https://doi.org/10.1002/APP.44164
47. Doost-Hoseini K, Taghiyari HR, Elyasi A (2014) Correlation between sound absorption coefficients with physical and mechanical properties of insulation boards made from sugar cane bagasse. Compos B Eng 58:10–15. https://doi.org/10.1016/J.COMPOSITESB.2013.10.011
48. Araújo Junior CP, Coaquira CAC, Mattos ALA, de Souza FM, Feitosa JPA, Morais JPS, de Freitas RM (2018) Binderless fiberboards made from unripe coconut husks. Waste Biomass Valorizat 9:2245–2254. https://doi.org/10.1007/S12649-017-9979-9/TABLES/5
49. Sun Y-C, Lin Z, Peng W-X, Yuan T-Q, Xu F, Wu Y-Q, Yang J, Wang Y-S, Sun R-C (2014) Chemical changes of raw materials and manufactured binderless boards during hot pressing: lignin isolation and characterization. BioResources 9:1055–1071
50. Mancera C, El Mansouri NE, Pelach MA, Francesc F, Salvadó J (2012) Feasibility of incorporating treated lignins in fiberboards made from agricultural waste. Waste Manag 32:1962–1967. https://doi.org/10.1016/J.WASMAN.2012.05.019
51. Reeb CW, Hays T, Venditti RA, Gonzalez R, Kelley S (2014) Supply chain analysis, delivered cost, and life cycle assessment of oil palm empty fruit bunch biomass for green chemical production in Malaysia. BioResources 9:5385–5416
52. Culbertson C, Treasure T, Venditti R, Jameel H, Gonzalez R (2016) Life cycle assessment of lignin extraction in a softwood kraft pulp mill. Nord Pulp Paper Res J 31:30–40
53. Alanya-Rosenbaum S, Bergman RD, Gething B (2021) Assessing the life-cycle environmental impacts of the wood pallet sector in the United States. J Clean Prod 320:128726
54. Ketkale H, Simske S (2023) A lifecycle analysis and economic cost analysis of corrugated cardboard box reuse and recycling in the United States. Resources 12:22
55. Baheramsyah A, Wardhana EM, Husein MA (2018) Analysis of palm fiber and coconut coir usage as purse seine cargo chamber insulator. Int J Mar Eng Innov Res 3:5462
56. Abdul Halim FA, Hashim IZ, Kamarozaman MNM (2023) Pallet from coconut husk
57. Aguda R, Bonilla S, Ben HJ, Revellame ED (2021) Challenges and opportunities in developing project management decision-making tools. J Eng Project Prod Manag 11:789
58. Aguda R (2020) Applying lean tools to chemical process development
59. Whittingham KL, Earle AG, Leyva-de la Hiz DI, Argiolas A (2023) The impact of the United Nations sustainable development goals on corporate sustainability reporting. BRQ Bus Res Quart 26:45–61

60. (2025) 2030 Sustainability goals report of Avery Dennison
61. (2024) 2024 Sustainability report of Amcor
62. (2024) 2024 Sustainability report of international paper
63. (2023) 2023 Responsibility report of packaging corporation of America
64. (2024) 2024 Sustainability report of Smurfit Westrock

Chapter 16
Integrated Pyrolysis-Gasification of Plastic and Polymers: Advancements, Pre-treatment Strategies, and Life Cycle Assessment

Rahul S. Raj, Vivek Kumar Sharma, Sai Parameshwar, and Akarsh Verma

1 Introduction

Plastic production has surged dramatically in recent decades, reaching 435 million tonnes in 2020, a significant increase from 234 million tonnes in 2000 [1, 2]. This escalation has been mirrored by a corresponding rise in plastic waste generation, with ~367 million tonnes produced globally in 2020, more than double the amount from 2000. Alarmingly, only about 10% of this plastic waste is recycled; the remainder is either incinerated (15%), landfilled (50%), or mismanaged, often ending up in the environment [3, 4]. Each year, an estimated 20 million tonnes of plastic waste leak into aquatic ecosystems, contributing to marine pollution and posing serious ecological and human health risks [5].

If current trends persist, global plastic production and waste generation are projected to increase by ~70% by 2040, worsening issues like marine litter, microplastic pollution, and greenhouse gas emissions [6, 7]. Plastics, primarily derived from fossil fuels, are responsible for about 5% of global greenhouse gas emissions, a figure expected to rise significantly in the coming decades. This situation underscores the urgent need for more effective plastic waste treatment and recycling methods to mitigate environmental harm and align with global sustainability goals [8, 9].

R. S. Raj (✉) · V. K. Sharma · A. Verma
Department of Mechanical Engineering, UPES, Dehradun, India
e-mail: rahul8348.raj@gmail.com

S. Parameshwar
Department of Petroleum Engineering and Earth Sciences, UPES, Dehradun, India

A. Verma et al. (eds.), *Polymers and Composite Materials for Packaging: Smart Food Packaging and Solutions*, Materials Horizons: From Nature to Nanomaterials,
https://doi.org/10.1007/978-981-95-0677-4_16

Traditional waste management strategies, such as mechanical recycling, have proven inadequate for GHG addressing the scale of plastic waste [10, 11]. Mechanical recycling is limited by the need for clean, well-sorted materials and is hindered by issues like polymer incompatibility and contamination [12]. Consequently, a significant portion of plastic waste ends up in landfills or is incinerated, both of which have detrimental environmental impacts [13].

In response to these challenges, thermochemical conversion technologies like pyrolysis and gasification have garnered attention as shown in Fig. 1 [14]. These high-temperature processes can convert mixed plastic waste, often in combination with biomass, into valuable energy-rich products [15–17]. Pyrolysis decomposes organic material in the absence of oxygen, yielding liquids, gases, and solid residues, while gasification involves partial oxidation to produce syngas, a mixture of hydrogen, carbon monoxide, and light hydrocarbons [18]. These processes offer several advantages over traditional methods, including better energy recovery and the ability to handle mixed or contaminated waste streams [14, 17].

Recent advancements have led to the development of integrated pyrolysis–gasification systems, which combine these processes to enhance efficiency and product yield [19]. By co-processing plastics with biomass, these integrated systems can lower activation energy requirements and significantly increase hydrogen output, transforming waste into clean fuel [20, 21]. Additionally, various process intensification techniques, such as microwave heating and catalytic tar cracking, further improve conversion efficiency and syngas quality [7, 8, 19].

These technological innovations are occurring alongside tightening regulations and ambitious sustainability targets. Internationally, the UN Sustainable Development Goals explicitly address plastic waste, with SDG 14.1 aiming to prevent and

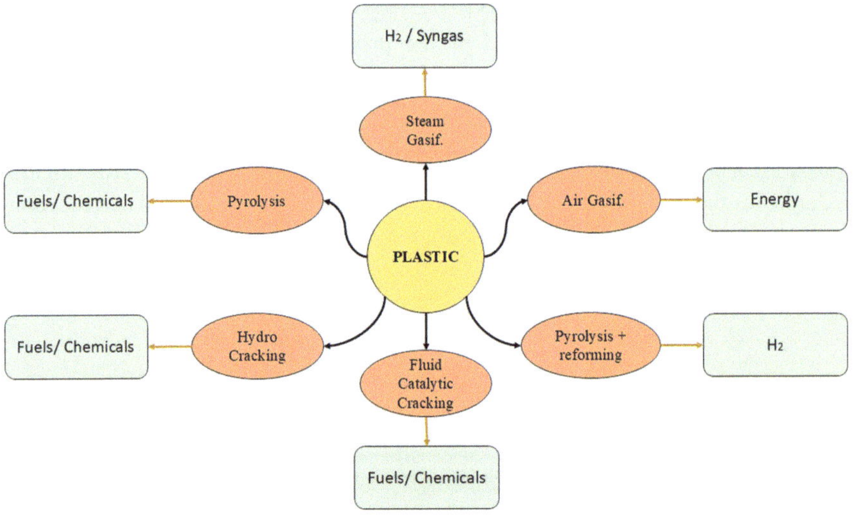

Fig. 1 Journey of plastic conversion into clean energy and fuel

reduce marine pollution by 2025 and SDG 12.5 targeting a significant reduction in waste through reuse, recycling, and recovery by 2030 [19, 22]. In 2022, the UN Environment Assembly adopted a historic resolution to develop a legally binding global treaty on plastic pollution, covering the full plastics life cycle. Regionally, initiatives like the EU's Circular Economy Action Plan and Green Deal have led to stringent plastic policies, including bans on single-use plastics and mandates for increased recycling rates [23–25].

Given the escalating plastic waste crisis, the limitations of traditional waste management methods, and the promising potential of thermochemical conversion technologies, this paper examines recent advancements in integrated pyrolysis–gasification of plastics [26, 27]. It explores strategies for process optimization and compares this approach with conventional treatment options, all supported by the latest research findings and their comprehensive LCA.

2 Fundamentals of Pyrolysis and Gasification

2.1 Pyrolysis of Plastics

Pyrolysis is a thermochemical decomposition process carried out in the absence of oxygen, wherein complex polymeric chains in plastic waste are broken down into smaller hydrocarbon compounds [28, 29]. The process yields three primary product streams: condensable liquids (commonly referred to as pyrolysis oil), non-condensable gases such as methane (CH_4), hydrogen (H_2), carbon monoxide (CO), and carbon dioxide (CO_2), and a residual solid char [30, 31]. Pyrolysis is particularly advantageous for treating heterogeneous plastic waste streams, as it can accommodate various polymer types—including polyolefins such as polyethylene (PE) and polypropylene (PP), polystyrene (PS), and polyethylene terephthalate (PET), without strict pre-sorting requirements. This makes it a flexible and scalable option for plastic waste valorization [15, 32].

The nature and yield of pyrolysis products are strongly influenced by several operational parameters. Temperature plays a critical role, with typical operating ranges between 350 and 600 °C [15]. The heating rate also affects the product distribution; fast pyrolysis, with heating rates exceeding 100 °C per minute, tends to maximize liquid product formation, whereas slow pyrolysis (1–10 °C/min) promotes the formation of solid char. Vapour residence time is another important factor; short residence times (<2 s) help preserve condensable vapours and minimize secondary cracking, thereby favouring higher liquid yields, whereas longer residence times typically enhance gas production [32–34]. Moreover, the type of plastic significantly influences the product outcome. Polyolefins generally produce high liquid yields, while plastics such as PVC and PET may emit chlorine or oxygenated compounds, necessitating pre-treatment or neutralization to mitigate environmental and equipment-related issues [18, 19, 25].

Recent technological advancements have focused on improving the selectivity and efficiency of the pyrolysis process [35, 36]. Catalytic pyrolysis, for instance, involves the use of catalysts such as zeolites and metal oxides to reduce reaction temperatures, suppress tar formation, and tailor the chemical composition of the pyrolysis products [37]. Additionally, microwave-assisted pyrolysis has been explored to achieve rapid and uniform heating, leading to enhanced energy efficiency [38]. Co-pyrolysis of plastics with biomass is also being investigated as a synergistic approach to improve the quality of pyrolysis oil and syngas while balancing the elemental composition of the feedstock [24, 32].

2.2 *Gasification of Plastics*

Gasification is a high-temperature thermochemical process, typically operating in the range of 700–1200 °C, which involves the partial oxidation of carbon-rich materials to generate synthesis gas, or syngas [36, 39, 40]. This gaseous product is primarily composed of carbon monoxide (CO), hydrogen (H_2), methane (CH_4), carbon dioxide (CO_2), and trace hydrocarbons [41, 42]. For plastic waste, gasification offers a significant advantage due to its ability to process mixed and contaminated waste streams while producing a clean-burning gas suitable for power generation, heating, or as a feedstock for chemical synthesis [43].

In contrast to pyrolysis, gasification introduces a controlled amount of an oxidizing agent typically air, oxygen, or steam, which initiates a series of complex reactions, including drying, devolatilization, oxidation, and reduction [44, 45]. The quality and composition of the resulting syngas are governed by several interdependent factors. Temperature is a key determinant, with higher operating temperatures above 900 °C favouring tar cracking and increasing the H_2/CO ratio, thus enhancing the calorific value and usability of the gas [42]. The choice of gasifying agent also plays a vital role; using steam instead of air can significantly improve hydrogen production and increase the heating value of the syngas. Another crucial factor is the equivalence ratio (ER), defined as the ratio of the actual air supplied to the stoichiometric air requirement [43, 45]. An ER in the range of 0.2–0.4 is typically considered optimal, as it maximizes syngas yield while minimizing complete combustion and the associated energy losses. Additionally, residence time influences the extent of conversion; while longer residence times may promote more complete breakdown of the feedstock, they can also lead to undesirable side reactions if not carefully managed [46–48].

Plastics generally possess a higher hydrogen-to-carbon (H/C) ratio compared to biomass, resulting in a syngas with a superior calorific value [24]. Despite these advantages, gasification of plastics presents challenges, particularly with the formation of tar and halogenated compounds, especially from feedstocks containing PVC and PET [49, 50]. These by-products can lead to operational difficulties and environmental concerns. Consequently, advanced treatment strategies such as secondary reforming, catalytic cracking, and plasma gasification have been developed to address these issues and enhance the quality and stability of the produced syngas [51, 52].

Table 1 Recent advances in pyrolysis and gasification of plastic waste

Process type	Feedstock	Key highlights	References
Co-pyrolysis	Municipal solid waste (MSW) and plastic waste	Comprehensive review on co-pyrolysis for renewable fuel and carbon materials; emphasizes environmental and economic sustainability aspects	[53]
Co-pyrolysis with machine learning	Biomass and polymeric wastes	Utilized evolutionary machine learning to optimize co-pyrolysis conditions; achieved high prediction accuracy ($R^2 > 0.9$)	[54]
Two-stage pyrolysis	HDPE, PP, PS, PET	Achieved hydrogen yields of up to 37.3/100 g of HDPE through pyrolysis followed by catalytic steam reforming	[55]
Gasification	Plastic waste	Reviewed mechanisms of plastic gasification; discussed integration with carbon capture for low-carbon hydrogen production	[19, 56]
Co-gasification	Plastic waste and biomass	Demonstrated that steam co-gasification enhances hydrogen production and syngas quality while reducing tar formation	[57]
Hydrothermal processing (HydroPRS)	Mixed plastic waste	Developed supercritical water-based recycling technology achieving high conversion efficiencies (~90%) with lower emissions	[58]
Integrated energy systems	Incineration power plant and P2G	Proposed a coordinated operation strategy integrating waste heat utilization and power-to-gas for low-carbon energy systems	[5]
Photo reforming	Plastic waste	Utilized high-entropy oxide catalysts for simultaneous plastic degradation and hydrogen production under light irradiation	[59]

These studies collectively advance the field of plastic waste management through innovative thermochemical processes as shown in Table 1, integration with renewable energy systems, and the application of machine learning for process optimization.

3 Integrated Pyrolysis-Gasification Process

3.1 *Process Flow and Operational Principles*

The integrated pyrolysis-gasification process is a two-stage thermochemical conversion technique designed to optimize the valorization of plastic waste as shown in Fig. 2 [24]. In the first stage, pyrolysis, plastic feedstocks are thermally decomposed in an oxygen-deficient environment, typically at temperatures ranging from 350 to

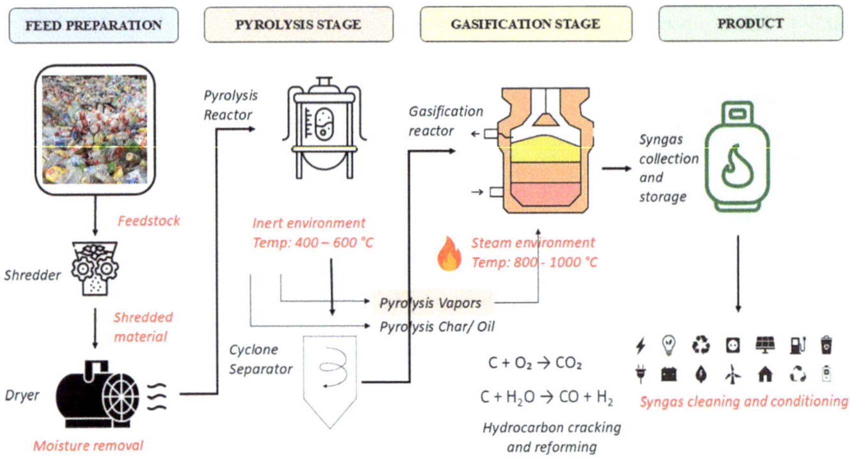

Fig. 2 Integrated pyrolysis-gasification pathway: converting plastic waste into syngas for a sustainable energy future

600 °C [60]. This process breaks down long-chain polymers into smaller molecules, yielding a mixture of condensable vapours (which can condense into pyrolysis oil), non-condensable gases (such as CH_4, H_2, CO, CO_2), and solid char [61]. The resultant volatile products and char are then introduced into the gasification stage, where they undergo partial oxidation at higher temperatures (700–1200 °C) in the presence of controlled amounts of oxidizing agents like air, oxygen, or steam. This subsequent stage converts the pyrolysis products into synthesis gas (syngas), primarily composed of carbon monoxide (CO), hydrogen (H_2), methane (CH_4), and carbon dioxide (CO_2) [60, 62].

This sequential approach allows for better control over the thermal decomposition and oxidation processes, enhancing the overall efficiency and quality of the syngas produced [61, 63]. The integration of these two processes enables the handling of a wide range of plastic wastes, including mixed and contaminated plastics, without the need for extensive pre-treatment or sorting [62].

3.2 Advantages of Integration

Integrating pyrolysis and gasification into a single waste-to-energy system offers distinct advantages over employing each process individually [64, 65]. One of the foremost benefits is the enhancement of syngas quality. The pyrolysis stage decomposes plastic polymers into simpler volatile compounds, which are more suitable to conversion in the subsequent gasification phase. This sequential breakdown improves the efficiency of carbon conversion and results in a cleaner syngas with elevated concentrations of hydrogen and carbon monoxide [66]. Another significant advantage is the reduction in tar formation, a common issue in direct gasification processes. Since complex hydrocarbons are already fragmented during pyrolysis, the tar load entering the gasifier is substantially decreased, mitigating risks such as clogging, corrosion, and operational downtime. Additionally, the integrated system promotes superior energy efficiency through heat integration [66, 67]. Heat generated during gasification can be harnessed to sustain the endothermic pyrolysis reactions, thereby minimizing the need for external energy inputs and improving the overall thermal performance of the system. Flexibility in feedstock handling is another notable benefit. The integrated approach can accommodate a diverse range of plastic wastes including mixed, contaminated, or non-sorted streams without strict pre-treatment requirements, making it a more adaptable and robust solution for heterogeneous plastic waste management [64, 65, 67].

3.3 Reactor Design Considerations

Developing reactors for an integrated pyrolysis-gasification system entails addressing multiple engineering and operational challenges [62, 68]. Material selection is critical, as the reactors must endure elevated temperatures and potentially corrosive environments. In particular, the presence of halogenated plastics such as polyvinyl chloride (PVC) can result in the formation of corrosive gases like hydrogen chloride, necessitating the use of corrosion-resistant alloys or ceramic linings. Thermal management between the two stages is another core consideration [68]. Efficient heat transfer mechanisms such as heat exchangers or regenerative systems are essential to maintain the required temperature gradients and improve energy recovery across the process. Consistent feed delivery can be facilitated through pre-treatment steps such as shredding and drying to ensure uniform particle size and composition. Scalability also plays a vital role in the reactor design process [69].

3.4 Process Integration Challenges

Despite its numerous advantages, the integration of pyrolysis and gasification presents a range of technical and economic challenges that must be carefully addressed for successful implementation [68]. Process synchronization is paramount; maintaining the optimal operational parameters such as temperature, residence time, and feedstock throughput for both stages require advanced control systems and responsive feedback loops [64]. Fluctuations in feedstock composition can disrupt the balance between the pyrolysis and gasification phases, negatively impacting system stability and product yield. Product consistency, especially in terms of syngas composition, also poses a challenge. The integrated process can potentially release hazardous pollutants such as dioxins, furans, or acid gases, particularly when halogenated plastics are involved. Consequently, effective gas cleaning units such as wet scrubbers, filters, and catalytic converters must be incorporated to meet environmental compliance and reduce health risks [64, 65].

3.5 Recent Developments

Recent studies have underscored the suitability of various polymers for integrated pyrolysis-gasification systems, with Table 2 presenting a comparative overview of key physicochemical properties that influence conversion efficiency [70, 71]. The ultimate and proximate analyses of these polymers reveal significant heterogeneity in elemental composition, thermal behaviour, and energy content. High-carbon and low-oxygen polymers such as polyethylene (PE), polypropylene (PP), and polystyrene (PS) exhibit elevated higher heating values (HHVs), which indicate strong potential for energy recovery. These polymers also have negligible ash content and extremely low moisture levels, characteristics that are advantageous for thermal conversion. These plastics show high volatile matter (VM) which promotes rapid devolatilization during pyrolysis and enhances syngas production in the subsequent gasification phase [72, 73].

Polymers such as polyethylene terephthalate (PET) and polyvinyl chloride (PVC) exhibit lower HHVs and higher oxygen content. PET, with a significantly lower carbon content and higher oxygen content, yields a much lower energy value, making it less efficient for thermochemical conversion. PVC, while comparable in volatile matter, presents operational challenges due to its chlorine content, which can form corrosive compounds like hydrogen chloride during thermal degradation. Tyres, though rich in carbon and energy, have relatively high fixed carbon and ash content, which may complicate reactor performance and syngas quality if not properly managed [71, 74].

The minimum operational temperatures for pyrolysis and gasification also vary across polymers. Thermoplastics such as PS require the lowest pyrolysis initiation temperature around 240 °C, whereas others like HDPE and LDPE require higher thresholds of 400–450 °C. Gasification temperatures for most polymers range between 600 and 750 °C, aligning with the optimal conditions for converting volatile compounds into syngas with favourable hydrogen yields. Among the few studies reporting hydrogen production metrics, PE and PP show promising yields, reinforcing their suitability for hydrogen-focused energy applications [75].

4 Reactor Designs and Operational Parameters

The design and operation of reactors are critical in optimizing the efficiency and output of pyrolysis and gasification processes, particularly when integrated to manage complex feedstocks such as mixed plastics. Recent advancements have led to the

Table 2 Pre- and post-treatment characteristics of polymers in integrated pyrolysis-gasification

Polymer	Ultimate analysis (wt.%)					Proximate analysis (wt.%)				Min. pyrolysis temp. (°C)	Min. gasifying temp. (°C)	HHV (MJ/kg)	H_2 yield (mmol/g)	References
	C	H	N	S	O	M	VM	FC	Ash					
PE	85.81	13.86	400	700	0	0.02	99.85	0	0.15	400	700	42.9	37	[73, 76]
PP	85.88	14.47	450	700	0	0	99.67	0.10	0.21	450	700	43.4	20	[77]
PET	61.87	4.35	450	650	33.78	0.7	89	10.2	0.1	450	650	10.1–13.2	9	[72]
LDPE	85.58	14.43	400	750	0	0.06	97.81	0.93	1.2	400	750	41.742	–	[78]
HDPE	85.49	14.52	450	750	0	0.09	98.12	0.97	0.82	450	750	40.614	–	[71]
PVC	37.78	4.83	350	600	0.3	0.01	96.5	3.4	0.7	350	600	–	16	[71]
PS	92.20	7.93	240	600	0	0	99.72	0.12	0.16	240	600	41.7	14	[75]
Tyres	85.2	7.3	400	600	0.5	1.2	61.3	33.5	5.2	400	600	38.5	–	[70]

development of innovative reactor designs and operational strategies that enhance product yields, reduce emissions, and improve overall process efficiency.

4.1 Reactor Types

The choice of reactor type significantly influences the product distribution in pyrolysis-gasification systems [74]. Fixed-bed reactors are widely recognized for their simplicity and low cost but face challenges related to limited heat transfer and batch-mode constraints, reducing their scalability for industrial applications. In contrast, fluidized-bed reactors offer superior heat and mass transfer, enabling uniform temperature distribution and continuous operation, though they require precise control to mitigate issues like bed agglomeration. Spouted-bed reactors, characterized by a central gas jet spouting mechanism, facilitate effective heat transfer and are well-suited for processing coarse or irregular feedstocks; however, they often exhibit poor mixing and high sensitivity to particle size variations. Rotary kilns, known for their ability to process large volumes of heterogeneous materials continuously, demand significant energy input and meticulous thermal regulation to avoid process inefficiencies [79–81].

To overcome the limitations of conventional designs, recent innovations have led to the development of hybrid reactors that integrate the strengths of multiple systems. Notable advancements include the incorporation of catalytic reformers within gasification units, which enhance syngas quality by breaking down tars and complex hydrocarbons into lighter, more valuable components [81–83]. Similarly, the emergence of circulating fluidized-bed reactors has further improved heat and mass transfer dynamics, promoting more efficient and homogeneous reactions. These innovations support the advancement of integrated pyrolysis-gasification technologies, making them more scalable, efficient, and suitable for processing diverse waste feedstocks [84–86].

4.2 Operational Conditions

The performance of pyrolysis and gasification processes is highly sensitive to operational parameters such as temperature, pressure, and the presence of catalysts [87]. Pyrolysis typically occurs between 350 and 600 °C, while gasification requires higher temperatures, ranging from 700 to 1200 °C. The temperature influences the composition of the products, with higher temperatures favouring the production of syngas over liquids and solids [81]. Operating pressures can vary depending on the reactor type and desired products. Higher pressures can enhance the yield of certain products and improve the efficiency of gasification reactions. For example, increasing pressure during pyrolysis in a hydrogen atmosphere has been reported to increase the yield of methane and ethane, while decreasing the yield of acetylene. The use of catalysts can significantly affect the product distribution. For instance, iron oxide has been used to suppress tar formation and increase hydrogen yield during gasification [80, 88]. Similarly, platinum-based catalysts can aid in the reforming of tars into lighter hydrocarbons, improving syngas quality.

Recent developments have focused on optimizing these operational parameters to enhance process efficiency. Machine learning algorithms have been employed to analyze vast amounts of data from historical and real-time sensor readings, enabling dynamic adjustments to process parameters and ensuring optimal performance under various conditions [88]. These algorithms can predict the outcomes of complex pyrolysis reactions, assisting in fine-tuning process parameters to achieve desired product yields and compositions [89]. Additionally, advancements in reactor design, such as the development of microwave-assisted pyrolysis systems, have shown promise in reducing reaction times and improving energy efficiency.

4.3 Recent Innovations in Reactor Designs

Advancements in reactor technology have focused on improving efficiency, scalability, and product quality. Innovations include the development of circulating fluidized-bed reactors that offer better heat and mass transfer properties, and the integration of catalytic reformers within gasification systems to enhance syngas quality. Table 3 provides a comparative overview of different reactor types used in pyrolysis and gasification processes. It highlights the mode of operation, advantages, limitations, and typical applications of each reactor type, aiding in the selection of the most appropriate reactor design based on specific process requirements and feedstock characteristics. The integration of recent innovations into reactor designs has led to more efficient and sustainable pyrolysis and gasification processes. By optimizing operational conditions and incorporating advanced technologies, these advancements contribute to the development of integrated systems that can effectively manage complex feedstocks, reduce emissions, and produce valuable products.

5 Comparative Analysis: Direct Gasification vs. Integrated Pyrolysis-Gasification

5.1 Syngas Composition

The composition of syngas generated from direct gasification and integrated pyrolysis-gasification processes plays a critical role in determining its suitability for specific end-use applications such as power generation, chemical synthesis, or fuel production. Direct gasification typically produces syngas rich in carbon monoxide (30–60%) and hydrogen (25–30%), with moderate levels of carbon dioxide (5–15%)

Table 3 Comparative overview of reactor types

Reactor type	Mode of operation	Advantages	Limitations	Applications	References
Fixed-bed	Batch/ semi-batch	Simple design, low cost, high char yield	Lower heat transfer, scalability issues	Small-scale pyrolysis and gasification	[74]
Fluidized-bed	Continuous	High heat and mass transfer, uniform temperature	Bed agglomeration, sensitive to gas flow rates	Medium to large-scale operations	[69]
Spouted-bed	Continuous	High heat and mass transfer, effective for coarse feedstocks	Poor mixing, sensitive to particle size distribution	Specialized applications	[79]
Rotary kiln	Continuous	Suitable for heterogeneous feedstocks, large capacity	High energy consumption, requires precise control	Large-scale industrial applications	[90]

and minimal methane content (0–5%) as shown in Table 4. This composition is particularly favourable for synthesis processes that require a high CO/H_2 ratio.

In contrast, integrated pyrolysis-gasification tends to yield a more balanced syngas profile, characterized by moderately lower concentrations of CO (20–40%) and H_2 (20–30%), but significantly elevated levels of methane (10–20%) and CO_2 (15–25%). The increased methane content enhances the energy density of the syngas, making it advantageous for combustion-based applications and direct fuel use. The lower CO and H_2 levels may, however, limit its effectiveness in synthesis applications unless further upgrading is performed.

Table 4 Syngas composition in direct gasification and integrated pyrolysis-gasification

Component (%)	Direct gasification	Integrated pyrolysis-gasification
CO	30–60	20–40
H_2	25–30	20–30
CO_2	5–15	15–25
CH_4	0–5	10–20

5.2 *Energy Efficiency*

Energy efficiency is a critical parameter in evaluating the technical viability and environmental sustainability of biomass conversion technologies. Both direct gasification and integrated pyrolysis-gasification represent thermochemical approaches with distinct operational characteristics and energy performance profiles.

Direct gasification involves the partial oxidation of biomass in a controlled environment, yielding a syngas composed predominantly of hydrogen (H_2), carbon monoxide (CO), carbon dioxide (CO_2), and methane (CH_4). As an exothermic process, it generates thermal energy, a portion of which sustains the reaction, while the remainder can be harnessed for electricity or heat generation. However, the energy efficiency of direct gasification systems remains relatively modest, typically ranging between 10 and 20%. Several factors contribute to this limitation. Incomplete combustion often results in unconverted residues or heat losses. Additionally, the syngas produced generally contains impurities such as tars, particulates, and acid gases, necessitating downstream cleaning and conditioning processes that further consume energy. Moreover, thermodynamic constraints, hinder the full utilization of thermal gradients due to challenges in maintaining high reaction temperatures without significant external input.

Despite these limitations, direct gasification continues to be a practical and scalable option, especially when coupled with combined heat and power (CHP) systems. Such integration can enhance overall system efficiency by simultaneously utilizing the thermal and electrical outputs of the process.

Integrated pyrolysis-gasification combines two complementary thermochemical steps such as pyrolysis and gasification to achieve higher energy conversion efficiencies. In this configuration, biomass or plastic is initially subjected to pyrolysis. The resultant gases, char and volatiles, are subsequently gasified to further extract chemical energy. This two-stage approach enhances overall system efficiency through synergistic utilization of both primary and secondary products. Notably, integrated systems have demonstrated energy conversion efficiencies of more than 70% when geared towards the production of methanol or dimethyl ether (DME), significantly surpassing the efficiency range observed in direct gasification.

The efficiency gains in integrated systems stem from several advantages. First, the utilization of by-products such as bio-oil and biochar adds value and minimizes waste, thereby improving net energy yield. Second, these systems offer improved feedstock flexibility, enabling the processing of materials with high moisture content or heterogeneous composition that would otherwise be unsuitable for conventional gasification. Third, the calorific value and quality of syngas produced in integrated systems tend to be higher, with lower impurity levels, rendering it more suitable for advanced applications such as chemical synthesis or high-efficiency power generation.

6 Techno-Economic and Environmental Assessments

Integrated pyrolysis-gasification systems present a promising pathway for sustainable biomass conversion, offering enhanced energy efficiency and reduced environmental impacts compared to traditional methods. This section delves into the techno-economic aspects, environmental performance focusing on Global Warming Potential (GWP) as a key Life Cycle Assessment (LCA) category and the commercial viability of these integrated systems [63].

6.1 Cost Analysis

The economic feasibility of integrated pyrolysis-gasification systems is influenced by both capital expenditures (CAPEX) and operational expenditures (OPEX). CAPEX encompasses costs associated with plant construction, equipment procurement, and infrastructure development. OPEX includes expenses related to feedstock acquisition, energy consumption, maintenance, labour, and waste management.

A techno-economic analysis conducted on a hybrid biomass-to-power system integrating pyrolysis and gasification indicated that while the initial investment is substantial, the combined system offers improved energy efficiency and reduced fuel costs over its operational lifetime. The integration allows for better utilization of energy, leading to lower overall operational costs compared to standalone systems.

6.2 Life Cycle Assessment

LCA evaluates the environmental impacts of a product or system from cradle to grave, encompassing all stages from raw material extraction to end-of-life disposal. Among various impact categories, Global Warming Potential (GWP) is a critical metric, quantifying the amount of greenhouse gases emitted, expressed in terms of CO_2 equivalents [63]. Studies have shown that integrated pyrolysis-gasification systems exhibit a significantly lower GWP compared to traditional pyrolysis and incineration methods. Table 5 compares the environmental impacts of standalone pyrolysis, standalone gasification, and integrated pyrolysis-gasification (IPG) processes. It highlights how IPG offers significant environmental advantages, including the lowest greenhouse gas emissions, minimal tar formation, and reduced solid residue generation due to its enhanced thermal integration and in-situ vapour cracking. Additionally, IPG simplifies pollution control requirements, making it a cleaner and more efficient waste-to-energy approach.

Furthermore, the integration of pyrolysis and gasification allows for the production of multiple valuable products, including biochar, bio-oil, and syngas, thereby enhancing the overall sustainability of the system. The ability to process mixed

Table 5 Comparative environmental impacts of pyrolysis, gasification, and integrated pyrolysis-gasification processes

Impact parameter	Pyrolysis	Gasification	Integrated pyrolysis-gasification (IPG)	References
GHG emissions	Moderate emissions; CO_2 and CH_4 from oil/gas combustion	Lower CO_2 due to partial oxidation; higher H_2/CO ratio in syngas	*Lowest emissions* due to higher efficiency and integrated heat recovery	[5, 91, 92]
Tar formation	*High tar content* in pyrolysis vapours; needs extensive scrubbing/cracking	Moderate tar; still requires reforming or high-temp cleanup	*Lowest tar* due to in-situ cracking during gasification of pyrolysis vapours	[5, 93]
Solid waste residue	*High char generation*; contains unconverted carbon and potential toxins	Ash and slag; lower in volume and carbon content	*Minimal residue* as char is further gasified, improving carbon conversion	[31, 82, 87, 94, 95]
Pollution control needs	Complex: oil condensation, tar removal, acid gas treatment, char handling	Syngas cleaning (H_2S, HCl, dust); needs temperature-resistant filters	*Simplified system*: no oil stream, fewer contaminants, reduced treatment requirements	[96–99]

and contaminated feedstocks further contributes to the reduction of environmental impacts, as it diverts waste from landfills and incineration.

7 Conclusion and Future Perspectives

This paper aimed to provide a comprehensive comparative analysis of direct gasification and integrated pyrolysis-gasification technologies for biomass conversion, focusing on their energy efficiency, environmental impact particularly Global Warming Potential (GWP) and techno-economic viability. Through detailed examination, it has been established that integrated pyrolysis-gasification systems generally offer superior performance across these parameters. These integrated systems not only achieve higher energy conversion efficiencies but also demonstrate significantly lower GWP, indicating a reduced environmental footprint. From an economic perspective, while the initial capital investment may be higher, the long-term operational savings and potential for producing multiple valuable by-products enhance their commercial viability.

Looking ahead, the development of integrated pyrolysis-gasification technologies holds considerable promise for advancing sustainable biomass conversion. Future research should focus on optimizing the integration of pyrolysis and gasification processes to further enhance energy efficiency and product yield. Advancements in catalyst development and reactor design could lead to more efficient conversion processes, reducing energy consumption and operational costs. Additionally, exploring the utilization of diverse feedstocks, including agricultural and municipal waste, can improve feedstock flexibility and contribute to waste reduction. The incorporation of carbon capture and storage (CCS) technologies within these systems could further mitigate greenhouse gas emissions, aligning with global climate goals. Furthermore, scaling up these technologies and demonstrating their feasibility in various geographical and economic contexts will be crucial for their widespread adoption.

Acknowledgements This research work was conducted by the financial support, provided by UPES, Dehradun, India.

Declaration of Competing Interest The authors declare that they have no known competing financial interests or personal relationships that could have appeared to influence the work reported in this paper.

References

1. Kumar H, Yoshida A (2024) Sustainable development growth in the plastic industries is promising or just a hoax: past, present, and future aspects on global perspective
2. Häder D-P (2021) Dumping of toxic waste into the oceans. Anthrop Pollut Aquat Ecosyst 54:353–371
3. Liu X, Lei T, Boré A, Lou Z, Abdouraman B, Ma W (2022) Evolution of global plastic waste trade flows from 2000 to 2020 and its predicted trade sinks in 2030. J Clean Prod 376:134373
4. Haba B, Djellali S, Abdelouahed Y, Boudjelida S, Faleschini F, Carraro M (2025) Transforming plastic waste into value: a review of management strategies and innovative applications in sustainable construction. Polymers 17(7):881
5. Demetrious A, Crossin E (2019) Life cycle assessment of paper and plastic packaging waste in landfill, incineration, and gasification-pyrolysis. J Mater Cycles Waste Manag 21:850–860
6. Lebreton L, Andrady A (2019) Future scenarios of global plastic waste generation and disposal. Palgrave Commun 5(1):1–11
7. Sharma S, Sharma V, Chatterjee S (2023) Contribution of plastic and microplastic to global climate change and their conjoining impacts on the environment: a review. Sci Total Environ 875:162627
8. Qureshi MS et al (2020) Pyrolysis of plastic waste: opportunities and challenges. J Anal Appl Pyrolysis 152:104804
9. Jain S (2019) The production of biodiesel using Karanja (*Pongamia pinnata*) and Jatropha (*Jatropha curcas*) oil. Biomass, biopolymer-based materials, and bioenergy. Elsevier, Amsterdam, pp 397–408
10. Ragaert K, Delva L, Van Geem K (2017) Mechanical and chemical recycling of solid plastic waste. Waste Manag 69:24–58
11. Singh AR, Singh SK, Jain S (2022) A review on bioenergy and biofuel production. Mater Today Proc 49:510–516
12. Vollmer I et al (2020) Beyond mechanical recycling: giving new life to plastic waste. Angew Chem Int Edn 59(36):15402–15423
13. Wojnowska-Baryła I, Bernat K, Zaborowska M (2022) Plastic waste degradation in landfill conditions: the problem with microplastics, and their direct and indirect environmental effects. Int J Environ Res Public Health 19(20):13223

14. Nanda S, Berruti F (2021) Thermochemical conversion of plastic waste to fuels: a review. Environ Chem Lett 19(1):123–148
15. Li N, Liu H, Cheng Z, Yan B, Chen G, Wang S (2022) Conversion of plastic waste into fuels: a critical review. J Hazard Mater 424:127460
16. Dwivedi G, Jain S, Shukla AK, Verma P, Verma TN, Saini G (2022) Impact analysis of biodiesel production parameters for different catalyst. Environ Dev Sustain 65:1–21
17. Singh VP, Jain S, Gupta JML (2022) Performance assessment of double-pass parallel flow solar air heater with perforated multi-V ribs roughness: part B. Exper Heat Transf 35(7):1059–1076
18. Manickavelan K, Ahmed S, Mithun K, Sathish P, Rajasekaran R, Sellappan N (2022) A review on transforming plastic wastes into fuel. J Niger Soc Phys Sci 74:64–74
19. Sharuddin SDA, Abnisa F, Daud WMAW, Aroua MK (2016) A review on pyrolysis of plastic wastes. Energy Convers Manag 115:308–326
20. Jain S, Sharma MP (2012) Oxidation, thermal, and storage stability studies of *J. curcas* biodiesel. Int Sch Res Notices 2012(1):861293
21. Dwivedi G, Jain S, Sharma MP (2011) Impact of biodiesel and its blends with diesel and methanol on engine performance. IJES 1(2):105–109
22. Kumar R et al (2021) Impacts of plastic pollution on ecosystem services, sustainable development goals, and need to focus on circular economy and policy interventions. Sustainability 13(17):9963
23. Wang S (2023) International law-making process of combating plastic pollution: status Quo, debates and prospects. Mar Policy 147:105376
24. Shah HH et al (2023) A review on gasification and pyrolysis of waste plastics. Front Chem 10:960894
25. Rahman MH, Bhoi PR, Menezes PL (2023) Pyrolysis of waste plastics into fuels and chemicals: a review. Elsevier, New Jersey
26. Saini M, Sharma A, Singh VP, Dwivedi G, Jain S (2022) Solar thermal receivers—a review. In: Advancement in materials, manufacturing and energy engineering: select proceedings of ICAMME 2021, pp 311–325
27. Singh VP, Jain S, Gupta JML (2023) Analysis of the effect of perforation in multi-v rib artificial roughened single pass solar air heater: part A. Exper Heat Transf 36(2):163–182
28. Jain S, Sharma MP (2012) Correlation development between the oxidation and thermal stability of biodiesel. Fuel 102:354–358
29. Jain S, Sharma MP (2011) Study of oxidation stability of *J. curcas* biodiesel/diesel blends. Int J Energy Environ 2(3):533–542
30. Gug J, Cacciola D, Sobkowicz MJ (2015) Processing and properties of a solid energy fuel from municipal solid waste (MSW) and recycled plastics. Waste Manag 35:283–292
31. Ciuta S, Tsiamis D, Castaldi MJ (2017) Gasification of waste materials: technologies for generating energy, gas, and chemicals from municipal solid waste, biomass, nonrecycled plastics, sludges, and wet solid wastes. Academic Press, New York
32. Peng Y et al (2022) A review on catalytic pyrolysis of plastic wastes to high-value products. Energy Convers Manag 254:115243
33. Tiwari R, Azad N, Dutta D, Yadav BR, Kumar S (2023) A critical review and future perspective of plastic waste recycling. Sci Total Environ 54:163433
34. Xue Y, Zhou S, Brown RC, Kelkar A, Bai X (2015) Fast pyrolysis of biomass and waste plastic in a fluidized bed reactor. Fuel 156:40–46
35. Kumar S, Jain S, Kumar H (2018) Performance evaluation of adaptive neuro-fuzzy inference system and response surface methodology in modeling biodiesel synthesis from jatropha–algae oil. Energy Sour A Recov Utilizat Environ Effects 40(24):3000–3008
36. Goyal P, Sharma MP, Jain S (2013) Optimization of transesterification of *J. curcas* oil to biodiesel using response surface methodology and its adulteration with kerosene. J Mater Environ Sci 4(2):277–284
37. Li Q, Faramarzi A, Zhang S, Wang Y, Hu X, Gholizadeh M (2020) Progress in catalytic pyrolysis of municipal solid waste. Elsevier, New Jersey

38. Chen X et al (2019) Recent developments in lignocellulosic biomass catalytic fast pyrolysis: strategies for the optimization of bio-oil quality and yield. Fuel Process Technol 196:106180
39. Jain S, Sharma MP (2013) Engine performance and emission analysis using oxidatively stabilized *J. curcas* biodiesel. Fuel 106:152–156
40. Goyal P, Sharma MP, Jain S (2012) Optimization of esterification and transesterification of high FFA *J. curcas* oil using response surface methodology. J Petrol Sci Res 1(3):36–43
41. Kong G, Wang K, Zhang X, Li J, Han L, Zhang X (2022) Torrefaction/carbonization-enhanced gasification-steam reforming of biomass for promoting hydrogen-enriched syngas production and tar elimination over gasification biochars. Bioresour Technol 363:960. https://doi.org/10.1016/j.biortech.2022.127960
42. Arena U (2012) Process and technological aspects of municipal solid waste gasification: a review. Waste Manag 32(4):625–639
43. Baruah D, Baruah DC (2014) Modeling of biomass gasification: a review. Renew Sustain Energy Rev 39:806–815
44. Susastriawan AAP, Saptoadi H, Purnomo S (2017) Small-scale downdraft gasifiers for biomass gasification: a review. Elsevier, New Jersey
45. Kumar A, Jones DD, Hanna MA (2009) Thermochemical biomass gasification: a review of the current status of the technology. Energies 2(3):556–581
46. Ayub Y, Zhou J, Ren J, He C (2023) An innovative integration of torrefaction, gasification, and solid oxide fuel cell for carbon–neutral utilization of biomass waste: process development, economic, exergy, advanced exergy, and exergoeconomics analysis. Energy Convers Manag 292:426. https://doi.org/10.1016/j.enconman.2023.117426
47. Kirubakaran V, Sivaramakrishnan V, Nalini R, Sekar T, Premalatha M, Subramanian P (2009) A review on gasification of biomass
48. Safarian S, Unnþórsson R, Richter C (2019) A review of biomass gasification modelling. Renew Sustain Energy Rev 110:378–391
49. Kumar-Shukla A et al (2020) Advances of carbon capture and storage in coal-based power generating units in an Indian context. Energies 13(16):4124
50. Kumar S, Jain S, Kumar H (2020) Experimental study on biodiesel production parameter optimization of jatropha–algae oil mixtures and performance and emission analysis of a diesel engine coupled with a generator fueled with diesel/biodiesel blends. ACS Omega 5(28):17033–17041. https://doi.org/10.1021/acsomega.9b04372
51. Buentello-Montoya DA, Zhang X, Li J (2019) The use of gasification solid products as catalysts for tar reforming. Elsevier, Amsterdam
52. Narnaware SL, Panwar NL (2021) Catalysts and their role in biomass gasification and tar abetment: a review. Biomass Convers Biorefin 16:1–31
53. Razzak SA (2024) Municipal solid and plastic waste derived high-performance biochar production: a comprehensive review. J Anal Appl Pyrolysis 65:106622
54. Shahbeik H et al (2023) Using nanocatalysts to upgrade pyrolysis bio-oil: a critical review. J Clean Prod 413:137473
55. Barbarias I, Lopez G, Artetxe M, Arregi A, Bilbao J, Olazar M (2018) Valorisation of different waste plastics by pyrolysis and in-line catalytic steam reforming for hydrogen production. Energy Convers Manag 156:575–584
56. Bashir S, Lapshun AL (2025) E-learning future trends in higher education in the 2020s and beyond. Cogent Educ 12(1):2445331
57. Ajorloo M (2024) A novel approach in thermodynamic analysis of waste to energy technique. UNSW Sydney
58. Loppinet-Serani A, Aymonier C, Cansell F (2010) Supercritical water for environmental technologies. J Chem Technol Biotechnol 85(5):583–589
59. Nguyen TT, Edalati K (2024) Impact of high-pressure columbite phase of titanium dioxide (TiO_2) on catalytic photoconversion of plastic waste and simultaneous hydrogen (H_2) production. J Alloys Compd 1008:176722
60. Yek PNY et al (2024) Co-processing plastics waste and biomass by pyrolysis–gasification: a review. Environ Chem Lett 22(1):171–188

61. Chhabra M, Dwivedi G, Baredar P, Kumar-Shukla A, Garg A, Jain S (2021) Production and optimization of biodiesel from rubber oil using BBD technique. Mater Today Proc 38:69–73. https://doi.org/10.1016/j.matpr.2020.05.791
62. Li T et al (2023) Performance analysis of an integrated biomass-to-energy system based on gasification and pyrolysis. Energy Convers Manag 287:117085
63. Raj RS, Jain S, Kumar-Sharma A (2024) Transforming municipal solid waste management through material and substance flow analysis: conversion pathways for sustainable energy production. Energy Convers Manag 322:119164. https://doi.org/10.1016/j.enconman.2024.119164
64. Im-orb K, Arpornwichanop A (2020) Process and sustainability analyses of the integrated biomass pyrolysis, gasification, and methanol synthesis process for methanol production. Energy 193:116788
65. Im-orb K, Wiyaratn W, Arpornwichanop A (2018) Technical and economic assessment of the pyrolysis and gasification integrated process for biomass conversion. Energy 153:592–603
66. Ayub Y, Zhou J, Shen W, Ren J (2023) Innovative valorization of biomass waste through integration of pyrolysis and gasification: process design, optimization, and multi-scenario sustainability analysis. Energy 282:128417
67. Moldoveanu SC (2010) Pyrolysis of hydrocarbons. Techn Instrum Anal Chem 28:131–229
68. Im-orb K, Detchusananard T, Ponpesh P, Arpornwichanop A (2017) Investigation of integrated biomass pyrolysis and gasification process for green fuel production. Energy Proced 142:204–209
69. Chen G, Andries J, Spliethoff H, Fang M, Van de Enden PJ (2004) Biomass gasification integrated with pyrolysis in a circulating fluidised bed. Sol Energy 76(1–3):345–349
70. Singh RK, Ruj B, Sadhukhan AK, Gupta P (2020) Catalytic and non-catalytic thermolysis of waste polystyrene for recovery of fuel grade products and their characterization. In: Ghosh SK (ed) Energy recovery processes from wastes. Springer, Singapore, pp 23–32. https://doi.org/10.1007/978-981-32-9228-4_3
71. Al-Qadri AA, Ahmed U, Ahmad N, Jameel AGA, Zahid U, Naqvi SR (2024) A review of hydrogen generation through gasification and pyrolysis of waste plastic and tires: opportunities and challenges. Int J Hydr Energy 77:1185–1204
72. Ryberg M, Laurent A, Hauschild MZ (2018) Mapping of global plastic value chain and plastic losses to the environment: with a particular focus on marine environment
73. Chaves M, Torres C, Tenorio C, Moya R, Arias-Aguilar D (2024) Syngas characterization and electric performance evaluation of gasification process using forest plantation biomass. Waste Biomass Valoriz 15(3):1291–1308
74. Olaleye AK, Adedayo KJ, Wu C, Nahil MA, Wang M, Williams PT (2014) Experimental study, dynamic modelling, validation and analysis of hydrogen production from biomass pyrolysis/gasification of biomass in a two-stage fixed bed reaction system. Fuel 137:364–374
75. Yuan Z et al (2020) Synergistic effect and chlorine-release behaviors during co-pyrolysis of LLDPE, PP, and PVC. ACS Omega 5(20):11291–11298. https://doi.org/10.1021/acsomega.9b04116
76. Xu T et al (2024) Life cycle assessment of plastic waste in Suzhou, China: management strategies toward sustainable express delivery. J Environ Manage 360:121201
77. Qureshi T, Farooq M, Nizami A, Ali U, Siddiqi MH (2024) Enhancing sustainability: structural evaluation of biomass cellulosic matrix for optimized conversion
78. Herrero NJ, Azañon R, Alvaro AJ, Ocampo D (2024) Determining the preeminent plastic wastes in the production of petrol using pyrolysis method and its effectiveness as an alternative fuel. Acta Chim Asiana 7(1):396–406
79. Niksiar A, Nasernejad B (2017) Activated carbon preparation from pistachio shell pyrolysis and gasification in a spouted bed reactor. Biomass Bioenergy 106:43–50
80. Mahinpey N, Gomez A (2016) Review of gasification fundamentals and new findings: reactors, feedstock, and kinetic studies. Chem Eng Sci 148:14–31
81. Gao Y et al (2023) Syngas production from biomass gasification: Influences of feedstock properties, reactor type, and reaction parameters. ACS Omega 8(35):31620–31631

82. Luo S, Xiao B, Hu Z, Liu S, Guan Y, Cai L (2010) Influence of particle size on pyrolysis and gasification performance of municipal solid waste in a fixed bed reactor. Bioresour Technol 101(16):6517–6520. https://doi.org/10.1016/j.biortech.2010.03.060
83. Yahaya AZ, Somalu MR, Muchtar A, Sulaiman SA, Daud WRW (2019) Effect of particle size and temperature on gasification performance of coconut and palm kernel shells in downdraft fixed-bed reactor. Energy 175:931–940
84. Bhavanam A, Sastry RC (2011) Biomass gasification processes in downdraft fixed bed reactors: a review. Int J Chem Eng Appl 2(6):425–433
85. Beheshti SM, Ghassemi H, Shahsavan-Markadeh R (2015) Process simulation of biomass gasification in a bubbling fluidized bed reactor. Energy Convers Manag 94:345–352
86. Zoungrana L, Sidibé SDS, Herman B, Coulibaly Y, Jeanmart H (2021) Design of a gasification reactor for manufacturing and operation in West Africa. Designs 5(4):76. https://doi.org/10.3390/designs5040076
87. Luo S, Xiao B, Hu Z, Liu S, Guan Y, Cai L (2010) Influence of particle size on pyrolysis and gasification performance of municipal solid waste in a fixed bed reactor. Bioresour Technol 101(16):6517–6520
88. Ukwuoma CC et al (2024) Enhancing hydrogen production prediction from biomass gasification via data augmentation and explainable AI: a comparative analysis. Int J Hydr Energy 68:755–776
89. Haq ZU, Akintunde SB, Salaudeen SA (2025) Hydrogen production from the steam gasification of hydrochar: a multi-output machine learning approach integrated with metaheuristic algorithms. Int J Hydr Energy 116:230–241
90. Chun YN, Kim SC, Yoshikawa K (2011) Pyrolysis gasification of dried sewage sludge in a combined screw and rotary kiln gasifier. Appl Energy 88(4):1105–1112
91. Muresan M, Cormos C-C, Agachi P-S (2013) Techno-economical assessment of coal and biomass gasification-based hydrogen production supply chain system. Chem Eng Res Des 91(8):1527–1541
92. Rakesh N, Dasappa S (2018) A critical assessment of tar generated during biomass gasification-Formation, evaluation, issues and mitigation strategies. Renew Sustain Energy Rev 91:1045–1064
93. Han J, Kim H (2008) The reduction and control technology of tar during biomass gasification/pyrolysis: an overview. Renew Sustain Energy Rev 12(2):397–416
94. Munir MT, Mardon I, Al-Zuhair S, Shawabkeh A, Saqib NU (2019) Plasma gasification of municipal solid waste for waste-to-value processing. Renew Sustain Energy Rev 116:109461
95. Hameed Z et al (2021) Gasification of municipal solid waste blends with biomass for energy production and resources recovery: current status, hybrid technologies and innovative prospects. Renew Sustain Energy Rev 136:110375
96. Vaskalis I, Zabaniotou A (2024) Comparative feasibility and environmental life cycle assessment of cotton stalks gasification and pyrolysis. Biomass 4(1):23–48
97. Javed MH, Ahmad A, Rehan M, Musharavati F, Nizami A-S, Khan MI (2025) Advancing sustainable energy: environmental and economic assessment of plastic waste gasification for syngas and electricity generation using life cycle modeling. Sustainability 17(3):1277
98. Sikarwar VS et al (2016) An overview of advances in biomass gasification. Royal Soc Chem. https://doi.org/10.1039/c6ee00935b
99. Wang F et al (2024) Insight into staged gasification of biomass waste: essential fundamentals and applications. Sci Total Environ 64:175954

Chapter 17
Recycled Plastics in Smart Packaging Applications

Harrison Appiah, Nnaemeka Ewurum, Sodiq Babatunde Yusuf, and Jovale Vincent Tongco

1 Introduction

Smart packaging represents an innovative frontier in packaging technology that goes beyond traditional containment and protection functions. It encompasses active and intelligent systems that can monitor, communicate, and in some cases, respond to changes in the product or its environment [1]. These technologies include oxygen scavengers, moisture controllers, temperature indicators, freshness sensors, and digital connectivity features that enhance product quality, safety, and consumer engagement [2]. Simultaneously, the global plastic packaging industry faces mounting environmental scrutiny. With annual plastic production exceeding 380 million tons and only 9% being recycled globally, plastic pollution has reached crisis proportions [3]. Packaging accounts for ~40% of plastic demand, with most items designed for single-use applications that contribute significantly to waste accumulation in landfills and marine environments [4]. The intersection of sustainability imperatives and functional requirements creates both challenges and opportunities for packaging innovation. Recycled plastics often present variable properties and potential contamination issues that can affect performance in demanding applications [5]. However, technological advances in recycling processes, material characterization, and smart systems integration are enabling new possibilities for high-value applications of recycled content [6]. This chapter examines the convergence of two important packaging trends: the push toward circular plastic economies and the demand for enhanced functionality through smart technologies. It explores the technical considerations, manufacturing approaches, and market dynamics shaping this emerging field. By analyzing current capabilities and future potential, this chapter provides insights into how recycled plastics can serve as viable substrates for smart

H. Appiah · N. Ewurum · S. B. Yusuf · J. V. Tongco (✉)
Department of Forest, Rangeland and Fire Sciences, University of Idaho, Moscow, ID, USA
e-mail: jtongco@uidaho.edu

A. Verma et al. (eds.), *Polymers and Composite Materials for Packaging: Smart Food Packaging and Solutions*, Materials Horizons: From Nature to Nanomaterials,
https://doi.org/10.1007/978-981-95-0677-4_17

packaging applications while addressing sustainability goals [7]. The subsequent sections will detail various recycled plastic materials suitable for smart packaging, integration techniques for embedding active and intelligent components, performance evaluations, and case studies demonstrating successful implementations. The chapter concludes with an assessment of market adoption factors and regulatory considerations that will influence further development in this sector.

2 The Plastic Waste Crisis and Recycling Landscape

Plastic production has increased exponentially since the 1950s, reaching 368 million metric tons in 2019 [8]. Packaging represents the largest application sector, accounting for ~40% of all plastic produced [3]. Despite growing awareness, only 9% of all plastic waste ever generated has been recycled, while 12% has been incinerated and 79% has accumulated in landfills or the natural environment [3]. Annual plastic waste generation reached 350 million metric tons globally, with 4.8–12.7 million metric tons entering the oceans [9]. This figure is projected to increase dramatically, with some estimates suggesting that by 2050, there could be more plastic by weight than fish in the oceans [4]. The predominant method involves sorting, washing, shredding, melting, and reforming plastics. While established for PET and HDPE, mechanical recycling suffers from quality degradation over multiple cycles due to polymer chain scission, cross-linking, and additive degradation [5]. Each recycling cycle typically reduces material properties by 5–30%, limiting the number of possible recycling loops [10]. Technologies including pyrolysis, gasification, and solvolysis break down polymers into their chemical constituents. While promising for contaminated, mixed, or difficult-to-recycle plastics, chemical recycling remains commercially limited due to high energy requirements, process complexity, and economic viability constraints. Recent innovations in catalytic depolymerization show potential for creating virgin-quality feedstock from PET waste [11]. Emerging enzymatic degradation approaches offer specificity for certain polymers. Enzymes like PETase can break down PET into its monomers, though commercial-scale application remains challenging [12, 13]. The packaging industry utilizes a diverse range of polymers, additives, and multi-layer structures that complicate recycling efforts. Smart packaging, with embedded electronics, sensors, or multiple materials, further increases complexity [14]. Food residues, adhesives, labels, and other contaminants significantly reduce recycling efficiency and quality output. Post-consumer waste typically contains 5–20% contamination by weight, requiring extensive washing and purification [15]. Fragmented collection systems and varying regional policies create inconsistency in waste streams. Only 63% of EU citizens have access to separate plastic collection systems, while figures are substantially lower in developing nations [16]. Recycled plastics often cannot compete economically with virgin materials, especially during periods of low oil prices. Processing costs for recycled plastics can be 2.5 times higher than virgin production, without accounting for externalities [17]. Extended Producer Responsibility schemes have been implemented

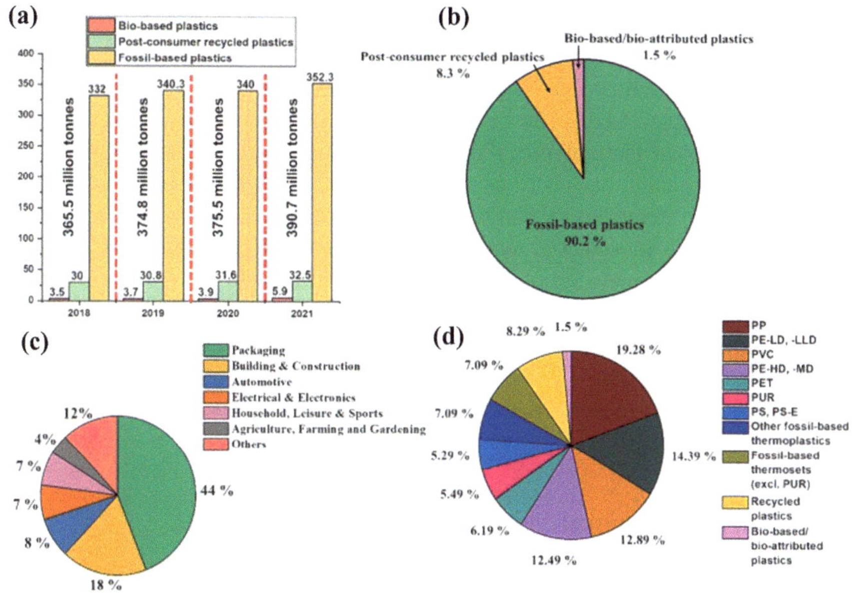

Fig. 1 **a** Global plastic production trends from 2018 to 2021. **b** Composition of global plastics production in 2021. **c** Breakdown of global plastics use by application in 2021. **d** Plastic-type breakdown in 2021 (reprinted with permission from reference [23])

in various regions, shifting waste management costs to producers [18]. The EU's Circular Economy Action Plan and Single-Use Plastics Directive mandate 25% recycled content in PET bottles by 2025, increasing to 30% for all plastic bottles by 2030 [19, 20]. The Basel Convention's 2019 amendment restricting global plastic waste trade has significantly altered international recycling dynamics [21]. Meanwhile, innovations in digital watermarking, artificial intelligence for sorting, and chemical recycling are beginning to address traditional barriers to plastics recycling [22]. Figure 1 shows the global plastic production trends, the composition of these plastics, and the breakdown of global plastics use by application.

3 Types of Recycled Plastics Suitable for Smart Packaging

Plastic polymers commonly used in packaging include polyethylene terephthalate (PET), polyethylene (PE), polypropylene (PP), polystyrene (PS), and polyvinyl chloride (PVC). Each polymer possesses distinct properties determining its recyclability and suitability for smart packaging applications [5]. PET demonstrates excellent transparency, gas barrier properties, and chemical resistance, making it ideal for food packaging and beverage containers. Its high recycling rate and relatively simple recycling process make recycled PET (rPET) a preferred material for smart packaging applications [24]. Polyethylene, including high-density (HDPE) and low-density (LDPE) varieties, offers flexibility, moisture resistance, and chemical stability [25]. HDPE recycling yields materials suitable for rigid packaging components, while recycled LDPE can be incorporated into flexible packaging structures that may host smart technologies [26]. Polypropylene provides thermal resistance, chemical stability, and barrier properties, though its recycling presents more challenges than PET. Recycled PP can be utilized in semi-rigid packaging applications where temperature resistance is required for certain smart packaging functionalities [27]. Post-consumer recycled plastics PCR generally exhibit decreased mechanical properties compared to virgin materials. Recycled PET typically shows a 5–10% reduction in tensile strength and impact resistance due to polymer chain degradation during the recycling process [28]. These changes must be considered when designing smart packaging that requires specific mechanical performance.

4 Impact of Thermal Window Constraints of PCR Materials

Thermal properties of PCR materials often display higher crystallinity and lower processing windows, affecting their compatibility with certain smart components that require precise thermal management during integration [29]. Optical properties, particularly important for visual indicators in smart packaging, can be compromised in recycled materials. rPET typically exhibits increased yellowness index and decreased transparency, potentially affecting color-change indicators or visual freshness monitors [30]. Contamination remains a significant challenge, as PCR materials may contain residual additives, degradation products, or cross-contamination from other polymers. These contaminants can potentially interfere with active or intelligent components in smart packaging systems [7]. Polylactic acid (PLA) derived from renewable resources offers biodegradability and compostability. While technically recyclable, PLA currently lacks established recycling infrastructure. Its transparency and printability make it suitable for certain smart packaging applications, particularly those requiring biodegradability [31]. Polyhydroxyalkanoates (PHAs) represent microbially produced polyesters with biodegradable properties [32]. Their variable mechanical properties and water resistance make them potential candidates for specific smart packaging applications, though commercial availability remains limited [33]. Thermoplastic starch (TPS) blends combine natural starch with plasticizers to create biodegradable packaging materials. These materials show promise for integration with certain bio-based sensors and indicators, particularly in applications where complete biodegradability is required. Material conductivity affects the integration of printed electronics and sensors. Recycled plastics often require surface treatments or additives to achieve suitable conductivity levels for electronic components [34]. Gas permeability generally increases in recycled plastics due to microvoids and structural changes, potentially affecting active packaging components that rely on controlled gas exchange. This property must be carefully managed in oxygen scavenger systems or modified atmosphere packaging using recycled materials [7]. Surface functionality for sensor attachment or indicator integration may require additional treatments when using recycled materials. Plasma treatment, corona discharge, or chemical modification can improve adhesion and functionality of smart components on recycled polymer surfaces [35].

5 Smart Packaging Technology Overview

Smart packaging extends beyond traditional packaging functions of protection and containment to include interactive and responsive features that enhance functionality. It broadly divides into three main categories: active, intelligent, and connected packaging [2]. Active packaging interacts directly with the product to improve preservation by incorporating components that release or absorb substances. Intelligent packaging monitors product conditions and communicates information about quality or safety. Connected packaging facilitates digital interactions between the product and consumers or supply chain stakeholders [1, 36].

6 Current Applications in Food, Pharmaceutical, and Consumer Goods

6.1 Food Industry

Smart packaging has gained significant traction in food applications. Time–temperature indicators monitor cold chain integrity, while modified atmosphere packaging extends shelf life. Oxygen scavengers prevent oxidation in sensitive products, and ethylene absorbers delay ripening in fresh produce [37]. Antimicrobial films with incorporated natural compounds like essential oils or silver nanoparticles inhibit pathogen growth, particularly valuable for minimally processed foods [38]. Freshness indicators based on pH or metabolite detection alert consumers to spoilage before consumption [39].

6.2 Pharmaceutical Applications

In pharmaceuticals, smart packaging improves medication adherence through electronic compliance monitoring and reminder systems. Temperature-sensitive indicators ensure vaccine and biologic product integrity throughout distribution [40].

Tamper-evident and authentication features combat counterfeiting, a critical concern for pharmaceutical safety [41]. Moisture control systems preserve the stability of hygroscopic medications [42].

6.3 Consumer Goods

Consumer goods employ smart packaging primarily for brand engagement and authentication. Near-field communication (NFC) and QR code technologies provide product information, usage instructions, and brand storytelling [43]. Thermochromic inks indicate optimal consumption temperature for beverages, while smart packaging in cosmetics monitors product freshness and usage patterns [2].

7 Technical Requirements for Smart Packaging Functionality

Smart packaging systems must meet several technical requirements to function effectively [1]. These include:

1. Material compatibility between the packaging substrate and smart components.
2. Sufficient barrier properties to protect both the product and functional elements.
3. Stability throughout the intended shelf life.
4. Manufacturability at commercial scale.
5. Regulatory compliance, particularly for food contact applications.
6. Cost-effectiveness relative to the value added.

Sensor integration presents challenges, requiring durability, sensitivity, and in many cases, food-grade status. Printed electronics must maintain functionality despite flexing and environmental stresses [2]. Power requirements for active electronic components remain a significant hurdle, though advances in energy harvesting, thin-film batteries, and low-power electronics are addressing these limitations [44]. Recycled plastics present unique compatibility challenges when integrated with smart technologies. The heterogeneous nature of post-consumer recycled (PCR) materials introduces variable chemical compositions that can interfere with smart packaging functionalities [45]. Residual contaminants from previous use cycles—adhesives, inks, or food residues—may interact with sensors or active components, leading to degraded performance or false readings [46]. Polymer chain degradation during the recycling process reduces molecular weight and creates oxidized groups, potentially affecting electrical conductivity required for printed electronics [47]. Additionally, the presence of various additives like stabilizers, plasticizers, and colorants in recycled streams complicates the design of sensitive detection systems [48]. The mechanical properties of recycled plastics typically show greater variability compared to virgin materials, creating challenges for consistent smart packaging performance.

Reduced tensile strength and impact resistance may compromise the protection of embedded electronics or sensors [49]. Thermal stability issues in recycled materials can affect the performance of temperature-sensitive indicators or limit processing options during manufacturing [50, 51]. Barrier properties—crucial for active packaging functionalities—often deteriorate in recycled materials. This deterioration can lead to premature oxygen ingress or moisture transmission, potentially reducing the effectiveness of shelf-life extension technologies [52]. Surface roughness and inconsistencies in recycled films may impact printing quality for QR codes or conductive inks necessary for intelligent packaging functions [53]. Accelerated aging studies show that recycled plastics integrated with smart components may experience shorter functional lifespans, with potential degradation of both the packaging material and the smart technology elements over time [54]. Recycled materials face stringent regulatory hurdles when combined with smart technologies, particularly for food contact applications. The European Food Safety Authority (EFSA) and the U.S. Food and Drug Administration (FDA) have established specific requirements for recycled plastics in food packaging that become more complex when smart technologies are introduced [55, 56]. Migration concerns are heightened in smart packaging systems, where additional components like sensors, indicators, or active ingredients may interact with recycled substrates [57]. Required safety assessments must evaluate not only the recycled material but also potential synergistic effects between the material and smart components [58]. The lack of harmonized global standards specifically addressing recycled content in smart packaging creates market fragmentation and increases compliance costs for manufacturers [59]. Additionally, end-of-life considerations become more complex as smart components may impact the recyclability of the packaging itself, creating potential regulatory conflicts with extended producer responsibility legislation [60]. Smart packaging technologies often rely on precise signal transmission that can be compromised by the inherent variability in recycled materials [61]. Radio frequency identification (RFID) tags and near-field communication (NFC) systems show decreased read ranges when applied to substrates with inconsistent compositions [62]. Electromagnetic shielding effects from metallic contaminants in recycled plastics can block wireless signals entirely in severe cases [63]. Optical clarity variations in recycled materials affect the reliability of visual indicators, QR code readability, and spectroscopic sensors [64]. This necessitates additional calibration protocols and quality control measures throughout the production process to ensure consistent performance.

8 Cost-Effectiveness and Scalability Challenges

The integration of smart technologies with recycled plastics faces significant economic barriers. The additional processing required to prepare recycled materials for smart packaging applications increases manufacturing costs by 15–30% compared to virgin material alternatives [65]. Quality control requirements are more stringent and expensive due to the variable nature of recycled feedstock

[66, 67]. Achieving consistent performance at an industrial scale remains difficult, with production runs using recycled materials showing higher rejection rates for smart packaging components compared to their virgin material counterparts [68]. The establishment of specialized supply chains capable of delivering both high-quality recycled plastics and compatible smart technology components represents a significant infrastructure challenge [69].

9 Active Packaging Solutions Using Recycled Plastics

Recycled plastics can effectively incorporate oxygen scavenger technologies, though with certain modifications to overcome inherent limitations. PET-based recycled materials show promise when combined with iron-based oxygen scavengers, achieving comparable performance to virgin materials after appropriate processing [70, 71]. Recent innovations have focused on nonmetallic alternatives like enzymes and antioxidants that remain effective in recycled polymer matrices despite the presence of contaminants. Moisture regulation remains critical for many packaged products. Recycled polypropylene (rPP) and recycled polyethylene (rPE) can be successfully combined with desiccants like silica gel, calcium oxide, and clay minerals through compartmentalization designs that prevent direct contact between absorbers and food products [72]. These systems maintain structural integrity while providing humidity control comparable to virgin plastic equivalents. Multi-layer systems incorporating recycled plastics as non-contact layers show promise. Research demonstrates that recycled PET can constitute up to 70% of moisture control packaging when used as structural layers surrounding virgin material barrier layers [73]. Antimicrobial functionality can be effectively incorporated into recycled plastics through several approaches. Natural antimicrobial compounds like essential oils and plant extracts represent a growing area of innovation. Recycled PLA combined with cinnamon oil microencapsulation has demonstrated effective inhibition of common food pathogens while maintaining the sustainability benefits of recycled materials. Surface modification techniques, including plasma treatment and UV grafting, have proven effective in attaching antimicrobial compounds to recycled polymer surfaces. These methods overcome the reduced binding sites and irregular surface characteristics common in recycled materials. Nanocomposite approaches show promise, with silver, zinc oxide, and copper nanoparticles effectively inhibiting microbial growth when incorporated into recycled PE and PP matrices. These nanoparticles maintain their efficacy despite the presence of contaminants common in recycled streams.

10 Case Study

Several commercial implementations demonstrate the viability of active packaging solutions using recycled content. TerraCycle and Procter & Gamble launched packaging containing 25% ocean-bound plastic with integrated odor-neutralizing technology for household products. Performance testing showed equivalent functionality to virgin plastic alternatives with significant environmental benefits. In the food sector, Amcor's recycled PET trays incorporating oxygen scavenger technology have extended fresh meat shelf life by up to five days while using 50% post-consumer recycled content [74, 75]. This implementation required process modifications to ensure consistent dispersal of active compounds throughout the variable recycled material. The pharmaceutical industry has adopted moisture-controlling blister packs with up to 40% recycled content for non-contact layers. Novartis reported successful implementation with no compromise in drug stability or packaging integrity [76]. Despite successful applications, several technical challenges persist. Migration concerns remain prominent, as recycled materials may contain unknown substances that could interact with active compounds. Advanced sorting and cleaning technologies combined with robust analytical verification have addressed these concerns. Consistency issues in recycled feedstock affect active compound distribution and efficacy. This has been addressed through master batch approaches where active compounds are concentrated in virgin material portions of otherwise highly recycled packaging structures. Performance variability has been overcome through adaptive manufacturing processes that adjust active compound loading based on recycled material batch characteristics. This approach ensures consistent end-performance despite variable starting materials.

11 Advanced Manufacturing Techniques

Recycled plastics present unique manufacturing challenges due to inherent batch-to-batch variability in material properties. Unlike virgin plastics, recycled materials often contain contaminants, mixed polymer types, and degraded polymer chains that affect processability. Advanced manufacturing techniques have evolved to address these challenges. Real-time process monitoring and adaptive control systems represent a significant advancement in handling recycled plastic variability. In-line rheological measurements and spectroscopic techniques enable continuous adjustment of processing parameters to maintain consistent product quality despite fluctuating input materials [77]. Multi-sensor arrays equipped with machine learning algorithms can detect material inconsistencies and automatically modify temperature profiles, pressure settings, and cooling rates to optimize production [78]. Material pre-conditioning has proven essential for stabilizing recycled plastic performance. Specialized compatibilizers and chain extenders restore molecular weight and improve compatibility between different polymer fractions commonly found in recycled streams [79]. Advanced sorting technologies employing NIR spectroscopy, AI-powered optical sorting, and electrostatic separation have dramatically improved the purity of recycled plastic feedstocks, making them more suitable for high-performance applications like smart packaging [80].

Additive manufacturing offers unprecedented flexibility for incorporating smart components into recycled plastic packaging. Recent research demonstrates the successful formulation of recycled PET and PP filaments with consistent printability when properly processed and stabilized [81]. These filaments can be used to create customized sensor housings, RFID tag enclosures, and other functional components that integrate with conventional packaging.

12 Printing with Recycled Plastics for Smart Packaging Components

Multi-material 3D printing enables the simultaneous deposition of recycled plastics alongside conductive materials, creating embedded circuitry directly within packaging structures [82]. This technique eliminates the need for separate electronic components while utilizing recycled materials as the structural substrate. 4D printing—where materials change shape or function in response to environmental stimuli—has been demonstrated using recycled plastics combined with responsive additives, creating packaging that can indicate tampering or product freshness [82]. Distributed manufacturing models enabled by 3D printing technology are particularly advantageous for recycled plastic utilization. Local recycling and printing operations reduce transportation emissions and enable rapid customization of smart packaging solutions for regional markets. Surface modification technologies are critical for enhancing the compatibility between recycled plastics and smart packaging components. Plasma treatment has emerged as an effective method for improving adhesion properties of recycled polymers without altering bulk material properties [83]. Low-temperature plasma processes create reactive surface groups that form strong bonds with inks, coatings, and adhesives used in smart packaging applications. Figure 2 shows the advanced techniques for manufacturing value-added products from recycled plastics.

13 Surface Treatment and Functionalization Methods

Chemical functionalization through grafting specific molecules onto recycled plastic surfaces has enabled direct sensor integration. A study demonstrated successful immobilization of colorimetric freshness indicators onto recycled PET through silane coupling agents, creating packaging that changes color in response to food spoilage volatiles [84]. Advanced coating technologies address barrier property deficiencies

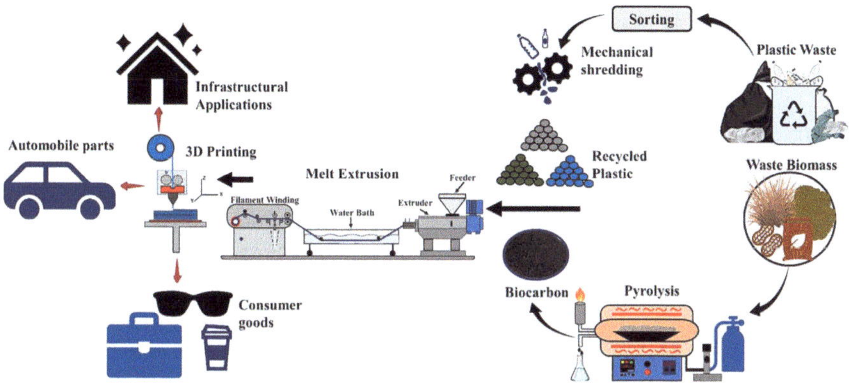

Fig. 2 Advanced techniques for manufacturing value-added products from recycled plastics (reprinted with permission from reference [23])

in recycled plastics. Atomic layer deposition (ALD) applies ultrathin, highly uniform barrier layers that prevent oxygen and moisture transmission while maintaining recyclability [85]. Similarly, sol–gel derived coatings provide enhanced barrier properties while serving as matrices for incorporating antimicrobial agents or oxygen scavengers into recycled plastics [86]. Laser surface texturing creates precise micropatterns on recycled plastic surfaces that enhance hydrophilicity, antimicrobial properties, or promote controlled release of active compounds [87]. These textured surfaces can be designed to gradually expose embedded sensors or indicators as needed during a product's shelf life.

14 Quality Control

Quality assurance systems specifically designed for recycled content manufacturing have become essential. Hyperspectral imaging combined with deep learning algorithms can detect contaminants and material inconsistencies at production speeds of up to 3 m/s, ensuring that smart packaging containing recycled content meets stringent performance and safety standards [88]. Integrated production lines that combine conventional plastic processing with printed electronics deposition represent the forefront of smart packaging manufacturing. These systems incorporate multiple quality checkpoints with feedback loops to ensure consistent performance despite material variability [89]. Real-time traceability systems track recycled content throughout the manufacturing process, providing quality assurance and enabling claim verification for regulatory compliance [90].

15 Life Cycle Assessment and Sustainability Metrics

Life cycle assessment (LCA) provides a systematic framework for evaluating the environmental impacts of recycled plastics in smart packaging applications from raw material extraction through production, use, and end-of-life management [91]. When comparing virgin and recycled plastics for smart packaging, comprehensive LCA addresses multiple impact categories such as global warming potential (carbon footprint), energy consumption, water usage, resource depletion, waste generation, ecotoxicity, and human health impacts. These metrics require careful boundary definition to ensure meaningful comparisons between packaging alternatives [92, 93]. For smart packaging incorporating electronic components, the system boundaries must include both the packaging substrate and the embedded technology components.

16 Environmental Impact Comparison: Virgin Versus Recycled Smart Packaging

Empirical evidence consistently demonstrates significant environmental advantages of recycled plastics over virgin materials in standard packaging applications, with typical reductions of 30–60% greenhouse gas emissions, 40–75% energy consumption, and 20–50% water usage [7, 94]. However, smart packaging introduces additional complexity. The incorporation of sensors, indicators, and other smart components can create trade-offs between material sustainability and functionality [95]. Key findings from recent studies include:

1. Recycled PET incorporating oxygen scavengers shows a 45% lower carbon footprint compared to virgin PET with identical functionality [96].
2. The environmental benefit of recycled content diminishes when embedded electronics require additional barrier layers or processing steps [97].
3. Smart packaging designs enabling proper end-of-life sorting through embedded identifiers can improve overall recycling rates by 15–30%, amplifying environmental benefits [98].

17 Market Adoption and Consumer Acceptance

Consumer acceptance represents a crucial factor in the widespread adoption of recycled plastics in smart packaging applications. Research shows evolving consumer attitudes toward recycled content, particularly in premium product categories where packaging quality expectations are high [99]. Studies indicate that while sustainability concerns drive positive intentions, actual purchasing behavior is more complex, with perceived quality and safety influencing final decisions [100]. A recent research demonstrated that transparent communication about recycling processes and quality control measures significantly improves consumer confidence in recycled smart packaging, with education level and prior sustainability engagement serving as key predictors of acceptance [101].

18 Outlook and Emerging Trends

Advancements in recycling technologies are rapidly transforming the potential for incorporating recycled plastics into smart packaging applications. Chemical recycling processes, including pyrolysis, solvolysis, and depolymerization, are enabling the conversion of plastic waste into virgin-equivalent monomers and feedstocks [5]. These technologies effectively address contamination issues and mixed plastic waste streams that conventional mechanical recycling cannot process efficiently. Artificial intelligence and machine learning algorithms are revolutionizing plastic sorting technologies. Advanced optical sorting systems now achieve purity levels exceeding 99% for certain polymer streams, greatly enhancing the quality of recycled content for smart packaging applications [102]. These technological improvements directly impact the feasibility of using recycled materials in high-performance packaging that incorporates smart technologies. Emerging biotechnological approaches show significant promise for plastic degradation and recycling. Engineered enzymes that can break down persistent polymer chains into their constituent monomers have demonstrated potential for PET recycling at industrial scales [103]. Similar enzymatic processes for polyolefins (PE, PP) are under development, potentially expanding the range of recycled materials suitable for smart packaging applications. The integration of biodegradable electronics with recycled substrates represents a significant advancement toward truly circular smart packaging systems. Organic thin-film transistors and biodegradable conductive inks compatible with recycled plastic substrates enable smart functionality without compromising end-of-life recyclability [104]. These developments address an important challenge in current smart packaging—the integration of electronic components that often render packages unrecyclable.

19 Summary and Conclusion

The integration of recycled plastics into smart packaging represents an important advancement in sustainable packaging innovation. Throughout this chapter, we've demonstrated how post-consumer and post-industrial recycled materials can be effectively incorporated into active, intelligent, and connected packaging systems despite challenges in material consistency and performance. The successful marriage of sustainability and functionality in packaging requires collaborative approaches across the value chain, from material scientists to brand owners. The convergence of recycled plastics and smart packaging technologies offers tremendous potential for addressing the simultaneous demands for sustainability, functionality, and consumer engagement. While technical and systemic challenges remain, collaborative innovation across disciplines and sectors presents a clear pathway toward packaging solutions that fulfill both environmental responsibilities and market requirements.

Conflicts of Interest There are no conflicts of interest to declare by the authors.

References

1. Yam KL, Takhistov PT, Miltz J (2005) Intelligent packaging: concepts and applications. J Food Sci 70:R1–R10. https://doi.org/10.1111/J.1365-2621.2005.TB09052.X
2. Vanderroost M, Ragaert P, Devlieghere F, De Meulenaer B (2014) Intelligent food packaging: the next generation. Trends Food Sci Technol 39:47–62. https://doi.org/10.1016/J.TIFS.2014.06.009
3. Geyer R, Jambeck JR, Law KL (2017) Production, use, and fate of all plastics ever made. Sci Adv 3:782. https://doi.org/10.1126/SCIADV.1700782/SUPPL_FILE/1700782_SM.PDF
4. Ellen MacArthur Foundation (2016) The new plastics economy: rethinking the future of plastics, pp 1–120
5. Ragaert K, Delva L, Van Geem K (2017) Mechanical and chemical recycling of solid plastic waste. Waste Manag 69:24–58. https://doi.org/10.1016/J.WASMAN.2017.07.044
6. Asgher M, Qamar SA, Bilal M, Iqbal HMN (2020) Bio-based active food packaging materials: sustainable alternative to conventional petrochemical-based packaging materials. Food Res Int 137:9625. https://doi.org/10.1016/J.FOODRES.2020.109625
7. Geueke B, Groh K, Muncke J (2018) Food packaging in the circular economy: overview of chemical safety aspects for commonly used materials. J Clean Prod 193:491–505. https://doi.org/10.1016/J.JCLEPRO.2018.05.005
8. Appiah H, Bar-Ziv E, Klinger JL, McDonald AG (2025) Exploring new applications of municipal solid waste. Sustainability 17:3719. https://doi.org/10.3390/SU17083719
9. Jambeck JR, Geyer R, Wilcox C, Siegler TR, Perryman M, Andrady A, Narayan R, Law KL (2015) Marine pollution: plastic waste inputs from land into the ocean. Science 347:768–771. https://doi.org/10.1126/SCIENCE.1260352
10. Hahladakis JN, Velis CA, Weber R, Iacovidou E, Purnell P (2018) An overview of chemical additives present in plastics: migration, release, fate and environmental impact during their use, disposal and recycling. J Hazard Mater 344:179–199. https://doi.org/10.1016/J.JHAZMAT.2017.10.014
11. Voss R, Lee RP, Fröhling M (2022) Chemical recycling of plastic waste: comparative evaluation of environmental and economic performances of gasification- and incineration-based treatment for lightweight packaging waste. Circul Econ Sustain 2:1369–1398. https://doi.org/10.1007/S43615-021-00145-7/FIGURES/5
12. Yoshida S, Hiraga K, Takehana T, Taniguchi I, Yamaji H, Maeda Y, Toyohara K, Miyamoto K, Kimura Y, Oda K (2016) A bacterium that degrades and assimilates poly(ethylene terephthalate). Science 351:1196–1199. https://doi.org/10.1126/SCIENCE.AAD6359
13. Austin HP, Allen MD, Donohoe BS, Rorrer NA, Kearns FL, Silveira RL, Pollard BC, Dominick G, Duman R, El-Omari K, Mykhaylyk V, Wagner A, Michener WE, Amore A, Skaf MS, Crowley MF, Thorne AW, Johnson CW, Lee Woodcock H, McGeehan JE, Beckham GT (2018) Characterization and engineering of a plastic-degrading aromatic polyesterase. Proc Natl Acad Sci USA 115:E4350–E4357. https://doi.org/10.1073/PNAS.1718804115
14. Wagner S, Schlummer M (2020) Legacy additives in a circular economy of plastics: current dilemma, policy analysis, and emerging countermeasures. Resour Conserv Recycl 158:104800. https://doi.org/10.1016/J.RESCONREC.2020.104800

15. Gall M, Wiener M, Chagas de Oliveira C, Lang RW, Hansen EG (2020) Building a circular plastics economy with informal waste pickers: recyclate quality, business model, and societal impacts. Resour Conserv Recycl 156:104685. https://doi.org/10.1016/J.RESCONREC.2020.104685
16. PlascticsEurope Plastics-the Facts (2020) An analysis of European plastics production, demand and waste data
17. Nielsen TD, Hasselbalch J, Holmberg K, Stripple J (2020) Politics and the plastic crisis: a review throughout the plastic life cycle. Wiley Interdiscip Rev Energy Environ 9:11007
18. OECD (2016) Extended producer responsibility: updated guidance for efficient waste management
19. European Commission (2019) CELEX_32019L0904_EN_TXT. J Eur Union 19:155–1199
20. Kasznik D, Łapniewska Z (2023) The end of plastic? The EU's directive on single-use plastics and its implementation in Poland. Environ Sci Policy 145:151–163. https://doi.org/10.1016/J.ENVSCI.2023.04.005
21. Brooks AL, Wang S, Jambeck JR (2018) The Chinese import ban and its impact on global plastic waste trade. Sci Adv 4:131. https://doi.org/10.1126/SCIADV.AAT0131
22. Ellen MacArthur Foundation (2019) Reuse rethinking packaging contents
23. Hassan M, Mohanty AK, Misra M (2024) 3D printing in upcycling plastic and biomass waste to sustainable polymer blends and composites: a review. Mater Des 237:112558. https://doi.org/10.1016/J.MATDES.2023.112558
24. Welle F (2011) Twenty years of PET bottle to bottle recycling—an overview. Resour Conserv Recycl 55:865–875. https://doi.org/10.1016/J.RESCONREC.2011.04.009
25. Tongco JV (2025) Effects of low concentration coupling agent addition on the physicomechanical behavior of wood fiber/HDPE composite. Mediter J Chem 15:187–195
26. Thaore V, Chadwick D, Shah N (2018) Sustainable production of chemical intermediates for nylon manufacture: a techno-economic analysis for renewable production of caprolactone. Chem Eng Res Des 135:140–152. https://doi.org/10.1016/J.CHERD.2018.05.026
27. Hahladakis JN, Iacovidou E (2018) Closing the loop on plastic packaging materials: what is quality and how does it affect their circularity? Sci Total Environ 630:1394–1400. https://doi.org/10.1016/J.SCITOTENV.2018.02.330
28. Nguyen PM, Berrard C, Daoud N, Saillard P, Peyroux J, Vitrac O (2024) Assessment of chemical risks and circular economy implications of recycled PET in food packaging with functional barriers. Resour Environ Sustain 17:100163. https://doi.org/10.1016/J.RESENV.2024.100163
29. Vilaplana F, Karlsson S (2008) Quality concepts for the improved use of recycled polymeric materials: a review. Macromol Mater Eng 293:274–297. https://doi.org/10.1002/MAME.200700393
30. De Santis F, Pantani R (2013) Optical properties of polypropylene upon recycling. Sci World J 2013:354093. https://doi.org/10.1155/2013/354093
31. Castro-Aguirre E, Iñiguez-Franco F, Samsudin H, Fang X, Auras R (2016) Poly(lactic acid)—mass production, processing, industrial applications, and end of life. Adv Drug Deliv Rev 107:333–366. https://doi.org/10.1016/J.ADDR.2016.03.010
32. Abbasi M, Coats ER, McDonald AG (2022) Green solvent extraction and properties characterization of Poly(3-hydroxybutyrate-co-3-hydroxyvalerate) biosynthesized by mixed microbial consortia fed fermented dairy manure. Bioresour Technol Rep 18:101065. https://doi.org/10.1016/J.BITEB.2022.101065
33. Dietrich K, Dumont MJ, Del Rio LF, Orsat V (2017) Producing PHAs in the bioeconomy: towards a sustainable bioplastic. Sustain Prod Consum 9:58–70. https://doi.org/10.1016/J.SPC.2016.09.001
34. Wu Y, Wang Z, Liu X, Shen X, Zheng Q, Xue Q, Kim JK (2017) Ultralight graphene foam/conductive polymer composites for exceptional electromagnetic interference shielding. ACS Appl Mater Interf 9:9059–9069. https://doi.org/10.1021/ACSAMI.7B01017
35. Netravali A, Mittal KL. (2024) Polymer surface modification to enhance adhesion: techniques and applications, p 573

36. Realini CE, Marcos B (2014) Active and intelligent packaging systems for a modern society. Meat Sci 98:404–419. https://doi.org/10.1016/J.MEATSCI.2014.06.031
37. Robertson GL (2016) Food packaging: principles and practice. In: Food packaging: principles and practice, 3rd edn, pp 1–688. https://doi.org/10.1201/B21347/FOOD-PACKAGING-GORDON-ROBERTSON/RIGHTS-AND-PERMISSIONS
38. Ghaani M, Cozzolino CA, Castelli G, Farris S (2016) An overview of the intelligent packaging technologies in the food sector. Trends Food Sci Technol 51:1–11. https://doi.org/10.1016/J.TIFS.2016.02.008
39. Müller P, Schmid M (2019) Intelligent packaging in the food sector: a brief overview. Foods 8:16. https://doi.org/10.3390/FOODS8010016
40. Samant Y, Lanjewar H, Parker D, Block L, Tomar GS, Stein B (2007) Evaluation of the cold-chain for oral polio vaccine in a rural district of India. Public Health Rep 122:112. https://doi.org/10.1177/003335490712200116
41. Pegu L, Chasta P, Kaushal M, Chandrul K (2019) Pharmaceutical packaging technology. Publ Int J Trend Sci Res Develop 21:1747–1754
42. Almuzaini T, Choonara I, Sammons H (2013) Substandard and counterfeit medicines: a systematic review of the literature. BMJ Open 3:2923. https://doi.org/10.1136/BMJOPEN-2013-002923
43. Spence C, Puccinelli NM, Grewal D, Roggeveen AL (2014) Store atmospherics: a multisensory perspective. Psychol Mark 31:472–488. https://doi.org/10.1002/MAR.20709
44. Mohebi E, Marquez L (2014) Intelligent packaging in meat industry: an overview of existing solutions. J Food Sci Technol 52:3947. https://doi.org/10.1007/S13197-014-1588-Z
45. Soares CTM, Ek M, Östmark E, Gällstedt M, Karlsson S (2022) Recycling of multi-material multilayer plastic packaging: current trends and future scenarios. Resour Conserv Recycl 176:105905. https://doi.org/10.1016/J.RESCONREC.2021.105905
46. Palanisamy Y, Kadirvel V, Ganesan ND (2025) Recent technological advances in food packaging: sensors, automation, and application. Sustain Food Technol 3:161–180. https://doi.org/10.1039/D4FB00296B
47. Reis Carneiro M, de Almeida AT, Tavakoli M, Majidi C (2023) Recyclable thin-film soft electronics for smart packaging and e-skins. Adv Sci 10:2301673. https://doi.org/10.1002/ADVS.202301673
48. Aguiar MIS, Sousa AF, Teixeira G, Tavares APM, Ferreira AM, Coutinho JAP (2024) Enhancing plastic waste recycling: evaluating the impact of additives on the enzymatic polymer degradation. Catal Today 429:114492. https://doi.org/10.1016/J.CATTOD.2023.114492
49. Zakaria Z, Osman H, Rahim NAA, Munusamy Y, Ismail H (2023) Recycled polyethylene terephthalate blends and composites: impact of PET waste, engineering design, and their applications. Recycl Polym Blends Compos Process Proper Appl 31:417–435. https://doi.org/10.1007/978-3-031-37046-5_21
50. Han JW, Ruiz-Garcia L, Qian JP, Yang XT (2018) Food packaging: a comprehensive review and future trends. Compr Rev Food Sci Food Saf 17:860–877. https://doi.org/10.1111/1541-4337.12343
51. Wang S, Liu X, Yang M, Zhang Y, Xiang K, Tang R (2015) Review of time temperature indicators as quality monitors in food packaging. Packag Technol Sci 28:839–867. https://doi.org/10.1002/PTS.2148
52. Seier M, Archodoulaki VM, Koch T (2024) The morphology and properties of recycled plastics made from multi-layered packages and the consequences for the circular economy. Resour Conserv Recycl 202:107388. https://doi.org/10.1016/J.RESCONREC.2023.107388
53. Hu B, Serranti S, Fraunholcz N, Di Maio F, Bonifazi G (2013) Recycling-oriented characterization of polyolefin packaging waste. Waste Manag 33:574–584. https://doi.org/10.1016/J.WASMAN.2012.11.018
54. Luoma E, Välimäki M, Immonen K (2023) Effect of accelerated aging on properties of biobased polymer films applicable in printed electronics. J Appl Polym Sci 140:e53414. https://doi.org/10.1002/APP.53414

55. Bolognesi C, Castle L, Cravedi J-P, Engel K-H, Fowler P, Franz R, Grob K, Gürtler R, Husøy T, Mennes W, Milana MR, Penninks A, Roland F, Silano V, Smith A, De Fátima M, Poças T, Tlustos C, Toldrá F, Wölfle D, Zorn H, Beausoleil C, Benford D, Bradley E, Brantsaeter AL, Calamandrei G, Doerge D, Greaves P, Schlatter JR, Simoneau C, Slob W, Van Engelen J, Van LH, Von Götz N, Woutersen RA, Alexander J, Mosesso P, Siani A, Arcella D, Castoldi AF, Croera C, Theobald A (2015) Scientific opinion on the risks to public health related to the presence of bisphenol A (BPA) in foodstuffs. EFSA J 13:3978. https://doi.org/10.2903/J.EFSA.2015.3978
56. Fda use of recycled plastics in food packaging (chemistry considerations): guidance for industry
57. Bhunia K, Sablani SS, Tang J, Rasco B (2013) Migration of chemical compounds from packaging polymers during microwave, conventional heat treatment, and storage. Compr Rev Food Sci Food Saf 12:523–545. https://doi.org/10.1111/1541-4337.12028
58. Undas AK, Groenen M, Peters RJB, van Leeuwen SPJ (2023) Safety of recycled plastics and textiles: review on the detection, identification and safety assessment of contaminants. Chemosphere 312:137175. https://doi.org/10.1016/J.CHEMOSPHERE.2022.137175
59. Thapliyal D, Karale M, Diwan V, Kumra S, Arya RK, Verros GD (2024) Current status of sustainable food packaging regulations: global perspective. Sustainability 16:5554. https://doi.org/10.3390/SU16135554
60. Kang Z, Huang Z, Peng Q, Shi Z, Xiao H, Yin R, Fu G, Zhao J (2023) Recycling technologies, policies, prospects, and challenges for spent batteries. iScience 26:108072. https://doi.org/10.1016/J.ISCI.2023.108072
61. Matt Reynolds (2024) Active, intelligent, connected, or otherwise "smart" packaging hits its stride|packaging world. https://www.packworld.com/sustainable-packaging/recycling/article/22885246/active-intelligent-connected-or-otherwise-smart-packaging-hits-its-stride. Accessed 21 April 2025
62. Paredes F, Herrojo C, Mata-Contreras J, Moras M, Núñez A, Ramon E, Martín F (2018) Near-field chipless radio-frequency identification (RFID) sensing and identification system with switching reading. Sensors 18:1148. https://doi.org/10.3390/S18041148
63. Sankaran S, Deshmukh K, Ahamed MB, Khadheer Pasha SK (2018) Recent advances in electromagnetic interference shielding properties of metal and carbon filler reinforced flexible polymer composites: a review. Compos A Appl Sci Manuf 114:49–71. https://doi.org/10.1016/J.COMPOSITESA.2018.08.006
64. Gazeau B, Minunno R, Zaman A, Shaikh F (2024) Elevating recycling standards: global requirements for plastic traceability and quality testing. Sustainability 16:5122. https://doi.org/10.3390/SU16125122/S1
65. Alaghemandi M (2024) Sustainable solutions through innovative plastic waste recycling technologies. Sustainability 16:10401. https://doi.org/10.3390/SU162310401
66. Schmidt Rivera XC, Leadley C, Potter L, Azapagic A (2019) Aiding the design of innovative and sustainable food packaging: integrating techno-environmental and circular economy criteria. Energy Proced 161:190–197. https://doi.org/10.1016/J.EGYPRO.2019.02.081
67. Kassab A, Al Nabhani D, Mohanty P, Pannier C, Ayoub GY (2023) Advancing plastic recycling: challenges and opportunities in the integration of 3D printing and distributed recycling for a circular economy. Polymers 15:3881. https://doi.org/10.3390/POLYM15193881
68. Stramarkou M, Boukouvalas C, Koskinakis SE, Serifi O, Bekiris V, Tsamis C, Krokida M (2022) Life cycle assessment and preliminary cost evaluation of a smart packaging system. Sustainability 14:7080. https://doi.org/10.3390/SU14127080
69. Shekarian E, Ijadi B, Zare A, Majava J (2022) Sustainable supply chain management: a comprehensive systematic review of industrial practices. Sustainability 14:7892. https://doi.org/10.3390/SU14137892
70. Gupta P (2023) Role of oxygen absorbers in food as packaging material, their characterization and applications. J Food Sci Technol 61:242. https://doi.org/10.1007/S13197-023-05681-8
71. Gaikwad KK, Singh S, Lee YS (2018) Oxygen scavenging films in food packaging. Environ Chem Lett 16:523–538. https://doi.org/10.1007/S10311-018-0705-Z/FIGURES/5

72. Sängerlaub S, Kucukpinar E, Müller K (2019) Desiccant films made of low-density polyethylene with dispersed silica gel—water vapor absorption, permeability (H_2O, N_2, O_2, CO_2), and mechanical properties. Materials 12:2304. https://doi.org/10.3390/MA12142304
73. Santomasi G, Todaro F, Petrella A, Notarnicola M, Thoden van Velzen EU (2024) Mechanical recycling of PET multi-layer post-consumer packaging: effects of impurity content. Recycling 9:93. https://doi.org/10.3390/RECYCLING9050093/S1
74. TerraCycle (2022) Procter and gamble launches bottle made from 100% recycled and ocean plastic|2017-10-05|brand packaging|packaging strategies. https://www.packagingstrategies.com/articles/94306-procter-gamble-launches-bottle-made-from-100-recycled-and-ocean-plastic. Accessed 30 April 2025
75. Amcor sustainability report (2023) The future of packaging is here
76. Novartis AG (2024) Novartis in society integrated report 2024
77. Hao P, Siebers C, Ragaert K, Gilabert FA (2024) Advancing the characterization of recycled polyolefin blends with a combined experimental and numerical approach to thermomechanical behavior. Polymers 16:1153. https://doi.org/10.3390/POLYM16081153
78. Krantz J, Licata J, Raju MA, Gao P, Ma R, Masato D (2025) Machine learning-based process control for injection molding of recycled polypropylene. Polymers 17:940. https://doi.org/10.3390/POLYM17070940
79. Logan H, Astrup TF, Damgaard A (2024) Additive inclusion in plastic life cycle assessments part I: review of mechanical recycling studies. J Ind Ecol 28:1582. https://doi.org/10.1111/JIEC.13542
80. Vollmer I, Jenks MJF, Roelands MCP, White RJ, van Harmelen T, de Wild P, van der Laan GP, Meirer F, Keurentjes JTF, Weckhuysen BM (2020) Beyond mechanical recycling: giving new life to plastic waste. Angew Chem Int Edn 59:15402–15423. https://doi.org/10.1002/ANIE.201915651
81. Sánchez-Rivera KL, Granger C, Appiah H, Nelson K, Grey S, Sun DJ, Estela-García JE, Chen E, Xu Z, Osswald TA, Turng L-S, McDonald AG, Van Lehn RC, Bar-Ziv E, Huber GW (2024) Cast film production with polyethylene recycled from a post-industrial printed multilayer film by solvent-targeted recovery and precipitation. ACS Mater Lett 21:4042–4050. https://doi.org/10.1021/acsmaterialslett.4c01048
82. Shaukat U, Rossegger E, Schlögl S (2022) A review of multi-material 3D printing of functional materials via vat photopolymerization. Polymers 14:2449. https://doi.org/10.3390/POLYM14122449
83. Dufour T (2023) From basics to frontiers: a comprehensive review of plasma-modified and plasma-synthesized polymer films. Polymers 15:3607. https://doi.org/10.3390/POLYM15173607
84. Makvandi P, Iftekhar S, Pizzetti F, Zarepour A, Zare EN, Ashrafizadeh M, Agarwal T, Padil VVT, Mohammadinejad R, Sillanpaa M, Maiti TK, Perale G, Zarrabi A, Rossi F (2020) Functionalization of polymers and nanomaterials for water treatment, food packaging, textile and biomedical applications: a review. Environ Chem Lett 19:583–611. https://doi.org/10.1007/S10311-020-01089-4
85. Vähä-Nissi M, Hirvikorpi T, Sievänen J, Matilainen K, Salo E, Harlin A (2012) Atomic layer deposition of thin inorganic coatings onto renewable packaging materials. Solid State Phenom 185:12–14. https://doi.org/10.4028/WWW.SCIENTIFIC.NET/SSP.185.12
86. Sfameni S, Rando G, Plutino MR, Sfameni S, Rando G, Plutino MR (2023) Perspective chapter: functional sol–gel based coatings for innovative and sustainable applications. Sol-Gel Method Recent Adv 72:514. https://doi.org/10.5772/INTECHOPEN.110514
87. Ozdemir M, Sadikoglu H (1998) A new and emerging technology: laser-induced surface modification of polymers. Trends Food Sci Technol 9:159–167. https://doi.org/10.1016/S0924-2244(98)00035-1
88. Taneepanichskul N, Hailes HC, Miodownik M (2025) Using hyperspectral imaging and machine learning to identify food-contaminated compostable and recyclable plastics. https://doi.org/10.14324/111.444/ucloe.3237

89. Välimäki MK, Sokka LI, Peltola HB, Ihme SS, Rokkonen TMJ, Kurkela TJ, Ollila JT, Korhonen AT, Hast JT (2020) Printed and hybrid integrated electronics using bio-based and recycled materials—increasing sustainability with greener materials and technologies. Int J Adv Manufact Technol 111:325–339. https://doi.org/10.1007/S00170-020-06029-8/TABLES/4
90. Gazeau B, Zaman A, Minunno R, Shaikh F (2024) Developing traceability systems for effective circular economy of plastic: a systematic review and meta-analysis. Sustainability 16:9973. https://doi.org/10.3390/SU16229973/S1
91. Hahladakis JN, Iacovidou E (2019) An overview of the challenges and trade-offs in closing the loop of post-consumer plastic waste (PCPW): Focus on recycling. J Hazard Mater 380:120887. https://doi.org/10.1016/J.JHAZMAT.2019.120887
92. Bohlmann GM (2004) Biodegradable packaging life-cycle assessment. Environ Prog 23:342–346. https://doi.org/10.1002/EP.10053
93. Papong S, Malakul P, Trungkavashirakun R, Wenunun P, Chom-In T, Nithitanakul M, Sarobol E (2014) Comparative assessment of the environmental profile of PLA and PET drinking water bottles from a life cycle perspective. J Clean Prod 65:539–550. https://doi.org/10.1016/J.JCLEPRO.2013.09.030
94. Hopewell J, Dvorak R, Kosior E (2009) Plastics recycling: challenges and opportunities. Philos Trans Royal Soc B Biol Sci 364:2115. https://doi.org/10.1098/RSTB.2008.0311
95. Schaefer D, Cheung WM (2018) Smart packaging: opportunities and challenges. Proced CIRP 72:1022–1027. https://doi.org/10.1016/J.PROCIR.2018.03.240
96. Maga D, Hiebel M, Aryan V (2019) A comparative life cycle assessment of meat trays made of various packaging materials. Sustainability 11:5324. https://doi.org/10.3390/SU11195324
97. Yokokawa N, Kikuchi-Uehara E, Sugiyama H, Hirao M (2018) Framework for analyzing the effects of packaging on food loss reduction by considering consumer behavior. J Clean Prod 174:26–34. https://doi.org/10.1016/j.jclepro.2017.10.242
98. CEAP (2020) Circular economy action plan (CEAP) 2020 summary for business Implications and next steps
99. (Somewhat) Conflicting views on consumer attitudes toward sustainable packaging—WhatTheyThink. https://whattheythink.com/articles/118218-somewhat-conflicting-views-consumer-attitudes-toward-sustainable-packaging/. Accessed 5 May 2025
100. Coelho PM, Corona B, ten Klooster R, Worrell E (2020) Sustainability of reusable packaging: current situation and trends. Resour Conserv Recycl X 6:100037. https://doi.org/10.1016/J.RCRX.2020.100037
101. Young E, Mirosa M, Bremer P (2020) A systematic review of consumer perceptions of smart packaging technologies for food. Front Sustain Food Syst 4:521813. https://doi.org/10.3389/FSUFS.2020.00063/BIBTEX
102. Guarda C, Caseiro J, Pires A (2024) Machine learning to enhance sustainable plastics: a review. J Clean Prod 474:143602. https://doi.org/10.1016/J.JCLEPRO.2024.143602
103. Maurya A, Bhattacharya A, Khare SK (2020) Enzymatic remediation of polyethylene terephthalate (PET)—based polymers for effective management of plastic wastes: an overview. Front Bioeng Biotechnol 8:602325. https://doi.org/10.3389/FBIOE.2020.602325/BIBTEX
104. Rodrigues C, Souza VGL, Coelhoso I, Fernando AL (2021) Bio-based sensors for smart food packaging—current applications and future trends. Sensors 21:2148. https://doi.org/10.3390/S21062148

Chapter 18
Insights from Battery Waste Recycling for Advancing Polymers and Composite Materials in Smart Food Packaging

Vivek Kumar Sharma, Divya Sharma, Sai Parmeshwar, Rahul Raj, and Akarsh Verma

1 Introduction

Smart food packaging revolutionizes food preservation, safety, and consumer involvement by incorporating active and intelligent functions into conventional packaging [1, 2]. Active packaging uses oxygen scavengers, antimicrobials, and moisture regulators to improve shelf life, while intelligent packaging uses indicators and sensors to check freshness and provide real-time quality data. These technologies enhance food safety, reduce waste, and raise customer awareness [3–5].

Smart packaging creation and deployment are difficult despite their benefits. Functional materials like high-performance sensors and bioactive agents are difficult and expensive to integrate into scalable and flexible systems [4, 6, 7]. Many smart packaging materials are petrochemical-based, posing sustainability and end-of-life concerns. Electronics and nanomaterials hinder recyclability and biodegradability, requiring sustainable alternatives.

Future solutions attempt to balance functionality with environmental responsibility [8–10]. They include biodegradable polymers, bio-based composites, and recyclable multilayer architectures with smart characteristics. Circular economy ideas and lifecycle assessments increasingly shape next-generation packaging, stressing

V. K. Sharma (✉) · R. Raj · A. Verma
Department of Mechanical Engineering, UPES (University of Petroleum and Energy Studies), Dehradun, India
e-mail: vivekmech007@gmail.com

D. Sharma
Physics and Material Science and Engineering Department, Jaypee Institute of Information Technology, Noida, Uttar Pradesh, India

S. Parmeshwar
Petroleum Engineering Department, School of Advanced Engineering, UPES, Dehradun, India

A. Verma et al. (eds.), *Polymers and Composite Materials for Packaging: Smart Food Packaging and Solutions*, Materials Horizons: From Nature to Nanomaterials,
https://doi.org/10.1007/978-981-95-0677-4_18

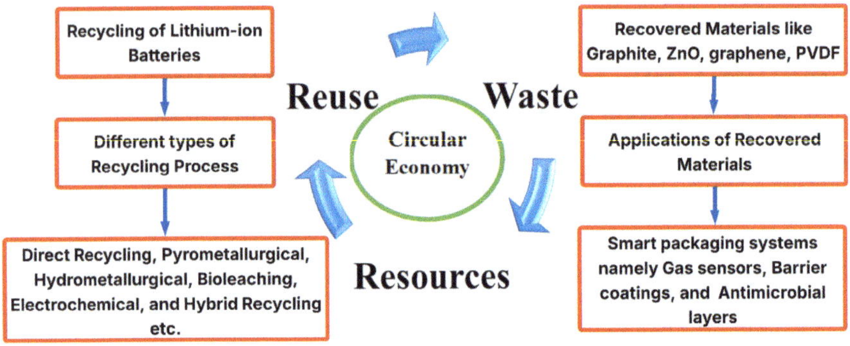

Fig. 1 Conceptual framework illustrating the integration of lithium-ion battery (LIB) recycling processes and the valorization of recovered materials—such as graphite, ZnO, graphene, and PVDF—into smart food packaging applications These recovered materials

environmental performance without compromising innovation [8, 11–13] as shown in Fig. 1.

Polymers and composites are essential to modern food packaging due to their versatility, lightweight, barrier qualities, and functional modification [14]. Polyethylene (PE), polypropylene (PP), polyethylene terephthalate (PET), and polystyrene (PS) are used for their mechanical strength, transparency, and chemical resistance. Polymers have moved from passive confinement to active food preservation and safety [4–15].

Innovative packaging solutions use polymers to integrate active agents (antimicrobials, antioxidants, moisture absorbers) and intelligent components (time–temperature indications). Biodegradable polymers including PLA, PHA, and starch-based variations are rising along with sustainability. These materials are versatile and degradable under controlled or industrial composting, supporting cyclic material flows [11, 16].

Inorganic fillers, nanoparticles, and natural fibers improve mechanical, thermal, and barrier properties of polymer-based composites [17]. For gas permeability reduction and shelf life, nanoclays and graphene oxide can be used. In addition, these composites can include sensor elements or active coatings for intelligent packaging. Polymers and composites are driving package innovation and environmental stewardship as consumer needs shift toward sustainability, safety, and interaction [18, 19].

Battery waste management, especially from lithium-ion batteries (LIBs), is a major environmental issue alongside packaging. LIB consumption has increased due to demand for portable devices, electric cars, and renewable energy storage, increasing end-of-life battery waste [20]. Batteries with lithium, cobalt, nickel, manganese, volatile organic electrolytes, and complicated polymeric components are dangerous. Poor disposal causes groundwater contamination, air pollution, and landfill fires [21, 22].

LIBs are sophisticated systems with electrodes, separators, electrolytes, and casings. When mismanaged, each ingredient causes environmental damage [23]. Cobalt and nickel are carcinogenic and bioaccumulative, and organic solvent breakdown produces persistent organic contaminants. Battery raw material mining has large environmental impacts, emphasizing the need for closed-loop material cycles and appropriate disposal [24, 25].

Effective LIB recycling reduces environmental damage and recovers resources. Lithium, cobalt, and graphite are extracted using hydrometallurgical and pyrometallurgical methods, supporting circular economy goals and minimizing dependence on virgin resources. Scalable, sustainable recovery technologies are needed to meet global LIB demand, affecting environmental policy and material science [26, 27].

Interdisciplinary research shows battery recycling and sustainable smart packaging are similar. Both systems involve complicated metal, polymer, adhesive, and coating assemblies that require creative disassembly, recovery, and reuse methods [28]. Battery recycling methods, especially those for separating multilayer components, can inform recyclable smart packaging. Material recovery hierarchies connect concepts [29, 30].

LIB recycling focuses chemical and thermal recovery of high-value materials. Similar to structured deconstruction, smart packaging could recover polymers, bioactive compounds, and integrated sensors. They promote material efficiency and reduce environmental impact, using circular design principles [31]. Fragmentation of composite structures is another convergence point. Battery recycling delaminates tightly bonded layers like current collector electrodes via mechanical, chemical, or thermal techniques [24]. Smart packaging can address multilayer film recycling issues with this skill. Smart and sustainable systems could be created using separation-friendly packaging and recycling-informed layering [27, 32].

In conclusion, smart food packaging and battery recycling can advance material circularity and sustainable engineering. LIB recovery expertise can enable smart packaging designs that balance technical performance and environmental concerns.

2 Polymers and Composite Materials in Smart Packaging

Smart food packaging relies on polymers for structural support and flexibility. Performance, cost, and environmental impact determine whether to use conventional or bio-based polymers. Due to their barrier qualities, mechanical strength, and processing adaptability, synthetic polymers like PE and PET are frequently employed [20, 33]. Due to its clarity and durability, PET is appropriate for rigid containers and low-density polyethylene (LDPE) for flexible films. However, their petrochemical origins and weak biodegradability raise sustainability problems.

Table 1 shows, polylactic acid (PLA), manufactured from renewable feedstocks, is a popular biodegradable option. PLA has strong mechanical strength and transparency but is brittle and low-barrier, necessitating blending or coatings. Microbial fermentation produces polyhydroxy alkanoates (PHA), which are suitable for smart

Table 1 Comparison of conventional versus bio-based polymers for smart packaging applications

Polymer type	Source	Key functional properties	Environmental characteristics	Typical applications in smart packaging	References
Polyethylene (PE)	Petrochemical	High flexibility, good moisture barrier, mechanical robustness	Non-biodegradable; widely recycled	Films, pouches, and multilayer flexible packaging	[29]
Polypropylene (PP)	Petrochemical	High thermal resistance, chemical stability, transparency	Non-biodegradable; partially recyclable	Closures, containers, and thermoformed trays	[30]
Polyethylene terephthalate (PET)	Petrochemical	High clarity, good oxygen barrier, structural rigidity	Recyclable (rPET); not biodegradable	Bottles, rigid containers, and smart labels	[34]
Polylactic acid (PLA)	Renewable (starch/sugar)	Transparent, stiff, biodegradable under industrial composting conditions	Bio-based; requires controlled degradation environments	Biodegradable films and intelligent packaging matrices	[32]
Polyhydroxyalkanoates (PHA)	Microbial fermentation	Flexible, biodegradable, good compatibility with bioactive agents	Fully biodegradable in natural conditions	Active packaging, antimicrobial layers	[33]
Starch-based polymers	Renewable (plants)	Cost-effective, biodegradable, low mechanical strength (improved via blending)	Biodegradable; limited water resistance	Compostable films and packaging inserts	[35]
Cellulose derivatives	Renewable (plant fibers)	Excellent oxygen barrier, printable, biodegradable	Compostable under natural conditions	Coatings, films, and sensor platforms	[36]

packaging due to their environmental degradation and active agent compatibility. Starch-based polymers and cellulose derivatives are biodegradable and inexpensive, but they need composite production to meet mechanical and barrier requirements [37, 38]. Nanomaterials and functional additives have enhanced polymer matrices in smart packaging [39, 40]. Nanoclays, graphene oxide, and metal oxides improve barrier and mechanical resilience, while bioactive chemicals add antibacterial, antioxidant, or self-healing capabilities [4, 5].

These advances enable multifunctional packaging that interacts dynamically with its contents and environment [41, 42]. Active and intelligent functions define smart packaging. Sensors that measure food quality, freshness barriers, and biodegradability are examples [43]. Food condition is monitored in real time by embedded TTIs, gas sensors, and spoiling detectors. Responsive inks, printed electronics, and nanostructured films are used in packaging. Flexible electronics allow for affordable biodegradable sensor creation [1, 39].

Food preservation requires barrier integrity to prevent oxidation, microbial growth, and moisture loss. Low-permeability high-performance composites use nanofillers such as metal oxides or layered silicates in polymer matrices. Active packaging solutions that regulate interior atmosphere without preservation need these advances [5, 43].

Smart packaging involves including environmental response. PLA and PHA can be made to breakdown in industrial composting or soil exposure, matching packaging life cycles to food perishability. Moisture or microbiological activity can cause degradation, integrating sustainability into package performance [28, 44–46]. Despite these advances, smart packaging has major end-of-life management issues. Recycling polymers, coatings, and electronics in multilayer systems is difficult. Material heterogeneity hinders sorting, resulting in high garbage or incineration rates [47, 48]. Biodegradable polymers require specific composting conditions that are not universally available, raising issues regarding their practicality [33, 49].

Functional additives like nanoparticles, conductive inks, and sensors can contaminate recycling streams and impair material quality, hindering recyclability [50, 51]. Adhesives and coatings typically prevent mechanical recycling. Chemical recycling can recover high-purity monomers, but its energy demand and scalability limit its adoption [52, 53].

Smart packaging components may be toxic. Antimicrobials, gas scavengers, and nanomaterials improve functionality but may leach, persist, or bioaccumulate. The environmental and health effects of these compounds are poorly understood, and present regulatory frameworks typically lag behind technology advances, generating safety and approval uncertainties.

Sustainable design is needed to address these issues [6, 38]. Eco-design principles recommend streamlined, modular packaging systems consisting of compatible or single-material constructions, non-toxic and removable additives, and end-of-life compatibility. Compostable electronics, green synthesis, and recycling-friendly materials can help intelligent packaging meet global sustainability targets [49].

By integrating utility with ecological responsibility, smart packaging materials—particularly polymers and composites—can revolutionize food systems by improving safety, quality, and sustainability. Sustainable progress in this subject requires a materials-centered, circular design, lifecycle thinking approach.

3 Battery Waste Recycling: An Overview

The direct recycling, pyrometallurgy, and hydrometallurgy are three main approaches to recycle spent lithium-ion batteries (LIBs), each with pros and cons in terms of material recovery, energy efficiency, and environmental impact [54–56]. These methods focus on recovering lithium, cobalt, and nickel, which are needed to make advanced energy storage devices. Direct recycling tries to prolong the life of battery components, especially cathode materials like $LiCoO_2$ or NMC. Re-lithiating intact cathodes after dismantling cells saves processing energy and material value. Direct recycling is sensitive to battery chemistry and design, restricting its use in mixed-stream or end-of-life battery scenarios [50–52].

Pyrometallurgical recycling recovers metals through reduction and roasting at >1000 °C. Cobalt, nickel, and copper are recovered as metallic alloys, but lithium is often found in slag and requires further treatment [49, 52]. Although compatible with mixed battery chemistries, this energy-intensive method may cause significant CO_2 emissions and material losses, especially for lighter elements like lithium and aluminum [53–55].

Hydrometallurgical recycling dissolves and recovers precious metals using water chemistry. Leaching agents like sulfuric acid or organic solvents enable metal extraction, followed by solvent extraction, precipitation, or electrochemical recovery for high-purity salts (e.g., Li_2CO_3, $CoSO_4$). Hydrometallurgy operates at lower temperatures and allows greater battery type selectivity and adaptability, but it requires strict chemical conditions and liquid waste disposal.

Hybrid techniques are being investigated to maximize recovery rates, reduce emissions, and boost economic viability [56–59]. These methods as shown in Table 2, can also help recycle smart packaging, which has multiple material classes and functions.

LIB components include metals, ceramics, and polymers. Lithium, cobalt, nickel, and copper are most economically valuable. These are obtained by hydrometallurgical leaching or pyrometallurgical alloying [60]. Secondary elements like manganese and aluminum can be recovered depending on battery chemistry, although they are worth less. The residual "black mass" after battery shredding often contains ceramic materials, especially lithium metal oxides [61]. In direct recycling, these ceramics may retain sufficient structural integrity to be re-lithiated and reused, reducing the need for complete resynthesis and lowering emissions [62]. This highlights the potential for reusing ceramic components in other applications requiring functional oxides.

Polymers in LIBs, including binders like polyvinylidene fluoride (PVDF) and separators composed of polyethylene or polypropylene, present additional recovery opportunities. Traditionally discarded during thermal or chemical treatment, PVDF is increasingly recognized for its economic and environmental significance [36].

Table 2 Comparative summary of battery recycling methods

Recycling method	Representative process steps	Key advantages	Key limitations	Primary recovered materials	References
Direct recycling	Cell disassembly, cathode re-lithiation	Preserves structural integrity of active materials; low energy requirement	Limited to homogenous chemistries; requires battery design standardization	Cathode materials (e.g., $LiCoO_2$)	[55, 56]
Pyrometallurgical	High-temperature smelting and slag separation	Can treat mixed-stream batteries; relatively simple equipment	High CO_2 emissions; lithium recovery inefficient	Metallic alloys of Co, Ni, Cu	[47, 48]
Hydrometallurgical	Chemical leaching, solvent extraction	High metal recovery rates with good purity; selective separation possible	Requires handling of hazardous chemicals; generates wastewater	Li_2CO_3, $CoSO_4$, $NiSO_4$	[57, 58]

Techniques for its recovery, such as solvent dissolution or selective precipitation, are being developed to enhance process sustainability. These concepts parallel the recovery challenges in multilayered food packaging, where polymer components may be contaminated or degraded during use [63].

The parallels between LIB recycling and smart packaging recovery are particularly notable in their handling of composite materials. Both systems consist of tightly integrated layers—metals, polymers, ceramics, and adhesives—that complicate disassembly. Efficient separation is essential to preserve material quality and facilitate downstream purification [64]. In LIBs, this begins with mechanical or thermal pretreatment, followed by advanced techniques like as ultrasonic delamination or solvent-based debonding that allow for selective disassembly without destroying reusable materials. Similar advances could facilitate the separation of multilayer sheets in smart packaging, boosting their recyclability [65].

Purification ensures the recovered materials fulfill quality standards for reintroduction into industry. Solvent extraction, ion exchange, and precipitation isolate battery-grade materials like nickel sulfate and lithium carbonate from LIBs. These chemical refinement methods can be utilized in packaging to recover useful ingredients and ensure food-contact safety for reused materials [60, 66].

End goal of recycling systems is material reuse. Cathode materials are reformulated or re-lithiated for new batteries, creating a closed-loop manufacturing cycle [67–69]. This could allow pure polymers or functional fillers to be reused in smart packaging, minimizing virgin material use and supporting circular economy aims [70]. The environmental and economic benefits of LIB recycling are compelling. Mining is energy-intensive and environmentally destructive. Recycling lessens demand [71–73]. Instead of virgin material manufacture, it reduces carbon emissions by 60% and landfill fires and harmful leachates. Smart packaging can assist reduce microplastic pollution and landfill burden by recovering polymers and additives [74].

Recovering cobalt and lithium lowers production costs and strengthens domestic supply chains, especially as energy storage and electric vehicle demand rise. In packaging, variable raw material costs and regulatory pressure on single-use plastics encourage reuse and upcycling. Additionally, recovery activities can drive employment development in green technology and support closed-loop business models for both battery and packaging businesses [75].

Overall, LIB recycling's advanced tactics and infrastructure can help smart packaging manage sustainability issues. By focusing on efficient separation, purification, and reuse of composite materials, these approaches offer a roadmap toward closing material loops and establishing more sustainable production-consumption systems across varied sectors.

4 Cross-Application Potential: From Battery Recycling to Smart Packaging

Innovative lithium-ion battery (LIB) recycling methods can improve polymer recovery and reuse in smart food packaging [76]. These techniques, originally developed to disassemble and manage complex multi-material systems with precision and minimal degradation, address smart packaging's multilayer composite and functional component separation and recycling difficulties [77]. Selective dissolving is used in hydrometallurgical recycling to remove battery components. This procedure can be used to isolate polymers in multilayer packaging by dissolving select layers like PE, PET, or PVA while maintaining others. This preserves polymer quality and reduces separation contamination for closed-loop recycling [78].

Thermal delamination, used in pyrometallurgical recycling to deconstruct electrode layers or volatilize binders, may be used in packaging. Modified low-temperature thermal techniques can break down smart packaging adhesives or thermosensitive layers without damaging reusable materials. These approaches are effective for recovering barrier layers and functional coatings [67, 79]. Density-based flotation and electrostatic separation, used in LIB material sorting, can recover polymers from packaging waste with fillers, metallic sensors, or inks. These methods improve sorting precision and prevent cross-contamination, preserving polymer quality.

The direct regeneration of battery cathodes without reverting to raw materials provides a blueprint for physically cleaning, remolding, or refunctionalizing packaging polymers. This upcycling pathway is important for high-performance or bio-based polymers because it offers a sustainable alternative to depolymerization [80].

Composite separation issues in LIBs and smart packaging are very similar. LIB cathodes consist of active layers like $LiCoO_2$ on aluminum collectors, bonded with polyvinylidene fluoride. PE, PET, EVOH, adhesives, and smart sensors are also used in packaging films. LIB techniques like solvent delamination, ultrasonic separation, and supercritical CO_2 dissolution can be used to isolate and recover packaging layers. During recycling, both domains favor functionality. Cathodes regenerate, so packaging polymers and barrier layers can be reused for high value. "Design for disassembly" and material circularity are increasingly important in next-generation material development [46].

Polymer binder recovery overlaps. PVDF is an essential but chemically resistant binder in LIBs, and its recovery often needs powerful solvents like N-methyl-2-pyrrolidone. However, cleaner solvents and supercritical fluids enable more sustainable recovery [75]. Packaging systems with resilient adhesives and coatings can leverage these approaches to extract and reuse binders, especially multilayer constructions. The search for biodegradable or detachable binders in battery systems may inspire packaging that is easy to disassemble and recycle [79].

Last, case studies and speculative applications demonstrate the potential for smart packaging made from battery-derived materials. Time–temperature indicators can be printed with conductive inks made from battery anode carbon black and graphite. In active packaging, transition metal oxides as MnO_2 or TiO_2 from cathode waste can serve as antibacterial or photocatalytic additives. Al_2O_3 and ZnO nanoscale coatings, found in LIB leftovers, improve gas-barrier properties in packaging films [41, 71].

Despite their early phases, some applications show promising cross-sector synergies. The confluence of recycling technologies between energy and food sectors could drive breakthroughs in sustainable material recovery, supporting a broader circular economy model.

5 Innovative Materials from Battery Recycling with Packaging Potential

As per Table 3, graphite, graphene, and carbon black, recovered from spent lithium-ion batteries, have intriguing physicochemical qualities for smart food packaging sensors and barrier layers. Their strong electrical conductivity, thermal stability, vast surface area, and mechanical flexibility enable multifunctionality and improve real-time monitoring and preservation packaging methods [81].

Graphite from battery anodes can be refined into conductive composites or inks for printed electronics. These sensors are useful for detecting gases (e.g., CO_2, ethylene),

Table 3 Recovered materials (graphite, PVDF, ZnO, etc.) and their corresponding smart packaging applications

Recovered material	Recovery source	Potential application in smart packaging	References
Graphite	Spent battery anode	Integration in printed sensor technologies	[81]
Graphene	Graphite refinement process	Development of high-performance barrier layers	[82]
Carbon black	Spent battery anode	Formulation of electrically conductive inks	[83]
Zinc oxide (ZnO)	Cathode material waste	Application in antimicrobial surface coatings	[15]
Titanium dioxide (TiO_2)	Cathode material waste	Production of photocatalytic functional films	[75]
Polyvinylidene fluoride (PVDF)	Binder recovery process	Fabrication of moisture and gas-barrier films	[75]
Recycled polyethylene terephthalate (rPET)	Battery casing waste	Manufacture of multilayer film structures	[46]

temperature changes, and microbial activity, providing real-time food freshness and safety indicators. Functionalizing graphite with certain chemical groups improves resistive or capacitive sensor sensitivity and selectivity [81].

Graphene, derived from the exfoliation or chemical treatment of graphite, improves polymer matrices like PLA and PET by decreasing gas and moisture permeability while preserving transparency and flexibility. This advancement allows for the creation of multifunctional films that exhibit high thermal and electrical conductivity, suitable for protective coatings and integrated sensors [82]. Carbon black serves as an economical conductive additive derived from battery anodes, combining with biodegradable polymers to produce printable sheets for smart indicators and sensors. This innovation fosters sustainability and minimizes landfill waste [84]. Metal oxides derived from spent batteries, such as zinc oxide, titanium dioxide, silver oxide, and copper oxide, demonstrate extensive antibacterial properties by generating reactive oxygen species or causing membrane disruption, thereby enhancing the shelf life of food when integrated into polymer films [85].

Furthermore, aluminum oxide (Al_2O_3) and silicon dioxide (SiO_2) improve barrier properties by forming tortuous diffusion pathways in multilayer films, while the sustainable recovery of ZnO and MnO_2 through hydrometallurgical methods promotes circular material utilization. Recycled polymers like PVDF and PET present innovative packaging solutions. PVDF can be recovered from lithium-ion batteries for use in barrier coatings, while PET can be subjected to both chemical and mechanical recycling processes to produce high-purity rPET. This supports multi-layer structures that incorporate biodegradable components without compromising performance standards [75].

Despite the complexities associated with recycling adhesives such as acrylics, EVA, and polyurethane due to cross-linking, investigations into recyclable binder systems suggest innovative strategies for reuse. Utilizing materials derived from batteries for packaging promotes resource conservation and the creation of innovative, sustainable solutions, merging the valorization of electronic waste with advancements in circular food packaging.

6 Sustainability and Circular Economy Considerations

A battery recycling and food packaging waste management share sustainability problems and potential in a comparative life cycle assessment (LCA) [86]. Both sectors use complex, multi-material systems and are under growing environmental scrutiny, especially during end-of-life.

In lithium-ion battery recycling, LCAs show that closed-loop recovery of essential elements like lithium, cobalt, and nickel benefits the environment. Hydrometallurgical recycling can cut CO_2 emissions by 60–70% and energy usage by over 50% compared to primary resource extraction. These benefits are offset by chemical use, wastewater creation, and battery collection, disassembly, and treatment logistics [39]. Thus, collection efficiency, material sorting accuracy, and recovery technology effectiveness determine battery recycling sustainability.

Conversely, food packaging LCAs—especially smart and multilayer variants—show material complexity and low recyclability as sustainability constraints [41].

Unrecycled PET and PE plastics cause landfill overflow and marine pollution. Smart packaging with sensors, adhesives, and coatings complicates mechanical recycling. Bioplastics and biodegradable materials have fewer lifespan impacts, but landfilling or a lack of industrial composting infrastructure reduces their sustainability [87]. Recycling, mono-material, and locally processed packaging systems outperform those with many components or specialized functions in LCA. As end-of-life design becomes more important, the two areas converge. Circular design ideas that prioritize material compatibility, modularity, and disassembly are being applied to battery systems and packaging. Manufacturers can improve lifecycle performance and comply with waste reduction and resource conservation policies by anticipating downstream processing needs during design [56].

This change enables closed-loop systems, which are growing in both industries. Advanced recovery methods can recycle important metals supporting components like graphite and PVDF into new batteries. Recycling in supply chains reduces primary extraction and improves material security by feeding spent batteries directly into fresh cell manufacturing. Direct regeneration of cathode materials is very efficient and resource-saving.

Material heterogeneity, food residue contamination, and the lack of integrated collecting infrastructure make closed-loop packaging systems harder to create [53]. However, design-for-recyclability, mono-material packaging, and compostable biopolymers are advancing [88]. Modular components like sensors and coatings are being added to smart packaging to make it more recyclable or compostable. Chemical recycling of PET plastics is producing food-grade materials from post-consumer waste, creating a circular packaging economy. Battery and packaging industries benefit from cross-sectoral learning [87]. Battery management traceability and labeling technologies can be applied to smart packaging for sorting and recovery. Sustainable package design ideas like minimizing material use and optimizing recyclability can inform battery design [87].

Extended producer responsibility (EPR) regulations demonstrate common sustainability pathways. EPR requires battery manufacturers to manage product recovery and safe disposal, encouraging recyclable and non-toxic design. EPR could improve smart package modularity, reduce hazardous chemicals, and match waste processing infrastructure [89]. Material traceability and labeling—now routine in battery recycling—also have applications. For automatic sorting and safety, batteries are marked with composition data. Similar methods might be used on smart packaging to communicate embedded material content, disposal directions, and recycling compatibility using standardized labels or digital identifiers like QR codes [90].

The EU's Battery Regulation, RoHS, and REACH provide guidelines for hazardous substance management and eco-design [91]. These would govern potentially harmful sensor elements, antimicrobial agents, and adhesives in smart packaging, assuring safety and sustainability.

Finally, battery recycling frameworks and legislation help improve smart food packaging sustainability [92]. The packaging industry as shown in Table 4, may improve its environmental performance and support a resilient circular economy by implementing circular design, material traceability, and regulatory best practices.

Table 4 Environmental impact metrics from LCA studies (e.g., CO_2 savings, energy use, recycling efficiency)

Evaluation metric	Battery recycling	Packaging recycling	References
Reduction in CO_2 emissions	Approximately 60–70%	Low to moderate reduction	[76]
Energy conservation potential	Greater than 50% energy savings	Limited energy savings	[89]
Material recovery efficiency	High, particularly for cobalt (Co) and lithium (Li)	Low to moderate efficiency	[90]
Water utilization impact	Moderate water usage	Low water usage	[92]

7 Challenges and Research Gaps

Smart food packaging using recycled materials faces significant technical, regulatory, and economic challenges, particularly in achieving the purity and compatibility required for food-contact applications under stringent standards from the U.S. FDA, EFSA, and GB. Battery recycling introduces risks of contamination from heavy metals and electrolytes, necessitating purification steps such as solvent extraction, membrane filtration, and thermal treatment, while post-consumer packaging waste demands the removal of inks, adhesives, and biological residues. Recyclates must integrate with polymers like PLA, PE, and PET without compromising integrity or sensory properties, often requiring surface modification of additives like carbon black and metal oxides to prevent leaching, validated through migration testing. Regulatory frameworks may also require functional barriers to isolate active packaging components from food. Economically, scaling such recycling remains difficult due to costly processes like thermal delamination and solvent extraction, which are rarely viable without incentives like carbon pricing or virgin plastic taxes. However, innovations such as adapting PVDF binder recovery for packaging and repurposing battery-derived materials can reduce reliance on synthetics and improve cost efficiency. Aligning these efforts with extended producer responsibility (EPR) schemes and sustainable branding may enhance economic viability and promote a circular economy, though cross-disciplinary research and pilot-scale demonstrations are essential, given the traditionally separate domains of electrochemical materials science and food-grade polymer engineering.

R&D collaboration across chemistry, engineering, toxicology, and industrial design is needed to close this gap as shown in Table 5. Materials compatibility, migratory behavior, and toxicological safety must be studied when adapting battery recycling methods for packaging, such as to extract smart coatings or sensors.

Transitioning from lab-scale feasibility to commercial deployment necessitates pilot-scale studies to optimize processing, identify operational bottlenecks, and assess industrial integration. These trials determine the viability of battery-derived graphene and metal oxides in packaging films while evaluating solvent-based delamination for handling real-world waste streams. Cross-sectoral studies further examine logistical, preprocessing, and regulatory considerations for repurposing battery-derived materials in consumer applications, fostering industrial confidence and informing material circularity policies. Despite technical and economic challenges, battery recycling and smart packaging technologies represent a promising avenue for sustainable materials innovation, requiring a systems-level approach that integrates collaborative research, regulatory alignment, and practical validation for safe and scalable reuse in food-grade applications.

Table 5 Key challenges in battery and packaging sectors along with proposed mitigation strategies to enhance sustainability, safety, and regulatory alignment

Identified challenge	Relevant sector	Proposed mitigation strategy
Material purity	Battery and packaging	Implementation of advanced purification technologies
Recyclability of multilayer films	Packaging	Adoption of design-for-recyclability principles
Presence of toxic additives	Battery and packaging	Substitution with environmentally benign functional agents
Economic viability	Battery and packaging	Utilization of subsidies and extended producer responsibility (EPR) mechanisms
Regulatory compliance	Packaging	Development and enforcement of harmonized international standards

The strategies focus on improving material purity, recyclability, and economic feasibility while addressing toxicology and compliance concerns

8 Conclusion and Future Outlook

Battery recycling and smart food packaging innovations provide a unique opportunity for cross-sectoral collaboration to enhance environmental sustainability and resource efficiency. Lithium-ion battery recycling methods, material management strategies, and circular economy principles contribute to the development of sustainable intelligent packaging systems, benefiting complex multi-material structures and challenging end-of-life scenarios. Synergies include repurposing carbon-based residues and metal oxides from spent batteries for sensors and antibacterial agents in packaging, while solvent-based separation, thermal delamination, and direct regeneration techniques facilitate the dismantling, purification, and reuse of functional packaging layers.

Circular battery value chains, incorporating extended producer responsibility (EPR), traceability, and design-for-recyclability, establish a framework for sustainable post-consumer packaging waste management. Advancing eco-friendly smart packaging necessitates interdisciplinary collaboration across materials science, process engineering, electronics, and food safety, emphasizing renewable and recycled materials, modular packaging design, and scalable recovery methods. Regulatory improvements are essential to enable safe cross-sectoral material reuse and incentivize eco-design through certifications, tax advantages, and EPR initiatives.

To achieve this future, three strategic recommendations are suggested:

1. Prioritize funding for multidisciplinary studies on battery recycling, food packaging science, and public health, focusing on material compatibility, contamination management, and lifecycle sustainability.

2. Cross-sector Pilot Programs: Pilot-scale collaborations involving battery recyclers, package makers, and regulatory authorities can evaluate recovered materials' technical integration, food-contact safety, and economic feasibility.
3. Improve regulatory frameworks to harmonize material standards across industries, encourage producer responsibility, and reward circular design techniques with fiscal incentives or green certifications.

Battery recycling and smart packaging sectors can create a closed-loop materials ecosystem that balances functionality, safety, and ecological resilience by combining technological innovation with sustainable design and forward-thinking legislation.

Declaration of Competing Interest The authors declare that they have no known competing financial interests or personal relationships that could have appeared to influence the work reported in this paper.

References

1. Khan S, Prasad A, Filipe CDM, Didar TF (2025) Smart food packaging for monitoring food quality. Antimicrobial food packaging. Elsevier, New Jersey, pp 175–192
2. Sundaresan J, Gupta A, Suadamara R (2025) Advances in smart food packaging for a sustainable future. In: IOP conference series: earth and environmental science. IOP Publishing, p 012117
3. Said NS, Lee WY (2025) Pectin-based active and smart film packaging: a comprehensive review of recent advancements in antimicrobial, antioxidant, and smart colorimetric systems for enhanced food preservation. Molecules 30(5):1144
4. Du L et al (2025) Application of smart packaging in fruit and vegetable preservation: a review. Foods 14(3):447
5. El Guerraf A et al (2024) Smart conducting polymer innovations for sustainable and safe food packaging technologies. Compr Rev Food Sci Food Saf 23(6):e70045
6. Edo GI et al (2025) Advancing sustainable food packaging: the role of green nanomaterials in enhancing barrier properties. Food Eng Rev 16:1–35
7. Arruda TR et al (2025) An overview of starch-based materials for sustainable food packaging: recent advances, limitations, and perspectives. Macromolecules 5(2):19
8. Cordeiro A et al (2025) Advancements in packaging materials: trends, sustainability, and future prospects. Circ Econ Sustain 32:1–32
9. Zhang M et al (2022) Recent advances in polymers and polymer composites for food packaging. Mater Today 53:134–161
10. Thul PP, Jagtap DN, Gosavi RA, Khatal MM, Labade P, Sawant AA (2025) Biotechnological innovations in packaging and sensors in the food processing
11. Oliver-Cuenca V et al (2024) Bio-based and biodegradable polymeric materials for a circular economy. Polymers 16(21):3015
12. Castagna A et al (2025) Transforming agricultural waste from mediterranean fruits into renewable materials and products with a circular and digital approach. Materials 18(7):1464
13. Cordeiro A et al (2025) Advancements in packaging materials: trends, sustainability, and future prospects. Circul Econ Sustain 65:1–32
14. Johari A, Agarwal P (2025) Multifunctional polymeric materials. Multifunct Mater Eng Biol Appl 85:47–79
15. Dejene BK (2024) Advancing natural fiber-reinforced composites through incorporating ZnO nanofillers in the polymeric matrix: a review. J Nat Fibers 21(1):2356015

16. Verma SK, Prasad A, Katiyar V (2024) State of art review on sustainable biodegradable polymers with a market overview for sustainability packaging. Mater Today Sustain 32:100776
17. Mishra T, Mandal P, Rout AK, Sahoo D (2022) A state-of-the-art review on potential applications of natural fiber-reinforced polymer composite filled with inorganic nanoparticle. Comp C Open Access 9:100298
18. Antony-Jose S et al (2025) A comprehensive review on cellulose nanofibers, nanomaterials, and composites: manufacturing, properties, and applications. Nanomaterials 15(5):356
19. Bin-Rashid A, Haque M, Islam SMM, Labib KMRU (2024) Nanotechnology-enhanced fiber-reinforced polymer composites: recent advancements on processing techniques and applications. Heliyon 10(2):785
20. Yi CQ, Bojeng MNBHBH, Kamis SKBH, Mubarak NM, Karri RR, Azri H (2024) Production of hydrogen using plastic waste via aspen hysys simulation. Sci Rep 14(1):4934
21. Philippot M, Alvarez G, Ayerbe E, Van Mierlo J, Messagie M (2019) Eco-efficiency of a lithium-ion battery for electric vehicles: influence of manufacturing country and commodity prices on GHG emissions and costs. Batteries 5(1):23. https://doi.org/10.3390/batteries5010023
22. Ren G, Liao C, Liu Z, Xiao S (2020) Lithium and manganese extraction from manganese-rich slag originated from pyrometallurgy of spent lithium-ion battery. Trans Nonferrous Metals Soc China 32(8):2746–2756. https://doi.org/10.1016/S1003-6326(22)65981-8
23. Luo X, Wang J, Dooner M, Clarke J (2015) Overview of current development in electrical energy storage technologies and the application potential in power system operation. Appl Energy 137:511–536
24. Jain S (2020) A review on recent trends in microwave-assisted catalytic biodiesel production using waste cooking oil and algal oil. ACS Omega 5(11):5438–5460. https://doi.org/10.1021/acsomega.9b04372
25. Wang X, Gu Z, Ang EH, Zhao X, Wu X, Liu Y (2022) Prospects for managing end-of-life lithium-ion batteries: present and future. Interdiscip Mater 1(3):417–433
26. Dalton B, Bhagabati P, De Micco J, Padamati RB, O'Connor K (2022) A review on biological synthesis of the biodegradable polymers polyhydroxyalkanoates and the development of multiple applications. Catalysts 12(3):319
27. Esterhuizen M, Kim YJ (2022) Effects of polypropylene, polyvinyl chloride, polyethylene terephthalate, polyurethane, high-density polyethylene, and polystyrene microplastic on *Nelumbo nucifera* (Lotus) in water and sediment. Environ Sci Pollut Res 29(12):17580–17590
28. Jain S (2019) The production of biodiesel using Karanja (*Pongamia pinnata*) and Jatropha (*Jatropha curcas*) oil. Biomass Biopolym Based Mater Bioenergy 78:397–408
29. Karan H, Funk C, Grabert M, Oey M, Hankamer B (2019) Green bioplastics as part of a circular bioeconomy. Trends Plant Sci 24(3):237–249. https://doi.org/10.1016/j.tplants.2018.11.010
30. Dospatliev L, Koyundzhiyska-Vasileva R, Stefanova V, Ivanov I, Ivanova I (2023) Biodegradable polymers in active and intelligent packaging systems. Polymers 15(4):879. https://doi.org/10.3390/polym15040879
31. Tao R, Xing P, Li H, Cun Z, Sun Z, Wu Y (2022) In situ reduction of cathode material by organics and anode graphite without additive to recycle spent electric vehicle $LiMn_2O_4$ batteries. J Power Sources 520:827. https://doi.org/10.1016/j.jpowsour.2021.230827
32. Ma X, Jian R, Zhang Y, Wang X, Sun Q (2022) Advances in starch-based materials for food packaging: structure, properties and functions. Trends Food Sci Technol 120:52–66. https://doi.org/10.1016/j.tifs.2022.01.025
33. Peelman N et al (2019) Application of bioplastics for food packaging. Trends Food Sci Technol 91:65–72. https://doi.org/10.1016/j.tifs.2019.06.001
34. Nofar M, Tabatabaei A (2020) Polylactide-based blends: a review on morphology development, properties, and applications. Polymers 12(6):1234. https://doi.org/10.3390/polym12061234
35. Montalbán MG, Cobos A, Víllora G (2021) Smart packaging: materials and applications. Materials 14(23):7069. https://doi.org/10.3390/ma14237069
36. Siracusa V, Blanco I (2020) Bio-based and biodegradable polymers for use in food packaging: a review. Int J Mol Sci 21(21):7075. https://doi.org/10.3390/ijms21217075

37. Yadav R, Singh M, Shekhawat D, Lee S-Y, Park S-J (2023) The role of fillers to enhance the mechanical, thermal, and wear characteristics of polymer composite materials: a review. Compos A Appl Sci Manuf 175:107775
38. Dejene BK, Gudayu AD (2024) Eco-friendly packaging innovations: integrating natural fibers and ZnO nanofillers in polylactic acid (PLA) based green composites: a review. Polym Plast Technol Mater 63(12):1645–1681
39. Jain S, Sharma MP (2012) Correlation development between the oxidation and thermal stability of biodiesel. Fuel 102:354–358
40. Jain S, Sharma MP (2013) Engine performance and emission analysis using oxidatively stabilized *J. curcas* biodiesel. Fuel 106:152–156
41. Jain S, Sharma MP (2012) Oxidation, thermal, and storage stability studies of *J. curcas* biodiesel. Int Sch Res Notices 2012(1):861293
42. Jain S, Sharma MP (2011) Study of oxidation stability of *J. curcas* biodiesel/diesel blends. Int J Energy Environ 2(3):533–542
43. Tariq M, Khan S, Nazir A, Ahmad W, Ali H (2023) Recent trends in biodegradable smart packaging materials: fundamentals, applications, and environmental impact. J Clean Prod 405:136828. https://doi.org/10.1016/j.jclepro.2023.136828
44. Dhal SB, Kar D (2025) Leveraging artificial intelligence and advanced food processing techniques for enhanced food safety, quality, and security: a comprehensive review. Discover Appl Sci 7(1):1–46
45. Drishya PK et al (2025) Advances in microbial and plant-based biopolymers: synthesis and applications in next-generation materials. Macromolecules 5(2):21
46. Singh N, Demirsöz R (2022) Recycling of traditional plastics: PP, PS, PVC, PET, HDPE, and LDPE, and their blends and composites. Nanomaterials in manufacturing processes. CRC Press, New York, pp 235–258
47. Jain S, Sharma MP (2012) Optimization of biodiesel production from Jatropha oil using response surface methodology. Fuel 103:572–578. https://doi.org/10.1016/j.fuel.2012.06.069
48. Kumar S, Jain S, Kumar H (2020) Experimental study on biodiesel production parameter optimization of jatropha–algae oil mixtures and performance and emission analysis of a diesel engine coupled with a generator fueled with diesel/biodiesel blends. ACS Omega 5(28):17033–17041
49. Raj RS, Jain S, Sharma AK (2024) Transforming municipal solid waste management through material and substance flow analysis: conversion pathways for sustainable energy production. Energy Convers Manag 322:119164
50. Kumar S, Jain S, Kumar H (2018) Performance evaluation of adaptive neuro-fuzzy inference system and response surface methodology in modeling biodiesel synthesis from jatropha–algae oil. Energy Sour A Recov Utilizat Environ Effects 40(24):3000–3008
51. Kumar U, Joseph-Paul S, Jain S (2021) Biochar: a source of nano catalyst in transesterification process. Mater Today Proc 46:5501–5505. https://doi.org/10.1016/j.matpr.2020.09.248
52. Singh VP, Jain S, Gupta JML (2023) Analysis of the effect of perforation in multi-v rib artificial roughened single pass solar air heater: part A. Exper Heat Transf 36(2):163–182
53. Kumar-Shukla A et al (2020) Advances of carbon capture and storage in coal-based power generating units in an Indian context. Energies 13(16):4124
54. Agnihotri M, Chamola R, Bhan U, Jain S (2024) Analyzing the performance of synthesized nano-catalysts for transesterification of waste cooking oil as environment friendly drilling fluid. Environ Res 244:117948. https://doi.org/10.1016/j.envres.2023.117948
55. Kumar Sharma V, Jain S, Singh VP (2020) A comparative analysis of recycling technologies for sustainable extraction of cathodic materials from battery waste
56. Singh AR, Singh SK, Jain S (2022) A review on bioenergy and biofuel production. Mater Today Proc 49:510–516
57. Wang T et al (2024) Direct recycling of spent nickel-rich cathodes in reciprocal ternary molten salts. J Power Sources 593:798. https://doi.org/10.1016/j.jpowsour.2023.233798
58. Gao Y, Li Y, Li J, Xie H, Chen Y (2020) Direct recovery of $LiCoO_2$ from the recycled lithium-ion batteries via structure restoration. J Alloys Compd 845:156234. https://doi.org/10.1016/j.jallcom.2020.156234

59. Jose SA et al (2024) Critical review of lithium recovery methods: advancements, challenges, and future directions. Processes 12(10):2203
60. Swain B, Jeong J, Lee J-C, Lee G-H (2008) Development of process flow sheet for recovery of high pure cobalt from sulfate leach liquor of LIB industry waste: a mathematical model correlation to predict optimum operational conditions. Sep Purif Technol 63(2):360–369. https://doi.org/10.1016/j.seppur.2008.05.022
61. Rostami T, Khoshandam B, Maroufi S (2022) Recovery of lithium, cobalt, nickel, and manganese from spent lithium-ion batteries through a wet-thermal process. Mater Res Bull 153:1118. https://doi.org/10.1016/j.materresbull.2022.111897
62. Xu J et al (2023) High-energy lithium-ion batteries: recent progress and a promising future in applications. Energy Environ Mater 6(5):e12450
63. Lee S, Koo H, Kang HS, Oh K-H, Nam KW (2023) Advances in polymer binder materials for lithium-ion battery electrodes and separators. Polymers 15(23):4477
64. Sharmili N, Nagi R, Wang P (2023) A review of research in the Li-ion battery production and reverse supply chains. J Energy Storage 68:107622
65. Li L, Duan Y (2023) Engineering polymer-based porous membrane for sustainable lithium-ion battery separators. Polymers 15(18):3690
66. Djoudi N, Mostefa MLP, Muhr H (2021) Hydrometallurgical process to recover cobalt from spent li-ion batteries. Resources 10(6):58
67. Singh VP, Jain S, Gupta JML (2022) Performance assessment of double-pass parallel flow solar air heater with perforated multi-V ribs roughness—part B. Exper Heat Transf 35(7):1059–1076
68. Saini M, Sharma A, Singh VP, Dwivedi G, Jain S (2022) Solar thermal receivers—a review. Adv Mater Manufact Energy Eng II Select Proceed ICAMME 2021:311–325
69. Chhabra M, Dwivedi G, Baredar P, Shukla AK, Garg A, Jain S (2021) Production and optimization of biodiesel from rubber oil using BBD technique. Mater Today Proc 38:69–73
70. Istrițeanu S (2021) Introduction to environmental challenges and waste in the automotive industry
71. Pratap Singh V, Author D, Kumar Sharma V, Jain S, Zhang X, Lecturer S (2020) Journal of energy storage a comparative analysis of recycling technologies for sustainable extraction of cathodic materials from battery waste powered by editorial manager® and ProduXion manager® from aries systems corporation. www.sun.ac.za
72. Dwivedi G, Jain S, Shukla AK, Verma P, Verma TN, Saini G (2022) Impact analysis of biodiesel production parameters for different catalyst. Environ Dev Sustain 74:1–21
73. Chhabra M et al (2025) Investigation of the shelf life of the optimized neem biodiesel and its execution and excretion characteristics on automotive diesel engine. Energy Sour A Recov Utilizat Environ Effects 47(1):4750–4772. https://doi.org/10.1080/15567036.2021.1881658
74. Kalungi P, Yao Z, Huang H (2024) Aspects of nickel, cobalt and lithium, the three key elements for Li-ion batteries: an overview on resources, demands, and production. Materials 17(17):4389
75. He J et al (2024) Electrospun PVDF-based polymers for lithium-ion battery separators: a review. Polymers 16(20):2895
76. Kühn S, Bartl A (2023) Packaging waste recycling: an overview of environmental benefits. Waste Manag 156:16–25. https://doi.org/10.1016/j.wasman.2023.01.012
77. Li X, Zhang Y, Wu F (2023) Material recovery and water utilization in LIB recycling processes. J Clean Prod 402:138423. https://doi.org/10.1016/j.jclepro.2023.138423
78. Beyerlein G-S, Zeinolebadi A, Aboud N, Jaeger P (2024) Supercritical fluid decontamination (SCFD) of poly (vinylidene fluoride) (PVDF) and further upcycling into highly porous aerogels with a well defined microstructure. J Clean Prod 446:141426
79. Bin-Rashid A, Haque M, Islam SMM, Uddin-Labib KMR, Chowdhury P (2024) Breaking boundaries with ceramic matrix composites: a comprehensive overview of materials, manufacturing techniques, transformative applications, recent advancements, and future prospects. Adv Mater Sci Eng 2024(1):2112358
80. Abdalla AM et al (2023) Innovative lithium-ion battery recycling: sustainable process for recovery of critical materials from lithium-ion batteries. J Energy Storage 67:107551

81. Saikam L, Arthi P, Senthil B, Shanmugam M (2023) A review on exfoliated graphite: synthesis and applications. Inorg Chem Commun 152:110685
82. Pusty M, Shirage PM (2022) Insights and perspectives on graphene-PVDF based nanocomposite materials for harvesting mechanical energy. J Alloys Compd 904:164060
83. Qin Y, Ouyang X, Lv Y, Liu W, Liu Q, Wang S (2023) A review of carbon-based conductive inks and their printing technologies for integrated circuits. Coatings 13(10):1769
84. Chen K-T, Yang Y-C, Yi Y-H, Zheng X-T, Tuan H-Y (2021) A carbon ink for use in thin, conductive, non peelable, amphiphilic, antioxidant, and large-area current collector coating with enhanced lithium ion battery performance. J Colloid Interface Sci 598:155–165
85. Zhou B, Su H, Liu W, Zhu Z, Wang L, Qi T (2024) Solvent extraction of metal ions from the leaching solutions of waste lithium-ion battery materials: a review. Sep Purif Technol 124:129173
86. Jain S (2019) The production of biodiesel using Karanja (*P. pinnata*) and Jatropha (*J. curcas*) Oil. Biomass Biopolym Based Mater Bioenergy 89:397–408
87. Doose S, Mayer JK, Michalowski P, Kwade A (2021) Challenges in ecofriendly battery recycling and closed material cycles: a perspective on future lithium battery generations. Metals 11(2):291
88. Fitriani F, Bilad MR, Aprilia S, Arahman N (2023) Biodegradable hybrid polymer film for packaging: a review. J Nat Fibers 20(1):2159606
89. https://www.acquiscompliance.com/blog/eu-battery-regulation-replace-battery-directive
90. https://www.plasticstoday.com/packaging/digital-product-passport-optimizes-recycling-traceability
91. https://en.wikipedia.org/wiki/CLP_Regulation
92. https://environment.ec.europa.eu/topics/waste-and-recycling/rohs-directive

Chapter 19
Applications of Biochar Composites in Packaging: Benefits and Limitations

Virendra Singh, Rajat Kumar Sharma, Neeraj Bisht, Neeraj Singh, and Akarsh Verma

1 Introduction

The packaging sectors encounter enormous challenges related to reducing their impact on the environment, improving safety, and increasing shelf life for their products. The conventional packing materials consist almost exclusively of petroleum-derived plastic materials, which cause environmental pollution and are non-biodegradable in nature. These non-biodegradable petrochemical-derived synthetic materials like polyethylene (PE), polypropylene (PP), polyethylene terephthalate (PET), and polystyrene (PS) are ideal packaging materials on account of their good mechanical properties, durability, and low cost. Yet they are non-biodegradable and also a contributing factor in plastic pollution worldwide and account for over 40% of total plastic production for packaging purposes [12]. Biochar, obtained by pyrolysis of biomass material, presents a sustainable option as a packaging material when used as a composite or filler with bio-based plastic materials. The special characteristics of biochar like a high surface area, porosity, and functional groups may be modified according to application requirements.

V. Singh · N. Singh
Department of Physics, G. B. Pant University of Agriculture and Technology, Pantnagar, India

R. K. Sharma (✉)
Department of Farm Machinery and Power Engineering, G. B. Pant University of Agriculture and Technology, Pantnagar, India
e-mail: rajat_sharma@hotmail.com

N. Bisht
Department of Mechanical Engineering, G. B. Pant University of Agriculture and Technology, Pantnagar, India

A. Verma
Department of Mechanical Engineering, UPES (University of Petroleum and Energy Studies), Dehradun, India

A. Verma et al. (eds.), *Polymers and Composite Materials for Packaging: Smart Food Packaging and Solutions*, Materials Horizons: From Nature to Nanomaterials,
https://doi.org/10.1007/978-981-95-0677-4_19

As a response to environmental issues, a shift has taken place toward biodegradable and renewable alternative packaging materials in the packaging industry. The bio-polyester materials like poly lactic acid (PLA), polyhydroxyalkanoate (PHAs), starch and cellulose derivatives have become popular and efficient alternatives due to their good properties like mechanical strength, compostability etc. Nevertheless, bio-polyester materials also contain a fewdrawbacks to their characteristics which involve brittleness, low melt strength, susceptibility to moistures and gases and lower gas and moisture barrier properties compared to plastic etc. [23, 33]. That's where biochar comes in and can be utilized as a filler material so as to counteract the aforementioned disadvantages.

2 Challenges in Packaging Materials

The newer packaging systems must perform a series of duties: physical damage resistance, microbial contamination inhibition, humidity resistance, oxygen exclusion, and light exclusion, and must be environmentally friendly. The main difficulties are [19, 31]:

Limited Barrier Properties of Biodegradable Materials

- Mechanical fragility in stress or humidity.
- Inconsistency in biodegradability under actual conditions.
- Bio-polymers are expensive and have complex processing needs.
- Additive or nanomaterial migration to food, requiring extensive toxicological assessments.

These requirements have resulted in the quest for functional fillers and additives capable of improving the performance characteristics of bio-matrices without sacrificing their environmental compatibility.

3 Biochar Composites for Packaging

Biochar, a carbonaceous material obtained from the pyrolysis of biomass, has also found application in a myriad of areas ranging from soil amendment and carbon sequestration and treatment of water and pollution remediation to construction material and energy storage, etc. [1, 9, 16, 21, 28–30]. In recent times, biochar and biochar-based composite materials have found a promising application as a sustainable packaging additive. Biochar can be utilized in a polymeric matrix for packaging purposes (Fig. 1). The tailorable physicochemical properties of biochar in terms of high surface area, porosity, electrical conductivity, and the presence of functional surface groups are to be leveraged to improve the functionality of biodegradable polymeric matrices.

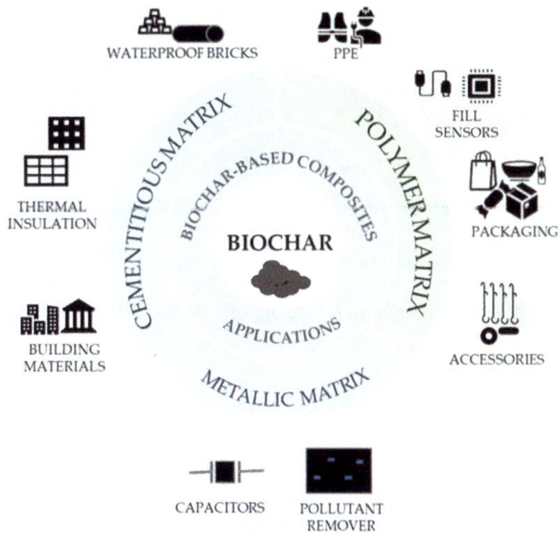

Fig. 1 Various application of biochar composites [10]

4 Advantages of Biochar in Packaging Composites

- **Mechanical Reinforcement**: The biochar particles enhance the biopolymer films' tensile strength, stiffness, and dimensional stability.
- **Barrier Properties**: Its hydrophobic and porous characteristics lower water vapor and oxygen permeability, increasing food shelf life.
- **Antioxidant and Antimicrobial Properties**: Bioactive compounds left behind and surface properties in biochar can slow down oxidative deterioration and hinder microorganisms from multiplying.
- **Sustainability**: Being made from agricultural or food waste, biochar promotes carbon sequestration and the principles of a circular economy.
- **Cost-Effectiveness**: Since biochar is a waste-derived filler material, it will lower the material cost and also value biomass wastes.

When integrated into matrices such as PLA, chitosan, or starch-based polymers, biochar forms composites that address key limitations of bioplastics, enabling their application in active and intelligent packaging systems. However, further research is required to assess the safety, effectiveness, and scalability of biochar composites in food-contact materials.

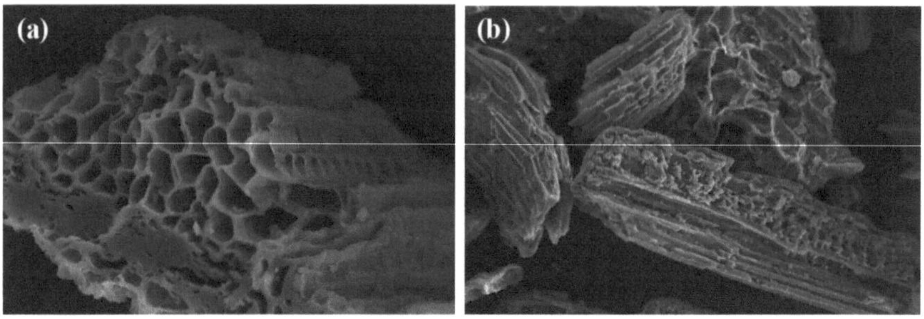

Fig. 2 SEM images of a pine needle derived biochar at **a** 10 μm scale, and **b** 50 μm scale

5 Biochar Production and Properties

Biochar, a carbonaceous material from the thermal pyrolysis of biomass under conditions of low oxygen availability, has gained popularity based on applications in environmental remediation and pollution reduction, energy storage, agriculture, and now food packaging. The characteristics of biochar, ranging from surface area and distribution of pores and functional groups, depend on the type of feedstock and the conditions of pyrolysis [3, 17, 18, 22, 24–27]. These characteristics play a key role in biochar's performance in food packaging applications (Fig. 2).

The influence of various factors on the properties of biochar is discussed below.

6 Influence of Feedstock Type

Feedstock material determines biochar's elemental makeup and porosity as well as ash fractions. Lignocellulosic biomass (e.g., straw, wood, husks) produces biochars having high carbon fixation and low ash content and is preferred in the production of biochars used in construction and composite materials [13]. The biochars from manure and sewage contain high concentrations of nutrients and minerals and thus find application in soil conditioning but tend to impart heavy metals and diminish carbon stability [32]. Biomasses having high concentrations of lignin are utilized for producing packaging materials as they form thermally stable aromatics carbon structures [8]. The various sources of biomass used in producing biochar and valuable properties of biochar for packaging applications are illustrated in Fig. 3.

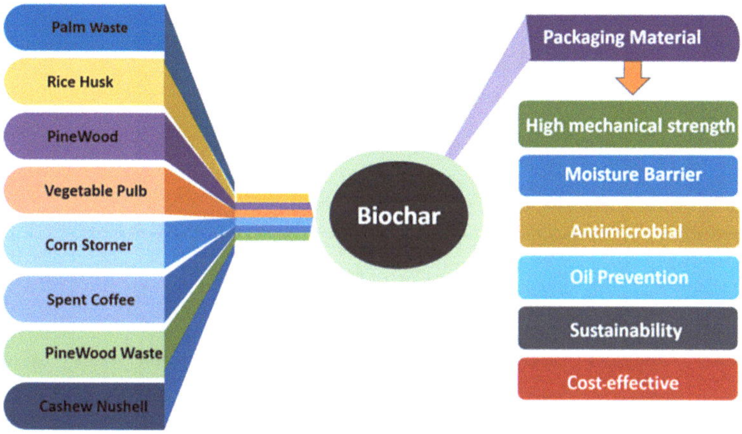

Fig. 3 Feedstocks types, as the raw materials, to produce biochar, and significant properties of packaging materials made from biochar composites

7 Effect of Pyrolysis Temperature

Pyrolysis temperature is the most critical process parameter regulating biochar properties. At lower pyrolysis temperatures (300–400 °C), biochars contain a larger percentage of volatiles and oxygen-containing functional groups, which make them highly hydrophilic and reactive but less thermally stable [5]. As the pyrolysis temperature rises (500–700 °C), volatiles decline progressively and biochars become more aromatic, specific surface area increases, and electrical conductivity grows. High-temperature biochars make excellent composite materials for packaging where barrier and strength properties may be enhanced [13]. The heating temperature and method classification categories are shown in Fig. 4.

8 Surface Chemistry and Porosity

The surface functional groups of biochar—hydroxyl, carboxyl, and carbonyl moieties—are highly variable depending on both pyrolysis conditions and the feedstock. The functionalities take important roles in numerous composite applications [31]. The functional groups highly depend on pyrolysis temperature, and generally, the functional groups decreased at higher pyrolysis temperature levels. Papadopoulou et al. [20] investigated the softwood biochar's functional group at two pyrolysis temperature levels of 550 °C and 700 °C through IR analysis (Fig. 5). The –OH, C–H, –C = O as well as C = C bonds were present in the biochar's IR spectra [20]. We can evidently see the decreasing trend of –C = O and other functionalities when pyrolysis temperature goes up to 700 °C. It may be caused by the decarboxylation

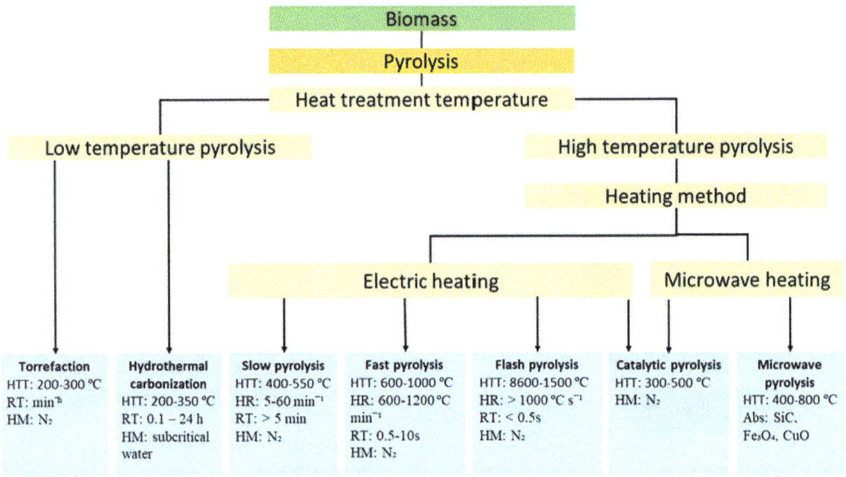

Fig. 4 Low temperature and high-temperature pyrolysis and heating methods [32]

and temperature-induced dehydration [4]. As discussed above, the biochar prepared at higher temperature levels are ideal for use in packages.

Also, porosity (micro- vs. meso- vs. macropores) influences the capacity of the material to capture gases, moisture and contaminants, which has application in food packaging systems for longer shelf life. Materials may be light in weight but high porosity materials will be weak and prone to breakage and highly damage sensitive.

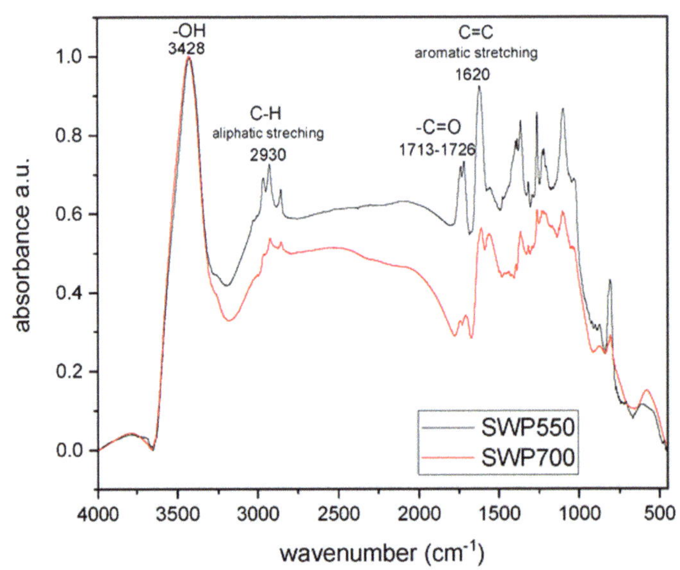

Fig. 5 Presence of functional groups on biochar surface and effect of pyrolysis temperature

Table 1 Proximate and elemental properties of pine needles and its char

Properties	*Pinus roxburghii*	Pine needles char
Water content (% wb)	7.78	4.7
Volatile matter (% db)	71.58	21.1
Ash content (% db)	2.08	7.6
Fixed carbon (% db)	26.34	71.3
Elemental analysis (wt %)		
C	44.99	77.7
H	5.46	3.2
N	0.99	2.4
O	48.55	9.1
H/C	1.46	0.50
O/C	0.81	0.09
Empirical formula	$CH_{1.46}N_{0.02}O_{0.81}$	$CH_{0.50}N_{0.03}O_{0.09}$
HHV (MJ kg^{-1})	17.67	28.1
pH	–	8.4
EC (dSm^{-1})	–	2.15
Iodine number	–	291

Materials of high porosity also tend to absorb the water and can capture the gases which will be unfavorable for the packaging application.

9 Optimization Considerations

Optimization of pyrolysis conditions is necessary in order to tailor biochar for targeted applications. Slow pyrolysis at longer residence time at moderate or elevated temperature maximizes yield and order of structure and fast pyrolysis process yields liquid products, resulted in reduced biochar yield [8]. Various post-treatments like activation (physical or chemical) can also enhance the surface area and make biochar useful for high-performance applications. The elemental and proximate characteristics of biochar obtained from pine needles at 500 °C pyrolysis temperature are shown in Table 1 [30].

10 Applications in Food Packaging

The incorporation of biochar in food packaging systems offers a diverse method in enhancing the performance of packages, promoting food preservation, and achieving sustainability. The unique properties of biochar such as high surface area, inherent porosity, and chemically active surface functionalities. provide numerous ways wherein it enhances the functionality of food packaging materials.

The addition of biochar in biopolymer films enhances the mechanical strength by overcoming the low tensile strength and flexibility usually affecting bioplastics. When added to PLA or PHA or starch-derivative polymers as reinforcing material, biochar behaves as a reinforcing agent at a nano- or micro- scale level. Polylactic Acid (PLA) is a highly used biodegradable plastic employed in packages; however, it has a few demerits as well. One of the key concerns when employing PLA is the high oxygen and moisture vapor transmission through it which may lead to food spoilage of the packaged food items. Its low ductility and heat resistance are also challenges. These can, however, be overcome by the addition of the right kind of reinforcing material. Different biomass fillers have been used in association with PLA as reinforcing material. Biochar happens to be a fiber which has also been used in association with PLA keeping in mind the packaging applications. Biochar greatly enhances the barrier properties of packaging films. Oxygen and water vapor transmission through them are key parameters related to the shelf life of perishable food materials. Because of the porous yet tortuous nature of the biochar material, biochar lengthens the path for the diffusion of gases and vapors through it. The physical barrier thus lowers the transmission of water vapor and oxygen through it and consequently reduces the oxidation and microbial decay processes in packaged foods gradually. Many researches report biochar-composite materials show as much as 50% reduction in oxygen and WVTR comparing with neat bio-polymers [5]. The interface between functional groups on the biochar surface and the matrix polymer results in enhanced load transfer and crack resistance under load conditions. Consequently, biochar-composite materials exhibit higher tensile strength, elongation at break and dimensional stability. Botta et al. [6] prepared biochar-PLA composite materials containing 1, 2.5 and 5 wt% of BC prepared by processing through an extruder employing a twin-co-rotating twin screw extruder machine. The tensile strength tests were conducted to estimate Young's Modulus as well as the extent of elongation at break. The Young's Modulus has been seen increasing as a function of increasing biochar levels by about 12.5, 17 and 20% for biochar 1%, biochar 2.5% and biochar 5% by weight when compared to pure PLA. However, a reduction in elongation before break has been noted as about 21%. This can be explained by intermolecular forces between biochar particles and PLA molecules, as a result of which the elastic modulus of composite material increased possibly due to biochar particles hindering the movement of PLA molecules [2]. TGA measurement indicate that the thermal erosion of PLA can be retarded by the incorporation of biochar. For all three types of plastics, addition of biochar lowers the global warming potential of the material and a linear correlation exists between amount of biochar added and reduction in the global warming potential compared to neat PLA by 3.3 kg CO_2 eq./kg of composite. Addition of biochar in the plastic material gives not only a reduction of a global warming potential related to production of plastic but also a net removal of CO_2 in the biochar. Both effects combined result in about twice the percentage reduction in GWP as percentage addition of biochar (e.g., a 20% addition of biochar gives a reduction in global warming potential of about 40%) [15]. Huang et al. [14] found that at 1% biochar filler, the tensile strength at load was enhanced from 56.2 MPa to 65.98 and 79.79 MPa at 100 and 200 mesh size of biochar, respectively, which indicates enhancement of about 16 and 43%.

Besides physical reinforcement, biochar also conveys active functional properties. Depending on the feedstock material and pyrolysis conditions, biochar preserves bioactive compounds showing antimicrobial and antioxidant activities. For example, biochars from hardwood, coconut shells, or herbaceous crops may hold leftover phenolic compounds, contributing to suppression of microbial growth. Feng et al. [11] reported antioxidant and antimicrobial properties of grain shell biochar-polymer film for packaging of blueberries. The composite film of C–Ag@PVA/CS had shown the superior antioxidant activity and antimicrobial activity against Gram-negative bacteria by killing the bacterial cell membrane. The composite film packaging has also facilitated delayed deterioration of blueberries and extended the storage duration. The above results pointed toward biochar-composite potential as active packaging material for food commodities for improved quality and safety.

Another promising area of application is biochar's capacity to adsorb ethylene gas, a plant hormone primarily involved in fruit ripening. Addition of biochar in packaging sachets or packaging materials maintains the delaying of climacteric fruits such as bananas, mangoes, and tomatoes and extends marketable life. Its adsorbent capacity can be employed similarly for odor and humidity scavenging and thus used in packaging dried foods or bakery stuffs. Charoensuk et al. [7] have studied agriculture waste based biochars for ethylene adsorber in delaying the ripening in bananas. The biochar from rice husk has shown considerable ethylene adsorption and has reduced the concentration of ethylene down to 69% in 144 h in closed system.

Furthermore, biochar films are investigated for their applicability in intelligent packaging systems, those able to sense and react to variations in the packaged surroundings. As a candidate for integration as a sensor on account of its electrical conductance and surface reactivity, biochar has the sensitivity and responsiveness for monitoring markers of spoilage (e.g., pH levels, gas concentrations) through color or electronic changes and giving immediate feedback on the quality of food.

These multiple functionalities make biochar a highly versatile food packaging additive. Successful application relies on optimization of loading concentration, particle size, dispersion homogeneity, and compatibility with the host polymer. The biochar's hydroxyl group can be a problem if biochar must be used as a filler in the PLA matrix. This can be overcome through chemical or physical modification as the situation demands. Agglomeration and loss of transparency can also be caused by loading levels > 10 wt% and could be detrimental for consumer acceptability at the retail level.

In summary, biochar application in food packaging encompasses passive, active, and intelligent functions. It enhances mechanical and barrier properties and also introduces bioactivity and benefits the cause of sustainability through biodegradability and waste valorization. These properties make biochar a next-generation material for designing sustainable and multi-functional food packaging.

11 Environmental and Economic Considerations

The application of biochar in the packaging sector fulfills the conditions of a circular economy by utilizing agricultural waste and decreasing the dependency on non-renewable resources. From an economic aspect, biochar application can give added value to agricultural by-products and lower the cost of waste management. The incorporation of biochar in packaging materials has significant environmental and economic benefits based on sustainability and the standards and regulations of a circular economy. The second key environmental benefit of biochar application stems from its sources: from the pyrolysis of agricultural residue, forest by-products, and food processing wastes. The biomass valorization embedded in biochar not only manages organic waste but also creates long-term carbon sequestration. The stable molecular skeleton of biochar-containing aromatic carbon resists microbial breakdown and can store carbon for hundreds of years if not millennia and hence serves as a carbon sink and a means of reducing climate change [12].

For packaging purposes, the addition of biochar in biopolymer matrices diminishes the reliance on plastics from fossil fuels and lessens the carbon footprint for production of the package. As compared to traditional fillers, biochar is a biodegradable material under certain conditions and a renewable material when used in tandem with compostable resins like PLA or starch plastics. Additionally, production through pyrolysis can be tailored so as to recover secondary by-products like bio-oil and gases as a source of renewable energies as well as improve the energy efficiency of the process [32].

From a cost viewpoint, adding biochar can noticeably reduce material costs for biodegradable polymers, which tend to be costly because they are produced on a small scale and synthesized from resources intensively. Low-cost and readily available through waste biomass, biochar serves as economical filler material replacing part of the base polymer without sacrificing—even in some instances improving—material properties like strength and barrier performance [8]. Accordingly, this substitution has the effect of making environmentally friendly packaging increasingly competitive and cost-effective compared with petroleum-derived packaging materials.

Additionally, decentralized biochar production at or near the sites of agricultural or industrial wastes would reduce transportation costs and boost local economies. This local production system may create several value-added commodities ranging from packaging materials.

Overall, the application of biochar in packaging presents a synergistic solution for mitigating the environmental impact and enhancing the economic viability of green materials. By waste valorization, carbon sequestration, and the reduction of the cost of polymers, biochar has the potential to become a transformative agent in converting the packaging sector toward a low-carbon, circular economy [5, 13]

12 Challenges and Future Perspectives

With all the possible benefits of biochar, however, a number of challenges face the application of biochar in food uses on a large scale. These challenges involve variability in biochar properties due to differences in production methods and biochar sources as well as lack of regulations for food-contact materials and an incomplete understanding of long-term consequences on food quality and safety. The variability in biochar production affects compatibility with the matrix of a composite material, making the final composite material's mechanical, thermal, or barrier properties inconsistent [5]. Future investigation should aim toward standardization of production processes, extensive assessment of safety, and examining large-scale application. Standardization of biochar properties must be achieved if performance is to be predictable and repeatable in packaging materials.

Another important area of concern revolves around regulatory compliance and safety. Being a packaging material and a food-contact material at times, biochar has to be under strict safety and health regulations and thus has to be carefully evaluated for possible contaminants like heavy metals, polycyclic aromatic hydrocarbons (PAHs), and persistent organic pollutants (POPs) that might be introduced through pyrolysis of some biomass sources. It is paramount that biochar be free from toxic contaminants and be safe for consumption by humans, and without a clear regulatory framework for the application of biochar in packages, uncertainty prevails [32].

From a processing level, uniform distribution of biochar particles in the polymeric matrix continues to be technically challenging. Inadequate dispersion may cause agglomeration and adversely affect the mechanical properties and clarity of the packaging film. Moreover, compatibility between hydrophilic biochar and hydrophobic polymeric matrices may demand surface modification of biochar through oxidation or grafting of a coupling agent. These processes enhance dispersion and adhesion but contribute to both production cost and complexity [13].

The economic scaling-up of biochar packaging production remains underdeveloped. While biochar itself is cheap, production through pyrolysis may be energy-intensive. It is important to develop low-energy, scalable, and cost-effective production processes for high-grade biochar for packaging purposes. Integration of biochar production through existing agricultural or industrial systems (e.g., biorefinery) has the potential for reducing the cost and environmental effects by sharing resources and minimizing waste.

Future research directions will involve optimizing biochar properties for targeted packaging functionalities like gas barrier, humidity resistance, or antimicrobial action. Emerging material functionalization and composite engineering can give rise to biochar-based packaging films having multi-functionality as active and intelligent packaging materials. Additionally, life cycle assessments (LCA) and techno-economic analyses (TEA) will be necessary in order to compare the general environmental and economic advantages of biochar packaging versus traditional and other bio-alternatives [5]. Interdisciplinary coordination between material science, environmental engineering, polymer chemistry, and food safety will be key to overcoming existing challenges. Through ongoing innovation and regulatory clarity as well as investments in infrastructure, biochar may be at the center of next-generation sustainable packaging material development.

13 Conclusion

Biochar is a groundbreaking material in the development of sustainable food packaging. As a carbonaceous by-product of biomass pyrolysis, its application goes from environmental remediation and agriculture to the essential area of food packaging and safety, shelf-life extension, and material sustainability. By reinforcing biodegradable polymers, promoting barrier and bioactive properties, and being included in active and intelligent packaging systems, biochar provides a path toward high-performance food packaging replacements for traditional plastics. But realizing this value at a commercial level demands a concerted and interdisciplinary effort. There has to be progress on a number of fronts-standards of production parameters for biochar, exploration of diversity in feedstocks, and rigorous safety assessments. Regulators must establish and harmonize standards for biochar as a food-contact material, and industry must commit to scalable, affordable processing technology. At the same time and in parallel with this development, consumer understanding and acceptability of biochar-containing packaging must be encouraged by transparency and education. Furthermore, biochar's function in a circular economy must be stressed. Its production from farming and industrial wastes gives it an intrinsic sustainability and also results in waste reduction, material recovery, and carbon reduction upon application in packaging. Techno-economic and life cycle analyses will be key to estimating benefits in numerical form and making it competitive against conventional packaging materials. Finally, biochar is not just a fill or an additive but a multi-functional platform for redefining the packaging of the future. Its integration may be a turning point toward environmentally friendly materials without sacrificing performance or safety. With ongoing innovation and concerted effort, biochar might become a core part of tomorrow's food packaging solutions, balancing the twin imperatives of sustainability and food security.

Declaration of Competing Interest The authors declare that they have no known competing financial interests or personal relationships that could have appeared to influence the work reported in this paper.

References

1. Anamika Singh V, Yadav BK (2020) Adsorption study of heavy metals from aqueous solutions using magnetite nanoparticles. J Phys Conf Ser 1504:012011. https://doi.org/10.1088/1742-6596/1504/1/012011
2. Anwer MAS, Naguib HE, Celzard A, Fierro V (2015) Comparison of the thermal, dynamic mechanical and morphological properties of PLA-Lignin & PLA-Tannin particulate green composites. Compos B Eng 82:92–99. https://doi.org/10.1016/j.compositesb.2015.08.028
3. Azad D, Pateriya RN, Sharma RK (2023) Chemical activation of pine needle and coconut shell biochar: production, characterization and process optimization. Int J Environ Sci Technol. https://doi.org/10.1007/s13762-023-04913-w
4. Bartoli M, Arrigo R, Malucelli G, Tagliaferro A, Duraccio D (2022) Recent advances in biochar polymer composites. Polymers (Basel) 14:2506. https://doi.org/10.3390/polym14122506
5. Bhattacharya T, Khan A, Ghosh T, Kim JT, Rhim J-W (2024) Advances and prospects for biochar utilization in food processing and packaging applications. Sustain Mater Technol 39:e00831. https://doi.org/10.1016/j.susmat.2024.e00831
6. Botta C, Grottola CM, Amato D, Acocella MR (2024) Biochar as sustainable filler of recycled polylactic acid (PLA): a new generation of processable biocomposites. Polymers (Basel) 16:3347. https://doi.org/10.3390/polym16233347
7. Charoensuk P, Chaiwong S, Suwunwong T, Halley PJ, Suwantong O (2024) Preparation and utilization of biochar from agricultural wastes as ethylene absorber for 'Gros Michel' banana ripening. Ind Crops Prod 222:119860. https://doi.org/10.1016/j.indcrop.2024.119860
8. Das O, Sarmah AK, Bhattacharyya D (2015) A novel approach in organic waste utilization through biochar addition in wood/polypropylene composites. Waste Manage 38:132–140. https://doi.org/10.1016/j.wasman.2015.01.015
9. Dash AK, Tripathy S, Naveenkumar A, Bhoi TK, Kumari A, Divya, Latare AM, Nandipamu TMK, Singh V, Raza MdB, Saraswat A, Bhadha J (2024) Biochar for pollutants bioremediation from soil and water ecosystem. In: Biochar production for green economy. Elsevier, pp 433–452. https://doi.org/10.1016/B978-0-443-15506-2.00026-2
10. Feliz Florian G, Ragoubi M, Leblanc N, Taouk B, Abdelouahed L (2024) Biochar production and its potential application for biocomposite materials: a comprehensive review. J Compos Sci 8:220. https://doi.org/10.3390/jcs8060220
11. Feng Q, Fan B, He Y-C, Ma C (2024) Antibacterial, antioxidant and fruit packaging ability of biochar-based silver nanoparticles-polyvinyl alcohol-chitosan composite film. Int J Biol Macromol 256:128297. https://doi.org/10.1016/j.ijbiomac.2023.128297
12. Geyer R, Jambeck JR, Law KL (2017) Production, use, and fate of all plastics ever made. Sci Adv 3. https://doi.org/10.1126/sciadv.1700782
13. Giorcelli M, Bartoli M (2019) Development of coffee biochar filler for the production of electrical conductive reinforced plastic. Polymers (Basel) 11:1916. https://doi.org/10.3390/polym11121916
14. Huang Y, Zhao S, Chen K, Huang B, Jin R (2024) A review of persulfate-based advanced oxidation system for decontaminating organic wastewater via non-radical regime. Front Environ Sci Eng 18:134. https://doi.org/10.1007/s11783-024-1894-2
15. Kane S, Van Roijen E, Ryan C, Miller S (2022) Reducing the environmental impacts of plastics while increasing strength: biochar fillers in biodegradable, recycled, and fossil-fuel derived plastics. Composites Part C: Open Access 8:100253. https://doi.org/10.1016/j.jcomc.2022.100253
16. Liu J, Zhang W, Jin H, Li Z, Liu G, Xing F, Tang L (2023) Exploring the carbon capture and sequestration performance of biochar-artificial aggregate using a new method. Sci Total Environ 859:160423. https://doi.org/10.1016/j.scitotenv.2022.160423
17. Mandal S, Sharma RK (2023) Application of gasification technology in agriculture for power generation. In: Handbook of energy management in agriculture. Springer Nature Singapore, Singapore, pp 1–22. https://doi.org/10.1007/978-981-19-7736-7_14-1

18. Mandal S, Sharma RK, Bhattacharya TK, Tanna H, Haydary J (2022) Charring of pine needles using a portable drum reactor. Chem Pap 76:1239–1252. https://doi.org/10.1007/s11696-021-01893-4
19. Marsh K, Bugusu B (2007) Food packaging—roles, materials, and environmental issues. J Food Sci 72. https://doi.org/10.1111/j.1750-3841.2007.00301.x
20. Papadopoulou K, Klonos PA, Kyritsis A, Tarani E, Chrissafis K, Mašek O, Tsachouridis K, Anastasiou AD, Bikiaris DN (2025) Synthesis and characterization of PLA/biochar bio-composites containing different biochar types and content. Polymers (Basel) 17:263. https://doi.org/10.3390/polym17030263
21. Rawat S, Wang C-T, Lay C-H, Hotha S, Bhaskar T (2023) Sustainable biochar for advanced electrochemical/energy storage applications. J Energy Storage 63:107115. https://doi.org/10.1016/j.est.2023.107115
22. Mandal S, Sharma RK, Bhattacharya TK (2023) Deriving fuel from pine needles through pyrolysis, charring and briquetting and their GHG emission potential. Curr Sci 124:1210–1215
23. Sha L, Chen Z, Chen Z, Zhang A, Yang Z (2016) Polylactic acid based nanocomposites: promising safe and biodegradable materials in biomedical field. Int J Polym Sci 2016:1–11. https://doi.org/10.1155/2016/6869154
24. Sharma RK, Bhattacharya TK, Kumain A, Chand P, Mandal S, Azad D (2020) Energy use pattern in wheat crop production system among different farmer groups of the Himalayan Tarai region. Curr Sci 118:448. https://doi.org/10.18520/cs/v118/i3/448-454
25. Sharma RK, Haydary J, Singh TP, Nazari MA, Mandal S, Verma A (2024) Fractional distillation of pine needle bio-oil: comprehensive characterization and rheological properties across temperature gradients. Biomass Convers Biorefin. https://doi.org/10.1007/s13399-024-06044-9
26. Sharma RK, Nazari MA, Haydary J, Singh TP, Mandal S (2023) A review on advanced processes of biohydrogen generation from lignocellulosic biomass with special emphasis on thermochemical conversion. Energies (Basel) 16:6349. https://doi.org/10.3390/en16176349
27. Sharma RK, Singh TP, Mandal S, Azad D, Kumar S (2022) Chemical treatments for biochar modification: opportunities, limitations and advantages. In: Engineered biochar. Springer Nature Singapore, Singapore, pp 65–84. https://doi.org/10.1007/978-981-19-2488-0_5
28. Singh N, Singh V, Bisht N, Negi P, Dhyani A, Sharma RK, Tewari BS (2025) A comprehensive review on supercapacitors: Basics to recent advancements. J Energy Storage 121:116498. https://doi.org/10.1016/j.est.2025.116498
29. Singh V, Joshi GC, Bisht D (2017) Energy dispersive x-ray fluorescent analysis of soil in the vicinity of industrial areas and heavy metal pollution assessment. J Appl Spectrosc 84:306–311. https://doi.org/10.1007/s10812-017-0468-5
30. Singh V, Pant N, Sharma RK, Padalia D, Rawat PS, Goswami R, Singh P, Kumar A, Bhandari P, Tabish A, Deifalla AM (2023) Adsorption studies of Pb(II) and Cd(II) heavy metal ions from aqueous solutions using a magnetic biochar composite material. Separations 10:389. https://doi.org/10.3390/separations10070389
31. Siracusa V, Rocculi P, Romani S, Rosa MD (2008) Biodegradable polymers for food packaging: a review. Trends Food Sci Technol 19:634–643. https://doi.org/10.1016/j.tifs.2008.07.003
32. Tengku Yasim-Anuar TA, Yee-Foong LN, Lawal AA, Ahmad Farid MA, Mohd Yusuf MZ, Hassan MA, Ariffin H (2022) Emerging application of biochar as a renewable and superior filler in polymer composites. RSC Adv 12:13938–13949. https://doi.org/10.1039/D2RA01897G
33. Trivedi AK, Gupta MK, Singh H (2023) PLA based biocomposites for sustainable products: a review. Adv Ind Eng Polym Res 6:382–395. https://doi.org/10.1016/j.aiepr.2023.02.002

Chapter 20
Integrating Carbon Nanotubes into Packaging Materials: An Overview of Enhanced Properties and Applications

Sai Parameshwar, Vivek Kumar Sharma, Rahul S. Raj, and Akarsh Verma

1 Introduction

Modern science and engineering benefit much from composite materials since their unusual structural composition and adaptable chemical characteristics [1, 2]. These materials two or more stages mix a supporting matrix with a catalytically active molecule. Integration helps with surface area, reusability, chemical and thermal stability, and surface area. By combining many components into a single system, composites may precisely change physical and chemical characteristics; hence, they are perfect for catalytic uses like industrial chemical manufacturing, energy transformation, and environmental cleanup [2, 3]. Their efficiency comes from synergistic phase interactions, which often improve reaction rates, selectivity, and durability over single-phase catalysts [4]. Figure 1 shows composite material market share worldwide.

Catalytic research has focused on carbon-based composites due to their high electrical conductivity, large surface area, and tunable surface chemistry. Graphene oxide, CNTs, and biochar are used as support matrices for metal nanoparticles or oxides, creating very effective catalytic systems. These composites excel in electrocatalysis and photocatalysis, where fast electron transport is crucial. Carbon nanotubes with platinum or palladium nanoparticles are widely used in fuel cells due to their catalytic activity and toxin resistance. Combining graphene oxide with photocatalysts like TiO_2 or ZnO effectively degrades organic pollutants in wastewater treatment [5–7].

S. Parameshwar (✉)
Department of Petroleum Engineering and Earth Sciences, UPES, Dehradun, Uttarakhand, India
e-mail: saiparameswars7@gmail.com

V. K. Sharma · R. S. Raj · A. Verma
Department of Mechanical Engineering, UPES (University of Petroleum and Energy Studies), Dehradun, Uttarakhand, India

A. Verma et al. (eds.), *Polymers and Composite Materials for Packaging: Smart Food Packaging and Solutions*, Materials Horizons: From Nature to Nanomaterials,
https://doi.org/10.1007/978-981-95-0677-4_20

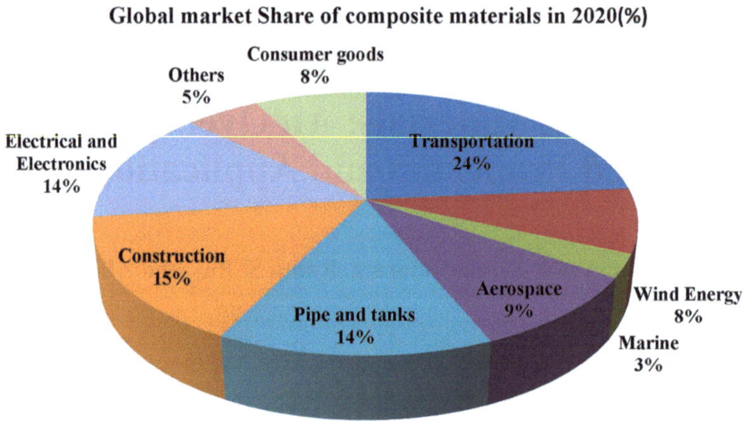

Fig. 1 Global market share of composite materials 2020 [15]

Metal oxide composites are key to catalysis. Secondary elements are often added to primary oxide systems like TiO_2, ZnO, or Fe_2O_3 to improve their photocatalytic and redox characteristics. TiO_2 compounds are valued for their strong oxidizing power, chemical stability, and environmental safety [8]. Their performance can be raised by combining them with graphene, CdS, or g-C_3N_4 by enhancing charge separation and light absorption minerals applied in water splitting, CO_2 reduction, color and hazardous chemical breakdown [9]. Particularly in food, environmental issues, and the necessity of superior materials are changing the global packaging sector. Packaging that guarantees product safety and durability while reducing environmental impact is required as consumer awareness grows and environmental laws tighten [10]. The durability of fossil fuel-based plastic packaging in ecosystems and the growing plastic waste problem are under study. Researchers and business experts are investigating sustainable materials that improve mechanical strength, heat resistance, and barrier properties. Nanotechnology has become a powerful tool [11–13].

Nanomaterials' nanoscale characteristics can greatly increase packaging performance. Carbon nanotubes are studied for their high aspect ratio, thermal and electrical conductivity, and mechanical strength. Carbon nanotubes in polymer matrices improve film strength, gas barrier efficacy, and performance [14]. Their antimicrobial, sensory, and mechanical features make them crucial to intelligent, adaptable packaging solutions. This chapter delves into the utilization of carbon nanotubes in polymer packaging. Priority are sustainable production techniques include biochar utilization and microwave-assisted synthesis. The talk stresses the importance of CNT-based nanocomposites in sustainable packaging and the requirement of balancing material efficiency with ecological responsibility.

2 Carbon Nanotubes: Types, Composition, and Properties

Formulated by the smooth rolling of graphene sheets into cylindrical forms, carbon nanotubes (CNTs) represent a special class of nanostructured carbon materials. Carbon nanotubes (CNTs) have drawn great attention since their basic discovery by Sumio Iijima in 1991 because of their amazing mechanical, chemical, and physical properties. Single-Walled Carbon Nanotubes (SWCNTs) and Multi-Walled Carbon Nanotubes (MWCNTs) are two examples of structurally based classification for Carbon Nanotubes (CNTs) [16].

Usually measuring between 0.8 and 2 nm, single-walled carbon nanotubes (SWCNTs) are made of a single sheet of graphene painstakingly rolled into a cylindrical form. Their electrical properties allow them to operate as either metals or semiconductors since they are highly sensitive to changes in chirality—the angle of graphene curvature—and size. Especially useful in fields including electronics, sensing technologies, and advanced nanoelectronic devices, SWCNTs have unique ability to control their electrical characteristics [17].

2.1 Multi-walled Carbon Nanotubes (MWCNTs)

Multi-walled carbon nanotubes (MWCNTs) comprise several concentric graphene cylinders, exhibiting interlayer spacing akin to that of graphite (~0.34 nm). These structures span around 2 nm to over 100 nm. Owing to their comparatively straightforward synthesis, mechanical durability, and scalability, MWCNTs are extensively employed in structural applications and composite materials [18].

2.2 Essential Physical and Chemical Characteristics of CNTs

The exceptional characteristics of carbon nanotubes stem from sp^2-hybridized carbon atoms organized in a hexagonal lattice and the quasi-one-dimensional structure. These features lead to several outstanding characteristics, summarized below. A comparative assessment of the environmental impacts of SWCNTs and MWCNTs is presented in Table 1 [19, 20].

2.2.1 Mechanical Strength

CNTs are among the strongest known materials, exhibiting tensile strengths estimated to be 50–100 times greater than steel on a per-weight basis. Their elastic modulus can reach up to 1 TPa, making them ideal reinforcements in polymer matrices for packaging and structural applications.

Table 1 Comparative properties of multi-walled and single-walled carbon nanotubes [21]

Feature	Multi-walled carbon nanotubes (MWCNTs)	Single-walled carbon nanotubes (SWCNTs)
Structure	Comprised of multiple concentric cylinders of rolled graphene sheets	Consist of a single cylindrical graphene layer
Diameter	Typically ranges from 2 to 100 nm	Typically ranges from 0.8 to 2 nm
Length	Extend from several micrometers up to millimeters	Extend from several micrometers up to millimeters
Synthesis complexity	Easier to synthesize; more suitable for large-scale production	More complex synthesis with sensitivity to conditions
Cost	Generally lower due to scalable fabrication techniques	Relatively higher, owing to intricate and resource-intensive production methods
Electrical properties	Predominantly metallic behavior	Can exhibit either metallic or semiconducting properties based on chirality
Mechanical strength	High mechanical strength, though typically lower than SWCNTs	Exceptional tensile strength, among the highest for any known material
Thermal conductivity	High (~3000 W/mK)	Very high (~3500 W/mK)
Purity control	Often contains more structural defects; lower purity	

2.2.2 Thermal Conductivity

With thermal conductivities exceeding 3000 W/mK, CNTs provide superior heat dissipation, which is particularly valuable in applications requiring thermal stability, such as the packaging of temperature-sensitive goods.

2.2.3 Electrical Conductivity

Depending on their chirality, CNTs can function as either conductors or semiconductors. Their electrical conductivity often surpasses that of traditional metals like copper, making them suitable for electromagnetic interference (EMI) shielding, antistatic coatings, and sensor-integrated packaging systems.

2.2.4 Barrier Properties

When integrated into polymer matrices, the high aspect ratio and tortuous diffusion pathways of CNTs significantly impede the transmission of gases such as oxygen and water vapor. This improvement in barrier efficacy results in prolonged shelf life and enhanced preservation of packaged goods. Carbon nanotubes possess a substantial surface-to-volume ratio, facilitating efficient interaction with various materials and chemical agents. Their surfaces can be modified—either covalently or non-covalently—to enhance dispersion within polymer matrices and optimize compatibility with biodegradable or food-grade ingredients.

2.3 *Methods for Synthesizing Carbon Nanotubes*

Carbon nanotubes (CNTs) are cylindrical nanostructures formed of carbon atoms organized in a hexagonal lattice, akin to rolled sheets of graphene. Since their discovery, various synthesis methods have been developed to produce CNTs with tailored properties suitable for a range of applications. The primary synthesis methods include arc discharge, laser ablation, chemical vapor deposition (CVD), and high-pressure carbon monoxide (HiPCO) processes (Fig. 2). Table 2 offers a comprehensive comparison of the advantages and disadvantages [4].

2.3.1 Arc Discharge Technique

This method entails generating an electric arc between two graphite electrodes within an inert gas environment, usually helium or argon. The elevated temperature produced by the arc results in the vaporization of carbon, which subsequently facilitates the formation of carbon nanotubes. Arc discharge is recognized for its ability to generate high-quality multi-walled carbon nanotubes (MWCNTs) and, under certain conditions, single-walled carbon nanotubes (SWCNTs). Nonetheless, this approach

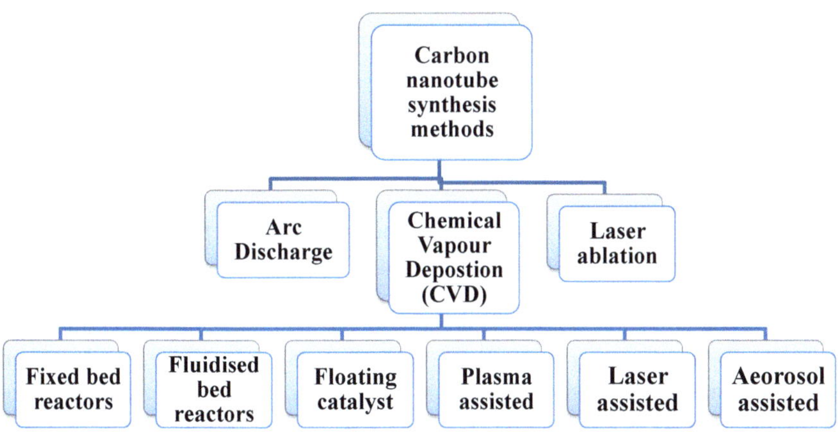

Fig. 2 Carbon nanotube synthesis methods

Table 2 Comparison of common carbon nanotube synthesis methods, including their advantages and limitations [4, 22, 24, 30, 31]

Synthesis method	Advantages	Disadvantages
Arc discharge	• Capable of synthesizing both SWCNTs and MWCNTs • Produces high-quality carbon nanotubes with relatively fewer structural defects	• Requires extremely high temperatures • Yields a mixture of CNTs and amorphous carbon, necessitating post-synthesis purification • Limited control over morphology and alignment
Laser ablation	• Enables the synthesis of high-purity SWCNTs with uniform diameters. Facilitates the formation of ordered CNT arrays	• Entails high operational costs due to the use of high-energy laser systems. Characterized by low production rates, rendering it less viable for industrial-scale applications
Chemical vapor deposition (CVD)	• Offers scalability and economic feasibility for large-scale production • Provides tunable control over parameters such as length, diameter, and alignment • Operates at comparatively lower temperatures	• May result in CNTs with higher defect densities • Requires stringent control over synthesis parameters to ensure product consistency and quality
High-pressure carbon monoxide (HiPCO)	• Produces high-purity SWCNTs with narrow diameter distributions. Amenable to scalable synthesis	• Involves complex and specialized high-pressure, high-temperature systems • Utilizes hazardous gases, thereby necessitating comprehensive safety protocols

frequently leads to a combination of CNTs and various carbon-based substances, requiring subsequent purification processes [4, 22, 23].

2.3.2 Method of Laser Ablation

A high-powered laser beam is precisely aimed at a graphite target embedded with metal catalysts like nickel or cobalt, within a high-temperature reactor that is saturated with inert gas. The laser effectively vaporizes the graphite, leading to the condensation of carbon atoms into carbon nanotubes as the material cools. Laser ablation is capable of producing high-purity single-walled carbon nanotubes (SWCNTs) with uniform diameters; however, its practical application is limited by low production yields and high operational costs [24, 25].

2.3.3 Chemical Vapor Deposition (CVD)

Chemical vapor deposition (CVD) is widely regarded as one of the most effective and scalable techniques for synthesizing carbon nanotubes. This method relies on the thermal decomposition of hydrocarbon gases—such as methane or acetylene—at high temperatures in the presence of transition metal catalysts (commonly iron or nickel) supported on substrates. Carbon atoms from the gaseous phase adhere to catalyst particles throughout the synthesis process, hence starting the creation of carbon nanotubes. Both single-walled carbon nanotubes (SWCNTs) and multi-walled carbon nanotubes (MWCNTs) are synthesized using the common method chemical vapor deposition (CVD). CVD is a flexible and scalable technique for nanotube construction since it offers exact control of fundamental structural properties including length, width, and alignment [22, 24, 26, 27].

2.3.4 High-Pressure Carbon Monoxide (HiPCO) Methodologies

Under high pressures and temperatures, the HiPCO process uses iron carbonyl catalysts to thermally decompose carbon monoxide gas. Generating single-walled carbon nanotubes with high purity and constant diameters calls for this gas-phase synthesis approach very effectively. HiPCO is respected for its ability to generate large volumes of SWCNTs with consistent structural characteristics. Still, it depends on highly specialized equipment and strict control of reaction parameters, which could limit its scalability and viability for broad industrial use. Regarding purity, yield, structural accuracy, scalability, and cost-effectiveness every CNT synthesis method has different benefits and constraints. The ideal technique selection frequently hinges on the particular application and the necessary equilibrium between performance requirements and production viability. Ongoing research aims to develop these technologies to improve production efficiency, refine control over CNT properties, and broaden their integration into other technological applications [28, 29].

2.4 Significance in Packaging Applications

The inherent characteristics of CNTs lead to enhanced functionality of composite films utilized in food packaging. They play a role:

1. In improving mechanical durability, which minimizes the necessity for multilayer structures.
2. Enhanced barrier characteristics, restricting oxidation, and microbial proliferation.
3. Incorporation of dynamic functionalities, including antimicrobial properties, spoilage detection sensors, or responsive behavior to stimuli.

The versatility and performance of CNTs position them as highly promising nanomaterials in the shift toward smart, sustainable, and high-performance food packaging systems [32, 33].

3 Composites Utilizing Carbon Nanotubes in Packaging

The integration of carbon nanotubes (CNTs) into polymer matrices signifies a groundbreaking method for developing high-performance, multifunctional packaging materials. The mechanical strength, barrier characteristics, and general performance of packing films can be effectively enhanced by including carbon nanotubes (CNTs) as nanofillers. These improvements include active packaging features including antibacterial characteristics and environmental monitoring capabilities in addition to passive protection, that is, increased resistance to gas penetration and physical stress. The several methods for including Carbon nanotubes into packaging materials, the resulting property enhancements, and the relevant applications of CNT-based nanocomposites in modern packaging technologies are discussed in this part [34, 35].

3.1 Polymer Matrices for Carbon Nanotube Integration

Many polymer matrices, including conventional synthetic plastics as well as biodegradable and bio-based choices, can include carbon nanotubes. Synthetic polymers, such as Polyethylene, Polypropylene, Polyethylene Terephthalate, Polystyrene, rule commercial packaging because of their strength and simplicity of production. Environmental issues, however, have hastened the move toward sustainable biopolymers including PLA, PHA, films based on starch, chitosan, and cellulose derivatives. Carbon nanotube-polymer nanocomposites' performance depends on their interaction with the polymer matrix. Carbon nanotubes are carboxylated, hydroxylated, or aminated to increase compatibility and dispersion. These changes add reactive groups to interfacial bonding, improving stress transfer, mechanical behavior, filler dispersion, and functional performance [36, 37].

3.2 Mechanical and Barrier Improvement

Carbon nanotubes are used to packaging films to increase mechanical strength. Carbon nanotubes increase tensile strength and modulus due to their load-bearing capability. Carbon nanotubes composition and orientation can be adjusted to tailor flexibility and toughness to packaging needs. Carbon nanotubes enhance gas barrier qualities, especially against oxygen, carbon dioxide, and water vapor, in addition to mechanical benefits. Gas diffusion is blocked by their convoluted, tortuous channel and high aspect ratio. Permeability decreases, prolonging package shelf life and removing the requirement for multilayer barrier constructions [38].

3.3 Enhancing Intelligent and Active Packaging

As functional components with unique capabilities, CNTs provide intelligent and active packaging solutions beyond passive protection.

3.3.1 Antimicrobials

Multi-walled carbon nanotubes (MWCNTs) are perfect for food packaging due to their antibacterial qualities. Their antibacterial effectiveness comes from physically disrupting bacterial cell membranes, causing cell leakage and death [39]. Carbon nanotubes in polymer films limit bacterial development on packing surfaces, which benefits perishables including meat, dairy, and fresh fruit. Carbon nanotubes can synergistically boost antibacterial effectiveness with silver nanoparticles. For food safety, nanoparticle migration must be prevented by uniform dispersion and stable integration in the polymer matrix. CNT-based films with proper engineering can extend shelf life and minimize microbiological contamination [40].

3.3.2 Sensors and Monitors

Carbon nanotubes help construct chemiresistive sensors due to their high electrical conductivity and environmental sensitivity. These sensors can be inserted directly into packing sheets to monitor food quality and safety, including spoiling gasses like ammonia and hydrogen sulfide, temperature, and humidity. This real-time monitoring helps customers and retailers maintain product integrity across the supply chain, supporting intelligent packaging and composition of gases in sealed containers [41].

3.4 Practical Illustrations and Case Analyses

A multitude of studies has evidenced the efficacy of Carbon nanotubes based composites in several packaging applications:

Polylactic acid/Carbon nanotubes nanocomposites exhibit enhanced thermal stability and oxygen barrier properties, making them suitable for biodegradable food containers [42].

1. Chitosan/CNT films demonstrate antimicrobial properties and enhanced tensile strength, making them suitable for packaging fresh produce [43].
2. Polyethylene/CNT composites have been investigated for their improved resistance to moisture and gases in flexible packaging applications [44].

Moreover, the development of hybrid nanocomposites that integrate CNTs with various nanomaterials, such as nanoclays, graphene oxide, and metal nanoparticles, is underway to harness synergistic effects, thereby advancing the limits of functionality.

3.5 Obstacles in Commercial Implementation

Although the laboratory results are encouraging, there are numerous challenges that need to be tackled for broader implementation:

1. The expense associated with the production of CNTs, although sustainable synthesis methods are mitigating this challenge.
2. Concerns about nanoparticle migration create regulatory difficulties especially for food-contact uses.

Carbon nanotube (CNT)-based composites are enabling a new generation of smart and sustainable packaging technologies; they are becoming more and more relevant given higher food safety criteria and continuous development of green synthesis techniques. Their great promise comes from their outstanding ability to raise strength, barrier performance, and functioning. Still, one should act carefully when including them into food-related uses. Notwithstanding their many benefits, there are serious questions about the use of carbon nanotubes in food packaging concerning probable health dangers, environmental effect, and compatibility with regulatory standards. Responsible commercialization of carbon nanotubes based nanocomposites demands minimizing environmental effects while guaranteeing customer safety [45, 46].

4 Food Packaging Health Hazards

In food packaging, carbon nanotubes (CNTs) cause major questions regarding their possible toxicity and consequences on human health. Though uncommon, nanoparticle migration into food is feasible even when CNTs are included into polymer matrix. This is particularly true under high temperatures, mechanical strain, or long storage times. These materials influence aspects including CNT size, surface properties, aggregation degree, presence of residual metal catalysts, and toxicological profile. Studies suggest that breathing or swallowing large amounts of unmodified CNTs especially in their pure form—may induce cytotoxic effects, oxidative stress, and inflammation. Nevertheless, these dangers are significantly reduced when CNTs are functionalized on the surface and evenly distributed in synthetic or biopolymer matrices. This highlights the significance of careful material processing and design for practical and safe packaging. Recent studies emphasize the critical need for migration testing and toxicity evaluations tailored for practical food-contact scenarios. To prioritize consumer safety, regulatory authorities worldwide—and with growing emphasis in India—are advocating for more thorough assessments prior to permitting CNT-enabled packaging materials to be introduced into the market. Therefore, although CNTs offer significant potential for enhancing packaging performance, their safe use must be supported by thorough scientific validation and conscientious material design [47, 48].

4.1 Assessment of Environmental Effects and Considerations for End-of-Life

The environmental implications of carbon nanotubes encompass their production, utilization, and disposal stages:

1. Production Phase: Traditional synthesis techniques often require significant energy and utilize hazardous substances. Nonetheless, sustainable methods—like the microwave-assisted biochar-based synthesis demonstrated in your study—can significantly lessen environmental impacts.
2. Use Phase: The likelihood of environmental release throughout the packaging lifecycle (e.g., mechanical wear, incineration) remains minimal when CNTs are effectively integrated. Nonetheless, the possibility of ecotoxicity resulting from leaching into soil or water systems must be acknowledged and necessitates comprehensive long-term studies.
3. Disposal Phase: The non-biodegradable nature of CNTs raises potential concerns regarding their accumulation in ecosystems. Packaging systems that integrate CNTs with biodegradable polymer matrices and are intended for composting or controlled incineration provide a more sustainable end-of-life option [49].

4.2 Regulatory and Ethical Considerations in India

The regulatory framework for nanomaterials in food packaging in India is currently in a state of development, with the formal incorporation of nanotechnology into packaging regulations being less advanced than the more established systems found in the U.S. or Europe. Nonetheless, various important agencies and guidelines contribute indirectly to the promotion of safe practices regarding the use of nanomaterials like carbon nanotubes (CNTs) in packaging systems.

The Food Safety and Standards Authority of India (FSSAI) serves as the primary regulatory entity responsible for the oversight of food safety. Although FSSAI has not published detailed guidelines specifically addressing nanomaterials in food packaging, it tends to align with global best practices and has demonstrated a keen interest in nano-enabled packaging through public consultations and partnerships with academic institutions.

The Bureau of Indian Standards (BIS) has released specific technical standards concerning nanotechnology, including IS/ISO/TS 80004, which outlines the terminologies associated with nanomaterials. While these measures are not currently enforceable for food packaging, they establish a foundation for potential regulatory frameworks in the future. The Department of Science and Technology (DST), through the nano mission, promotes research and risk assessment in nanotechnology. The aim of these programs is to ensure that nanomaterials produced in India adhere to safety-by-design standards and foster environmental sustainability. Considering environmental factors, it is essential to acknowledge that regulations set forth by the Central Pollution Control Board (CPCB) and the Ministry of Environment, Forest and Climate Change (MoEFCC) may pertain to the disposal and environmental consequences of waste containing nanomaterials. Nevertheless, certain rules pertaining to the utilization of carbon nanotubes (CNTs) remain in the developmental stage [50, 51].

5 Anticipated Insights and Challenges

The use of carbon nanotubes (CNTs) into packaging materials represents a notable progression in the development of high-performance, intelligent, and ecological food packaging solutions. As nanotechnology advances, CNT-based composites are anticipated to move from experimental research to extensive commercial utilization. Nonetheless, actualizing this promise on a substantial scale necessitates traversing a multifaceted terrain of scientific, technical, regulatory, and socio-economic obstacles.

5.1 Expansion Eco-friendly Carbon Nanotube Manufacturing

A significant challenge is the scalability of sustainable production processes for carbon nanotubes. Traditional methods, including chemical vapor deposition (CVD) and arc discharge, necessitate significant financial commitment and pose environmental sustainability issues. In contrast, environmentally friendly alternatives like CNT synthesis via biochar and microwave irradiation present promising pathways by utilizing agricultural waste and reducing energy consumption.

Future directions in this area include:

- Developing standardized synthesis protocols to ensure consistent quality and yield.
- Automating and scaling microwave-assisted synthesis systems for industrial production.

- Developing resilient supply networks for biomass feedstocks, especially in agriculturally focused nations such as India.

5.2 *Domains for Continued Inquiry and Prospects*

To fully realize the potential of CNTs in packaging, numerous study domains necessitate additional investigation:

- Long-term performance assessments under actual storage and transportation environment.
- Investigating the biodegradability of CNT-polymer composites, particularly in composting environments.
- Designing and developing multifunctional packaging systems that combine enhanced barrier, mechanical, and sensing capabilities.
- Conducting comprehensive life cycle assessments (LCA) to evaluate environmental impacts from production to disposal.

6 Conclusion

Development of high-performance, sustainable, and intelligent food packaging systems gains a major stride forward when carbon nanotubes (CNTs) are included into polymer-based packaging materials. Particularly in the case of multi-walled carbon nanotubes (MWCNTs), CNTs can greatly improve the functioning of both conventional and biodegradable polymers because to their exceptional mechanical strength, thermal and electrical conductivity, and good gas barrier qualities. Beyond its use in structural reinforcement, CNTs allow active and smart packaging elements like enhanced shelf-life preservation, antibacterial action, and real-time environmental monitoring. Further enhancing their environmental credentials are green synthesis innovations include microwave-assisted conversion of fruit waste-derived biochar into CNTs. By transforming agricultural waste into high-value products, these environmentally friendly manufacturing techniques not only reduce the carbon footprint but also support circular economy aims.

Especially MWCNTs qualify for broad packaging applications since they offer advantages in terms of scalable and reasonably priced technologies. Single-walled carbon nanotubes (SWCNTs) are valuable for particular uses needing exceptional sensitivity or perfect electrical performance even if they are more costly and challenging to manufacture. Still, there remain challenges in the manner CNT-based packaging will be generally embraced. Issues concerning toxicity, the lack of comprehensive regulatory standards, and cost-related hurdles have to be taken under consideration if we are to ensure responsible and safe deployment. More research on their biocompatibility, migratory potential, and environmental impact is needed to enable the ethical integration of CNTs into the food packaging industry—especially in countries like India—especially in developing regulatory frameworks in those countries. Combining complex materials science, sustainability, and packaging innovation, carbon nanotube-based nanocomposites provide as synthesis technologies develop and improved awareness of safety concerns expands, CNTs—which offer durable, smart, and ecologically responsible alternatives—have the power to transform food packaging.

Declaration of Competing Interest The authors declare that they have no known competing financial interests or personal relationships that could have appeared to influence the work reported in this paper.

References

1. Abramovich H (2017) Introduction to composite materials. In: Stability and vibrations of thin walled composite structures. Elsevier, pp 1–47. https://doi.org/10.1016/B978-0-08-100410-4.00001-6
2. Amaechi CV, Agbomerie CO, Orok EO, Ye J (2020) Economic aspects of fiber reinforced polymer composite recycling. In: Encyclopedia of renewable and sustainable materials. Elsevier, pp 377–397. https://doi.org/10.1016/B978-0-12-803581-8.10738-6
3. Dwivedi G, Jain S, Shukla AK, Verma P, Verma TN, Saini G (2022) Impact analysis of biodiesel production parameters for different catalyst. Environ Dev Sustain. https://doi.org/10.1007/s10668-021-02073-w
4. Emamhadi MA, Sarafraz M, Akbari M, Thai VN, Fakhri Y, Linh NTT, Mousavi Khaneghah A (2020) Nanomaterials for food packaging applications: a systematic review. Food Chem Toxicol 146:111825. https://doi.org/10.1016/j.fct.2020.111825
5. Hsissou R, Seghiri R, Benzekri Z, Hilali M, Rafik M, Elharfi A (2021) Polymer composite materials: a comprehensive review. Compos Struct 262:113640. https://doi.org/10.1016/j.compstruct.2021.113640
6. Goyal P, Sharma MP, Jain S (2013) Optimization of transesterification of Jatropha curcas oil to biodiesel using response surface methodology and its adulteration with kerosene. J Mater Environ Sci 4(2):277–284
7. Jain S, Sharma MP (2012) Correlation development between the oxidation and thermal stability of biodiesel. Fuel 102:354–358. https://doi.org/10.1016/j.fuel.2012.06.110
8. Goyal P, Sharma MP, Jain S (2012) Optimization of esterification and transesterification of high FFA Jatropha curcas oil using response surface methodology. J Petrol Sci Res 1(3)
9. Ibrahim KS (2013) Carbon nanotubes-properties and applications: a review. Carbon Lett 14(3):131–144. https://doi.org/10.5714/CL.2013.14.3.131
10. Mamalis AG, Robinson M, Manolakos DE, Demosthenous GA, Ioannidis MB, Carruthers J (1997) Crashworthy capability of composite material structures. Compos Struct 37(2):109–134. https://doi.org/10.1016/S0263-8223(97)80005-0
11. Okpalugo TIT, Papakonstantinou P, Murphy H, McLaughlin J, Brown NMD (2005) High resolution XPS characterization of chemical functionalised MWCNTs and SWCNTs. Carbon 43(1):153–161. https://doi.org/10.1016/j.carbon.2004.08.033
12. Jain S (2019) The production of biodiesel using Karanja (Pongamia pinnata) and Jatropha (Jatropha curcas) oil. In: Biomass, biopolymer-based materials, and bioenergy. Elsevier, pp 397–408. https://doi.org/10.1016/B978-0-08-102426-3.00017-5
13. Jain S, Sharma MP (2011) Study of oxidation stability of Jatropha curcas biodiesel/diesel blends. Int J Energy Environ 2(3):533–542

14. Pal N (2020) Nanoporous metal oxide composite materials: a journey from the past, present to future. Adv Coll Interface Sci 280:102156. https://doi.org/10.1016/j.cis.2020.102156
15. Popov V (2004) Carbon nanotubes: properties and application. Mater Sci Eng R Rep 43(3):61–102. https://doi.org/10.1016/j.mser.2003.10.001
16. Rudakiya D, Patel Y, Chhaya U, Gupte A (2019) Carbon nanotubes in agriculture: production, potential, and prospects. In: Nanotechnology for agriculture. Springer Singapore, pp 121–130. https://doi.org/10.1007/978-981-32-9370-0_8
17. Siciliano S, Lopresto CG, Lamonaca F (2024) From traditional packaging to smart bio-packaging for food safety: a review. Euro-Mediterr J Environ Integr 9(4):1971–1986. https://doi.org/10.1007/s41207-024-00627-8
18. Soni SK, Thomas B, Kar VR (2020) A comprehensive review on CNTs and CNT-reinforced composites: syntheses, characteristics and applications. Mater Today Commun 25:101546. https://doi.org/10.1016/j.mtcomm.2020.101546
19. Virgil'ev YuS, Kalyagina IP (2004) Carbon–carbon composite materials. Inorganic Mater 40:S33–S49. https://doi.org/10.1023/B:INMA.0000036327.90241.5a
20. Jain S, Sharma MP (2012) Oxidation, thermal, and storage stability studies of Jatropha curcas biodiesel. ISRN Renew Energy 2012:1–15. https://doi.org/10.5402/2012/861293
21. Wu Z-S, Zhou G, Yin L-C, Ren W, Li F, Cheng H-M (2012) Graphene/metal oxide composite electrode materials for energy storage. Nano Energy 1(1):107–131. https://doi.org/10.1016/j.nanoen.2011.11.001
22. Sharma R, Sharma AK, Sharma V (2015) Synthesis of carbon nanotubes by arc-discharge and chemical vapor deposition method with analysis of its morphology, dispersion and functionalization characteristics. Cogent Eng 2(1):1094017. https://doi.org/10.1080/23311916.2015.1094017
23. Jain S, Sharma MP (2013) Engine performance and emission analysis using oxidatively stabilized Jatropha curcas biodiesel. Fuel 106:152–156. https://doi.org/10.1016/j.fuel.2012.11.076
24. Shoukat R, Khan MI (2021) Carbon nanotubes: a review on properties, synthesis methods and applications in micro and nanotechnology. Microsyst Technol 27(12):4183–4192. https://doi.org/10.1007/s00542-021-05211-6
25. Kumar Shukla A, Ahmad Z, Sharma M, Dwivedi G, Nath Verma T, Jain S, Verma P, Zare A (2020) Advances of carbon capture and storage in coal-based power generating units in an Indian context. Energies 13(16):4124. https://doi.org/10.3390/en13164124
26. Kumar S, Jain S, Kumar H (2018) Performance evaluation of adaptive neuro-fuzzy inference system and response surface methodology in modeling biodiesel synthesis from Jatropha–algae oil. Energy Sources Part A Recover Util Environ Eff 40(24):3000–3008. https://doi.org/10.1080/15567036.2018.1515277
27. Kumar S, Jain S, Kumar H (2020) Experimental study on biodiesel production parameter optimization of Jatropha-algae oil mixtures and performance and emission analysis of a diesel engine coupled with a generator fueled with diesel/biodiesel blends. ACS Omega 5(28):17033–17041. https://doi.org/10.1021/acsomega.9b04372
28. Yahyazadeh A, Nanda S, Dalai AK (2024) Carbon nanotubes: a review of synthesis methods and applications. Reactions 5(3):429–451. https://doi.org/10.3390/reactions5030022
29. Raj RS, Jain S, Kumar Sharma A (2024) Transforming municipal solid waste management through material and substance flow analysis: conversion pathways for sustainable energy production. Energy Convers Manage 322:119164. https://doi.org/10.1016/j.enconman.2024.119164
30. Raj Singh A, Kumar Singh S, Jain S (2022) A review on bioenergy and biofuel production. Mater Today Proc 49:510–516. https://doi.org/10.1016/j.matpr.2021.03.212
31. Singh VP, Jain S, Gupta J (2023) Analysis of the effect of perforation in multi-v rib artificial roughened single pass solar air heater: - Part A. Exp Heat Transf 36(2):163–182. https://doi.org/10.1080/08916152.2021.1988761
32. Zhang H-B, Lin G-D, Zhou Z-H, Dong X, Chen T (2002) Raman spectra of MWCNTs and MWCNT-based H2-adsorbing system. Carbon 40(13):2429–2436. https://doi.org/10.1016/S0008-6223(02)00148-3

33. Singh VP, Jain S, Gupta J (2022) Performance assessment of double-pass parallel flow solar air heater with perforated multi-V ribs roughness—Part B. Exp Heat Transf 35(7):1059–1076. https://doi.org/10.1080/08916152.2021.2019147
34. Gangoli VS, Godwin MA, Reddy G, Bradley RK, Barron AR (2019) The state of hipco single-walled carbon nanotubes in 2019. C—J Carbon Res 5(4):65. https://doi.org/10.3390/c5040065
35. Saini M, Sharma A, Singh VP, Dwivedi G, Jain S (2022) Solar thermal receivers—a review, pp 311–325. https://doi.org/10.1007/978-981-16-8341-1_25
36. Pal G, Kumar S (2016) Modeling of carbon nanotubes and carbon nanotube–polymer composites. Prog Aerosp Sci 80:33–58. https://doi.org/10.1016/j.paerosci.2015.12.001
37. Raul PK, Thakuria A, Das B, Devi RR, Tiwari G, Yellappa C, Kamboj DV (2022) Carbon nanostructures as antibacterials and active food-packaging materials: a review. ACS Omega 7(14):11555–11559. https://doi.org/10.1021/acsomega.2c00848
38. Xie X, Mai Y, Zhou X (2005) Dispersion and alignment of carbon nanotubes in polymer matrix: a review. Mater Sci Eng R Rep 49(4):89–112. https://doi.org/10.1016/j.mser.2005.04.002
39. David ME, Ion R-M, Grigorescu RM, Iancu L, Holban AM, Nicoara AI, Alexandrescu E, Somoghi R, Ganciarov M, Vasilievici G, Gheboianu AI (2021) Hybrid materials based on multi-walled carbon nanotubes and nanoparticles with antimicrobial properties. Nanomaterials 11(6):1415. https://doi.org/10.3390/nano11061415
40. Saleemi MA, Kong YL, Yong PVC, Wong EH (2022) An overview of antimicrobial properties of carbon nanotubes-based nanocomposites. Adv Pharm Bull 12(3):449–465. https://doi.org/10.34172/apb.2022.049
41. Tang R, Shi Y, Hou Z, Wei L (2017) Carbon nanotube-based chemiresistive sensors. Sensors 17(4):882. https://doi.org/10.3390/s17040882
42. Carson L, Kelly-Brown C, Stewart M, Oki A, Regisford G, Luo Z, Bakhmutov VI (2009) Synthesis and characterization of chitosan–carbon nanotube composites. Mater Lett 63(6–7):617–620. https://doi.org/10.1016/j.matlet.2008.11.060
43. Gonçalves C, Gonçalves I, Magalhães F, Pinto A (2017) Poly(lactic acid) composites containing carbon-based nanomaterials: a review. Polymers 9(7):269. https://doi.org/10.3390/polym9070269
44. Raja M, Ryu SH, Shanmugharaj AM (2013) Thermal, mechanical and electroactive shape memory properties of polyurethane (PU)/poly (lactic acid) (PLA)/CNT nanocomposites. Eur Polymer J 49(11):3492–3500. https://doi.org/10.1016/j.eurpolymj.2013.08.009
45. Gupta N, Gupta SM, Sharma SK (2019) Carbon nanotubes: synthesis, properties and engineering applications. Carbon Lett 29(5):419–447. https://doi.org/10.1007/s42823-019-00068-2
46. Rahman G, Najaf Z, Mehmood A, Bilal S, Shah A, Mian S, Ali G (2019) An overview of the recent progress in the synthesis and applications of carbon nanotubes. C 5(1):3. https://doi.org/10.3390/c5010003
47. Bangar SP, Whiteside WS, Kajla P, Tavassoli M (2025) A review of advancements, properties, and challenges of carbon nanotubes in food packaging. J Food Meas Charact 19(4):2172–2194. https://doi.org/10.1007/s11694-025-03127-7
48. Dhara D, Biswas S, Das SK, Biswas O (2021) Status of food safety and food security in India in the perspective of FSSAI. Indian J Anim Health 60(2-Spl). https://doi.org/10.36062/ijah.2021.spl.01821
49. Köhler AR, Som C, Helland A, Gottschalk F (2008) Studying the potential release of carbon nanotubes throughout the application life cycle. J Clean Prod 16(8–9):927–937. https://doi.org/10.1016/j.jclepro.2007.04.007
50. Liu Y, Zhao Y, Sun B, Chen C (2013) Understanding the toxicity of carbon nanotubes. Acc Chem Res 46(3):702–713. https://doi.org/10.1021/ar300028m
51. Chhabra M, Dwivedi G, Baredar P, Kumar Shukla A, Garg A, Jain S (2021) Production & optimization of biodiesel from rubber oil using BBD technique. Mater Today Proc 38:69–73. https://doi.org/10.1016/j.matpr.2020.05.791

Chapter 21
Valorization of Agricultural Crop Residues for Biodegradable and Eco-friendly Food Packaging Applications

Rajat Kumar Sharma, Virendra Singh, Deepshikha Azad, Devendra Kumar, and Akarsh Verma

1 Introduction

The expanding rate of population growth worldwide is elevating pressures on food systems, natural resources, and waste management. At the same time, the accumulation of non-biodegradable waste, especially from synthetic plastic materials used in the packaging industry in vast quantities is also increasing several environmental problems [1, 2]. The food packaging industry, which holds a considerable proportion of global plastic usage, is the major contributory factor to environmental degradation through its use of petroleum-based polymers that are naturally not susceptible to biological decay. It is projected that more than 400 million tons of plastics are manufactured every year, with a significant portion used for single-packaging applications [3]. Disposal of such post-consumer waste usually ends in their dumping in

R. K. Sharma (✉) · D. Azad
Department of Farm Machinery and Power Engineering, Govind Ballabh Pant University of Agriculture and Technology, Pantnagar, Uttarakhand, India
e-mail: rajat_sharma@hotmail.com

V. Singh
Department of Physics, Govind Ballabh Pant University of Agriculture and Technology, Pantnagar, Uttarakhand, India

D. Kumar
Department of Electrical Engineering, Govind Ballabh Pant University of Agriculture and Technology, Pantnagar, Uttarakhand, India

A. Verma
Department of Mechanical Engineering, UPES (University of Petroleum and Energy Studies), Dehradun, India

A. Verma et al. (eds.), *Polymers and Composite Materials for Packaging: Smart Food Packaging and Solutions*, Materials Horizons: From Nature to Nanomaterials,
https://doi.org/10.1007/978-981-95-0677-4_21

waste dumps, burning, or unintentional dispersal in terrestrial and aquatic systems, thus generating long-term ecological contamination, microplastic spread, and related human health hazards. In the face of increased environmental, regulatory, and societal demands, innovation in sustainable, biodegradable substitutes to conventional plastic packaging has attracted significant interest. Biodegradable packaging obtained from renewable sources of biomass is a promising route toward the reduction of plastic waste while helping to achieve the goals of a circular economy [4]. In this regard, lignocellulosic crop-residue material has proven to be a valuable and underutilized resource. These crop-residue materials, such as rice straw, wheat straw, corn husks, sugarcane bagasse, and barley husks, are produced in enormous amounts as by-products of crop harvesting and processing. Worldwide, many billion tons of such crop-residue biomass are being produced every year, of which a significant amount is wasted or is being disposed of in environmentally harmful manners, such as open-field burning. These waste management practices not only waste a potentially useful resource but also generate aggravating air contamination, greenhouses gases, and loss of nutrients to the soils [5–13].

Valorizing agricultural crop residues as the conversion of low-value or waste biomass to more valued functional materials is an ecologically and economically beneficial approach [14]. The structural make-up of these residues in the form of cellulose, hemicellulose, and lignin as their main constituents makes them good substrates for the manufacture of environmentally friendly films. The most common natural polymer, cellulose, has good mechanical properties and the ability to form films, hemicellulose has good oxygen barrier properties; and lignin provides the films with hydrophobicity as well as UV stability. New processing technologies of the biomass such as chemical pretreatments (alkali and acid hydrolysis), enzymatic hydrolysis, mechanical defibrillation, and the extraction of nanocellulose have opened the way to convert the lignocellulosic biomass to films, composites, and molded packaging articles with their properties being increasingly on par with those of conventional plastics.

There have been many empirical studies to prove the efficiency of lignocellulosic-derived materials as food packaging solutions. For instance, cellulose nanocrystals isolated from rice straw have been used as reinforcement additives in poly(lactic acid) and polyvinyl alcohol matrices to produce films with enhanced strength in terms of resistance to stretching and barrier properties [15]. Sugarcane bagasse has also been used successfully to manufacture biodegradable foam trays and molded articles for the packaging of dry food [16]. Lignocellulosic matrices with the addition of bioactive entities like essential oils and silver nanoparticles have also enabled the preparation of functional packages with antimicrobial and antioxidant activity, thus enabling food preservation and shelf life enhancement [17]. Although these materials enjoy tremendous technological potential, their wider commercial deployment is hampered by a number of impediments. These include logistical issues related to the gathering, transport, and warehousing of crop residues, geographical and seasonal variability in the nature of the crop-residue biomass; the absence of standard protocols of processing technology adopted nationwide; and the inability to scale-up the processes of manufacture. Moreover, the functional limitations of lignocellulosic materials in terms of sensitivity to water and lower thermal stability as opposed to their conventional plastic counterparts also require additional work on the modification of the material as well as formulations of the composites. Regulatory requirements and issues related to safety aspects, especially in the light of migration behavior and biodegradation in real life, also need to go through rigorous testing to ascertain compliance with food contact material specifications [18, 19].

From a sustainability point of view, the valorization of agricultural residue waste for biodegradable packaging is in line with numerous United Nations Sustainable Development Goals (SDGs), such as SDG 12 (responsible consumption and production), SDG 13 (climate action), and SDG 15 (life on land). It represents a strategic way of lowering the environmental impact of the packaging sector in addition to solving the problem of agricultural waste management [20]. Furthermore, the development of value chains of biomass-based packaging has the potential to create new rural and agrarian community-based economic opportunities, thus supporting inclusive and resilient economic development.

Against the international urgency of shifting to sustainable packaging solutions and the yet untapped potential of lignocellulosic biomass, this chapter explores the current state of crop-residue valorization for food packaging applications from a biodegradable packaging material point of view. It gives an overview of the physicochemical properties of major crop residues, summarizes the recent technological developments in the processing of the biomass, and compares the functional properties of the obtained packaging materials. It also delves into the issues related to industrial-scale implementation and the prospects for further research and development. In that way, this work emphasizes the significance of valorizing agricultural biomass as the keystone to the development of environmentally sound food packaging technologies.

2 Availability of Various Agriculture and Forest Biomass in India

India has immense strength in agricultural and forest biomass, providing a sustainable and renewable resource pool for power generation, industrial purposes, and climate-resilient development. The country's diverse agro-climatic regions offer a vast range of crop residue, forest by-products, and organic waste, which remain untapped in most places. State-wise variations in the total production of biomass along with surplus availability reveal the scope for localized biomass-based projects. Punjab, Uttar Pradesh, and Maharashtra indicate high instances of both the production of and surpluses in biomass, indicating the scope for scaled-up bioenergy projects and green rural enterprise. Areas like Haryana and Gujarat also reveal significant levels of surplus notwithstanding moderate output, indicating well-functioning residue management and collection facilities. Northeastern and hilly states like Mizoram, Sikkim, and Arunachal Pradesh offer little in terms of their contribution of biomass due to terrain and low intensity of agriculture. Mapping this pattern of biomass is important to plan the decentralized system of renewables, drive circular economy approaches, and make India carbon neutral. Beyond power, the surplus of biomass presents a nascent opportunity to manufacture biocomposites for green eco-packaging material. Rice husk, wheat straw, and sugarcane bagasse from agriculture can be transformed to make biodegradable substitutes of plastics, helping in sustainable packaging industries, curbing plastic pollutions, and generating green enterprise in rural and peri-urban areas of the country [5] (Table 1).

Table 1 Agriculture and forest biomass availability in India

State	Biomass production (MT/year)	Surplus biomass (MT/year)
Punjab	53.30	22.2
Maharashtra	47.6	14.8
Uttar Pradesh	60.3	13.7
Haryana	29.0	11.3
Gujarat	29.0	11.1
Madhya Pradesh	33.3	9.6
Tamil Nadu	22.5	8.9
Rajasthan	29.9	8.6
Andhra Pradesh & Telangana	43.9	7.0
Assam	11.4	6.3
Bihar	25.8	5.5
Orissa	20.1	5.7
West Bengal	36.0	4.3
Karnataka	34.2	3.4
Kerala	11.6	3.4
Chhattisgarh	11.3	2.1
Himachal Pradesh	2.9	1.0
Uttarakhand	2.1	0.5
Jharkhand	3.1	0.4
Jammu and Kashmir	1.6	0.3
Manipur	0.9	0.2
Meghalaya	0.6	0.2
Tripura	0.1	0.2
Goa	0.7	0.1
Nagaland	0.5	0.1
Arunachal Pradesh	0.4	0.1
Mizoram	0.1	0.0
Sikkim	0.1	0.0
India (Total)	511.0	145.0

3 Composition of Different Agriculture and Forest Biomass

The compositional properties of different lignocellulosic biomass sources largely dictate their feasibility in the development of composites and packaging material. Rice straw consists of ~ 25–35% cellulose, 32–37% hemicellulose, and 6–10% lignin. High cellulose and hemicellulose contents contribute to strength and film-forming potential, but the latter's hydrophilous behavior restricts water resistance potential [21]. Rice husk, with a similar profile of cellulose, is silica-enriched, which ensures thermal stability and barrier function. Presence of significant amounts of lignin and silica also contributes to improved durability and environmental resistance [22]. Wheat straw has moderate to high cellulose and hemicellulose contents with a well-balanced lignin content. It helps in the preparation of composites with good tensile properties but with moisture susceptibility, which calls for chemical or physical treatments [23]. Sugarcane bagasse is a good source with its good cellulose content of ~ 51.8% and ~ 21.7% lignin [24]. Its availability as a suitable raw material for the production of nanocellulose is significant. Lignin content adds natural antioxidant value to its functional properties in the development of packaging applications. Barley husk and barley straw consist of 30–35% cellulose and contain major contents of arabinoxylan-rich hemicellulose with 17–25% lignin content [24]. These contribute to the development of oxygen barrier films as well as structural strength in the development of composites. Barley cob as well as corn husk also retains cellulose contents of ~ 45% and ~ 31–39%, respectively. Their moderate hemicellulose and the low lignin content enable their acceptance in the development of biocomposites, but their water resistance is a limitation. These types of biomass altogether provide feasible substitutes as biodegradable composites with functional value in industrial applications.

The compositional variations of different biomass sources have a critical influence on the functional properties of their resulting derived biocomposites. Rice husk's high silica content is greatly responsible for the consequential enhancement in moisture and barrier to gases, thus ensuring thermal stability and durability. Barley husk, which is rich in arabinoxylans, is favorable for developing oxygen-impermeable, firm films and is thus a good raw material in protective packaging applications. Sugarcane bagasse with its cellulose content of ~ 51.8% is of particular value in extracting the cellulose as nanocellulose and obtaining good mechanical strength and biodegradability in developed biocomposites. Corn husk is in contrast to the foregoing and has comparatively lower lignin content, which decreases its natural water resistance and may detract from its possible application in its original form. These differences emphasize the significance of the composition of the biomass in designing specific packaging functions of the biocomposite material, indicating the value of the adoption of targeted biomasses and treatments to optimize the functioning of the material in sustainable packaging (Table 2).

Table 2 Compositional analysis of different biomass

Biomass type	Cellulose (%)	Hemicellulose (%)	Lignin (%)	Other components	References
Rice straw	25.4–35.5	32.3–37.1	6.4–10.4	Small amounts of ash, wax, silica	Mandal et al. [21]
Rice husk	35	33	23	High silica, ash	Johar et al. [22]
Wheat straw	30–40	30–35	14–15	Minor waxes, ash	Liu et al. [23]
Sugarcane bagasse	~ 51.8	27.6	~ 21.7	~ 2.7% ash,	Pires et al. [24]
Barley husk/Straw	30–35	30–33 (mainly arabinoxylans)	17–25	Crystalline wax layer (in straw)	Pires et al. [24]
Corn cob	~ 45	~ 35	~ 15	Small % ash, trace proteins	Pires et al. [24]
Corn husk	31–39	Moderate	2–14	4–5% ash, trace fats and waxes	Pires et al. [24]
Pine needles	27.17	24.15	29.15	Ash	Mandal et al. [25]

4 Framework for Biomass-Based Composite Formation

The conversion of agricultural waste into biodegradable films is achieved through a number of well-planned steps. It begins with the selection of crop waste such as rice straw, wheat straw, sugarcane bagasse, barley husk, or corn husks. These waste materials have useful compounds such as cellulose, hemicellulose, and lignin. Initially, the biomass is cleaned. It is washed to remove dirt and dust, dried to reduce the moisture level, and then crushed or chopped to pieces so that further processing is simplified. It is possible that the dewaxing process is carried out with the application of hot water or alcohol to eliminate the natural waxes on the surface of the fiber. After cleaning and processing, the biomass is chemically treated to isolate and cleanse the cellulose. A common treatment is alkaline treatment with the use of such chemical agents as sodium hydroxide to break up the lignin. This breaks up the cellular arrangement of the vegetable material. Acid treatment is another process, usually with the use of sulfuric acid or hydrochloric acid, which dissolves the hemicellulose and enables the extraction of minute cellulose crystals. Bleaching can also occur to remove residual color and other contaminations, thereby yielding a whiter and purer cellulose product.

To make the material even finer, the processed biomass is also put through physical or mechanical processes like high-pressure mixing, ultrasound treatment, steam explosion, or microwaving. These processes disintegrate the cellulose particles to create smaller units known as nanofibrils or nanocrystals. These extremely tiny fibers are robust and possess good film-casting properties. Then the nanocellulose is blended with the biodegradable compounds such as polylactic acid (PLA), polyvinyl alcohol (PVA), or starch to form a composite. Plasticizers such as glycerol are added to enhance the flexibility and strength of the films. Additional ingredients such as chitosan, silver nanoparticles, or vegetable oils may also be added to confer additional functions such as the prevention of the growth of microbes or enhancement of the shelf life. The film is subsequently created with techniques including solvent casting, where the liquid solution is filled in a shallow mold and dried, or through the process of extrusion, where the melted substance is pushed out of the mold in thinly spread layers. The last steps include the application of treatments such as creating a waterproof coating layer or the use of crosslinking agents to improve the resistance of the film to water and heat. The resulting material is a biodegradable packaging film, creating a green option to conventional plastics and minimizing agricultural loss and contamination.

5 Comparison of Various Agricultural Biomass Sources for Nanocellulose-Based Biocomposites

Table 3 presents a comprehensive overview of the experimental studies on the development of the biocomposites from various lignocellulosic crop residues. The studies highlight the key role of the biomass composition, extraction process, and matrix compatibility in determining the structural and functional behaviors of the resultant composites, as is specifically the case with packaging applications. In the context of rice straw, Perumal et al. [15] and Agustin et al. [26] demonstrated the extraction of cellulose nanofibers (CNFs) and their reinforcement in polymer matrices like polyvinyl alcohol (PVA) and polylactic acid (PLA) [15, 26]. Mechanical improvements in these composites are largely due to the high crystallinity and aspect ratio of the nanocellulose, which enable effective stress transfer and enhanced interfacial bonding between the matrix and the fibers. The addition of chitosan, a natural polycationic polymer, facilitated additional crosslinking in the matrix, adding to enhanced biodegradability as well as antimicrobial activity. However, the high water absorption due to the highly hydrophilic nature of cellulose—owing to its numerous hydrogen bond-forming hydroxyl functional groups—requires surface treatments or blending with hydrophobic additives to increase moisture resistance.

Johar et al. [22] and Biswas et al. [27] used rice husk as a feedstock, highlighted by its silica and lignin content [22, 27]. The addition of cellulose nanocrystals (CNCs) from rice husk to starch-based films enhanced the tensile strength considerably. This is due to enhanced interfacial adhesion between the hydrophilic CNCs and the starch matrix and the intrinsic rigidity and crystallinity of CNCs as effective load-bearing fillers. Furthermore, the addition of silver nanoparticles (AgNPs) introduced robust antimicrobial activity due to AgNPs' well-established capacity to destabilize microbe membranes. Apart from these functional merits, the general application of rice husk is frustrated by its lower-than-favorable bulk density and abrasive nature,

Table 3 Comparison of various agricultural biomass sources for nanocellulose-based biocomposites

Biomass source	Study focus	Key findings	Result	Advantage	Limitation	References
Rice straw	Cellulose nanofiber extraction and composite reinforcement	Improved mechanical and barrier properties in PLA/PVA matrices	Enhanced strength and biodegradability	Abundant, good for reinforcement and eco-friendly films	High water absorption	Perumal et al. [15], Agustin et al. [26]
Rice husk	Use of CNC and silica in biocomposites	Effective filler for starch films and AgNPs antimicrobial films	Improved tensile strength and antimicrobial activity	High silica content provides barrier and thermal stability	Limited commercial use	Johar et al. [22], Biswas et al. [27]
Wheat straw	Nanocellulose and hemicellulose for PLA and PHBV composites	Improved Young's modulus, suitable for fruit packaging	Mechanical property enhancement	Cost-effective, easily available	Hydrophilic nature affects water barrier	Yang et al. [28], Berthet et al. (2015)
Sugarcane bagasse	CNC films and foam trays with starch	Good film-forming ability and mechanical strength	High tensile strength; edible films	Renewable, large volume waste	Hydroxyl groups reduce water resistance	Ghaderi et al. [29], Mandal and Chakrabarty [30]
Barley husk/straw	Arabinoxylan and CNC in barrier films	Formed strong oxygen barrier and reinforced nanocomposites	Improved oxygen resistance and film stiffness	Rich in arabinoxylans; useful for oxygen barrier films	Needs modification for moisture sensitivity	Fortunati et al. [31]; Höije et al. [32]
Corn cob	Reinforcement in chitosan and PVA films	Increased strength with glycerol and irradiation	Strengthened bioplastic structure	Good compatibility with polymers	Hydrophilic without treatment	de Andrade et al. [33], Silvério et al. [34]
Corn husk	Transparent nanofibrillated cellulose films	High transparency and moderate strength	Improved tensile and optical properties	Visually appealing films for packaging	Low water resistance	Xiao et al., [35], Ratanakamnuan et al. [36]

which make processing and logistics challenging. Wheat straw composites, studied by Yang et al. [28] and Berthet et al. (2015), were found to have better mechanical properties when mixed with PLA and polyhydroxybutyrate-co-valerate (PHBV) [28]. Solid-state shear milling enabled better dispersion and fiber-matrix adhesion with reduced fiber particle size, thus having enhanced Young's modulus and flexural strength. These results agree with the fibrous morphology of wheat straw, which is composed of cellulose and hemicellulose and is effective in reinforcing polymer matrices. Nevertheless, a high water vapor transmission rate (WVTR) still exists owing to the hydrophilic nature of the straw fiber, especially the hemicellulose component, which restricts their use in applications requiring good water barrier properties.

Sugarcane bagasse, with its cellulose content of ~52%, has also undergone a wide range of studies. Ghaderi et al. [29] and Mandal and Chakrabarty [30] prepared films and trays from bagasse and starch with the addition of nanocellulose, which displayed significant tensile strength and structural uniformity improvements [29, 30]. These were due to the pure cellulose and uniform dispersal of the nanofibers throughout the matrix, which provide mechanical strength. Furthermore, plasticizers and additives like magnesium stearate and guar gum aided in the flexibility and processability of the composites. However, the intrinsic hydrophilicity caused by the excess of –OH groups limits their uses in damp conditions. This can be overcome by chemical modification in the form of esterification or the addition of hydrophobic polymers.

Barley husk and straw were examined by Fortunati et al. [31] and Höjie et al. [32] with a view to the extraction of arabinoxylans and CNCs. The resultant films were found to possess good oxygen barrier properties as a result of the high molecular weight and film formation ability of arabinoxylans, which form close-packing structures with minimal permeability to gases. In addition, the combination of barley-derived CNCs with chitosan and PVA provided improved strength and rigidity due to synergy of interchain hydrogen bonding [31, 32]. In spite of such merits, the hydrophilic hemicellulose chain prevents such films from being water-resistant, and this requires subsequent structural development to make them more water-resistant. Composites from corn cob were tested by de Andrade et al. [33] and Silvério et al. [34], with emphasis put on the reinforcement of chitosan and PVA matrices. Glycerol plasticizer facilitated increased flexibility in films, while electron beam irradiation induced crosslinking in the matrix, resulting in enhanced tensile strength. High reinforcement efficiency of CNC from corn cob is attributed to their rigidity and high surface area for interfacial interactions to occur. Esterified nanocellulose fibers also gained improved hydrophobicity with the replacement of hydroxyl groups with non-polar entities, thus gaining moisture barrier properties [33, 34]. Similarly, Xiao et al. [35] and Ratanakamnuan et al. [36] examined the fabrication of clear nanofibrillated films from corn husk using ultrasonication. The films exhibited good light transmittance and reasonable mechanical strength as a result of fine fibrillation creating thin, uniform layers favorable toward transparency and structural stability. Unfortunately, the quite limited lignin content of corn husk restricts its native water resistance due to lignin contribution to the hydrophobic character of the cell walls of plants. Thus, the use of more reinforcement or hydrophobic treatment is needed for applications susceptible to moisture [35, 36].

In general, the agricultural residue-derived biocomposites can be successfully engineered for the use in packaging by utilizing the inherent structural and chemical characteristics of the biomass. Enhancements in mechanical, optical, and antimicrobial properties observed are closely related to the nanostructured morphology and surface chemistry of constituent cellulose and hemicellulose. However, the prevalence of issues of sensitivity to moisture highlights the necessity of targeted chemical modification and sophisticated composite design strategies to maintain consistent behavior across a wide range of environmental conditions. These works in total offer a conceptual framework for the development of sustainable, bio-based packaging solutions.

6 Scope and Limitations

The use of lignocellulosic agricultural biomass to create food packaging materials is a promising and environmentally sound alternative to petroleum-based polymers. From agro-residues such as rice straw, wheat straw, sugarcane bagasse, corn husk, and barley husk, not only is the material renewable and biodegradable but also provides a pragmatic solution to the issue of agro-waste accumulation. By processing biomass to create value-added packaging articles, the environmental problems of stubble burning, land and air contamination, and microplastic pollution can be avoided. This process also contributes to the global push toward a circular economy by allowing the reuse of agricultural by-products in high-value applications.

The lignocellulosic biomass is mainly made up of cellulose, hemicellulose, and lignin. Each of these contributes specific properties to the resultant composite materials. The nanostructured cellulose in its forms of nanocrystals or nanofiber provides the packaging films with high mechanical strength and rigidity and thus increases their structural integrity. Hemicellulose contributes to the barrier properties of the packaging material because of its film-forming ability and low oxygen permeability. Lignin, which is hydrophobic in nature, can increase the moisture resistance and heat stability of the material, especially as a coating or filler. Native cellulose and hemicellulose, which are hydrophilic in nature, however, cause high water uptake, restraining the use of the material in humid conditions. This has forced the development of modification methods such as esterification, blending with hydrophobic polymers, and the use of nanoparticles to enhance moisture resistance and functional properties.

The addition of natural additives such as chitosan, as well as inorganic compounds such as silver nanoparticles, gives the films antimicrobial activity, thus extending food product shelf life. They are also typically biodegradable and compostable, which means they can degrade in the natural environment without the formation of toxic residues. This property is highly appealing considering the recent ban on single-use plastics in many areas.

In spite of these benefits, there are still problems in terms of material harvesting, process efficiency, and reproducibility of the biomass fractionation. Additional problems in the areas of component migration to food, storage stability, and long-term stability also have to be analyzed before the introduction of these materials can become widespread in food packaging applications. Still, continued research and technological development from multiple disciplines are likely to overcome these limitations to make the use of biomass-based composites on a large scale in sustainable food packaging systems feasible.

7 Conclusion

The valorization of lignocellulosic crop residues represents a promising and sustainable route to the development of degradable food packaging material. Rice straw, wheat straw, sugarcane bagasse, corn husk, and barley husk are agro-residues that are rich in cellulose, hemicellulose, and lignin—components that are responsible for the mechanical strength, film-forming properties, and ability of packaging material to resist moisture. Improvements in the processing of biomass, including the extraction of nanocellulose, mixing with biopolymers, and chemical treatments, have made possible the manufacture of such biocomposites with improved tensile strength, barrier properties, and antimicrobial activity. These biomaterials not only address environmental concerns such as plastic pollution and stubble burning but also promote circular economy objectives by converting waste to value-added value. Furthermore, their composability and biodegradability also put them in line with worldwide sustainability requirements, especially with the rising regulatory push toward phasing out single-use plastics. Yet, there are still issues in the attainment of commercial scale-up. Biomass composition variability, water barrier limitations, as well as logistical issues of gathering and processing need to be overcome. In addition, thorough evaluations of food safety, migration behavior, and in-use biodegradation are necessary to confirm regulatory acceptance and safe consumer applications. In spite of such limitations, packaging based on lignocellulosic biomass has tremendous potential to transform the packaging sector. With concerted interdisciplinary research, enabling policy initiatives, and inter-sector collaboration, such biomaterials can become alternative sustainable solutions in place of synthetic polymers, significantly adding to the cause of environmental conservation and sustainable growth.

Conflicts of Interest There are no conflicts of interest to declare by the authors.

References

1. Bartoli M, Arrigo R, Malucelli G, Tagliaferro A, Duraccio D (2022) Recent advances in biochar polymer composites. Polymers 14(12):2506. https://doi.org/10.3390/polym14122506
2. Papadopoulou K, Klonos PA, Kyritsis A, Tarani E, Chrissafis K, Mašek O, Tsachouridis K, Anastasiou AD, Bikiaris DN (2025) Synthesis and characterization of PLA/biochar bio-composites containing different biochar types and content. Polymers 17(3):263. https://doi.org/10.3390/polym17030263
3. Lebreton L, Andrady A (2019) Future scenarios of global plastic waste generation and disposal. Palgrave Commun 5(1):6. https://doi.org/10.1057/s41599-018-0212-7
4. Liang Y, Tan Q, Song Q, Li J (2021) An analysis of the plastic waste trade and management in Asia. Waste Manage 119:242–253. https://doi.org/10.1016/j.wasman.2020.09.049
5. Mandal S, Sharma RK (2023) Application of gasification technology in agriculture for power generation. In: Handbook of energy management in agriculture. Springer Nature Singapore, pp 1–22. https://doi.org/10.1007/978-981-19-7736-7_14-1
6. Mandal S, Sharma RK, Bhattacharya TK, Tanna H, Haydary J (2022) Charring of pine needles using a portable drum reactor. Chem Pap 76(2):1239–1252. https://doi.org/10.1007/s11696-021-01893-4
7. Mandal S, Sharma RK, Bhattacharya TK (2023) Deriving fuel from pine needles through pyrolysis, charring and briquetting and their GHG emission potential. Curr Sci 124(10):1210–1215
8. Sharma RK, Bhattacharya TK, Kumain A, Chand P, Mandal S, Azad D (2020) Energy use pattern in wheat crop production system among different farmer groups of the Himalayan Tarai region. Curr Sci118(3):448. https://doi.org/10.18520/cs/v118/i3/448-454
9. Sharma RK, Haydary J, Singh TP, Nazari MA, Mandal S, Verma A (2024) Fractional distillation of pine needle bio-oil: comprehensive characterization and rheological properties across temperature gradients. Biomass Convers Biorefin. https://doi.org/10.1007/s13399-024-060 44-9
10. Sharma RK, Haydary J, Singh TP, Nazari MA, Mandal S, Verma A (2025) Fractional distillation of pine needle bio-oil: comprehensive characterization and rheological properties across temperature gradients. Biomass Convers Biorefin 15(8):12429–12440. https://doi.org/10.1007/s13399-024-06044-9
11. Sharma RK, Nazari MA, Haydary J, Singh TP, Mandal S (2023) A Review on advanced processes of biohydrogen generation from lignocellulosic biomass with special emphasis on thermochemical conversion. Energies 16(17):6349. https://doi.org/10.3390/en16176349
12. Sharma RK, Singh TP, Mandal S, Azad D, Kumar S (2022) Chemical treatments for biochar modification: opportunities, limitations and advantages. In: Engineered biochar. Springer Nature Singapore, pp 65–84. https://doi.org/10.1007/978-981-19-2488-0_5
13. Singh N, Singh V, Bisht N, Negi P, Dhyani A, Sharma RK, Tewari BS (2025) A comprehensive review on supercapacitors: basics to recent advancements. J Energy Storage 121:116498. https://doi.org/10.1016/j.est.2025.116498
14. Singh V, Pant N, Sharma RK, Padalia D, Rawat PS, Goswami R, Singh P, Kumar A, Bhandari P, Tabish A, Deifalla AM (2023) Adsorption studies of Pb(II) and Cd(II) heavy metal ions from aqueous solutions using a magnetic biochar composite material. Separations 10(7):389. https://doi.org/10.3390/separations10070389
15. Perumal AB, Sellamuthu PS, Nambiar RB, Sadiku ER (2018) Development of polyvinyl alcohol/chitosan bio-nanocomposite films reinforced with cellulose nanocrystals isolated from rice straw. Appl Surf Sci 449:591–602. https://doi.org/10.1016/j.apsusc.2018.01.022
16. Elkayaly D, Hazem N, Fahim IS (2022) Green and sustainable packaging manufacturing: a case study of sugarcane bagasse-based tableware in Egypt. Circ Econ Sustain 2(3):829–856. https://doi.org/10.1007/s43615-021-00136-8
17. Kapoor DU, Gaur M, Kumar A, Singh S, Chidrawar VR, Prajapati BG (2024) Value-added lignocellulose from waste biomass as a potential source in fabrication of lignin-based composite

for multifaceted applications. Regen Eng Transl Med. https://doi.org/10.1007/s40883-024-00357-x
18. Anamika, Singh, V., & Yadav, B. K. (2020). Adsorption study of Heavy Metals from Aqueous Solutions using Magnetite Nanoparticles. Journal of Physics: Conference Series, 1504(1), 012011. https://doi.org/10.1088/1742-6596/1504/1/012011
19. Dash AK, Tripathy S, Naveenkumar A, Bhoi TK, Kumari A, Divya, Latare AM, Nandipamu TMK, Singh V, Raza MdB, Saraswat A, Bhadha J (2024) Biochar for pollutants bioremediation from soil and water ecosystem. In: Biochar production for green economy. Elsevier, pp 433–452). https://doi.org/10.1016/B978-0-443-15506-2.00026-2
20. Hughes C (2020) SDG 12: responsible consumption and production. In: Jean Monnet sustainable development goals network policy brief series. RMIT University, Melbourne, Australia
21. Mandal S, Bhattacharya TK, Tanna H (2017) Energy harnessing routes of rice straw. Curr Sci 113(1):21–23
22. Johar N, Ahmad I, Dufresne A (2012) Extraction, preparation and characterization of cellulose fibres and nanocrystals from rice husk. Ind Crops Prod 37(1):93–99. https://doi.org/10.1016/j.indcrop.2011.12.016
23. Liu R, Yu H, Huang Y (2005) Structure and morphology of cellulose in wheat straw. Cellulose 12(1):25–34. https://doi.org/10.1007/s10570-004-0955-8
24. Pires JRA, Souza VGL, Fernando AL (2019) Valorization of energy crops as a source for nanocellulose production—current knowledge and future prospects. Ind Crops Prod 140:111642. https://doi.org/10.1016/j.indcrop.2019.111642
25. Mandal S, Bhattacharya TK, Verma AK, Haydary J (2018) Optimization of process parameters for bio-oil synthesis from pine needles (Pinus roxburghii) using response surface methodology. Chem Pap 72(3):603–616. https://doi.org/10.1007/s11696-017-0306-5
26. Agustin MB, Ahmmad B, Alonzo SMM, Patriana FM (2014) Bioplastic based on starch and cellulose nanocrystals from rice straw. J Reinf Plast Compos 33(24):2205–2213. https://doi.org/10.1177/0731684414558325
27. Biswas MC, Tiimob BJ, Abdela W, Jeelani S, Rangari VK (2019) Nano silica-carbon-silver ternary hybrid induced antimicrobial composite films for food packaging application. Food Packag Shelf Life 19:104–113. https://doi.org/10.1016/j.fpsl.2018.12.003
28. Yang S, Bai S, Wang Q (2018) Sustainable packaging biocomposites from polylactic acid and wheat straw: Enhanced physical performance by solid state shear milling process. Compos Sci Technol 158:34–42. https://doi.org/10.1016/j.compscitech.2017.12.026
29. Ghaderi M, Mousavi M, Yousefi H, Labbafi M (2014) All-cellulose nanocomposite film made from bagasse cellulose nanofibers for food packaging application. Carbohyd Polym 104:59–65. https://doi.org/10.1016/j.carbpol.2014.01.013
30. Mandal A, Chakrabarty D (2014) Studies on the mechanical, thermal, morphological and barrier properties of nanocomposites based on poly(vinyl alcohol) and nanocellulose from sugarcane bagasse. J Ind Eng Chem 20(2):462–473. https://doi.org/10.1016/j.jiec.2013.05.003
31. Fortunati E, Benincasa P, Balestra GM, Luzi F, Mazzaglia A, Del Buono D, Puglia D, Torre L (2016) Revalorization of barley straw and husk as precursors for cellulose nanocrystals extraction and their effect on PVA_CH nanocomposites. Ind Crops Prod 92:201–217. https://doi.org/10.1016/j.indcrop.2016.07.047
32. Höije A, Gröndahl M, Tømmeraas K, Gatenholm P (2005) Isolation and characterization of physicochemical and material properties of arabinoxylans from barley husks. Carbohyd Polym 61(3):266–275. https://doi.org/10.1016/j.carbpol.2005.02.009
33. de Andrade MR, Nery TBR, de Santana e Santana TI, Leal IL, Rodrigues LAP, de Oliveira Reis JH, Druzian JI, Machado BAS (2019) Effect of cellulose nanocrystals from different lignocellulosic residues to chitosan/glycerol films. Polymers11(4):658. https://doi.org/10.3390/polym11040658

34. Silvério HA, Flauzino Neto WP, Pasquini D (2013) Effect of incorporating cellulose nanocrystals from corncob on the tensile, thermal and barrier properties of poly(vinyl alcohol) nanocomposites. J Nanomater 2013(1). https://doi.org/10.1155/2013/289641
35. Xiao S, Gao R, Gao L, Li J (2016) Poly(vinyl alcohol) films reinforced with nanofibrillated cellulose (NFC) isolated from corn husk by high intensity ultrasonication. Carbohyd Polym 136:1027–1034. https://doi.org/10.1016/j.carbpol.2015.09.115
36. Ratanakamnuan U, Manorom W, Inthasai P (2013) Preparation of biodegradable film from esterified corn husk cellulose. Adv Mater Res 701:229–233. https://doi.org/10.4028/www.scientific.net/AMR.701.229

MIX
Papier aus verantwortungsvollen Quellen
Paper from responsible sources
FSC® C105338

If you have any concerns about our products,
you can contact us on
ProductSafety@springernature.com

In case Publisher is established outside the EU,
the EU authorized representative is:
Springer Nature Customer Service Center GmbH
Europaplatz 3, 69115 Heidelberg, Germany

Printed by Libri Plureos GmbH
in Hamburg, Germany